Lecture Notes in Mathematics

Edited by A. Dold and B. Eckmann

503

Applications of Methods of Functional Analysis to Problems in Mechanics

Joint Symposium IUTAM/IMU Held in Marseille, September 1–6, 1975

Edited by P. Germain and B. Nayroles

Springer-Verlag
Berlin · Heidelberg · New York 1976

Editors
Paul Germain
Faculté des Sciences
Mécanique Théorique-Tour 66
9 quai Saint Bernard
F-75005 Paris

Bernard Nayroles
Laboratoire de Mécanique
et d'Acoustique
31, chemin Joseph-Aiguier
F-13274 Marseille Cedex 2

Library of Congress Cataloging in Publication Data

Joint Symposium IUTAM/IMU on Applications of methods of functional analysis to problems in mechanics, Marseille, 1975.
Applications of methods of functional analysis to problems in mechanics.

(Lecture notes in mathematics ; 503)
1. Mechanics—Congresses. 2. Functional analysis—Congresses. I. Germain, Paul. II. Nayroles, Bernard, 1937- III. International Union of Theoretical and Applied Mechanics. IV. International Mathematical Union (Founded 1950) V. Title. VI. Series: Lecture notes in mathematics (Berlin) ; 503.
QA3.L28 no. 503 [QA801] 510'.8s [620.1'001'5157]
76-5454

AMS Subject Classifications (1970): 35A15, 35J55, 49H05, 73C99, 73E99, 73K25, 76-02

ISBN 3-540-07629-8 Springer-Verlag Berlin · Heidelberg · New York
ISBN 0-387-07629-8 Springer-Verlag New York · Heidelberg · Berlin

Printed in Germany
Printing and binding: Beltz Offsetdruck, Hemsbach/Bergstr.

AVANT-PROPOS

Il y a deux ans l'Union Internationale de Mécanique Théorique et Appliquée (I. U. T. A. M.) proposa à l'Union Internationale de Mathématiques (I. M. U.) d'organiser conjointement un symposium "Sur les applications de l'analyse fonctionnelle aux problèmes de mécanique". L'invitation fut acceptée. Les deux Unions nommèrent un Comité Scientifique International auquel elles confièrent la responsabilité scientifique de la rencontre. Le Laboratoire de Mécanique et d'Acoustique du C. N. R. S. accepta de se charger de son organisation à Marseille. Ce symposium vient de se tenir durant la première semaine de septembre. Ce sont ses actes que nous avons l'honneur de présenter ici.

Il y avait relativement longtemps qu'une rencontre entre mathématiciens et mécaniciens n'avait été patronnée par les deux Unions. Les relations entre la Mécanique et la mathématique qui furent si étroites dans le passé -au point qu'un progrès dans l'une des disciplines entraînait bien souvent immédiatement un progrès dans l'autre- s'étaient récemment quelque peu relachées. Chacune avait tendance à s'enfermer dans ses problèmes et à développer des modes de pensée autonomes et un langage propre . L'initiative prise par les deux Unions reposait sur la conviction que le moment était favorable pour tenter de remédier à une telle situation. L'expérience a montré qu'elles ont vu juste.

Un grand nombre de savants ont en effet manifesté leur désir de participer à cette rencontre et d'y être invités à présenter une communication. A titre d'exemple, les trente-huit communications retenues par le Comité Scientifique International ont été choisies parmi plus de soixante-quinze projets qui lui avaient été soumis. Par ailleurs, la répartition des origines des participants et des conférenciers couvre continûment un large secteur allant des départements de Mathématiques pures jusqu'aux départements de Mécanique et de Sciences pour l'ingénieur. On peut se féliciter enfin de la présence d'un grand nombre de jeunes scientifiques aussi bien parmi les auteurs que parmi les participants.

Ce succès est dû pour une large part au thème retenu pour cette rencontre. Le sujet est relativement neuf, surtout pour les mécaniciens dont la formation mathématique ne comporte pas, bien souvent, d'initiation à l'analyse fonctionnelle ; il est spécialement fascinant pour les mathématiciens auxquels la mécanique -et notamment la mécanique des milieux continus avec l'infinie variété des comportements des milieux étudiés et des conditions aux frontières rencontrées dans les situations concrètes- offre une gamme extraordinaire de problèmes nouveaux de types variés dont il convient d'étudier l'existence, l'unicité et les propriétés des solutions. Les intérêts des uns et des autres se rencontrent dans l'étude des résolutions numériques et, comme on l'a souvent remarqué, la mise en oeuvre des possibilités offertes par les ordinateurs modernes a fait beaucoup pour rapprocher mécaniciens et mathématiciens. Nous permettra-t-on d'avancer toutefois que la convergence de pensée des mathématiciens et des mécaniciens sur le thème de l'analyse fonctionnelle va plus profond : la formulation des inéquations variationnelles par exemple n'est souvent pas autre chose que la manière directe d'écrire le principe des puissances virtuelles ; au-delà de la dualité des langages -qui n'a pas été peut être encore parfaitement surmontée au cours de ce symposium- on retrouve, ou plutôt on devine, l'attrait et la fécondité d'une formulation globale, "quasi géométrique", toute chargée de significations physiques.

Point n'est besoin, nous semble-t-il, de présenter très en détail l'ensemble des travaux de ce symposium. Nous avons la conviction que le présent ouvrage sera un instrument de travail fort utile. Le spécialiste y trouvera une mise au point récente sur le sujet qu'il étudie. Les scientifiques -mathématiciens, mécaniciens, ingénieurs, physiciens- qui souhaitent aborder ou approfondir ce thème disposeront avec ce volume d'un document de base où chaque question est abordée sous les angles différents et complémentaires ; c'est une propriété essentielle qui caractérise cet ouvrage parmi ceux traitant de questions analogues. Ainsi chacun peut trouver, en plus de l'introduction aux idées majeures et à la bibliographie du sujet qui l'intéresse, le mode d'approche qui lui convient le mieux. Nous espérons donc que le lecteur ratifiera l'opinion des participants qui ont reconnu la remarquable qualité de la très grande majorité des communications.

Attirons enfin l'attention sur les six Conférences Générales ; le choix fait par le Comité Scientifique s'est révélé, en effet, particulièrement heureux. Leurs différents thèmes donnent une excellente idée des questions que l'on entendait traiter et elles furent effectivement bien souvent illustrées et complétées par les communications. Leurs auteurs ont fait un effort très remarqué et très réussi pour trouver le langage approprié à l'ensemble de l'auditoire sans manquer à la rigueur et à la précision voulues. Dans chacune d'elles on trouve évoquées des situations mécaniques concrètes et d'importance majeure abordées par des techniques mathématiques élaborées et conduisant parfois, au-delà de l'application de ces techniques, à un approfondissement et à un développement des concepts et des méthodes mathématiques.

Voici en effet ce que nous invitons le lecteur à découvrir dans ces Conférences Générales et dans l'ensemble des communications, pour son bénéfice professionnel et surtout pour sa joie personnelle : la fécondité du dialogue mathématique-mécanique que l'on a tenté ici de renouer, et -nous l'espérons- avec un premier succès chargé de promesses pour l'avenir.

Paul GERMAIN — Bernard NAYROLES

Septembre 1975

MEMBERS OF SCIENTIFIC COMMITTEE

Chairman : P. GERMAIN

Université Paris VI
Institut de Mécanique Théorique et Appliquée
Paris, France

J. LIGHTHILL

University of Cambridge
Department of Applied Mathematics and
Theoretical Physics
Cambridge, U.K.

K. KIRCHGÄSSNER

Universität Stuttgart
Mathematisches Institut A
Stuttgart, B.R.D.

J.L. LIONS

Collège de France
Paris, France

G. STAMPACCHIA

Instituto Tonelli
Universita di Pisa
Pisa, Italy

F. STUMMEL

J.W. Goethe Universität Frankfurt/Main
Fuchbereich Mathematik
Frankfurt, B.R.D.

LIST OF THE PARTICIPANTS

AGUIRRE-PUENTE J.	C.N.R.S. Laboratoire d'Aérothermique 4ter, route des Gardes 92190 MEUDON	France
AMIEL R.	Institut de Mathématiques et Sciences Physiques Département de Mathématiques Parc Valrose 06034 NICE - Cédex	France
ANDERSSON B.J.	Kungl. Tekniska Högskolan S-100 44 STOCKHOLM 70	Sweden
ANDRY J.R.	Department of Mechanical Engineering Michigan State University EAST LANSING, Michigan 48823	U.S.A.
ANTMAN S.S.	Department of Mathematics University of Maryland COLLEGE Park Maryland 20742	U.S.A.
ARANTES OLIVEIRA E.R.	Instituto Superior Técnico Laboratorio de Resistencia de materiais Av. Rovisco Pais LISBOA	Portugal
ARTOLA M.	Résidence des Rosiers Bellevue Bâtiment E 33170 GRADIGNAN	France
AUMASSON C.	O.N.E.R.A. 29, avenue de la Division Leclerc 92310 CHATILLON	France
BARDOS C.	Institut de Mathématiques et Sciences Physiques Dept. de Mathématiques-Parc Valrose 06034 NICE- Cédex	France
BAIOCCHI C.	Istituto di Matematica Università di Pavia PAVIA	Italia
BALL D.J.	Dept. of Electrical, Electronic and Control Engineering Sunderland Polytechnic Chester Road SUNDERLAND SR 1 3SD	England
BALL J.M.	Dept. of Mathematics Heriot-Watt University RICCARTON, MIDLOTTIAN	Scotland

BENACHOUR — Institut de Mathématiques et Sciences Physiques, Département de Mathématiques, Parc Valrose, 06034 NICE - Cédex — France

BENJAMIN T.B. — Fluid Mechanics Research Institute, University of Essex, COLCHESTER — England

BERGER M.S. — Belfer Graduate School, Yeshiva University, Amsterdam Ave. & 185-6th St., NEW-YORK , N.Y. 10033 — U.S.A.

BERNADOU M. — I.R.I.A., Domaine de Voluceau, 78150 ROCQUENCOURT — France

BESSONNET G. — Laboratoire de Mécanique, 40, avenue du Recteur Pineau, 86022 POITIERS — France

BISHOP R.E.D. — Dept. of Mechanical Engineering, University College London, Torrington Place, LONDON WCIE 7 JE — England

BLAKELEY W. — Department of Mathematics, Wolverhampton Polytechnic, WOLVERHAMPTON — England

BONA J.L. — Department of Mathematics, The University of Chicago, CHICAGO, Illinois 60637 — U.S.A.

BOSSAVIT A. — E.D.F., Service Informatique et Mathématiques Appliquées, 17, av. du Général de Gaulle, B.P. n° 27, 92140 CLAMART — France

BOUC R. — C.N.R.S., Laboratoire de Mécanique et d'Acoustique, 31, chemin Joseph-Aiguier, 13274 MARSEILLE Cédex 2 — France

BOUCHER M. — 75, voie de Châtenay, 91370 VERRIERES-le-BUISSON — France

BOUJOT Jacqueline — O.N.E.R.A., 29, av. Div. Leclerc, 92320 CHATILLON — France

BRANCHER J.-P. — 1, rue Saint-Antoine, 54250 BOUXIERES-AUX-DAMES — France

BRAUCHLI H. — Institut für Mechanik, Eidgenössiche Technische Hochschule Zürich, Rämistrasse 101, CH-8006 ZÜRICH — Schweiz

BRIERE T.	9, rue du Vert-Buisson 76000 ROUEN	France
BRUN L.	Centre d'Etudes de Limeil B.P. n° 27 94190 VILLENEUVE-St-GEORGES	France
BUDIANSKI B.	Pierce Hall, Harvard University CAMBRIDGE Massachussets 02138	U.S.A.
CHAVENT G.	I.R.I.A. Domaine de Voluceau 78150 ROCQUENCOURT	France
CHIPOT M.	L.E.M.T.A. 2, rue de la Citadelle B.P. n° 850 54011 NANCY- Cédex	France
CIMATTI G.	Istituto Elaborazione Informazione Via S. Maria 46 56100 PISA	Italia
CIMETIERE A.	"Le Studel" - App. 486 86000 POITIERS	France
CIORANESCU Doina-Maria	Université Paris VI Analyse Numérique Equipe de Recherche Associée 215 Tour 55 9, quai Saint-Bernard 75005 PARIS	France
COIRIER J.	Laboratoire de Mécanique 40, avenue du Recteur Pineau 86022 POITIERS	France
COLEMAN B.D.	Mellon Institute of Science Carnegie- Mellon University 4400 Fifth Avenue PITTSBURGH, Pennsylvania 15213	U.S.A.
COLLATZ L.	2000 HAMBURG 67 Eulenkrugstrasse 84	B.R.D.
COMO M.	Dipartimento di Strutture Università della Calabria COSENZA	Italia
COUTRIS Nicole	11, rue Mansard 92170 VANVES	France
DAFERMOS C.	Division of Applied Mathematics Brown University PROVIDENCE, Rhode Island 02912	U.S.A.
DEBORDES O.	C.N.R.S. Laboratoire de Mécanique et d'Acoustique 31, chemin Joseph-Aiguier 13274 MARSEILLE- Cédex 2	France

DESABAYE P.	C.E.A. Centre d'Etudes de Vaujours B.P. n°7 93270 SEVRAN	France
DESTUYNDER P.	C.E.A. Service MA- Centre d'Etudes de Limeil B.P. n° 27 94190 VILLENEUVE-St-GEORGES	France
DO C.	Université de Nantes (E.N.S.M.) 3, rue du Maréchal Joffre 44000 NANTES	France
DRUCKER D.C.	Dean, College of Engineering University of Illinois URBANA, Illinois 61801	U.S.A.
DUVAUT G.	Université P. et M. Curie Place Jussieu 75005 PARIS	France
FABRIZIO M.	Università di Bologna Istituto Matematico "Salvatore Pincherle" Piazza di Porta S. Donato, 5 40127 BOLOGNA	Italia
FILIPPI P.	C.N.R.S. Laboratoire de Mécanique et d'Acoustique 31, chemin Joseph-Aiguier 13274 MARSEILLE Cédex 2	France
FINN R.	Mathematics Department Stanford University STANFORD, California 94305	U.S.A.
FRAEIJS de VEUBEKE B.	Institut de Mécanique 75, rue du Val Benoît 4000 LIEGE	Belgique
FRAENKEL L.E.	Department of Applied Mathematics and Theoretical Physics Silver Street CAMBRIDGE CB3 9EW	England
FREMOND M.	Laboratoire Central des Ponts et Chaussées 58, boulevard Lefebvre 75732 PARIS	France
GAJEWSKI H.	Akademie der Wissenschaften der DDR Zentralinstitut für Mathematik und Mechanik D.D.R. 108 BERLIN Mohrenstrasse 39	D.D.R.
GERHARDT C.	F B Mathematik, Universität Postfach 3980 D-65 MAINZ	B.R.D.

GERMAIN P.	Université de PARIS VI Institut de Mécanique Théorique et Appliquée Tour 66, 4, Place Jussieu 75230 PARIS Cédex 05	France
GEYMONAT G.	Istituto Matematico Politecnico Corso Duca degli Abruzzi 24 10100 TORINO	Italia
GRIMALDI A.	Dipartimento di Strutture Università della Calabria COSENZA	Italia
HERNANDEZ J.	L.A. Numérique T. 55-65, 5e 4, Place Jussieu 75230 PARIS Cédex 05	France
HEWIT J.R.	Dept. of Mechanical Engineering University NEWCASTLE UPON TYNE	England
HEYDEN A.M.A.	Technische Hogeschool Delft Vakgroep Technische Mechanica Mekelweg 2 DELFT	Netherlands
HOGFORS C.	Chalmers University of Technology Division of Mechanics S. 402 GOTEBORG 5	Sweden
HSIAO G.C.	University of Delaware College of Art and Sciences Department of Mechanics 223 Sharp Laboratory NEWARK , Delaware 19711	U.S.A.
IOOSS G.	Institut de Mathématiques et Sciences Physiques Dépt. de Mathématiques, Parc Valrose 06034 NICE -Cédex	France
JANSSENS P.	Université de Bruxelles Faculté des Sciences Appliquées Service de Mécanique 50, avenue F. Roosevelt 1050 BRUXELLES	Belgique
JEAN M.	C.N.R.S. Laboratoire de Mécanique et d'Acoustique 31, chemin Joseph-Aiguier 13274 MARSEILLE Cédex 2	France
JULLIEN Y.	C.N.R.S. Laboratoire de Mécanique et d'Acoustique 31, chemin Joseph-Aiguier 13274 MARSEILLE Cédex 2	France

KALKER J.J.	Department of Mathematics Delft University of Technology Julianalaan 132 DELFT	Netherlands
KESTENS J.	Université Libre de Bruxelles Service d'Analyse des Contraintes Avenue Ad. Buyl 87 1050 BRUXELLES	Belgique
KIRCHGASSNER K.	Universität Stuttgart Mathematisches Institut A D-7 STUTTGART N Herdweg 23	B.R.D.
KOITER W.T.	Delft University of Technology Department of Mechanical Engineering Mekelweg 2 DELFT	Netherlands
KOTORYNSKI W.	University of Victoria Department of Mathematics P.O. Box 1700 VICTORIA BRITISH COLUMBIA Canada V8W 2Y2	Canada
KREYSZIG E.	Department of Mathematics University of Windsor WINDSOR, Ontario	Canada
LABISCH F.	Lehrstuhl für Mechanic II der Ruhr-Universität Bochum 4630 BOCHUM Universitätsstrasse 150 IA3-	B.R.D.
LACHAT J.-C.	CETIM 55, avenue Félix-Louat 60300 SENLIS	France
LADEVEZE P.	Université PARIS VI Institut de Mécanique Théorique et Appliquée - U.E.R. 49 Tour 66, 4, Place Jussieu 75230 PARIS Cédex 05	France
LANCHON Hélène	L.E.M.T.A. 2, rue de la Citadelle B.P. n° 850 54011 NANCY Cédex	France
LEE J.K.	Dept. of Aerospace Engr. & Engr. Mechanics ENS Bldg. 345 The University of Texas AUSTIN, Texas 78712	U.S.A.
LEFEBVRE J.P.	C.N.R.S. Laboratoire de Mécanique et d'Acoustique 31, chemin Joseph-Aiguier 13274 MARSEILLE Cédex 2	France

LENE Françoise	Département de Mécanique Théorique Tour 66- Université PARIS VI 4, Place Jussieu 75230 PARIS Cédex 05	France
LICHNEWSKY A.	Université Paris-Sud-Centre d'Orsay Mathématique- Bât. 425 91405 ORSAY	France
LIONS J.L.	Collège de France 11, place Marcelin Berthelot 75005 PARIS	France
MACERI F.	Istituto di Scienza delle Costruzioni Facoltà di Ingegneria Piazzale Tecchio 80125 NAPOLI	Italia
MAISONNEUVE O.	Institut de Mathématiques Université Montpellier II Place Eugène Bataillon 34060 MONTPELLIER Cédex	France
MARCOIN G.	D.R.M.E. 5bis, avenue de la Porte de Sèvres 75015 PARIS	France
MERCIER B.	I.R.I.A.-Domaine de Voluceau 78150 ROCQUENCOURT	France
MIKHAILOV G.K.	USSR National Committee on Theoretical and Applied Mechanics Leningrad Avenue 7 MOSCOW A-40	U.S.S.R.
MOREAU J.-J.	Université des Sciences et Techniques du Languedoc - Mathématiques Place Eugène Bataillon 34060 MONTPELLIER Cédex	France
NAPOLITANO L.G.	Istituto di Aerodinamica Università degli Studi di Napoli Facoltà d'Ingegneria P. le Tecchio 80125 NAPOLI	Italia
NASTASE Adriana	Aerodynamisches Institut der R.W.T.H. Tempelgraben 55 51 AACHEN	B.R.D.
NAYROLES B.	C.N.R.S. Laboratoire de Mécanique et d'Acoustique 31, chemin Joseph-Aiguier 13274 MARSEILLE Cédex 2	France
NEDELEC J.C.	Ecole Polytechnique Centre de Mathématiques Appliquées Route de Saclay 91120 PALAISEAU	France

NGUYEN Q.S.	Laboratoire de Mécanique des Solides Ecole Polytechnique Route d'Orsay 91120 PALAISEAU	France
NIORDSON F.	Technical University of Denmark Building 404, DK-2800 LYNGBY	Denmark
ODEN J.T.	Department of Aerospace Engr. and Engr. Mechanics ENS Bldg. 345 The University of Texas AUSTIN , Texas 78712	U.S.A.
ORTIZ E.L.	Mathematics Department Imperial College LONDON SW7 2RM	England
OVSJANNIKOV L.	Insitute of Hydrodynamics NOVOSIBIRSK 630090	U.S.S.R.
PLANCHARD J.	Electricité de France Direction des Etudes et Recherches 17, avenue du Général de Gaulle 92140 CLAMART	France
POTIER-FERRY M.	Université de Paris VI- Mécanique 4, place Jussieu 75005 PARIS	France
PRICE W.G.	University College London Torrington Place LONDON WCIE 7 JE	England
PRIMICERIO M.	Università degli Studi Istituto Matematico "Ulisse Dini" Viale Morgagni, 67/A 50134 FIRENZE	Italia
PRITCHARD A.J.	Department of Engineering University of Warwick COVENTRY CV 4 7 AL	England
RAOUS M.	C.N.R.S. Laboratoire de Mécanique et d'Acoustique 31, chemin Joseph-Aiguier 13274 MARSEILLE Cédex 2	France
RIEDER G.	Lehrstuhl und Institut für Technische Mechanik Technische Hochschule Aachen D-51 AACHEN Templergraben 64	B.R.D.
ROBERT J.	Université de Besançon Route de Gray-La Bouloie 25030 BESANÇON Cédex	France
ROGULA D.	Institute of Fundamental Technological Research Swietokrzyska 21 00-049 WARSAW	Poland

ROMANO G.	Dipartimento di Strutture Università della Calabria COSENZA	Italia
ROUGEE P.	58, avenue Schildge 91120 PALAISEAU	France
SATHER D.P.	Department of Mathematics University of Colorado BOULDER , Colorado 80302	U.S.A.
SATTINGER D.H.	School of Mathematics University of Minnesota MINNEAPOLIS, Minnesota 55455	U.S.A.
SAUT J-C.	Université PARIS VII U.E.R. de Mathématiques Tour 45-55 2, place Jussieu 75221 PARIS Cédex 05	France
SAYIR M.	Institut für Mechanik ETH, Rämistrasse 101 8006 ZÜRICH	Schweiz
SCHEURER B.	C.E.A. C.E.L. B.P. n° 27 94190 VILLENEUVE-St-GEORGES	France
SEDOV L.I.	Leninski Gory MGU, Zona II, kv. 84 MOSCOW B-234	USSR
SMOLLER J.A.	University of Michigan Department of Mathematics ANN ARBOR, Michigan 48104	U.S.A.
SOCOLESCU D.	Institut für Angewandte Mathematik Universität Karlsruhe 75 KARLSRUHE Englerstrasse 2	B.R.D.
SOCOLESCU Rodica	Institut für Angewandte Mathematik Universität Karlsruhe 75 KARLSRUHE Englerstrasse 2	B.R.D.
SOLOMON L.	Université de Poitiers Laboratoire de Mécanique 40, avenue du Recteur-Pineau 86022 POITIERS	France
STAMPACCHIA G.	Scuola Normale Superiore Piazza dei Cavalieri 56100 PISA	Italia
STUART C.A.	Institut Battelle 7, route de Drize CH-1227 Carouge GENEVE	Suisse

STUMMEL F.	J. W. Goethe Universität Frankfurt/Main Fachbereich Mathematik 6 FRANKFURT a. M. Robert Mayer Strasse 6-10	B. R. D.
STUMPF H.	Ruhr-Universität Bochum Lehrstuhl für Mechanik II 4630 BOCHUM Universitätstrasse 150 IA 3	B. R. D.
TARTAR L.	Université Paris-Sud Département Mathématique 91405 ORSAY	France
TEMAM R.	Université Paris-Sud Département Mathématique 91405 ORSAY	France
TERRIER M.	C. E. A. -Centre d'Etudes de Limeil B. P. n° 27 94190 VILLENEUVE-St-GEORGES	France
TING T. W.	Department of Mathematics University of Illinois URBANA, Illinois 61801	U. S. A.
TRIBILLON J-L.	D. R. M. E. 26, boulevard Victor 75996 PARIS ARMEES	France
VALLEE	Université de Poitiers Laboratoire de Mécanique 40, avenue du Recteur-Pineau 86022 POITIERS	France
WANG Y.	Université PARIS VI et C. N. R. S. Laboratoire Associé 189 Analyse Numérique- Tour 55-65 4, place Jussieu 75230 PARIS Cédex 05	France
WEISSGERBER V.	Technische Hochschule Fachbereich Mathematik Schlopgartenstrasse 7 D 61- DARMSTADT	B. R. D.
WENDLAND W.	Technische Hochschule Fachbereich Mathematik Schlopgartenstrasse 7-9 D 61- DARMSTADT	B. R. D.
ZANDBERGEN P. J.	Technische Hogeschool Twente afd. T. W. P. O. Box 217 ENSCHEDE	Netherlands
ZEIDLER E.	Sektion Mathematik Karl-Marx Platz 701 LEIPZIG	D. D. R.

CONTENTS

INEQUATIONS QUASI-VARIATIONNELLES DANS LES PROBLEMES A FRONTIERE LIBRE EN HYDRAULIQUE

Claudio BAIOCCHI

Istituto Matematico dell'Università

et

Laboratorio di Analisi Numerica del C.N.R.

Pavia, Italie

L'étude du mouvement des fluides à travers des matériaux poreux conduit en général à des problèmes mathématiques du type "frontière libre" , problèmes qui ont été intensivement traités dans la littérature spécialisée, à cause du grand intérêt des problèmes physiques correspondants (Cf. p. ex. {9}).

Pour fixer les idées, on considérera le problème suivant ([1]) (voir figure ci-dessous). Sur une base horizontale imperméable, deux bassins d'eau, de niveaux y_1 , y_2 , sont séparés par une digue en matériau poreux ; l'eau filtre du bassin plus élevé au bassin moins élevé, et on veut déterminer les grandeurs physiques (telles que pression, débit, vitesses, ...) et géométriques (partie mouillée de la digue, lignes de courant, ...) associées au mouvement. On va aussi supposer la digue en matériau incompressible, isotrope, homogène, et le flux incompressible, stationnaire, irrotationnel, bidimensionnel ; on suppose aussi négligeables les effets de capillarité.

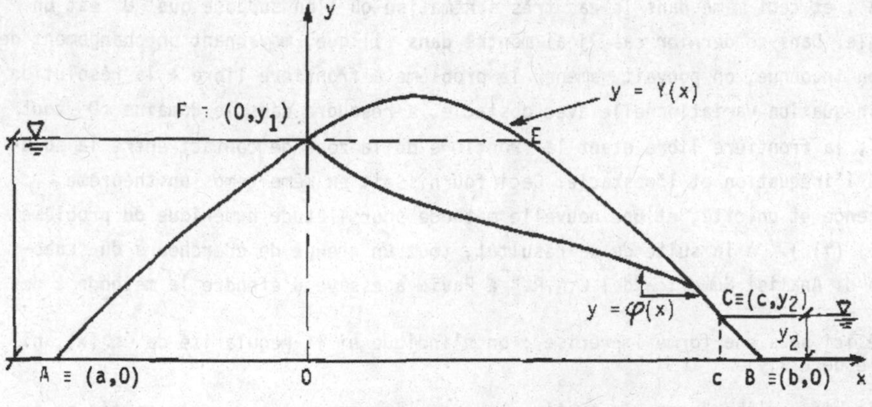

([1]) Pour des problèmes plus généraux, ainsi que pour une ample bibliographie sur le sujet, on renvoie à {6}, {7}.-

Si l'on désigne par D la section de la digue ; par Ω la partie mouillée de D ; par p(x,y) la pression au point (x, y) de D ; et par $y = \varphi(x)$ la "partie supérieure" du bord de Ω, moyennant la loi de DARCY, on aboutit au problème mathématique suivant [2] :

PROBLEME A - *Trouver un sous-ensemble* Ω *de* D *, délimité par* $\widehat{AF}$, $\widehat{AB}$, $\widehat{BC}$ *et une courbe "régulière"* $y = \varphi(x)$ *joignant* F *à* C *, tel que l'on puisse résoudre dans* Ω *le problème aux limites* [3] :

(1) $\Delta u = 0$ dans Ω $\quad (\Delta = \frac{\partial^2}{\partial x^2} + \frac{\partial^2}{\partial y^2})$;

(2) $u_{|\widehat{AF}} = y_1$; $u_{|\widehat{BC}} = y_2$; $u_{y|\widehat{AB}} = 0$; $u_{|y=\varphi(x)} = y$;

(3) $\frac{\partial u}{\partial n} = 0$ le long de la partie intérieure à D de la courbe $y = \varphi(x)$ $\quad (\frac{\partial}{\partial n}$ = dérivée normale).

Une fois le problème A résolu, on récupère aisément les autres grandeurs associées au mouvement ; par exemple (en prenant le poids spécifique du liquide et le coefficient de perméabilité égaux à 1) la vitesse est donnée par - grad u, le débit est donné par $-\int_0^{y_1} u_x(x,t)dt$, la pression est donnée par $p(x,y) = \begin{cases} u(x,y)-y & \text{dans } \Omega \\ 0 & \text{dans } D\setminus\Omega \end{cases}$

Bien qu'il s'agisse d'un problème très étudié dans la littérature spécialisée, on n'avait pas, à ma connaissance, un résultat d'existence et unicité pour le problème A ; et ceci même dans le cas très schématisé où l'on suppose que D est un rectangle. Dans ce dernier cas, j'ai montré dans {1} que, moyennant un changement de fonction inconnue, on pouvait ramener le problème à frontière libre à la résolution d'une inéquation variationnelle avec obstacle, à résoudre dans le domaine D tout entier ; la frontière libre étant la frontière de la zone de contact entre la solution de l'inéquation et l'obstacle. Ceci fournissait en même temps un théorème d'existence et unicité, et une nouvelle méthode pour l'étude numérique du problème (Cf. {5} [4]). A la suite de ce résultat, tout un groupe de chercheurs du "Laboratorio di Analisi Numerica del C.N.R." à Pavia a essayé d'étendre la méthode à des

---

[2] Posé ici sous une forme imprécise : on n'indique ni la régularité de $\varphi(x)$, ni celle de u(x,y).

[3] "à frontière libre" car une partie de $\partial\Omega$ est inconnue et sur cette partie on impose "une condition de trop" (voir (3)).

[4] Méthode qui, d'un côté, est parfaitement justifiée sur le plan théorique ; d'un autre côté, du point de vue pratique, par rapport aux autres méthodes jusqu'ici proposées, elle a apporté un gain sensible à la fois en simplicité de programmation et en temps d'exécution.

cas moins schématisés (géométrie générale, perméabilité variable, problèmes non stationnaires, présence de plusieurs fluides immiscibles, ...) en se heurtant toutefois à une grosse difficulté ; les résultats obtenus (cf. {6}) nécessitaient, grosso modo, la restriction suivante : la"paroi de droite" $\widehat{ECB}$ de la digue doit être verticale.

Il y avait deux raisons qui rendaient inévitable cette restriction : en effet on s'est aperçu maintenant que, premièrement, la géométrie générale nécessite un instrument plus puissant que les inéquations variationnelles ([5]) ; deuxièmement, comme on va le voir dans un moment, si la paroi $\widehat{ECB}$ n'est pas verticale, le problème A, tel qu'il est formulé, n'admet pas d'unicité ; il admet des solutions non acceptables du point de vue physique, et qu'il faut donc éliminer en imposant d'autres conditions.

Pour mieux dégager tout cela on va d'abord transformer le problème A en exploitant toutes les inconnues en termes de la pression $p(x,y)$, et en précisant les hypothèses de régularité.

On remarque d'abord que, si u est une fonction régulière satisfaisant (1),(2), d'après le principe du maximum (sous la forme de Hopf) on doit avoir $u(x,y) > y$ dans Ω([6]) et sur $\widehat{AB}$; en termes de la pression $p(x,y)$ (qui vaut $u(x,y)-y$ dans Ω et 0 hors de Ω) on a donc :

(4) $$p(x,y) \geqslant 0 \text{ dans } \bar{D} \quad ; \quad p_{|\widehat{AB}} > 0$$

(5) $$\Omega = \{(x,y) \in D \mid p(x,y) > 0\} .$$

Pour ne pas imposer a priori trop de régularité sur φ, on va supposer seulement que Ω est un "sous-graphe", à savoir si $(x_0, y_0) \in \Omega$ et si $y \in]0, y_0[$, on a $(x_0, y) \in \Omega$; soit, en termes de p :

(6) $$\{(x_0, y_0) \in D,\ p(x_0, y_0) > 0 ,\quad y \in]0, y_0[\} \Longrightarrow p(x_0, y) > 0 .$$

Ceci posé, on va montrer que le problème A, en termes de p(x,y) équivaut ([7]) à:

PROBLEME B - *On cherche une fonction* p *telle que :*

(7) $$p \in C^\circ(\bar{D}) \cap H^1(D) \quad ([8])$$

([5]) et précisément il faut utiliser les inéquations quasi-variationnelles. Les I.Q.V. ont été récemment introduites par Bensoussan et Lions {8} pour l'étude de problèmes de contrôle impulsionnel ; dans ce cas aussi elles traduisent des problèmes à frontière libre.

([6]) On suppose D, Ω ouverts ; $\bar{D}$, $\bar{\Omega}$ désigneront les fermetures correspondantes.

([7]) à part les précisions sur la régularité !

([8]) $H^1(D)$ est l'usuel espace de Sobolev ; (7) signifie que p est une fonction continue sur $\bar{D}$ et dont les dérivées p_x, p_y (au sens des distributions) sont dans $L^2(D)$.

qui satisfasse (au sens de $C^\circ(\bar{D})$*)* (4) *et :*

(8) $\quad p_{|\widehat{AF}} = y_1 - y \quad ; \quad p_{|\widehat{FEC}} = 0 \quad ; \quad p_{|\widehat{BC}} = y_2 - y \quad ,$

et telle que, si l'on définit Ω *par* (5), *on ait la validité de* (6) *et de :*

(9) $\left\{ \begin{array}{l} \text{pour tout } \psi \in C^1(\bar{D}) \text{ nul dans un voisinage} \\ \text{de } \widehat{AFECB} \text{ on a :} \\ \int_\Omega \operatorname{grad}(p+y) \cdot \operatorname{grad} \psi \, dx\, dy = 0 \, . \end{array} \right.$

En effet on remarquera que (puisque p=u-y dans Ω) (9) est l'usuelle formulation variationnelle de (1), (3) et $\frac{\partial u}{\partial y}\Big|_{\widehat{AB}} = 0$; les autres relations de (2) sont contenues dans (8) ([9]).

Maintenant il est immédiat de voir que le problème B admet toujours une solution en général "non physique" : il suffit de résoudre en p^* :

$$\left\{ \begin{array}{l} \Delta p^* = 0 \quad \text{dans } D \\ \text{conditions aux limites :} \\ (8) \quad \text{et} \quad \frac{\partial p^*}{\partial y}\Big|_{\widehat{AB}} = -1 \end{array} \right.$$

et on s'aperçoit aisément (principe du maximum) que $p^* > 0$ dans $D \cup \widehat{AB}$; donc (4) est satisfaite, et (5) fournit $\Omega = D$; (6) est alors valable ainsi que (9), donc p^* est une solution du problème B.

Désignant par $y = Y(x)$ le "bord supérieur de D" on aura donc, en correspondance à $p = p^*$, $\varphi(x) = Y(x)$ pour $0 \leqslant x \leqslant c$; et on remarquera que cette solution "non physique" ne peut pas être éliminée en ajoutant des hypothèses du type :

(10) la fonction φ ([10]) est décroissante ,

car, si par exemple $y_1 = \max Y(x)$ et $Y(x)$ est convexe, en correspondance à $p = p^*$, la condition (10) est remplie (et $\Omega = D$ n'est pas la "vraie" solution physique) ; on ne peut pas non plus imposer a priori $\Omega \neq D$, car dans certains cas ([11]) la solution physique correspond effectivement à $\Omega \equiv D$ (et $p \equiv p^*$).

Dans {2} j'ai proposé de remplacer le problème B par le problème C suivant :

([9]) En ce qui concerne $u = y$ le long de $y = \varphi(x)$ (à savoir p = 0 le long de la partie supérieure de $\partial\Omega$), elle est contenue dans (8) pour la partie qui est sur ∂D ; et découle de (7) et de la définition de Ω en ce qui concerne $\partial \Omega \cap D$.

([10]) On remarquera que, sous la formulation B du problème, on peut définir φ par la formule :
pour $x \in [0,c]$, $\varphi(x) = \sup \{y \mid p(x, y) > 0\}$

([11]) Par exemple si D est un triangle, de sommets A, B, F, avec l'angle en F aigu.

PROBLEME C - *On cherche* $p(x,y)$ *solution du problème* B *et qui de plus vérifie :*

$$(11) \quad \begin{cases} \text{la fonction } x \longmapsto \int_0^{Y(x)} p_x(x,t)\,dt \\ \text{est non décroissante pour } x \in [o,c] \end{cases}$$

Il s'agit d'une hypothèse très naturelle du point de vue physique ; elle traduit le fait que le débit de la digue à travers la section verticale $\{x = x_0\}$ (débit qui, à une constante près, vaut $-\int_0^{Y(x_0)} p_x(x_0,t)\,dt$) est une fonction non croissante de x_0, pour $x_0 \in [o,c]$ ([12]).

Pour l'étude du problème C, on va d'abord effectuer un changement de fonction inconnue semblable à celui effectué dans {1} pour le cas où D est un rectangle ; précisément on pose :

$$(12) \qquad U(x,y) = \int_0^y p(x,t)\,dt \qquad \forall\ (x,y) \in \bar{D} \ ;$$

Evidemment si l'on connait U on évalue p par :

$$(13) \qquad p(x,y) = U_y(x,y) \qquad \forall\ (x,y) \in \bar{D}\ .$$

Les relations suivantes sont immédiates :

$$(14) \qquad U_{|\widehat{AB}} = 0 \quad ; \quad U_{y|\widehat{AF}} = y_1 - y \quad ; \quad U_{y|\widehat{FEC}} = 0 \quad ; \quad U_{y|\widehat{BC}} = y_2 - y \quad ;$$

et d'ailleurs de (9) (qui entraîne en particulier $\Delta p = -D_y\,\chi_\Omega$ ([13])), on déduit :

$$(15) \qquad -\Delta U = \chi_\Omega \quad \text{dans } D\ .$$

En ce qui concerne Ω on peut bien sûr écrire (grâce à (5),(13)) :

$$\Omega = \{\ (x,y)\ |\ U_y(x,y) > 0\ \}\ ;$$

toutefois cette relation est "peu commode" par rapport à l'équation (15). D'ailleurs de (6), on a aussi :

$$(16) \qquad U(x,y) \leq U(x, Y(x)) \qquad \forall\ (x,y) \in \bar{D}$$

$$(17) \qquad \Omega = \{(x,y) \in D\ |\ U(x,y) < U(x, Y(x))\}$$

([12]) En termes de u on peut aussi montrer que (11) équivaut à la relation (elle aussi évidente du point de vue physique , et qui est plus commode pour traiter le problème tridimensionnel) :

$\frac{\partial u}{\partial n} \leq 0$ le long de la partie commune à $y = \varphi(x)$ et $y = Y(x)$ ($\frac{\partial}{\partial n}$ dérivée normale extérieure) .

([13]) χ_Ω désigne la fonction caractéristique de Ω dans D, à savoir

$$\chi_\Omega(x,y) = \begin{cases} 1 & \text{pour } (x,y) \in \Omega \\ 0 & \text{pour } (x,y) \in D \setminus \Omega\ . \end{cases}$$

En ce qui concerne la régularité de U, on remarquera que U résout un problème aux limites de type mêlé (Dirichlet - dérivée oblique ; cf. (14)) avec un second membre dans $L^\infty(D)$ (cf. (15)) ; à la suite d'une étude assez fine de la régularité de la solution d'un problème de ce type ([14]) on peut montrer que l'on a :

(18) $\forall\, r \in [\,1, +\infty\,[\;,\quad U \in W^{2,r}(D)$ ([15]) .

Finalement un calcul immédiat montre que (11) s'exprime, en termes de U, sous la forme :

(19) $x \longmapsto U(x,Y(x))$ est convexe sur $[\,o, c\,]$.

De (1.14), ..., (1.19) on peut déduire plusieurs types d'inéquations quasi - variationnelles. Le type le plus simple ([16]) consiste à poser, pour v(x,y) fonction "régulière" sur $\bar{D}$:

$$(Mv)(x,y) = \begin{cases} v(x,y) + 1 & \text{pour } (x,y) \in \bar{D},\quad x \notin [\,o,c\,] \\ \text{conv}\,[\,v(x,Y(x))^+] & \text{pour } (x,y) \in \bar{D},\quad x \in [\,o,c\,] \end{cases}$$ ([17])

On a, grâce à (15), (16), (17), (19) :

(20) $U \leqslant MU \;;\; \Delta U \geqslant -1 \;;\; (U - MU)(\Delta U + 1) = 0$ dans D

et le problème (14), (20) est un problème typique quasi-variationnel (avec opérateur M "non local", comme dans {8}).

Pour un problème "très proche" de (14), (20), moyennant la technique des I.Q.V., j'ai montré dans {3} que :

Il existe deux fonctions, U_{min} , U_{max} *telles que :*

i) la formule (13) *avec* $U = U_{min}$ *donne une solution du problème* C *; et de même pour* $U = U_{max}$ ([18]) ;

ii) toute solution p(x,y) *du problème* C *est telle que la fonction* U *définie par* (12) *vérifie :*

(20) $U_{min}(x,y) \leqslant U(x,y) \leqslant U_{max}(x,y) \qquad \forall\,(x,y) \in \bar{D}$.

([14]) Les "points difficiles" sont évidemment A et B, où l'on a à la fois saut de conditions aux limites et points singuliers du bord (y=Y(x) est supposée régulière). Pour des raisonnements semblables cf. {3}.

([15]) à savoir $U, U_x, U_y, U_{xx}, U_{yy}, U_{xy} \in L^r(D)$.

([16]) même s'il n'est pas le plus commode ; cf. le n°1 de {3} où l'on construit toute une famille de problèmes quasi-variationnels satisfaits par U .

([17]) où $t^+ = \max(t, 0)$ et conv f = enveloppe convexe de f .

([18]) On a donc existence pour le problème C.

La démonstration de ce théorème est d'ailleurs "de type constructif", à savoir : on peut évaluer (à la machine) des approximations de U_{min}, U_{max} ; et les résultats numériques obtenus au L. A. N. à Pavie suggèrent la conjecture que l'on a toujours :

$$U_{min}(x,y) \equiv U_{max}(x,y) \tag{21}$$

donc (cf. (20)) unicité pour le problème C.

En effet on sait démontrer la validité de (21) si $\widehat{ECB}$ est verticale (cf.{4}); dans le cas général la validité de (21) est un problème ouvert ; de même ouvert est le problème de la "régularité" de la frontière libre : par exemple dans le cas où D est un rectangle on sait démontrer que $y = \varphi(x)$ (définie dans (^{10})) est une fonction continue, strictement décroissante, analytique sur $]o, c[$; des propriétés analogues dans le cas général ne sont pas connues.

BIBLIOGRAPHIE -

{1} BAIOCCHI (C.), *Su un problema a frontiera libera connesso a questioni di idraulica*, Ann. Mat. pura e appl. XCII (1972).

{2} BAIOCCHI (C.), C.R.Acad.Sc.Paris, 278 (1974), 1201-1204 ; et Conférence au Congrès international des Mathématiciens, Vancouver, 1974.

{3} BAIOCCHI (C.), *Studio di un problema quasi variazionale connesso a problemi di frontiera libera*, A paraître au Boll.U.M.I. (1975).

{4} BAIOCCHI (C.), Travail en cours de rédaction.

{5} BAIOCCHI (C.), COMINCIOLI (V.), GUERRI (L.), VOLPI (G.), *Free boundary problems in the theory of fluid flow through porous media : a numerical approach*, Calcolo, X (1973).

{6} BAIOCCHI (C.), COMINCIOLI (V.), MAGENES (E.), POZZI (G.), *Free boundary problems in the theory of fluid flow through porous media : existence and uniqueness theorems*, Ann. di Mat. pura e appl. XCVII (1973).

{7} BAIOCCHI (C.), MAGENES (E.), *Sur les problèmes à frontière libre* ... (en russe), Ouspeki Mat. Nauk. 1974.

{8} BENSOUSSAN (A.), LIONS (J.L.), C.R.Acad.Sc.Paris, 276 (1973), 1189-1193 et 1333-1337.

{9} HARR (M.E.), *Groundwater and seepage*, Mc.Graw Hill, New-York, 1967.

THE ALLIANCE OF PRACTICAL AND ANALYTICAL INSIGHTS INTO THE NONLINEAR PROBLEMS OF FLUID MECHANICS

T. Brooke Benjamin

Fluid Mechanics Research Institute,
University of Essex, Colchester, UK

The field of this Symposium is evidently one of growing importance and popularity, and various good reasons can be given for its vitality. My special intention in this lecture is to emphasize its status as a branch of applied science, a status well established by existing achievements. That is, I shall in all my remarks represent the standpoint taken by the user of functional analysis for practical scientific aims, rather than the standpoint of the mathematician concerned with the intellectual discipline of functional analysis for its own sake. Everybody who has even a superficial acquaintance with functional analysis must be impressed by the great sweep and power of the ideas, and must respect the pioneers of the subject and those contemporary mathematicians who continue to advance it fundamentally. The subject abounds with brilliant and deeply satisfying accomplishments - say, the Hahn-Banach theorem and its prolific range of applications, the classical Leray-Schauder theory of topological degree and its modern generalizations, the Sobolev classifications of function spaces, the Lyusternik-Shnirel'man theory of critical points on infinite-dimensional surfaces - and everybody will have his own choices for particular esteem. But many people, like myself, are excited by these mathematical resources not only for their intrinsic splendour but also for their potentiality as means to another end, namely the solution of theoretical problems posed by the physical world. Such problems make, of course, nice exercises for the mathematical machinery originally developed in abstract, but my claim is much more than this. The standpoint I wish to represent recognizes that the tools available in functional analysis can sometimes be supremely expedient in their applications to physical problems, winning ground that is genuinely valuable by the criteria of good science. In this respect the powerful mathematical methods highlighted in this Symposium are complementary to other methods of investigation that are generally no less productive, and the interaction amongst the different styles of research is the aspect I hope particularly to illustrate. I shall refer presently to specific examples in fluid mechanics, a subject rich in challenging nonlinear problems some of which have already yielded to the methods of functional analysis and of other branches of pure mathematics, but a subject still posing many mysteries.

In the present company there is hardly need to defend the proposition that interplay between mathematics and the physical sciences is mutally profitable. We may defer to overwhelming historical evidence of its value if any argument is necessary. The essence of the benefit was beautifully expressed by Jacques Hadamard in his monograph 'The Psychology of Invention in the Mathematical Field' (1945). After noting that many important discoveries - in physics as well as mathematics - are made without any possible applications being foreseen, he went on to say, "We must add, however, that, conversely, application is useful and eventually essential to theory by the very fact that it opens new questions for the latter. One could say that application's constant relation to theory is the same as that of the leaf to the tree: one supports the other, but the former feeds the latter." Hadamard's aphorism sums up the case perfectly, but let us particularize it in terms of people. Nature, the subject of the applications, provides magnificently serious problems for mathematicians, whose resources of logical precision can sometimes achieve refined insights into the workings of natural phenomena. But practical scientists may be better equipped to point out the most significant problems, also to recognize and interpret the clues that nature offers towards their solution. The power of mathematics becomes transcendent when a problem is abstracted from its original natural setting, cleared of the inexactnesses inevitable in practice and posed tidily in mathematical language. But to reach this stage meaningfully, also sometimes to point the way on, physical intuition and proper appreciation of the experimental facts are the crucial implements. It is well to remember that many of the most significant advances in the science of continuum mechanics have been made, and continue to be made, by potent combinations of intuition and comparatively elementary mathematics - by means that might perhaps seem intolerably crude to some of us here. This remark does not disparage the work of applied mathematicians who trouble to uphold high standards of mathematical rigour: rather, it underlines the great potentiality of allying styles of research that can be supremely powerful on their respective grounds. Good science comes in many guises, and the only universal test of its value is that it opens up new realms of understanding.

To make a brief appreciation of work applying rigorous mathematics to problems in fluid mechanics, most of it may be classified in one or other of three broad categories as follows:

(i) Justification and perhaps generalization of tentative descriptions - such as linearized or asymptotic approximations - which may already have extracted the main scientific content of a given problem. Here the value of sound mathematics is to consolidate and refine theories. Work of this kind seldom wins the enthusiasm of the applied scientists who first solved the problems in question tentatively but correctly. It is vital to long-term progress, however, which is bound sooner or later to depend on the secure foundations thus established.

(ii) Confirmation that idealized models of physical processes are free from extraneous catastrophes. Here the accomplishment is to give solid support to the craft of mathematical model-making, particularly for evolutionary processes. Natural phenomena are generally too complicated for comprehensive description, so, guided by intuition, simplified models are propounded in the hope that the study of them - analytical or numerical - will provide essential explanations of real events. The procedure forms the mainstream of all applied science, and is its most exciting intellectual exercise, but the rationale is, of course, inherently speculative. The mathematician's responsibility is to check rigorously the behaviour of such models, which function in an abstract universe where mischievous, quite unnatural singularities of behaviour may occur and where physical intuition may be quite unreliable. The intuitional model-builder will hope that the model problem is well posed in some way acceptable as a simulation of nature, but often only the mathematician can verify this hope.

(iii) Provision of useful qualitative information that is inaccessible by any other means. This is perhaps the most rewarding role for applied functional analysis. The methods available in functional analysis - and its intersections with topology, differential geometry and global analysis - have, as their supreme potentiality from the standpoint of applied science, the capacity to establish definite conclusions about the solutions of nonlinear problems that are beyond the scope of old-fashioned constructive methods of anlaysis. A few specific nonlinear problems in fluid mechanics have already been elucidated in this way, and it seems to be one of the most promising avenues for future progress, in particular with regard to the very difficult problems of turbulence. Clearly, the new methods are tools for exploration as well as consolidation in theoretical fluid mechanics.

To complete these introductory remarks, here are two general reflections on the technicalities of applied functional analysis. First, let us recall a common feature of the methods in question that distinguishes them from the tentative and descriptive methods still serving much practical research in mechanics. It is that, when seeking to establish solutions with regularity properties essential to the original physical problem, the need usually arises to broaden the class of mathematical objects under consideration, generally functions. Typically, the abstract notion of a weak solution is introduced as a vital step in the argument, whereas the ultimate aim is to pin down a more prosaic mathematical object. Otherwise expressed, the need is to reason first in terms of coarse topologies, because the finer topologies seemingly appropriate to descriptions of the physical context are intractable for conclusive arguments. The matter is, of course, tied up with the conception of comleteness of function classes upon which most existence theories rely. This characteristic logical device is easily wielded by pure mathematicians, but it is notably foreign to the intuitive thinking applied by, for instance, aerodynamicists and meteorologists to problems of fluid mechanics. Note, however, that the latter style of thinking has something in common superficially with global analysis, which too is largely concerned with vector fields whose smoothness is taken for granted. (Some relevant work of Arnold and others in this area will be mentioned later.)

The second point concerns the efficiency of functional-analytic methods in providing for generalizations of a given problem. Pure mathematicians are, by the ethos of their subject, disposed to establish any proposition with the utmost generality possible, and the same tendency is naturally prominent in applied-mathematics research where abstract methods are used. Obvious practical advantages may be achieved thereby: for example, it may be useful to know that certain qualitative properties of the solution of a hydrodynamic problem are common to a wide range of boundary geometries. The criteria of mathematical and practical interest are generally different, however, and in applied mathematics it is often true that broadly scientific considerations rather than narrowly technical ones indicate what elaborations of a theory may be significant and worth pursuing. In particular, there is always good scientific sense in distinguishing clearly between, on the one hand, situations that are generic or dense in the ensemble of possibilities and, on the other, situations that are possible yet extraordinary. The former usually deserve priority, being the heart of the scientific problem, even if the latter may be specially challenging and interesting by virtue of their exceptionality. This philosophic principle in fact has a respectable place in other branches of mathematics, notably topology, and it is important also in human affairs. For instance, it is perfectly expressed by the well-known legal maxim, 'Hard cases make bad law'.

I now turn to two specific problems which have been studied recently by myself and colleagues in the Fluid Mechanics Research Institute, and which may in a modest way exemplify a few of the foregoing general ideas.

I. VORTEX RINGS: AN APPLICATION OF VARIATIONAL METHODS

To bring out the point of this example, the dynamical problem will have to be explained in some detail before functional analysis is evoked. The example illustrates that appreciation of the physical aspects can be the key to an expedient theory, indicating the most profitable uses of rigorous mathematics, and that in turn analysis can suggest interesting physical interpretations.

Perfect-fluid theory on the lines now to be summarized seems adequate to explain the main characteristics of vortex rings, which comprise an impressive class of fluid motions often observed in nature. (Smoke rings are the most familiar example. A beautiful example on a small scale may be observed by letting a drop of ink fall from a height of about 3 cm into a glass of still water.) Vortex rings manifest a remarkable property of permanence, and so theoretical questions about their stability are no less interesting scientifically than questions about their existence as special solutions of the dynamical equations.

A global existence theory for steady vortex rings in an ideal fluid has recently been given by Fraenkel & Berger (1974), and they used a variational method. Their approach presents formidable technical difficulties, however, and though they were admirably resourceful in overcoming these difficulties it seems fair to say that the complexity of their analysis is somewhat disproportionate to the status of the results achieved. One of the functionals crucially involved in the analysis has no physical significance, and for this reason no relation can be seen between the particular steady motions established and neighbouring time-dependent motions. The alternative approach now proposed keeps much closer conceptually to the original physical problem, and consequently the results have greater potentiality for explaining what happens when vortex rings are realized. Also, many of the analytical difficulties seem to be obviated, and, as the truly outstanding advantage, the result concerning existence has an immediate implication concerning stability.

In respect of the stability problem it is very relevant to refer to the work of V.I. Arnold who, in an important series of papers (e.g. 1965, 1966a, 1966b), introduced several new ideas about the hydrodynamics of perfect fluids. His deepest contributions were presented in the language of differential geometry, and the ideas in question have since been developed by other global analysts including J. Marsden, D.G. Ebin and R. Abraham. The basic consideration is that a perfect fluid with rigid boundaries, on which the motion is therefore tangential everywhere, is a system whose configuration space is an infinite-dimensional Lie group, namely the group of volume-preserving diffeomorphisms. And the possible motions of the fluid, that is, solutions of the Euler dynamical equation, are geodesics on this group. Much profound theory has been worked out on this basis, but a deficiency still remaining is the lack of a global existence theorem for the initial-value problem in three space dimensions. Only a local, small-time result is so far available. [We should note here, incidentally, that the present time-dependent problem is free from this difficulty, being in effect a problem of two-dimensional flow for which special case global results have been obtained (e.g. Kato 1967).] On the whole this class of mathematical work on hydrodynamics does not appear to have much practical bearing, but one of Arnold's original discoveries is definitely valuable in respect of applications. This is a general variational principle for steady flows, and the principle offers a means of proving Liapunov stability in certain cases. The general formalism of Arnold's method is very difficult to apply, however, and from a practical point of view the perhaps most valuable aspect of Arnold's discovery is simply that, for any specific problem, there will always be an underlying variational principle whose useful form may be worked out ad hoc. In fact the following analysis closely follows an application discussed in one of Arnold's papers (1965).

DYNAMICAL EQUATIONS. We consider axisymmetric motions, without swirl, of an incompressible inviscid fluid with unit density. The velocity $\underline{u}$ of the fluid everywhere satisfies the condition of mass conservation div $\underline{u} = 0$; accordingly, with respect to cylindrical polar coordinates (r,θ,z), it can be represented by

$$\underline{u} = (-\psi_z/r,\ 0,\ \psi_r/r), \qquad (1)$$

where $\psi = \psi(r,z,t)$ is the Stokes stream-function. The fluid being supposed unbounded, the motion is represented by the evolution with time t of the scalar function ψ defined on the half-plane $r \geqslant 0$, $-\infty < z < \infty$ (henceforth denoted by HP). The kinematical boundary and asymptotic conditions are

$$\psi = O(r^2) \text{ as } r \to 0, \qquad \psi, \psi_r, \psi_z \to 0 \text{ as } r^2+z^2 \to \infty. \qquad (2)$$

The vorticity of the fluid is

$$\text{curl}\ \underline{u} = (0, \omega, 0) \quad \text{with} \quad \omega = -\frac{1}{r}\psi_{zz} - \left(\frac{1}{r}\psi_r\right)_r,$$

and we shall particularly need to consider the function

$$\zeta = \omega/r = \mathcal{L}\psi, \quad \text{say.}$$

It is in many respects convenient to take $y = \frac{1}{2}r^2$ instead of r as the first independent variable. In terms of this,

$$\mathcal{L}\psi = -\frac{1}{2y}\psi_{zz} - \psi_{yy},$$

and the volume element is $d\mu = 2\pi dy dz$.

Taking the curl of the Euler equation of motion, one obtains

$$\zeta_t + \partial(\psi, \zeta) = 0, \tag{3}$$

where the shorthand $\partial(\psi,\zeta)$ is introduced for $\partial(\psi,\zeta)/\partial(y,z)$. In view of the representation (1) of the velocity components, (3) means that the total derivative $d\zeta/dt$ following the motion is zero. Thus, the value of ζ for any infinitesimal ring of fluid particles remains constant, and the μ-measure of any set of values assumed by ζ is therefore an invariant of the motion. In other words, the functions ζ of (y,z) evolved during any time interval are rearrangements of the initial function. For the motions in question the support of ζ is bounded, so that the integral

$$I = \int_{HP} y\zeta\, d\mu = \int_{\text{supp}(\zeta)} y\zeta\, d\mu$$

is meaningful, and by use of (3) and (2) it appears that I is invariant with t. In fact I is identifiable with the total impulse of the motion, and its invariance accords with the principle of momentum conservation expressed in a form covering the case of infinite fluids (cf. Benjamin 1970, §3.8).

For a motion that is steady in a frame of reference translating at constant speed c in the axial direction, so that $\psi = \psi_0(y,z')$ and $\zeta = \zeta_0(y,z')$ with $z' = z - ct$, equation (3) reduces to

$$\partial(\psi_0 - cy, \zeta_0) = 0.$$

This implies that ζ_0 is a function of $\psi_0 - cy$ alone, thus

$$\zeta_0 = \mathcal{L}\psi_0 = F(\psi_0 - cy), \tag{4}$$

and we note that $\psi_0 - cy = \Psi(y,z')$, say, is the stream-function for the steady motion observed in the reference frame (y,z'). A vortex ring is a motion of this kind for which $F(\Psi)$ is a non-decreasing function, posit-

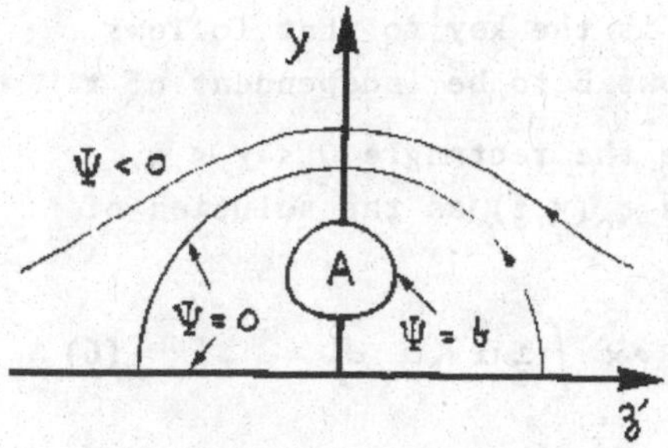

ive for $\Psi > b \geq 0$ and zero for $\Psi \leq b$. The accompanying diagram is sketch of streamlines Ψ = const. The region A = supp(ζ_0), which we shall call the vortex core, is usually connected and separate from the z'-axis, so corresponding to a torus in the physical domain, and outside it the motion is irrotational. It is possible, however, that A includes the axis, and one such case (Hill's vortex) will be discussed later.

The existence problem for vortex rings is to show that the nonlinear equation (4) has a solution complying with the conditions (2). Note that it is a free-boundary problem in the respect that A cannot be prescribed. In the method of treatment adopted by Fraenkel & Berger (1974), ψ_0 is taken as dependent variable and difficulties arise because the support of this function is unbounded. For just this reason, standard 'direct' methods of the calculus of variations - which concern weakly continuous functionals - are unavailing for direct application to the problem. The difficulty was obviated by considering a restricted form of the problem, on a bounded domain with zero boundary conditions for ψ_0: this provides the essential compactness property, and, on the basis of estimates for the solution thus established, the arbitrary restriction is finally removed. A variational principle involving an isoperimetrical constraint was used, which relates to a form of (4) with right-hand side $\lambda F(\Psi)$, where F is a prescribed function and λ an unknown Lagrange multiplier.

The alternative approach exploits the fact that the identity $\zeta = \mathcal{L}\psi$ is invertible, thus

$$\psi = \int_{HP} k(r, z; \hat{r}, \hat{z})\, \zeta(\hat{r}, \hat{z})\, d\hat{\mu} = \int_{supp(\zeta)} k\zeta\, d\hat{\mu} = K\zeta, \quad say. \tag{5}$$

Here the Green function k, corresponding to the conditions (2), is well known: it is just the normalized stream-function for an infinitesimal vortex ring, and several expressions for it are given in Lamb's textbook (1932, §161). For $y > 0$, $\hat{y} > 0$, k is positive and a symmetrically decreasing function of $z - \hat{z}$ (i.e. even and decreasing with $|z - \hat{z}|$). Now, by Green's theorem combined with (2) and the identity $\zeta = \mathcal{L}\psi$, the kinetic energy of the motion described by ψ has the alternative expressions

$$2E = \int_{HP} \frac{1}{r^2}\left(\psi_r^2 + \psi_z^2\right) d\mu = \int_{supp(\zeta)} \zeta\psi\, d\mu = \int_{supp(\zeta)} \zeta K\zeta\, d\mu,$$

and the last of these, a functional of ζ, is the key to what follows. Note that the Euler equation of motion shows E to be independent of t.

VARIATIONAL PRINCIPLE. Let D denote the rectangle $0 < y < \alpha$, $-\beta < z < \beta$. We characterize the function $\zeta_0(y,z)$ as the solution of the variational problem

$$E(\zeta_0) = \max E(u) = \max \int_D \tfrac{1}{2} u K u \, d\mu \tag{6}$$

subject to constraints

(i) $\int_D z u \, d\mu = 0$ (normalization to factor out translations with respect to z),

(ii) $\int_D y u \, d\mu = I$ fixed,

(iii) u belong to the set of rearrangements on D of a bounded non-negative function u_0, such that $\mathrm{supp}(u_0) \subset D$ is a connected set.

In particular, (iii) means that $E(\zeta_0)$ is maximal for perturbations satisfying (i) and (ii) in the semi-groups with parameter s that are determined by

$$u_s + \partial(f,u) = 0 \quad \forall\, s \geq 0, \qquad u|_{s=0} = \zeta_0, \tag{7}$$

where $f = f(y,z,s)$, $f|_{\partial D} = 0$, is chosen arbitrarily.

The following facts are noted without proof:

(a) As mentioned already below (5), the kernel k of the linear integral operator K is positive on D × D and symmetrically decreasing in $z - \hat{z}$. Hence, by use of a standard inequality in the theory of rearrangements (Hardy et al. 1952, Thm 380), it can readily be shown that the maximizing function ζ_0 must be a symmetrically decreasing function of z.

(b) If the dimensions α and 2β of D are chosen sufficiently large in comparison with the mean diameter of $\mathrm{supp}(u_0)$, then $\mathrm{supp}(\zeta_0)$ does not extend near the boundaries of D distant from origin. This fact, which is crucial to the existence theory on present lines, may be proved by considering a function u with the contrary property, though admissible by (i), (ii), (iii), and showing that E is increased by a rearrangement of u that brings the extremities of supp(u) closer to the origin.

(c) If the specified value of I is large enough, then $\mathrm{supp}(\zeta_0)$ does not extend to the z-axis (so we have a genuine, toroidal vortex ring). This property is in fact unnecessary to the theory, and its negation needs to be included to cover cases such as Hill's vortex, but for simplicity we now assume it to hold.

The system (7) may be used to express successive variations in order of smallness, on the lines of the classical calculus of variations. One proceeds formally as if ζ_0 and f were analytic functions, so that the variation $u - \zeta_0$ is calculable by means of the formula $u = \exp(\int_0^s Lds')\zeta_0$ with $Lu = -\partial(f,u)$, and the ordering is made in powers of s. The resulting expressions for conditional variations of E can then be extended by continuity to be meaningful over suitably wide function classes for ζ_0. The FIRST VARIATION complying with (iii) is thus $\dot{u} = -s\partial(f,\zeta_0)$, so that

$$\dot{E} = \int_D \tfrac{1}{2}(\zeta_0 K\dot{u} + \dot{u}K\zeta_0)\,d\mu = \int_D \dot{u}K\zeta_0\,d\mu$$

$$= -\int_D K\zeta_0\,\partial(f,\zeta_0)\,d\mu = \int_D f\,\partial(K\zeta_o,\zeta_o)\,d\mu$$

and

$$\dot{I} = -\int_D y\,\partial(f,\zeta_0)\,d\mu = \int_D f\,\partial(y,\zeta_0)\,d\mu.$$

Since f is arbitrary it follows by the Lagrange principle that

$$\partial(K\zeta_0 - cy, \zeta_0) = 0$$

for some constant c, and this confirms that ζ_0 satisfies (4) as a necessary condition for a maximum.

Remark. An equally serviceable variational principle is given by relaxing the constraint (ii), fixing c and asking for a maximum of E - cI. Thus the propagation speed of the vortex ring is used as parameter, but the right to specify its impulse is lost. As regards the physical realities, the above formulative seems more fundamental than this alternative, also than Fraenkel & Berger's formulation. The most natural conception of the problem, we suggest, is to specify two properties conserved by all axisymmetric motions, namely impulse and the measure composition of ζ.

The SECOND VARIATION complying with (iii) is $\ddot{u} = -\tfrac{1}{2}s\partial(f,u) - \tfrac{1}{2}s^2\partial(f_s,\zeta_0)$, the second part of which, having the form of the first variation, makes no contribution to E. Hence one obtains

$$2\ddot{E} = \int_D \dot{u}\{s^2K\dot{u} + s\,\partial(f, K\zeta_0 - cy)\}\,d\mu \tag{8}$$

where now f, and correspondingly u, is restricted to comply with

$$-\dot{I} = -\int_D \zeta_0\,\partial(f,y)\,d\mu = \int_D \zeta_0 f_z\,d\mu = 0. \tag{9}$$

This result may be used to establish the important property that the set $A = \mathrm{supp}(\zeta_0)$ is connected. The following is an outline of the proof.

It has already been pointed out (§(a) below (7)) that A cannot be disconnected in the z-direction. So we need only to consider the possibility of disconnections in the y-direction. Assume that $A = A_1 \cup A_2$ and A_2 is separate from A_1. Take $f = f(y,z) \in C^{\infty}(D)$ such that

$$\left.\begin{array}{l} f_y = 0,\ f_z = \gamma \text{ in } A_1 \\ f_y = 0,\ f_z = -\delta \text{ in } A_2 \end{array}\right\}$$

with the positive constants γ, δ chosen to satisfy (9). The respective variation is thus admissible, corresponding in fact to infinitesimal displacements of A_1 and A_2 towards each other parallel to the y-axis, such as to preserve the value $I(\zeta_0)$ of I. When this f is substituted in (8), a long but straightforward calculation lead to

$$2\ddot{E} = -(\gamma + \delta)^2 \int_{A_2} (\zeta_0)_y (\psi_1)_y \, d\mu$$

in which $$\psi_1 = \int_{A_1} k \zeta_0 \, d\hat{\mu}$$

is the 'potential' due to the part of ζ_0 in A_1. By hypothesis, $\mathcal{L}\psi_1 = 0$ on a set including A_2. Hence, by Green's theorem,

$$2\ddot{E} = (\gamma + \delta)^2 \int_{A_2} (\zeta_0)_z (\psi_1)_z \, d\mu$$

and the integral must be positive by virtue of the fact that ζ_0 is a symmetrically decreasing function of z. (Note that ψ_1 is C^{∞} in A_2, and so the integral remains meaningful if $(\zeta_0)_z$ is interpreted as a distribution in the case that ζ_0 is discontinuous.) Thus $E > 0$, contrary to the specification that $E(\zeta_0)$ is a maximum. This completes the proof.

An adaption of this line of reasoning establishes the following, equally helpful fact. For $\sigma \in]0, \max u_0[$, define $S_\sigma(\zeta_0) = \{(y,z) : \zeta_0 \geq \eta\}$. Then the sets $S_\sigma(\zeta_0)$ are all connected, which means that the contour lines (or plateaus) $\zeta_0 = \sigma$ are homeomorphic to concentric circles (or annuli). Accordingly, if u_0 is continuous, then so is ζ_0. For a proof one may assume $S_\sigma(\zeta_0)$ to be disconnected for some σ and consider the representation $\zeta_0 = \bar{\zeta}_0 + \zeta_0'$, such that $S_\sigma(\zeta_0) = \operatorname{supp}(\zeta_0')$ and $\bar{\zeta}$ is a plateau on this set. Clearly, $E(\zeta_0)$ must be a maximum for I-preserving rearrangements of ζ_0' alone, leaving $\bar{\zeta}_0$ unchanged, and the second variation is reducible to a form corresponding to that for a conditional maximum of $E(\zeta_0')$ plus positive terms. Hence a contradiction can be demonstrated as above. It is easy to show also that the sets $S_\sigma(\zeta_0)$ are all simply connected.

EXISTENCE. Now, at last, some functional analysis is needed; however, the problem has been put into a form where comparatively little of it is enough to achieve the main purpose. The existence of a solution to the problem of steady vortex-rings is proved by showing that a conditional maximum of the energy functional (6)is attained within an appropriate function space. In virtually all the standard spaces of functions defined on the finite domain D or its closure, this functional is bounded and moreover continuous with respect to weak convergence. Accordingly, following the standard argument, we can consider maximizing sequences of functions satisfying the given constraints, but, of course, only the convergence of such sequences in some weak sense can be asserted in the first place. However, the basic uncertainty about the identity of the weak limit can be resolved by the crucial a priori estimates just demonstrated - that the maximizing function ζ_0 must be a connected and 'concentric' rearrangement of the given function u_0.

Let us outline the argument using $L^2(D)$. Here we have no need for the concept of gradient operators which is a familiar advantage of Hilbert spaces in variational problems, but a property needed crucially is that, being reflexive, $L^2(D)$ is weakly compact. An absolutely standard argument shows the functional E to be weakly continuous in this space, and the linear functional I is obviously well defined on it. So we can certainly assume a sequence $\{u_n\}$ of I-preserving rearrangements (with, therefore, $\|u_n\| = \|u_0\|$) that maximizes E in the limit $n \to \infty$, and by the compactness property we know that this sequence, or subsequence of it, converges at least weakly to some element $\phi \in L^2(D)$. The existence of a weak limit being assured, the properties established a priori for the maximizing function now show that $\|\phi\| = \|u_0\|$, which implies that the sequence converges strongly (in the norm topology); and the identification of ϕ with a solution ζ_0 of the variational problem is confirmed.

It is interesting to reflect that, as weak compactness is the main logical element, the argument might be put in terms of many other Banach spaces. We essentially need a space that is the dual of another, so that in it, according to a fundamental theorem of functional analysis, all bounded sequences are compact in the weak* topology. Thus, for instance, $C(\overline{D})$ would be useless, but its dual, the space of normalized functions of bounded variation on $\overline{D}$, might perhaps be used. This suggestion seems quite attractive since this space seems a natural setting for rearrangements of a bounded function. As a last comment on the issue, to express the view of an occasional, non-expert user of functtional analysis for practical purposes, it is noted how the existence argument depends briefly but crucially on a logical device beyond intuitive mathematics. A brief sortie has to be made into the abstract territory of weak topology, but then one immediately steps back with an assurance that the problem has a solution.

STABILITY. Here we have a valuable bonus from the demonstration of existence, for we may infer that the vortex ring represented by the maximizing function ζ_0 is a stable motion. Consider any perturbation of ζ_0 satisfying the conditions of the variational problem to be the initial state $\zeta(y,z,0)$ for a free motion. The dynamical equation (3) shows that $\zeta(y,z,t)$ remains for all $t > 0$ in the class of rearrangements of ζ_0, and $I(\zeta)$ is an invariant of the motion, so equalling $I(\zeta_0)$. Moreover, the energy difference $\Delta E = E(\zeta_0) - E(\zeta)$ is invariant, and is non-negative since $E(\zeta_0)$ is a global maximum for the class of functions in which ζ remains. Hence ΔE may serve as a Liapunov functional, to establish stability for axisymmetric perturbations within the stated class.

To complete a demonstration of unconditional stability for axisymmetric perturbations, various details have to be supplied. In particular, a metric allowing translations in z to be factored out needs to be introduced, and allowance needs to be made for perturbations not satisfying the conditions of the variational problem, which wider class may be represented as conditional perturbations from vortex rings neighbouring on the given one. The arguments required are more or less standard, however (cf. Benjamin 1972), and the present simple considerations are virtually enough to show that the given vortex ring is stable in a practical sense, at least for axisymmetric perturbations. The question of stability for general three-dimensional disturbances is more difficult.

DISCONTINUOUS VORTICITY. Finally, we note an interesting, simplified version of the problem, presented when ζ_0 is a positive constant λ in the vortex core and zero elsewhere (i.e. F in (4) is a step function). The variational problem is then to find a connected set A in HP, symmetric about the y-axis, such that

$$2E = \lambda \int_A\int_A k \, d\mu \, d\hat{\mu} \quad \text{is maximum} \qquad (10)$$

$$\text{for} \int_A d\mu = |A| \quad \text{fixed and} \quad \int_A y \, d\mu = \frac{I}{\lambda} \quad \text{fixed} \qquad (11)$$

(i.e. the area and first moment of A are fixed). Vanishing of the conditional first variation evidently requires that

$$\psi\big|_{\partial A} = \int_A k\big|_{\partial A} \, d\mu = cy + b, \qquad (12)$$

which is precisely the dynamical condition that the boundary ∂A of the vortex core is a streamline $\Psi = b$ in the moving frame of reference (y,z'). This problem lends itself to solution by successive approximations, either in principle as the basis of an existence proof or by computer. Thus, a sutiably large D is divided into rectangles such that (11) can be satisfied by an integral number of them, and the integral (10) is evaluated in the form $\sum_{i,j} E_{i,j}$, where i,j range over the labelling numbers of each set of elements selected compatibly with (11). The finite optimization problem is obviously soluble. By successive subdivision of the elements and determination of the optimal set, the maximum will be increased at each stage and the series of approximations will converge.

If the first of the constraints (11) is relaxed, the necessary condition (12) for a maximum is replaced by

$$\psi\big|_{\partial A} = cy,$$

implying that in the moving frame the vortex boundary ∂A coincides with the streamline $\Psi = 0$ which also includes the axis $y = 0$. This is the only case for which an explicit solution of the problem is known, namely Hill's spherical vortex (Lamb, p.245) with $\Psi(\lambda/5)y(a^2 - z'^2 - 2y)$ in $z'^2 + 2y$ $(= z'^2 + r^2) \leq \underline{a}^2$ and irrotational motion outside the sphere of radius $\underline{a}$. It thus appears plausible that, among axisymmetric vortex flows with constant ζ, an interesting classification in physical terms can be made as follows. For a given impulse I, the maximum possible kinetic energy is realized by a Hill vortex for which $\underline{a}$ is determined by $I = (4\pi/15)\lambda a^5$ and correspondingly $|A| = (4\pi/3)a^3$. If $|A|$ is constrained to be less than this value, the maximum is realized by a vortex ring, whose mean-square radius $(2\bar{y})^{\frac{1}{2}} = (2I/\lambda|A|)^{\frac{1}{2}}$ is inversely proportional to $|A|^{\frac{1}{2}}$.

II. BIFURCATION OF VISCOUS FLOWS: AN APPLICATION OF DEGREE THEORY

The theoretical results to be summarized were worked out in a recent paper (Benjamin 1975), where the Leray-Schauder theory was applied to aspects of the Navier-Stokes equations. Its subject was the bifurcation and stability of solutions representing bounded steady flows. This work has had a practical outcome in several ways and, conversely, physical thinking has been a valuable guide to the mathematical study.

One practical effect of progress with the theoretical problem was to stimulate a search for new examples of bifurcation phenomena, on which the conclusions of the abstract theory might be tested experimentally. There are in fact very few specific examples about which much is known already, and the reason for this shortage is clear. All of them are examples that were first studied, long ago, by means of linearized stability analyses of the explicit kind, and this approach is tractable only for specially simple, idealized situations. But most of the recent, precise mathematical work on problems of hydrodynamic stability has also focused on the classical problems, taking advantage of the extensive background of information about them. The two prototype problems, about which an enormous literature has accumulated, are (i) G.I. Taylor's problem of flow between rotating cylinders and (ii) the problem named after Bénard concerning convection between horizontal planes. Truly, there are remarkably few other examples already known where the second state of the system after bifurcation is a steady flow. And, as will be explained presently, the familiar examples (i) and (ii) are both extraordinary in mathematical respects, due to the severe, unrealistic simplifications that had to be introduced originally to make analysis possible. With a general qualitative theory available, however, one is able to look boldly into more complicated and realistic situations, especially ones for which constructive analysis is virtually impossible.

The following are two such cases that have already been studied experimentally, confirming predictions of the qualitative theory. First, a modification of the classical Taylor experiment has been studied in which, contrary to the usual procedure, the length of the fluid-filled annulus is made comparable with its width. In the past the experiment has almost always been done with a long annulus, the object having been to minimize end-effects and so simulate the idealized theoretical model where the annulus is assumed infinitely long. But when end-effects are dominant the primary bifurcation is qualitatively different from the predictions of the idealized theory, in particular manifesting a strong hysteresis respective to gradual increases and decreases of speed around the bifurcation value.

The second case is shown in the accompanying photographs. A circular cylinder closed at the ends is partially filled with liquid and, with its axis horizontal, is rotated at constant speed N. The flow velocities are sufficiently small for the action of gravity to predominate on the free surface, which functions in effect as a fixed horizontal boundary spanning the cross-section. When N is less than a critical value, the flow is more or less axially uniform over the central region, departing from this state only in boundary layers of the Ekman type at the two ends. This situation is shown in the upper photograph. The lower photograph shows the situation arising when N is increased to a value above the critical. A cellular flow pattern then develops, presumably bifurcating from - and 'exchanging stability' with - the primary, nearly uniform flow.

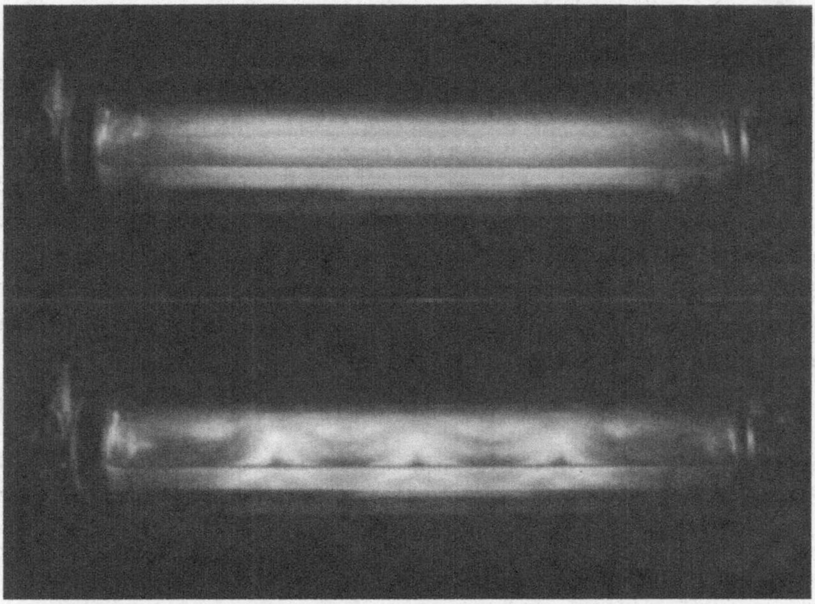

Bifurcation of steady flow of liquid partially filling a horizontal cylinder rotated at constant speed. The liquid is a weak aqueous solution of glycerol, with a pearly substance added for visualization of the flow.

It is worth emphasis that this phenomenon is too complex for constructive analysis. The range of Reynolds numbers is such that viscous and nonlinear inertial effects are comparable everywhere in the flow, and the boundary geometry precludes any simplifying approximation. Even the primary flow appears impossible to calculate, other than by computer solution of the Navier-Stokes equations. It is also worth emphasis that, unlike the Taylor bifurcation phenomenon, this one seems to have no intuitive explanation. There is no simple physical reason why the wavy flow should develop.

<u>OUTLINE OF THEORY</u>. We suppose that a viscous incompressible fluid fills a bounded domain D in $\mathbb{R}^3$. The boundary conditions, specifying the velocity of the fluid on the boundary ∂D, are taken to be independent of time t. According to the famous existence theory of Leray and Hopf (see Ladyzhenskaya 1969, Ch.5), the time-independent hydrodynamical problem has at least one solution: that is, there is a vector field U(x) satisfying the Navier-Stokes equations and the boundary conditions, so corresponding to the Eulerian velocity field of the fluid. Subject to mild assumptions about the smoothness of the boundary, it can

be shown that $U \in C_{2+\alpha}(D \to \mathbb{R}^3)$, so that U is a classical solution of the time-independent problem.

We shall consider perturbations from this primary solution, expressing the velocity field in the form $U(x) + v(x,t)$ which requires v to be solenoidal and to vanish on ∂D. As a function of the position vector x, v is considered to belong to a Hilbert space H which has standard uses in the theory of the Navier-Stokes equations. Namely, H is the subspace of $W_1^2(D \to \mathbb{R}^3)$ defined as the completion in the W_1^2 norm of the class of C^∞ solenoidal vector fields with compact support in D. In terms of operators defined on H, the boundary-value problem for v may be represented by the equation

$$Av_t = v - Bv - Cv. \tag{1}$$

The three operators are defined by the following expressions, in which the left-hand sides are inner products in H with an arbitrary element $\psi \in H$:

$$(\psi, Av) = -\int_D \psi \cdot v \, dx,$$

$$(\psi, Bv) = -R\int_D \psi \cdot \{(U\cdot\nabla)v + (v\cdot\nabla)U\} \, dx,$$

$$(\psi, Cv) = R\int_D v \cdot (v\cdot\nabla)\psi \, dx.$$

Here R is a positive parameter, the Reynolds number such that 1/R is the dimensionless viscosity of the fluid, and the linear operator B generally depends on R also through the function U(x). By appeal to standard embedding theorems it can be shown that A,B and C are compact operators $H \to H$. [This property of A follows from Rellich's theorem. For B and C, the argument may be found in Ladyzhenskaya (1969, pp. 116,117) or Sattinger (1973, pp. 159,160).] Note also that B is the strong Fréchet derivative of B + C at the zero point θ of H.

For a time-independent perturbation $v = V(x)$, the operator equation is

$$V = BV + CV, \tag{2}$$

which has, of course, the trivial solution $V = \theta$ corresponding to the primary flow. For any isolated solution V_m of (2) (i.e. for any isolated fixed point of the compact operator B + C), an index can be defined by $i_m = \deg(I - B - C, \Omega_m)$, where Ω_m is a neighbourhood of V_m small enough to include no other fixed point. An account of all solutions of (2) is made in the following theorem, whose proof is a by-product of the Leray-

Hopf existence theory:

THEOREM 1. Let (2) have only isolated solutions $V_m \in H$ $(m = 1,2,\ldots)$ which include the zero solution $(m = 1)$ and which have respective indices i_m. Then the total number k of solutions is finite and

$$\sum_{m=1}^{k} i_m = 1. \tag{3}$$

Note that the assumption of isolated solutions is very reasonable. A solution is necessarily isolated except at critical values of R where the Fréchet derivative, exemplified by B for the zero solution, has 1 as an eigenvalue. And even in the critical cases a solution remains isolated if the condition of Theorem 3 below is satisfied.

The result (3) is helpful in the interpretation of examples of bifurcation, and its main significance appears in the light of another theorem as follows, which relates to the time-dependent problem.

THEOREM 2. If the index i_1 of the zero solution of (2) equals - 1 and the operator B does not have 1 as an eigenvalue, then a positive number σ can be found such that the linear equation

$$-\sigma A\eta = \eta - B\eta \tag{4}$$

has a non-trivial solution $\eta \in H$.

The proof consists in showing that, if the positive number γ is chosen sufficiently large, the equations

$$u = s(Bu + \gamma Au) \qquad \text{with} \qquad 0 < s \leq 1$$

have no non-trivial solution. This fact follows straightforwardly from the definitions of A and B, on the justified assumption that U has bounded derivatives on D. Hence, by the homotopy invariance of degree

$$\deg(I - B - \gamma A, \mathcal{B}_1) = 1,$$

where $\mathcal{B}_1$ is the unit ball in H. But the conditions of the theorem imply that

$$\deg(I - B, \mathcal{B}_1) = i_1 = -1,$$

and so there can be no admissible homotopy between $I - B - \gamma A$ and $I - B$ on the unit sphere. The assertion of the theorem now follows.

The importance of Theorem 2 is that it shows the primary flow to be unstable whenever $i_1 = -1$. For (4) implies that $\eta(x)\, e^{\sigma t}$ is a solution of the linearized form of (1), and it has been proved by Sattinger (1970) (confirming what had been taken for granted by generations of intuitional applied mathematicians!) that an exponentially growing solution of the linearized perturbation problem is a sufficient condition of hydrodynamic instability. The conclusion respective to the zero solution, that $i_1 = -1$ implies instability, evidently extends to any solution V_m of (2).

The practical bearing of these facts is best illustrated by a bifurcation diagram, as sketched in the accompanying figure. Here f(V) is a linear functional with f(θ) = 0, and each point on the curves corresponds to a solution V possible at the respective value of the parameter R. In all such cases, however many bifurcations occur, the sum of the indices remains 1 independently of R. And, as is commonly supposed from knowledge of specific examples, the parts of such a diagram that are characterized here by i = - 1 always represent unstable states of the system.

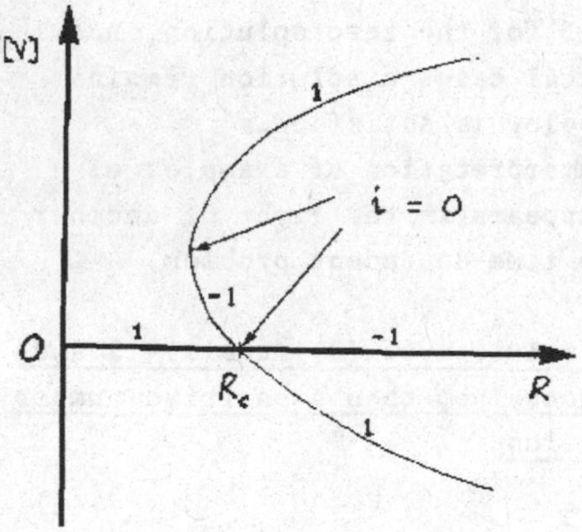

GENERIC FORM OF BIFURCATION. The diagram illustrates the case of transcritical bifurcation, where the branch of non-zero solutions in the product space H × ℝ extends locally both above and below the critical value R_c of R. The practical implications of this are well known, particularly that the primary flow is only locally stable for values of R somewhat less than R_c and that hyteresis in the manner already mentioned will be observed experimentally. So, it appears a particular significant indication of the theory that in practice bifurcation phenomena will be transcritical virtually always. This property is due essentially to the fact that the operator C is quadratic, as also is the corresponding remainder after linearizing (2) relative to any non-zero solution.

It may be asserted as a global proposition with regards to realistic applications that, under the influence of quadratic nonlinearities, the transcritical form of bifurcation is generic even if not universal. The truth of this proposition becomes clear when, for instance, one considers finite-dimensional examples. On the other hand, well-established results for the idealized Taylor problem show that the primary bifurcation has a supercritical form, where non-trivial solutions exist only for $R > R_c$. The task of reconciling these conflicting theoretical conclusions, and assessing the respective experimental evidence, has been a particularly interesting part of the present research.

If the linear operator B does not have 1 as an eigenvalue, then the index of the zero solution is a property of B alone and is calculable by a well-known formula due to Leray and Schauder (Krasnosel'skii 1964, p. 136). But for a critical value of R, at which 1 is an eigenvalue of B, the index depends on the non-linear part of the perturbation equation. The transcritical form of bifurcation requires that the index is then zero, as marked in the diagram. (For supercritical bifurcation, it is necessary that the index is 1 at $R = R_c$, if it is 1 for $R = R_c^-$ and - 1 for $R = R_c^+$.) The relevant abstract result is the following theorem, which is included in more general propositions given in Krasnosel'skii's book (1964, see p. 217, Thm 4.1, and p. 223):

THEOREM 3. Let $\lambda(R_c) = 1$ be a simple eigenvalue of B to which the eigenfunction ξ corresponds. Let

$$(\xi^*, C\xi) \equiv R_c \int_D \xi \cdot (\xi \cdot \nabla)\xi^* \, dx \neq 0, \tag{5}$$

where ξ^* is the eigenfunction of the adjoint operator B^*. Then the zero solution of (2) is isolated at $R = R_c$ and

$$i_1(R_c) = \deg(I - B - C, \mathcal{B}_\varepsilon) = 0.$$

(Here $\mathcal{B}_\varepsilon$ denotes a sufficiently small ball centred on the zero element in H.)

Note that if the additional condition $\lambda'(R_c) \neq 0$ is satisfied, then R_c is guaranteed to be a transcritical bifurcation point (cf. Krasnosel'skii, pp. 233,234).

The condition (5) appears to be generic to problems of bounded flows with natural boundary conditions. Note that the integral would vanish if B were self-adjoint, so that $\xi \equiv \xi^*$, but this is easily seen never to be the case in real examples. The possibility that otherwise the integral might vanish cannot be ruled out entirely, but it appears so exceptional as to be practically irrelevant.

There is, however, a fairly obvious reason why the non-degeneracy condition (5) is evaded by the idealized model of the Taylor problem. This is seen to follow from the fact that the idealized problem is invariant to translations along the axis of the concentric cylinders, so that the non-uniform flow after bifurcation can be assumed to satisfy periodic rather than natural boundary conditions. Moreover, for other hydrodynamic examples of supercritical bifurcation, such as the idealized Bénard problem, the attribute in common is invariance under some continuous transformation group, the invariance being non-trivial in the sense that the second solution appearing at bifurcation depends on the transformed co-ordinates. Ideal symmetries of this kind are not found in real, experimental situations, and thus the abstract theory reveals a significant point of principle as regards the interpretation of bifurcation phenomena in practice. For the Bénard problem, it has already been shown by Joseph (1971) that the effect of lateral boundaries is to change the bifurcation from the supercritical to the transcritical form. And we may now claim that the same qualitative effect is always present, even if small, in the Taylor experiment and in other bifurcation experiments such as the one illustrated earlier.

In conclusion, I wish to record my indebtedness to my colleague Dr. J.L. Bona, who contributed much to the research summarized in this lecture.

REFERENCES

Arnold, V.I. 1965 Conditions for nonlinear stability of stationary plane curvilinear flows of an ideal fluid. Dokl. Akad. Nauk SSSR 162, 975-978. (Soviet Math. Dokl. 6, 773.)

Arnold, V.I. 1966a Sur un principe variationnel pour les écoulements stationnaires des liquides parfaits et ses applications aux problèmes de stabilité non linéaires. J. Mécanique 5, 29.

Arnold, V.I. 1966b Sur la géométrie différentielle des groupes de Lie de dimension infinie et ses applications à l'hydrodynamique des fluides parfaits. Ann. Inst. Fourier (Grenoble) 16, fasc. 1, 319-361.

Benjamin, T.B. 1970 Upstream influence. J. Fluid Mech. 40, 49-79.

Benjamin, T.B. 1972 The stability of solitary waves. Proc. Roy. Soc. London, A 328, 153-183.

Benjamin, T.B. 1975 Applications of Leray-Schauder degree theory to problems of hydrodynamic stability. To appear in Math. Proc. Cambridge Phil. Soc.

Fraenkel, L.E. & Berger, M.S. 1974 A global theory of steady vortex rings in an ideal fluid. Acta mathematica 132, 13-51.

Hadamard, J. 1945 The Psychology of Invention in the Mathematical Field. Princeton University Press. (Dover reprint 1954.)

Hardy, G.H., Littlewood, J.E. & Pólya, G. 1952 Inequalities, 2nd ed. Cambridge Univeristy Press.

Joseph, D.D. 1974 Stability of convection in containers of arbitrary shape. J. Fluid Mech. 47, 257-282.

Kato, T. 1967 On classical solutions of the two dimensional non-stationary Euler equation. Arch. Rat. Mech. Anal. 25, 188-200.

Kransnosel'skii, M.A. 1964 Topological Methods in the Theory of Nonlinear Integral Equations. London: Pergamon.

Ladyzhenskaya, O.A. 1969 The Mathematical Theory of Viscous Incompressible Flow, 2nd ed. New York: Gordon and Breach.

Lamb, H. 1932 Hydrodynamics, 6th ed. Cambridge University Press.

Sattinger, G.H. 1970 The mathematical problem of hydrodynamic stability. J. Math. Mech. 19, 797-817.

Sattinger, G.H. 1973 Topics in Stability and Bifurcation Theory. Berlin: Springer-Verlag.

ASYMPTOTIC BEHAVIOUR OF SOLUTIONS OF VARIATIONAL INEQUALITIES WITH HIGHLY OSCILLATING COEFFICIENTS

J.L. Lions
Collège de France, Paris.

INTRODUCTION

We report in this lecture on the asymptotic behaviour of the solution of a number of boundary problems for operators with highly oscillating coefficients([1]).

These problems are mainly motivated by the study of various physical phenomenae connected with composite materials.

We present here some of the mathematical techniques which could be of some help in these matters. We study in this report :

(1) Stationary boundary value problems (Section 1).
(2) Stationary problems with obstacles (Section 2).
(3) Evolution problems, with or without obstacles (Section 3).

We also present in Section 3.5. some problems leading to non local effects.

The results for stationary equations are due to DE GIORGI-SPAGNOLO [1]; cf. for related topics, SPAGNOLO [1][2], SBORDONE [1], MARINO-SPAGNOLO [1] .

Stationary and evolution equations of various types have been considered by SANCHEZ-PALENCIA [1] . Formulaes giving higher order terms for stationary problems have been given by BABUSKA [1][2] .

([1]) Detailed proofs with many other problems and examples will be given in BENSOUSSAN-LIONS-PAPANICOLAOU [3] .

Variational inequalities of stationary type for operators with highly oscillating coefficients have been studied by BOCCARDO and MARCELLINI [1] , BOCCARDO and I. CAPUZZO DOLCETTA [1] , MARCELLINI [1] BENSOUSSAN-LIONS-PAPANICOLAOU [1] , and similar problems for variational inequalities of evolution in BENSOUSSAN-LIONS-PAPANICOLAOU [2] .

For the numerical analysis of these problems, we refer to BABUSKA, loc. cit., and BOURGAT [1] .

The Author wishes to thank Professors BABUSKA, DE GIORGI, LANCHON, SANCHEZ PALENCIA, SBORDONE, SPAGNOLO, for several interesting discussions and comments on the problems considered here.

Some questions related to the topics of this report but not considered here are : problems of optimal control where the control lies in the coefficients (cf. L. TARTAR [1]), problems of averaging, problems of optimal control of systems with highly oscillating coefficients (cf. BENSOUSSAN-LIONS-PAPANICOLAOU [3]).

The plan is as follows :

1. STATIONARY BOUNDARY VALUE PROBLEMS.

1.1. Setting of the problem.

Let $\mathcal{O}$ be an open bounded (to fix the ideas) set of $\mathbb{R}^n$ with smooth boundary Γ ; in $\mathcal{O}$ we consider a second order elliptic ([1]) operator A^ε with highly oscillating coefficients with period ε in all variables $x_1, \dots, x_n$, ε "small".

Remark 1.1.

Of course $n = 2$ or 3 in the applications. All what we are going to say readily extends to the case when the coefficients have different period $\varepsilon_1, \dots, \varepsilon_n$ along the different directions $x_1, \dots, x_n$. ■

Analytically, A^ε can be written as follows :

(1.1)
$$A^\varepsilon v = - \sum_{i,j=1}^{n} \frac{\partial}{\partial x_i}\left(a_{ij}\left(\frac{x}{\varepsilon}\right)\frac{\partial}{\partial x_j}\right)$$

where the a_{ij}'s are given functions in $\mathbb{R}^n$, with period 1 in all variables ; let us denote by Π^n the cube $]0,1[^n$; then we can assume :

(1.2)
$$\left|\begin{array}{l} a_{ij} \in L^\infty(\Pi^n), \quad (a_{ij} \text{ extended by periodicity to } \mathbb{R}^n), \\ a_{ij} = a_{ji} \quad \forall i,j \\ \sum_{i,j} a_{ij}(y)\xi_i\xi_j \geq \alpha \sum_i \xi_i^2 \quad \forall \xi_i \in \mathbb{R}, \ \alpha > 0, \\ \text{a.e for } y \in \Pi^n. \end{array}\right.$$

([1]) For systems, cf. BAHBALOV [1] .

We consider first <u>the Dirichlet problem for</u> A^{ε}.

Let u_{ε} be the unique solution of

$$(1.3)\qquad \begin{cases} A^{\varepsilon} u_{\varepsilon} = f, \quad f \text{ given in } L^2(\mathcal{O}), \\ u_{\varepsilon} = 0 \quad \text{on } \Gamma. \end{cases}$$

Remark 1.2.

The coefficients $a_{ij}(\frac{x}{\varepsilon})$ of A^{ε} are <u>not regular</u> but only L^{∞} functions ; therefore (1.3) should be understood in the <u>weak form</u> as follows : we introduce the usual <u>Sobolev spaces</u> :

$$H^1(\mathcal{O}) = \{ v \mid v, \frac{\partial v}{\partial x_i} \in L^2(\mathcal{O}) \},$$

$$H^1_o(\mathcal{O}) = \{ v \mid v \in H^1(\mathcal{O}), \quad v = 0 \text{ on } \Gamma \}$$

equipped with their usual Hilbertian structure.

For $u, v \in H^1(\mathcal{O})$, we define

$$(1.4)\qquad a^{\varepsilon}(u,v) = \sum_{i,j} \int_{\mathcal{O}} a_{ij}(\frac{x}{\varepsilon}) \frac{\partial u}{\partial x_j} \frac{\partial v}{\partial x_i}\, dx.$$

Then u_{ε} is <u>the unique solution</u> in $H^1_o(\mathcal{O})$ of

$$(1.5)\qquad a^{\varepsilon}(u_{\varepsilon}, v) = (f,v) \quad \forall v \in H^1_o(\mathcal{O}),$$

where $(f,v) = \int_{\Omega} f(x)v(x)dx$. ■

Remark 1.3.

The operator A^{ε} as given above corresponds to a highly non homogeneous material, with a periodic structure, as it is the case in many situations arising in Physics. ■

A natural problem, which has an interest from the purely mathematical viewpoint, but also from the numerical and from the physical view points, is now :

<u>What is the asymptotic behaviour of</u> u^{ε} <u>as</u> $\varepsilon \to 0$?

1.2. The homogeneized operator.

We are going now to construct an operator $\mathcal{A}$ with constant coefficients ; later on we shall show how one can "replace" A^ε by the much simpler operator $\mathcal{A}$.

We introduce the following notations :

$$W = \{ \varphi | \varphi \in H^1(\Pi^n), \quad \varphi(y_1,\dots,y_{i-1}, 0, y_{i+1},\dots, y_n) =$$

$$= \varphi(y_1, \dots, y_{i-1}, 1, y_{i+1}, \dots, y_n) \quad \forall i \}$$

(we define in this way a closed subspace of $H^1(\Pi^n)$) ;

for $\varphi,\psi \in W$, we set

$$\alpha(\varphi,\psi) = \sum_{i,j}\int_{\Pi_n} a_{ij}(y) \frac{\partial\varphi}{\partial y_j} \frac{\partial\psi}{\partial y_i}\, dy \; ; \tag{1.6}$$

we denote by y_i the function $y \to y_i$; we consider the equation :

$$\alpha(\chi^i,\psi) = \alpha(y_i,\psi) \quad \forall \psi \in W, \; \chi^i \in W . \tag{1.7}$$

Since $\alpha(\varphi,\varphi)$ is coercive on $\overset{\circ}{W} = W/R$, (1.7) uniquely defines $\chi^i \in \overset{\circ}{W}$. Therefore we uniquely define q_{ij} by

$$q_{ij} = \alpha(\chi^i - y_i, \chi^j - y_j). \tag{1.8}$$

It is a simple matter to check that

$$\left| \begin{array}{l} q_{ij} = q_{ji}, \\ \sum q_{ij}\, \xi_i \xi_j \geqslant q_o \sum \xi_i^2 , \quad q_o > 0. \end{array} \right. \tag{1.9}$$

We then define $\mathcal{A}$ by

$$\mathcal{A} v = - \sum_{i,j} q_{ij} \frac{\partial^2 v}{\partial x_i \partial x_j} \; ; \tag{1.10}$$

we call $\mathcal{A}$ the homogeneized operator associated to A^ε for reasons that will be given below.

1.3. An asymptotic theorem.

We shall give in Section 1.4. below some indications on the proof(s) of the following result (cd. DE GIORGI-SPAGNOLO [1]) :

Theorem 1.1. We assume that (1.2) holds true. Let A^ε be given by (1.1) and let $\mathcal{A}$ be given by (1.10). Let u_ε be the solution of (1.3) and let u be the solution of :

$$\mathcal{A}u = f , \quad u \in H_0^1(\mathcal{O}) . \tag{1.11}$$

Then , as $\varepsilon \to 0$,

$$u_\varepsilon \to u \quad \text{in} \quad H_0^1(\mathcal{O}) \quad \text{weakly.} \tag{1.12}$$

Remark 1.4

One can consider, more generally, operators A^ε of the form

$$A^\varepsilon v = - \sum_{i,j} \frac{\partial}{\partial x_i}\left(a_{ij}\left(x, \frac{x}{\varepsilon}\right)\frac{\partial v}{\partial x_j}\right) \tag{1.13}$$

where $a_{ij}(x,y) \in C^1(\bar{\mathcal{O}} \times \Pi^n)$, a_{ij} periodic in y , and such that

$$\begin{aligned} &|a_{ij}(x,y) - a_{ij}(x',y)| \leqslant C|x-x'|^\beta , \quad 0 < \beta \leqslant 1 \ \forall x,x' \in \bar{\mathcal{O}}, \\ &\forall y \in \Pi^n , \ a_{ij} = a_{ji} , \\ &\sum a_{ij}(x,y)\, \xi_i\, \xi_j \geqslant \alpha \sum \xi_i^2 \quad \forall x,y,\xi_i . \end{aligned} \tag{1.14}$$

Then one defines (cf. BENSOUSSAN-LIONS-PAPANICOLAOU [1]) an homogeneized operator $\mathcal{A}$ with variable coefficients :

$$\mathcal{A}v = - \sum \frac{\partial}{\partial x_i}\left(q_{ij}(x)\frac{\partial v}{\partial x_j}\right) , \tag{1.15}$$

where $q_{ij}(x)$ is defined as in (1.8) but with $\alpha(\varphi,\psi)$ replaced by

$$\alpha(x;\varphi,\psi) = \sum_{i,j} \int_{\Pi^n} a_{ij}(x,y) \frac{\partial \varphi}{\partial y_j} \frac{\partial \psi}{\partial y_i}\, dy . \tag{1.16}$$

One has then a result similar to Theorem 1.1. ∎

1.4. Remarks on the Proof of Theorem 1.1. :

The proof of Theorem 1.1. proceeds in two steps.

Step 1. Prove Theorem 1.1. when the a_{ij}'s are smooth functions (It is a simple matter to check that one can always assume that f is smooth).

Step 2. Prove the result in the general case . ∎

Of course Step 2 relies on a priori estimates.

Since, by virtue of (1.2), $a^{\varepsilon}(v,v) \geq \alpha\|v\|^2$, where $\|v\|$ = norm of v in $H_0^1(\mathcal{O})$, one obtains from (1.5) that

(1.17) $\|u_\varepsilon\| \leq C$ = constant which does not depend on ε.

This is not sufficient to complete Step 2 (assuming Step 1 solved); in order to complete Step 2, one has to use a deeper estimate, due to MEYERS [1] (which relies on the Calderon-Zygmund theory of singular integrals) : there exists a $p > 2$ such that

(1.18) $$\left\| \frac{\partial u_\varepsilon}{\partial x_i} \right\|_{L^p(\mathcal{O})} \leq C . \quad ∎$$

There are (at least) two possible approaches for Step 1 [1].

A first possibility consists in looking for an asymptotic expansion using two scales of variables.

Remark 1.5.

On this approach, one can obtain next terms in the asymptotic expansion of u_ε :

(1.19) $$u_\varepsilon = u + \varepsilon u_1 + \ldots$$

[Cf. also BABUSKA [1][2] for formulaes for u_1]. This method leads to the pointwise estimate

(1.20) $$|u_\varepsilon(x) - u(x)| \leq C\varepsilon$$

when all dataes are sufficiently smooth. ∎

A second possibility is to use the probabilistic interpretation

---

[1] A different proof is given by DE GIORGI-SPAGNOLO, loc. cit.

of u_ε. Cf. BENSOUSSAN-LIONS-PAPANICOLAOU [1], and [3] for detailed proofs.

Another (more direct) proof is given in TARTAR [2].

1.5. Convergence of "local energy".

It is in general not true that, under the hypothesis of Theorem 1.1. one has $u_\varepsilon \to u$ in $H_o^1(\mathcal{O})$ strongly.

Of course, it follows from Theorem 1.1. that

$$a^\varepsilon(u_\varepsilon,u_\varepsilon) = (f,u_\varepsilon) \to (f,u) = \mathcal{Q}(u,u)$$

if we set

$$\mathcal{Q}(u,v) = \sum_{i,j}\int_{\mathcal{O}} q_{ij}\frac{\partial u}{\partial x_j}\frac{\partial v}{\partial x_i}\,dx \tag{1.21}$$

But one can prove more (cf. DE GIORGI-SPAGNOLO [1]); let us define, for any given measurable set $S\subset\bar{\mathcal{O}}$:

$$a_S^\varepsilon(u,v) = \sum_{i,j}\int_S a_{ij}\left(\frac{x}{\varepsilon}\right)\frac{\partial u}{\partial x_j}\frac{\partial v}{\partial x_i}\,dx, \tag{1.22}$$

$$\mathcal{Q}_S(u,v) \quad \sum_{i,j}\int_S q_{ij}\frac{\partial u}{\partial x_j}\frac{\partial v}{\partial x_i}\,dx. \tag{1.23}$$

Then one has, for any set S:

$$a_S^\varepsilon(u_\varepsilon,u_\varepsilon) \to \mathcal{Q}_S(u,u), \text{ as } \varepsilon\to 0. \tag{1.24}$$

1.6. Other boundary value problems.

Let us now consider another function a_o, such that

$$\left|\begin{array}{l} a_o \in L^\infty(\Pi^n),\quad a_o \text{ periodic over } \mathbb{R}^n, \\ \\ a_o(y) \geq \alpha > 0 \text{ a.e.,}\end{array}\right. \tag{1.25}$$

and let us define the operator A^ε by

$$A^\varepsilon v = -\sum\frac{\partial}{\partial x_i}\left(a_{ij}\left(\frac{x}{\varepsilon}\right)\frac{\partial v}{\partial x_j}\right) + a_o\left(\frac{x}{\varepsilon}\right)v. \tag{1.26}$$

We consider the Neumann's boundary value problem ; if we set now

(1.27) $$a^{\varepsilon}(u,v) = \sum \int_{\mathcal{O}} a_{ij}(\frac{x}{\varepsilon}) \frac{\partial u}{\partial x_j} \frac{\partial v}{\partial x_i} dx + \int_{\mathcal{O}} a_o(\frac{x}{\varepsilon}) uv \, dx ,$$

then u_ε is the unique solution of

(1.28) $$a^{\varepsilon}(u_\varepsilon, v) = (f,v) \quad \forall v \in H^1(\mathcal{O}) \ , \quad u_\varepsilon \in H^1(\mathcal{O}) ;$$

(1.28) is the Neumann's boundary value problem in its weak or variational forms.

We define

(1.29) $$\overline{a_o} = \int_{\Pi^n} a_o(y)\, dy$$

and

(1.30) $$\mathcal{a}(u,v) = \sum \int_{\mathcal{O}} q_{ij} \frac{\partial u}{\partial x_j} \frac{\partial v}{\partial x_i} dx + \overline{a_o} \int_{\mathcal{O}} uv \, dx.$$

The homogeneized problem is now :

(1.31) $$\mathcal{a}(u,v) = (f,v) \quad \forall v \in H^1(\mathcal{O}) \ , \quad u \in H^1(\mathcal{O}).$$

One can prove ([1]):

(1.32) $$u_\varepsilon \to u \ \text{ in } \ H^1(\mathcal{O}) \ \text{ as } \ \varepsilon \to 0.$$

Remark 1.6.

One has similar results for mixed boundary value problems, i.e with a Dirichlet's condition on part of Γ and a Neumann's boundary condition on the remaining part of Γ. ■

([1]) S. SPAGNOLO ; personal communication, Pisa, April 1975.

1.7. Domains with holes.

Let Ω be an open set of the form

$$\Omega_\varepsilon = \mathcal{O} - \overline{\Omega'_\varepsilon} \tag{1.33}$$

Ω'_ε = open set contained in $\mathcal{O}$ such that $\overline{\Omega'_\varepsilon} \subset \mathcal{O}$ and where Ω'_ε consists in the union of "small" holes arranged with period ε in all variables.

Let the functions a_{ij}'s be given as in Section 1.1. and let us consider the problem

$$A^\varepsilon u_\varepsilon = f \quad \text{in } \Omega, \quad u_\varepsilon \in H^1_o(\Omega) . \tag{1.34}$$

We want again to study the asymptotic behaviour of u_ε when $\varepsilon \to 0$ and when Ω'_ε is changed by homothety as ε decreases to zero.

We "approximate" u_ε in two steps.

We define $\mathcal{Q} \subset]0,1[^n$ as the image by homothety $1/\varepsilon$ of that part of Ω'_ε which is in the fundamental period region of $\mathcal{O}$.

We consider next

$$b_{ij\eta}(y) = \begin{cases} a_{ij}(y) & \text{for } y \in \Pi^n - \mathcal{Q} , \\ \frac{1}{\eta}\delta_{ij} & \text{for } y \in \mathcal{Q} , \quad \eta > 0, \end{cases} \tag{1.35}$$

and

$$b_{o\eta}(y) = \begin{cases} 0 & \text{in } \Pi^n - \mathcal{Q} , \\ \frac{1}{\eta} & \text{in } \mathcal{Q} , \end{cases} \tag{1.36}$$

all these functions being extended to $\mathbb{R}^n$ by periodicity.

We then define, $\forall u, v \in H^1(\mathcal{O})$:

$$b^\varepsilon_\eta(u,v) = \sum_{i,j} \int_{\mathcal{O}} b_{ij\eta}\left(\frac{x}{\varepsilon}\right) \frac{\partial u}{\partial x_j} \frac{\partial v}{\partial x_i}\, dx + \int_{\mathcal{O}} b_{o\eta}\left(\frac{x}{\varepsilon}\right) uv\, dx \tag{1.37}$$

and we denote by $u_{\varepsilon\eta}$ the solution of

$$(1.38)\quad \begin{cases} b_\eta^\varepsilon(u_{\varepsilon\eta}, v) = (f,v)_\Omega = \int_\Omega fv\,dx \quad \forall v \in H_0^1(\mathcal{O}), \\ u_{\varepsilon\eta} \in H_0^1(\mathcal{O}). \end{cases}$$

One shows that

$$(1.39)\qquad \|u_\varepsilon - u_{\varepsilon\eta}\| \leq C\eta^{1/4} .$$

Then one can approximate $u_{\varepsilon\eta}$ by the solution of the homogeneized problem associated with b_η^ε for $\eta>0$ fixed.

One uses the methods of previous Sections for $b_\eta^\varepsilon(u,v)$.

In this manner, one has "replaced" the domain with holes Ω_ε by the much simpler domain $\mathcal{O}$, and the operator A^ε by an operator $\mathcal{B}_\eta$ with constant coefficients in $\mathcal{O}$.

The question which remains is : how to choose η ? We conjecture that a "good" choice of η consists in taking η of the order of ε^2.

2. STATIONARY PROBLEMS WITH OBSTACLES.

2.1. Homogeneization for Variational Inequalities.(1)

We use the notations of Section 1 and we consider the following V.I.

(2.1)
$$\begin{cases} u_\varepsilon \leq 0, \quad A^\varepsilon u_\varepsilon - f \leq 0, \; u_\varepsilon(A^\varepsilon u_\varepsilon - f) = 0 \text{ in } \mathcal{O}, \\ u_\varepsilon \in H_o^1(\mathcal{O}) . \end{cases}$$

Problem (2.1) is equivalent to finding u_ε such that

(2.2)
$$\begin{cases} a^\varepsilon(u_\varepsilon, v-u_\varepsilon) \geq (f, v-u_\varepsilon) \; \forall \, v \leq 0 \;, \; v \in H_o^1(\mathcal{O}) \;, \\ u_\varepsilon \in H_o^1(\mathcal{O}) \;, \; u_\varepsilon \leq 0 \text{ in } \mathcal{O} . \end{cases}$$

Remark 2.1.

We refer to LIONS-STAMPACCHIA [1] for V.I. of elliptic type ; in the present situation, (2.2) is equivalent to minimizing

(2.3)
$$J_\varepsilon(v) = \frac{1}{2} a^\varepsilon(v,v) - (f,v)$$

over
$$K = \{ v \mid v \in H_o^1(\mathcal{O}) \;, \quad v \leq 0 \quad \text{in } \mathcal{O} \},$$

hence the existence and uniqueness of u_ε immediately follows . ■

We want to study the behaviour of u_ε as $\varepsilon \to 0$.

Let $\mathcal{a}(u,v)$ be defined as in (1.21) and let u be the solution of the "homogeneized V.I." :

(1) We write V.I. in short.

(2.4) $\mathcal{a}(u,v-u) \geq (f,v-u) \quad \forall v \in K, \quad u \in K.$

One has the following :

Theorem 2.1. Let the hypothesis of Theorem 1.1. hold true. Let u_ε (resp. u) be the solution of the V.I. (2.2)(resp. (2.4)). Then

(2.5) $$u_\varepsilon \to u \quad \text{in } H_0^1(\Omega) \quad \text{weakly as} \quad \varepsilon \to 0.$$

2.2. Remarks on the proof of Theorem 2.1.

One possible proof of Theorem 2.1. consists in approximating u_ε uniformly in ε by the penalty method ; we denote by $u_{\varepsilon\eta}$ the unique solution of the penalized equation

(2.6) $$A^\varepsilon u_{\varepsilon\eta} + \frac{1}{\eta} u_{\varepsilon\eta}^+ = f \;, \; u_{\varepsilon\eta} \in H_0^1(\mathcal{O})$$

where in general $\varphi^+ = \sup(\varphi, 0)$.

One proves that

(2.7) $$\| u_{\varepsilon\eta} - u_\varepsilon \| \leq C \; \eta^{1/2} \quad \text{where } C \text{ does not depend on } \varepsilon.$$

Similarly if u_η denotes the solution of

(2.8) $$\mathcal{a} u_\eta + \frac{1}{\eta} u_\eta^+ = f \;, \; u_\eta \in H_0^1(\mathcal{O}) \;,$$

then one has

(2.9) $$\| u - u_\eta \| \leq C \; \eta^{1/2}.$$

Therefore in order to prove (2.5) one has only to prove that

(2.10) $$u_{\varepsilon\eta} \to u_\eta \text{ in } H_0^1(\mathcal{O}) \text{ weakly, for } \eta > 0 \text{ fixed.}$$

But (2.10) follows easily from the results of Section 1.

2.3. Free boundaries.

Let S_ε denote the free boundary in problem (2.1), i.e. the interface between the region where $u_\varepsilon = 0$ and the region where $u_\varepsilon < 0$.

Similarly, let S be the free boundary in problem (2.4).

The following question is open : in what sense (if any) does S give an "approximation" of S_ε?

2.4. Other V.I.

One can replace in (2.2) the condition $v \leq 0$, $u \leq 0$ by a "general" condition

(2.11) $$v \in K , \quad u_\varepsilon \in K$$

where K denotes a (non empty) closed convex subset of $H_0^1(\mathcal{O})$. (1)

Let u_ε be the solution of

(2.12) $$a^\varepsilon(u_\varepsilon, v-u_\varepsilon) \geq (f, v-u_\varepsilon) \quad \forall v \in K , \quad u_\varepsilon \in K$$

and let u be the solution of

(2.13) $$\mathcal{a}(u, v-u) \geq (f, v-u) \quad \forall v \in K , \quad u \in K.$$

When is it true that $u_\varepsilon \to u$ in $H_0^1(\mathcal{O})$ weakly as $\varepsilon \to 0$?

The general question seems to be open.
The answer is negative if K is finite dimensional (TARTAR).
The answer is positive for K defined by

(2.14) $$K = \{v \mid v \in H_0^1(\mathcal{O}) \quad , \quad \phi_1 \leq v \leq \phi_2 \}$$

where $\phi_i \in H^1(\mathcal{O})$, $\phi_1 \leq 0 \leq \phi_2$ on Γ.

It is also positive when the condition $\phi_1 \leq v \leq \phi_2$ should take place on a subset E of $\mathcal{O}$.

Cf. BOCCARDO-MARCELLINI [1] , BOCCARDO-CAPUZZO DOLCETTA [1] , MARCELLINI [1] , CRISTIANO [1] , BENSOUSSAN-LIONS-PAPANICOLAOU [1].

2.5. Homogeneization of Quasi-Variational Inequalities. (2)

The Q.V.I. have been introduced in BENSOUSSAN-LIONS [1] for the solution of optimal impulse control problems. C. BAIOCCHI [1][2] observed that some free boundary problems arising in hydrodynamics can be

(1) One can also consider convex subsets of $H^1(\mathcal{O})$.
(2) We write Q.V.I. in short.

reduced to Q.V.I. ; this leads to the general question : <u>are homogeneization results of the previous types still valid for some Q.V.I.</u> ?

This question is open in general; we present here briefly a particular case where the answer is affirmative.

We assume that $a^{\varepsilon}(u,v)$ is defined by (1.27) and we introduce

(2.15) $$V = H^1(\mathcal{O}) \times H^1(\mathcal{O}) ;$$

for $u = \{u_1, u_2\}$, $v = \{v_1, v_2\}$ in V ; we define :

(2.16) $$b^{\varepsilon}(u,v) = a^{\varepsilon}(u_1,v_1) + a^{\varepsilon}(u_2,v_2)$$

(2.17) $$M(v) = \{k_2 + v_2,\ k_1 + v_1\},\quad k_i > 0\ ,\quad i = 1,2\ ,$$

(2.18) $$(f,v) = (f_1,v_1) + (f_2,v_2).$$

We consider the Q.V.I. :

(2.19) $$\left| \begin{array}{l} b^{\varepsilon}(u_{\varepsilon}, v-u_{\varepsilon}) \geqslant (f, v-u_{\varepsilon}) \quad \forall v \in V \ \text{ such that} \\ v \leqslant M(u_{\varepsilon})^{(1)}\ ,\quad u_{\varepsilon} \leqslant M(u_{\varepsilon}). \end{array} \right.$$

This Q.V.I. <u>admits a unique solution</u> (2).

We define the <u>homogeneized</u> Q.V.I. as follows ; we define :

(2.20) $$\mathcal{B}(u,v) = \mathcal{A}(u_1,v_1) + \mathcal{A}(u_2,v_2)\ ,$$

where $\mathcal{A}$ is defined by (1.30) ; then we let u be the solution of the Q.V.I. :

(2.21) $$\mathcal{B}(u,v-u) \geqslant (f,v-u) \quad \forall v \leqslant M(u),\ u \leqslant M(u).$$

<u>Then one can prove that</u>

(2.22) $$u_{\varepsilon} \to \bar{u} \quad \text{in} \quad V \quad \text{weakly as} \quad \varepsilon \to 0.$$

---

(1) I.e. $v_1 \leqslant k_2 + u_{\varepsilon 2}$, $v_2 \leqslant k_1 + u_{\varepsilon 1}$.

(2) We assume that $f_i \in L^{\infty}(\mathcal{O})$, $f_i \geqslant 0$. One has $u_{\varepsilon i} \geqslant 0$.

3. EVOLUTION PROBLEMS.

3.1. Parabolic Equations.

Let us consider now functions $a_{ij}(y,t)$ defined in $\mathbb{R}^n \times]0,T[$, such that

$$(3.1)\quad \left| \begin{array}{l} a_{ij} \in L^\infty(\mathbb{R}^n_y \times]0,T[), \; a_{ij} \text{ is of period 1 in all the} \\ y_i\text{'s variables}, \; a_{ij} = a_{ji}, \; \sum a_{ij}(y,t)\xi_i\xi_j \geq \alpha \sum \xi_i^2, \; \alpha > 0, \\ \text{a.e. in } y \text{ and in } t. \end{array} \right.$$

We shall set :

$$(3.2)\quad A^\varepsilon(t)v = -\sum_{i,j} \frac{\partial}{\partial x_i}\left(a_{ij}\left(\frac{x}{\varepsilon},t\right)\frac{\partial v}{\partial x_j}\right),$$

$$(3.3)\quad a^\varepsilon(t;u,v) = \sum_{i,j} \int_{\mathcal{O}} a_{ij}\left(\frac{x}{\varepsilon},t\right)\frac{\partial u}{\partial x_j}\frac{\partial v}{\partial x_i}\,dx.$$

Let f be given satisfying

$$(3.4)\quad f \in L^2(Q), \; Q = \mathcal{O} \times]0,T[.$$

It is known that there exists a unique function u_ε such that

$$(3.5)\quad u_\varepsilon \in L^2(0,T;H^1_0(\mathcal{O})), \; \frac{\partial u_\varepsilon}{\partial t} = u'_\varepsilon \in L^2(Q),$$

$$(3.6)\quad (u'_\varepsilon(t),v) + a^\varepsilon(t;u_\varepsilon(t),v) = (f(t),v) \quad \forall v \in H^1_0(\mathcal{O}),$$

$$(3.7)\quad u_\varepsilon(0) = 0.$$

As in previous Sections, we want to study the behaviour of u_ε as $\varepsilon \to 0$.

For almost every fixed t we define

$$\mathcal{A}(t)v = - \sum_{i,j} q_{ij}(t) \frac{\partial^2 v}{\partial x_i \partial x_j} \tag{3.8}$$

where

$$\left| \begin{array}{l} q_{ij}(t) = \alpha(t;\chi_t^i - y_i, \chi_t^j - y_j), \\ \alpha(t;\varphi,\psi) = \sum \int_{\Pi^n} a_{ij}(y,t) \frac{\partial\varphi}{\partial y_j} \frac{\partial\psi}{\partial y_i} dy , \\ \alpha(t,\chi_t^i,\psi) = \alpha(t;y_i,\psi) \quad \forall \psi \in W. \end{array} \right. \tag{3.9}$$

We consider next the <u>"homogeneized" parabolic equation</u>

$$u \in L^2(0,T;H_o^1(\mathcal{O})) \ , \quad u' \in L^2(Q), \tag{3.10}$$

$$(u'(t),v) + \mathcal{A}(t;u(t),v) = (f(t),v) \quad \forall v \in H_o^1(\mathcal{O}) \ , \tag{3.11}$$

$$u(o) = 0 \tag{3.12}$$

(in (3.11) we have set :

$$\mathcal{A}(t;u,v) = \sum q_{ij}(t) \int_{\mathcal{O}} \frac{\partial u}{\partial x_j} \frac{\partial v}{\partial x_i} dx). \tag{3.13}$$

One can show :

<u>Theorem</u> 3.1. : <u>Let the hypothesis</u> (3.1)(3.4) <u>take place</u>. <u>Then if</u> u_ε (resp. u) <u>denotes the solution of</u> (3.5)(3.6)(3.7)(resp. of (3.10)(3.11)(3.12)) <u>one has</u>

$$\left| \begin{array}{l} u_\varepsilon \to u \ \ \underline{\text{in}}\ L^2(0,T;H_o^1(\mathcal{O})) \ \ \underline{\text{weakly}}, \\ u'_\varepsilon \to u' \ \ \underline{\text{in}}\ L^2(Q)\ \underline{\text{weakly}}. \end{array} \right. \tag{3.14}$$

<u>Remark</u> 3.1

Let $b(y)$ be given in $\mathbb{R}^n$ such that

$$\left| \begin{array}{l} b \in L^\infty(\mathbb{R}^n), \ b \text{ is periodic of period 1 in all variables,} \\ b(y) \geq b_o > 0 \ . \end{array} \right. \tag{3.15}$$

Let us consider the equation :

(3.16) $$(b^{\varepsilon} u'_{\varepsilon}, v) + a^{\varepsilon}(t; u_{\varepsilon}, v) = (f,v) \quad \forall v \in H_o^1(\mathcal{O}),$$

(3.5) and (3.7) being unchanged, where

(3.17) $$b^{\varepsilon}(x) = b(x/\varepsilon).$$

We set :

(3.18) $$\bar{b} = \int_{\Pi_n} b(y)dy ;$$

we consider the "homogeneized" parabolic equation :

(3.19) $$(\bar{b}\, u', v) + \mathcal{a}(t; u, v) = (f,v) ,$$

conditions (3.10)(3.12) being unchanged. Then one has still (3.14) [1] ∎

3.2. Parabolic V.I.

With the notations of Section 3.1., let now u_ε be the solution of the V.I. of evolution

(3.20) $$\left| \begin{array}{l} \left(\frac{\partial u_\varepsilon}{\partial t}, v-u_\varepsilon\right) + a^{\varepsilon}(t;u_\varepsilon, v-u_\varepsilon) \geqslant (f,v-u_\varepsilon) \ \forall v \leqslant 0, v \in H_o^1(\mathcal{O}), \\ u_\varepsilon \leqslant 0 \quad \text{a.e} \end{array} \right.$$

conditions (3.5) and (3.7) being unchanged.

The "homogeneized" V.I. is :

(3.21) $$\left| \begin{array}{l} \left(\frac{\partial u}{\partial t}, v-u\right) + \mathcal{a}(t;u,v-u) \geqslant (f,v-u) \ \forall v \leqslant 0 , \ v \in H_o^1(\mathcal{O}) , \\ u \leqslant 0 \quad \text{a.e.} \end{array} \right.$$

together with conditions (3.10)(3.12).

One has then a result similar to (3.14).

For the proof, one can use methods analogous to those of Section 2.2. ∎

Remark 3.2.

One has a similar result for V.I. associated to the operator introduced in (3.16). ∎

---

(1) Problems of this type have been considered by SANCHEZ-PALENCIA; cf. [1] and personal communication.

Remark 3.3.

One has problems and results entirely similar to those of Section 2.4. for evolution V.I. ∎

3.3. Homogeneization of the Stefan's free boundary problem.

Let us consider the following Stefan's problem : we are looking for a function $\theta_\varepsilon(x,t) > 0$ (temperature) which satisfies

$$\frac{\partial\theta_\varepsilon}{\partial t} + A^\varepsilon\theta_\varepsilon = 0 \quad \text{when } \theta_\varepsilon > 0, \tag{3.22}$$

where x belongs to a domain $\Omega(t)$ (for $t > 0$ fixed) bounded by a fixed boundary Γ' and a free boundary $S_\varepsilon(t)$; on $\Gamma'\times(0,T)$, θ_ε is given > 0 :

$$\theta_\varepsilon(x,t) = g_o(x,t) \quad x \in \Gamma', \tag{3.23}$$

and on $S_\varepsilon(t)$ we have the classical conditions

$$\left|\begin{array}{l} \theta_\varepsilon = 0 , \\ \sum a_{ij}(\frac{x}{\varepsilon})\frac{\partial\theta_\varepsilon}{\partial x_j} n_i = - L\ V.n \end{array}\right. \tag{3.24}$$

where $n = \{n_i\}$ = normal to $S_\varepsilon(t)$, L given > 0 constant, V = speed of $S_\varepsilon(t)$.

The initial temperature is given :

$$\theta_\varepsilon(x,o) = \theta_o(x) \ , \ x \in \Omega(o) \quad (\Omega(o) \text{ given}). \tag{3.25}$$

Remark 3.4.

We could consider other boundary conditions than (3.23). ∎

As it is now well known [1] one can transform problem (3.22)..(3.25) into a V.I. of evolution as follows : we extend θ_ε by zero outside the union of the domains $\Omega(t)$ in $\tilde{\theta}_\varepsilon$, so that $\tilde{\theta}_\varepsilon$ is now defined in $Q=\mathcal{O}\times]0,T[$, $\mathcal{O}$ fixed with boundaries Γ' and Γ'' ; we introduce next :

([1]) Cf. G. DUVAUT [1] and a general report from this A.: DUVAUT [2]. The idea of the transformation of free boundary problems into V.I. and in Q.V.I. is due to C. BAIOCCHI [1] [2] [3]. New applications have been given by BREZIS-STAMPACCHIA [1] [2] and by FRIEDMAN-KINDERLHERER [1]. Cf. also the bibliographies of these works and LIONS [1] .

(3.26) $$u_\varepsilon(x,t) = \int_0^t \tilde{\theta}_\varepsilon(x,s)\,ds.$$

Let us define :

$$g(x,t) = \int_0^t g_0(x,s)\,ds,$$

$$f(x) = \tilde{\theta}_0(x) - L\,\chi(x), \quad \chi(x) = \text{characteristic function of the set } \mathcal{O} - \Omega(o).$$

$$K(t) = \{v \mid v \in H^1(\mathcal{O}),\ v=0 \text{ on } \Gamma'',\ v(x)=g(x,t) \text{ on } \Gamma',\ v \geq 0 \text{ a.e. in } \mathcal{O}\}.$$

<u>Then</u> u_ε <u>is characterized as the unique solution of the V.I.</u>

(3.27)
$$\left|\begin{array}{l} \left(\frac{\partial u_\varepsilon}{\partial t}, v-u_\varepsilon\right) + a^\varepsilon(u_\varepsilon, v-u_\varepsilon) \geq (f, v-u_\varepsilon) \quad \forall v \in K(t), \\ u_\varepsilon \in K(t), \end{array}\right.$$

(3.28) $$u_\varepsilon \in L^2(0,T;H_0^1(\mathcal{O}))\ , \quad \frac{\partial u_\varepsilon}{\partial t} \in L^2(Q),$$

(3.29) $$u_\varepsilon(x,o) = 0.$$

This is, up to minor changes, a V.I. similar to (3.20). Therefore, one has the result :

(3.30)
$$\left|\begin{array}{l} u_\varepsilon \to u \text{ in } L^2(0,T;H_0^1(\mathcal{O})) \quad \text{weakly}, \\ \frac{\partial u_\varepsilon}{\partial t} \to \frac{\partial u}{\partial t} \text{ in } L^2(Q) \text{ weakly} \end{array}\right.$$

where u is the solution of the "homogeneized" problem :

(3.31)
$$\left|\begin{array}{l} \left(\frac{\partial u}{\partial t}, v-u\right) + \mathcal{a}(u, v-u) \geq (f, v-u) \quad \forall v \in K(t), \\ u \in K(t), \end{array}\right.$$

and conditions similar to (3.28)(3.29).

Since $\tilde{\theta}_\varepsilon = \frac{\partial u_\varepsilon}{\partial t}$ we see that

(3.32) $$\tilde{\theta}_\varepsilon \to \tilde{\theta} \text{ in } L^2(Q) \quad \text{weakly}$$

<u>where</u> $\tilde{\theta}$ <u>is the extension by</u> 0 <u>of the solution</u> θ <u>of the "homogeneized"</u>

Stefan's problem :

$$(3.33) \qquad \frac{\partial \theta}{\partial t} + \mathcal{A}\theta = 0,$$

with conditions similar to (3.23)(3.24)(3.25).

Remark 3.5.

Let S_ε denote the free boundary for the initial problem and let S be the free boundary for the "homogeneized" problem. As in the case of stationary problems (cf. Section 2.3) it is an open question to see in what sense S_ε gives an "approximation" of S. ∎

3.4. Various Remarks.

Remark 3.6.

One can prove a result similar to (3.14) for Q.V.I. of evolution which are the "parabolic" analogous to the Q.V.I. introduced in Section 2.5. ∎

Remark 3.7.

One can study in a way that is similar to what has been done in Section 1.7. parabolic equations and parabolic V.I. in domains with holes. ∎

Remark 3.8.

One has also similar results for second order hyperbolic equations and some V.I. for hyperbolic operators.

For this and also for other V.I. for parabolic operators, we refer to BENSOUSSAN-LIONS-PAPANICOLAOU [2] ∎

3.5. Non local limit operators.

We mention now a class of partial differential operators of evolution whose "homogeneized associates" are not partial differential operators. ([1])

([1]) Cf. BENSOUSSAN-LIONS-PAPANICOLAOU [2]. In the Note [1] of these A. there is a stationary example where the same kind of phenomenon appears.

We consider two operators of the type (1.1) :

(3.34) $$A_k^{\varepsilon} v = -\sum_{i,j} \frac{\partial}{\partial x_i}\left(a_{ijk}\left(\frac{x}{\varepsilon}\right) \frac{\partial v}{\partial x_j}\right) \quad , \; k = 1,2$$

where the functions a_{ij1} and a_{ij2} satisfy conditions analogous to (1.2).

Let u_ε be the solution of

(3.35) $$\frac{\partial}{\partial t} A_1^{\varepsilon} u_\varepsilon + A_2^{\varepsilon} u_\varepsilon = f \; , \qquad f \in L^2(Q) \; ,$$

(3.36) $$u_\varepsilon \in L^2(0,T;\, H_0^1(\mathcal{O})) \; , \; u_\varepsilon(o) = 0 \; .$$

Equations of this type have been considered by SHOWALTER and TING [1]. This problem admits a unique solution which belongs to $L^\infty(0,T;H_0^1(\mathcal{O}))$, and which is such that $\frac{\partial u_\varepsilon}{\partial t} \in L^2(0,T;H_0^1(\mathcal{O}))$.

One could think that if we denote by $\mathcal{A}_k$ the homogeneized operator associated to A_k^{ε} , then the homogeneized problem associated to (3.35) is :

$$\frac{\partial}{\partial t} \mathcal{A}_1 u + \mathcal{A}_2 u = f .$$

<u>But this is not true</u>. The homogeneized equation associated to (3.35) is a <u>convolution equation</u> :

(3.37) $$\mathcal{A}(t) \overset{*}{} u = f \quad , \quad u = 0 \text{ for } \; t < 0 \; ,$$

(where in (3.37) f is extended by 0 for $t < 0$), where $\mathcal{A}$ is defined by its <u>Laplace transform</u>

(3.38) $$\hat{\mathcal{A}}(p) = \int_0^\infty e^{-pt} \mathcal{A}(t)\, dt \; ;$$

$\hat{\mathcal{A}}(p)$ is given as follows; let us introduce

(3.39) $$\alpha(p;\varphi,\psi) = \sum_{i,j} \int_{\Pi^n} (a_{ij2}(y) + p\, a_{ij1}(y)) \frac{\partial \varphi}{\partial y_j} \frac{\partial \psi}{\partial y_i}\, dy \; ;$$

for every p such that $\operatorname{Re} p \geq 0$, we define $\chi_p^i \in \overset{\circ}{W}$ as the unique solution of

(3.40) $$\alpha(p;\chi_p^i,\psi) = \alpha(p;\, y_i,\psi) \qquad \forall\, \psi \in W \; ;$$

we then set

(3.41) $$q_{ij}(p) = \alpha(p;\ \chi_p^i - y_i\ ,\ \chi_p^j - y_j)\ ;$$

then $\hat{\mathcal{a}}(p)$ is given by

(3.42) $$\hat{\mathcal{a}}(p) = -\Sigma q_{ij}(p)\ \frac{\partial^2}{\partial x_i \partial x_j}.$$

We then have :

when $\varepsilon \to 0$, one has

$u_\varepsilon \to u$ in $L^\infty(0,T;H_o^1(\mathcal{O}))$ weak star and

$\frac{\partial u_\varepsilon}{\partial t} \to \frac{\partial u}{\partial t}$ in $L^2(0,T;H_o^1(\mathcal{O}))$ weakly.

<u>Remark</u> 3.9. :

The operator $\mathcal{a}^*$ is not a <u>local operator</u> but an integro-differential operator.

We shall return later to the study of these non local effects produced by highly oscillating coefficients.

REFERENCES

BABUSKA, I. [1] Solution of problems with interfaces and singularities. Inst. Fluid Dyn. Applied Math. April 1974.

[2] Solution of the interface problem by homogeneization Inst. Fluid.Dyn. Applied Math. March 1974.

BAHVALOV, N.S. [1] Doklady Akad. Nauk. USSR, 218 (1974), pp.1046-1048.

BAIOCCHI, C. [1] Su un problema di frontiera libera connesso a questioni di idraulica. Ann. Mat. Pura e Applic. XCII (1972), pp. 107-127. (C.R.A.S. 273 (1971), pp. 1215-1217).

[2] Problèmes à frontière libre en hydraulique. C.R.A.S.278, (1974), pp. 1201-1204.

[3] These Proceedings.

BENSOUSSAN, A., and LIONS,J.L. [1] Nouvelle formulation de problèmes de contrôle impulsionnel et applications. C.R.A.S. Paris, 276 (1973), pp. 1189-1192.

[2] Contrôle impulsionnel et systèmes d'Inéquations Quasi Variationnelles. C.R.A.S. Paris, 278(1974) pp. 747-751.

BENSOUSSAN, A., LIONS, J.L., and PAPANICOLAOU, G. [1] Sur quelques phénomènes asymptotiques stationnaires. C.R.A.S., Paris, (1975).

[2] Sur quelques phénomènes aymptotiques d'évolution. C.R.A.S., Paris (1975).

[3] Book in preparation.

BOCCARDO, L., and MARCELLINI, P. [1] Sulla convergenza delle soluzioni di disequazioni variazionali. Ist. Matematico U. Dini, Firenze, 1975.

BOCCARDO, L., and CAPUZZO DOLCETTA, I. [1] To appear.

BOURGAT, J.F. [1] To appear.

BREZIS,H., and STAMPACCHIA,G. [1] The hodograph method in fluid dynamics in the light of variational inequalities. To appear (C.R.A.S. 276,(1973), pp. 129-132).

[2] These Proceedings.

CRISTIANO [1] To appear.

DUVAUT, G. [1] Résolution d'un problème de Stefan, C.R.A.S. Paris, 276, (1973), pp. 1461-1463.

[2] Problèmes à frontière libre en théorie des milieux continus. Conférence Toulouse, 1975.

FRIEDMAN,A., and KINDERLHERER, D. [1] A class of parabolic quasi variational inequalities. To appear.

DE GIORGI, E. and SPAGNOLO, S. [1] Sulla convergenza degli integrali dell'energia per operatori ellittici del secondo ordine. Boll. UMI (4) 8 (1973), pp. 391-411.

LIONS, J.L. [1] Introduction to some aspects of free surface problems. Synspade. University of Maryland, May 1975.

LIONS, J.L., and STAMPACCHIA, G. [1] Variational Inequalities. C.P.A.M. (1967), pp. 493-519.

MARCELLINI, P. [1] Un teorema di passagio al limite per la somma di funzioni convesse. Boll. U.M.I. 11 (1975).

MARINO, A. and SPAGNOLO, S. [1] Un tipo di approssimazione dell'operatore ... Annali Scuola Normale Superiore di Pisa, XXIII (1969), pp. 657-673.

MEYERS, G. [1] An L^p-estimate for the gradient of solutions of second order elliptic divergence equations. Annali Scuola N. Sup. Pisa, 17 (1963), pp. 189-206.

SANCHEZ-PALENCIA, E. [1] Comportements local et macroscopique d'un type de milieux physiques hétérogènes. Int. J. Eng. Sci. (1974), Vol. 12, pp. 331-351.

SBORDONE, C. [1] Sulla G-convergenza di equazioni ellittiche e paraboliche. Ricerche di Mat. (1975).

SHOWALTER, R.E., and TING, T.W. [1] Pseudo-parabolic partial differential equations. SIAM J. Math. Anal. (1970), pp. 1-26.

SPAGNOLO, S. [1] Sul limite delle soluzioni di problemi di Cauchy relativi all'equazione del calore. Annali Scuola Normale Superiore di Pisa, XXI (1967), pp. 657-699.

[2] Sulla convergenza di soluzioni di equazioni paraboliche ed ellittiche. Annali Scuola NOrmale Superiore di Pisa, XXII (1968), pp. 571-597.

TARTAR, L. [1] Problèmes de contrôle des coefficients dans des équations aux dérivées partielles, in Lecture Notes in Economics and Mathematical Systems, Springer, 107, (1975), pp. 420-426.

[2] To appear.

APPLICATION OF CONVEX ANALYSIS
TO THE TREATMENT OF ELASTOPLASTIC SYSTEMS

J.J. MOREAU

0. INTRODUCTION

Convex analysis, i.e. the study of convex subsets or of convex numerical functions in topological linear spaces, has progressed greatly during the recent years. This was the work of people with quite diverse backgrounds : potential theory, the general theory of topological linear spaces, partial differential equations and the calculus of variations, approximation theory, optimization and optimal control, economics ... Such very alive topics as variational inequalities, monotone operators and nonlinear semigroups are also closely intermingled with convex analysis.

The necessary facts for understanding the sequel of this lecture are sketched in Sect. 1 below ; for more details, the reader could refer to [7], [12], [19], [33].

It was with definite mechanical motivations that the author took part in the general development of convex analysis (see the reference lists in [19] or [25]). The very concept of subgradient, as formalized in [16], was devised on mechanical purposes : it allowed to treat frictionless unilateral constraints in mechanical systems as a special case of the force-configuration relations which admit a "superpotential" (Cf. [20] [24] [25] ; as an example of dynamical problem with unilateral constraints, see [17], [18], devoted to the inception of cavitation in a liquid). The "subdifferential calculus" proved also perfectly adapted to the formulation and handling of resistance laws such as the Coulomb law of friction (when the normal component of the contact force is treated as a state variable) or the Prandtl - Reuss law of perfect plasticity (cf. [21] [25]) ; strain hardening can also be tackled in the same way (cf. [28] [29]).

A short note [22] outlined how the quasi-static evolution of an elastoplastic system could be studied by these methods. An essential step in solving the problem consists in what is called the sweeping process (cf. Sect. 6 below) associated with a moving convex set in some normed space. The sweeping process has been extensively investigated in numerous reports of the series "Travaux du Séminaire d'Analyse Convexe, Montpellier" : existence of solutions under various assumptions, constructive algorithms, asymptotic properties (for a basic exposition, see also [24] ; concerning the stochastic version of this process see [3]).

The method was first applied to an elastoplastic system whose configuration manifold is a Hilbert space (see [24]);the Hilbert structure is naturally associated with the elastic potential. However, the main assumption made in this case, involving a nonempty interior for the rigidity set, relatively to the Hilbert topology, is not satisfied by usual continuous elastoplastic systems, so that the practical interest of the theory at this stage is restricted to systems with only a finite number of degrees of freedom.

The purpose of the present lecture is to explain the general method by developing the yet impublished study of an <u>elastoplastic rectilinear rod with small longitudinal displacements</u>.

The occurence in that case of only one space variable brings much simplification . In further publications the author shall adapt the same ideas to two- or three - dimensional continuous systems. But in such more complicated situations, the elements of which the existence is obtained, with some approximation algorithms, can only be considered as "weak solutions" of the evolution problem. Determining some cases of "smoothness" still remains an open task.

The Hencky - Nadai model will not be considered here ; the phenomenon it describes appears more as an extreme case of nonlinear elasticity than proper plasticity. The reader could refer to [4] [5] [6] [11] [13].

1. SUBDIFFERENTIALS

1.a. DEFINITIONS

Let X and Y be a pair of real linear spaces placed in duality by a bilinear form $< .,. >$. For sake of simplicity it will be supposed that this duality is separating , i.e. the linear form defined on X by $x \mapsto < x,y >$ is identically zero only if y is the origin of Y and the symmetrical assumption is made regarding the linear form defined on Y by $y \mapsto < x,y >$.

Recall that a locally convex topology on X (resp. Y) is said compatible with the duality if the continuous linear forms relative to this topology are exactly those which can be expressed as above. The closed convex subsets of X or also the lower semicontinuous convex numerical functions on this space are the same for all these topologies ; therefore as soon as a dual pair of linear spaces is given, we shall refer to closed convex sets or to l.s.c. convex functions without specifying the topology.

In what follows, X and Y play symmetric roles. Let f be a function defined, for instance, on X with values in $] -\infty , +\infty]$. An element $y \in Y$ is called <u>a subgradient of</u> f <u>at the point</u> $x \in X$ <u>if the value</u> f(x) <u>is finite and if the affine function, taking the same value as</u> f <u>at the point</u> x,

$$u \mapsto < u - x, y > + f(x)$$

<u>is a minorant of</u> f <u>all over</u> X . The (possibly empty) set of these subgradients is called the <u>subdifferential</u> of f at the point x ; this is a subset of Y denoted by $\partial f(x)$.

For instance in the special case where f is convex on X and weakly differentiable at the point x with $y \in Y$ as gradient (or "Gâteaux differential") at this point, the subdifferential reduces to the singleton $\{y\}$.

The following concept is immediately connected with subdifferentials : The numerical function f^* defined on Y by

$$(1.1) \qquad f^*(y) = \sup_{x \in X} [< x,y > - f(x)]$$

is called the <u>polar</u> or <u>conjugate</u> function of f, relative to the duality $< X,Y >$. Then

(1.2) $\quad \partial f(x) = \{y \in Y \; : \; f(x) + f^*(y) - < x,y > \leq 0 \}$

(where the $\leq$ sign may equivalently be replaced by =). As f^* is the supremum of a collection of affine functions on Y which are continuous for the l.c. topologies compatible with the duality, this function is convex and l.s.c. . Then (2.2) proves that $\partial f(x)$ is closed and convex.

Iterating the process one may consider f^{**}, the polar function of f^*. Standard separation arguments show that <u>if</u> f <u>is convex</u> <u>l.s.c.</u>, <u>with values in</u> $]-\infty, +\infty]$, <u>then</u> $f^{**} = f$. If in addition it is specified that f is not the constant $+\infty$, this function is said <u>proper closed convex</u> and the same properties hold for $g = f^*$.

Suppose that f and g are, in that way, a pair of mutually polar proper closed function ; then for $x \in X$ and $y \in Y$ the four following properties are equivalent

$$(1.3) \qquad \begin{cases} y \in \partial f(x) \\ x \in \partial g(y) \\ f(x) + g(y) - < x,y > \leq 0 \\ f(x) + g(y) - < x,y > = 0 \; ; \end{cases}$$

then x and y are said <u>conjugate</u> relatively to f,g.

1.b. EXAMPLE.

Let C be a subset of X ; its <u>indicator function</u> ψ_C (i.e. $\psi_C(x) = 0$ if $x \in C$ and $+\infty$ if not) is proper closed convex if and only if C is nonempty closed and convex. The polar function

$$y \mapsto \psi_C^*(y) = \sup_{x \in X} [< x,y > - \psi_C(x)] = \sup_{x \in C} < x,y >$$

is classically called the <u>support function</u> of C (relative to the duality $< X,Y >$). The nonzero subgradients of ψ_C are obviously related to the <u>supporting hyperplanes</u> of the set C . Precisely, for any $x \in X$, the set $\partial\psi_C(x)$ is a closed conic convex subset of Y (with vertex at the origin) ; it is empty if and only if $x \notin C$; as soon as $x \in C$ this set contains at least the origin of Y . Generally speaking, $\partial\psi_C(x)$ is called the <u>normal outward cone to</u> C associated with x .

More specially let U be a closed linear subspace of X ; then $\psi_U^* = \psi_V$, where

V denotes the subspace of Y orthogonal to U . And

$$(1.4)\qquad \partial\psi_U(x) = \begin{cases} V & \text{if } x \in U \\ \emptyset & \text{if } x \notin U . \end{cases}$$

1.c. ADDITION RULE

Here is the most usual problem of the subdifferential calculus :

Let f_1 and f_2 be two numerical functions on X ; for every $x \in X$ one trivially has

$$(1.5)\qquad \partial f_1(x) + \partial f_2 \subset \partial\,(f_1 + f_2)\,(x)\,.$$

Various sufficient conditions have been established ensuring that this inclusion is actually an equality of sets. We shall only need the following one : <u>If f_1 and f_2 are convex and if there exists a point $x_0 \in X$ at which both functions take finite values, one of them being continuous at this point</u> (for some topology compatible with the duality $\langle X,Y \rangle$), <u>then</u> (1.5) <u>holds as an equality for every</u> $x \in X$.

As an illustration let C be a convex subset of X and f a convex function. If there exists a point x_0 at which f takes a finite value and which is interior to C relatively to some topology compatible with the duality this is a point of continuity for ψ_C so that one has

$$\partial(\psi_C + f)\,(x) = \partial\,\psi_C\,(x) + \partial\,f(x)$$

for every $x \in X$.

Consequently x is a minimizing point of the restriction of f to C if and only if $\partial\psi_C\,(x) + \partial f(x)$ contains the origin of Y , or equivalently

$$\partial f(x) \cap -\,\partial\psi_C(x) \neq \emptyset\,.$$

1.d. CONVEX INTEGRAL FUNCTIONALS

Many rules of the subdifferential calculus involve several pairs of dual spaces at the same time. The following situation is of primary importance in continuum mechanics when, starting from the local behavior of the medium, one generates functional formulations.

Let (T,μ) be a measure space and let X and Y be two spaces of measurable mappings of (T,μ) into $\mathbb{R}^n$, such that the bilinear pairing

$$(1.6) \qquad \langle x,y \rangle = \int_T x(t).\, y(t)\, \mu(dt)$$

is meaningful for every $x \in X$ and $y \in Y$ (the dot represents the natural scalar product in $\mathbb{R}^n$). Let $u \mapsto f(t,u)$ and $v \mapsto g(t,v)$ be a pair of convex numerical functions on $\mathbb{R}^n$, depending on $t \in T$ and which, for each t, are the polar of each other in the sense of the scalar product of $\mathbb{R}^n$. Under mild assumptions concerning the function spaces X and Y (they are satisfied in particular by the Lebesgue spaces $L^p(T, \mathbb{R}^n)$ and $L^q(T, \mathbb{R}^n)$, $1/p + 1/q = 1$) and simple measurability assumptions concerning the numerical functions $f,g : T \times \mathbb{R}^n \to]-\infty, +\infty]$, R.T. Rockafellar [30] [33] has established that the functionals F and G respectively defined on the spaces X and Y by the integrals (possibly taking the value $+\infty$)

$$F : x \mapsto \int_T f(t,x(t)\,)\,\mu(dt)$$

$$G : y \mapsto \int_T g(t,y(t))\,\mu(dt)$$

are mutually polar convex functions relatively to the pairing (1.6) ; furthermore x and y are conjugate relatively to F and G if and only if $x(t)$ and $y(t)$ are conjugate relative to $f(t,.)$ and $g(t,.)$, for almost every t in the sense of μ.

A very convenient account of this question may be found in [34]. Of course the case of a pair of functions $f(t,.)$ and $g(t,.)$ defined, instead of $\mathbb{R}^n$, on a pair of infinite dimensional dual spaces has also been investigated (see e.g. [7],[34] or various reports in "Travaux du Séminaire d'Analyse Convexe , Montpellier" by C. Castaing, P. Clauzure, M.F. Sainte - Beuve, M. Valadier).

1.e. THE $\mathcal{C}$, $\mathcal{M}$ DUALITY

In the same connection, we shall use in this lecture another result of R.T. Rockafellar [33]: Suppose now that K is a compact topological space, with no measure a priori given. Let $x \mapsto \gamma(x)$ be a <u>multifunction</u> of K into $\mathbb{R}^n$, with non-empty closed convex values (in other words, γ is a nonempty closed convex subset of $\mathbb{R}^n$, depending on $x \in K$). This multifunction is supposed <u>lower semicontinuous</u> in the classical sense that, for every open subset Ω of $\mathbb{R}^n$ the set $\{x \in K : \gamma(x) \cap \Omega \neq \emptyset\}$ is open. Let us consider the (Banach) space $\mathcal{C}$ of the continuous mapping of K into $\mathbb{R}^n$ and its dual, i.e. the space $\mathcal{M}$ of the n-dimensional Radon measures on K, with the natural pairing. The set of the <u>continuous selectors</u> of γ , i.e.

$$C = \{s \in \mathscr{C} : \forall x \in K,\ s(x) \in \gamma(x)\}$$

is a closed convex subset of $\mathscr{C}$ and it is nonempty by virtue of a theorem of E. Michael [15] . The statement is that the support function of C in the sense of the duality $\mathscr{C}, \mathscr{M}$ may be constructed as follows : For every $m \in \mathscr{M}$, there exists, non uniquely, a nonnegative (bounded) scalar Radon measure μ on K relatively to which m possesses a density $\frac{dm}{d\mu} \in L^1(K, \mu ; \mathbb{R}^n)$ and one has

$$\psi_C^*(m) = \int_K \psi_{\gamma(x)}^* \left(\frac{dm}{d\mu}(x) \right) \mu(dx) ;$$

changing the measure μ is clearly immaterial, because $\psi^*_{\gamma(x)}$, the support function of $\gamma(x)$ in the sense of the natural duality of $\mathbb{R}^n$, is positively homogeneous ; in particular, μ may be the "absolute value" of the n-dimensional measure m . In addition, for $s \in \mathscr{C}$ and $m \in \mathscr{M}$, the relation $m \in \partial\psi_C(s)$ is equivalent to : for every μ as above, the density function satisfies

$$\frac{dm}{d\mu}(x) \in \partial\psi_{\gamma(x)}(s(x))$$

for every $x \in K$, except possibly on a set whose μ-measure is zero ; here again the choice of μ is immaterial because $\partial\psi_{\gamma(x)}$ is a cone in $\mathbb{R}^n$.

2. FORCES AND VELOCITIES

2.a. THE $\mathcal{V}$, $\mathcal{F}$ FORMALISM

One is used, in classical mechanics, to associate with each possible configuration of a material system a pair of real linear spaces, infinite dimensional if the system has an infinite number of degrees of freedom, which will be denoted in the sequel by $\mathcal{V}$ and $\mathcal{F}$. The elements of $\mathcal{V}$ constitute, in a general sense, the possible values of the velocity of the system if it comes to pass through the considered configuration. The elements of $\mathcal{F}$ are the possible values of various forces which may be exerted on the system in that event. Forces, in such an abstract sense, are merely items of the code under which the available physical information about the considered material system is fed into the calculating machinery of Mechanics.

Denoting by $< v,f >$ the power of the force $f \in \mathcal{F}$ if the systems happens to have the velocity $v \in \mathcal{V}$, one places the spaces $\mathcal{V}$ and $\mathcal{F}$ in duality. The traditional method of virtual power (or of virtual work) precisely consists in exploiting this duality.

Observe that, in a given mechanical situation, there are usually several ways of applying this $\mathcal{V}$, $\mathcal{F}$ formalism (comparative examples are developed in [24] or [25]). For instance, if the mechanical system is a continuous medium occupying in the considered configuration a region Ω of the physical space, it may be convenient to take as $\mathcal{V}$ and $\mathcal{F}$ two spaces of tensor fields defined in Ω. The element $v \in \mathcal{V}$ will be the field $\omega \mapsto \dot{\varepsilon}^{ik}(\omega)$, the time-rate of strain of the medium, while every "force" $f \in \mathcal{F}$ must take the form of a field of strain tensors $\omega \mapsto \sigma_{ik}(\omega)$. The latter may be the proper internal strain of the medium, but such an f may also depict some external mechanical action according to the following rule : for every $v \in \mathcal{V}$, $< v,f >$ must equal the corresponding power.

Recall that, under integrability assumptions relative to the Lebesgue measure $d\omega$ of Ω, one has classically

$$(2.1) \qquad < v,f > \; = - \int_{\Omega} \dot{\varepsilon}^{ik}(\omega)\, \sigma_{ik}(\omega)\, d\omega \, .$$

The minus sign in this expression is a pure accident due to the sign conventions made when defining the components of stress in solid mechanics (while pressure, in fluid mechanics, is counted with the opposite convention). Sometimes it will be found simpler to place such spaces in duality by using the natural functional analytic scalar product : then one shall remember that it represents the negative of the power.

2.b. RESISTANCE LAWS.

In this general $\mathcal{V}$, $\mathcal{F}$ formalism, let us call a <u>resistance law</u> some relation between the possible velocity $v \in \mathcal{V}$ of the system and the value $f \in \mathcal{F}$ of some of the forces it undergoes.

The most elementary case is that of a linear of <u>viscous</u> resistance. Then the relation has the form $-f = Lv$, where $L : \mathcal{V} \to \mathcal{F}$ is a linear mapping, self-adjoint with regard to the power pairing $<.,.>$ and monotone, i.e.

$$\forall v \in \mathcal{V} : \qquad < v, L v > \geq 0 . \tag{2.2}$$

Trivially, the numerical function $Q : v \mapsto < v, L v > / 2$ is a quadratic form on the space $\mathcal{V}$ (the Rayleigh function) and $L v$ is its weak gradient at the point v. By (2.2) the quadratic form Q is nonnegative, thus convex and the preceding relation may equivalently be written as

$$- f \in \partial Q (v) . \tag{2.3}$$

The advantage of the subdifferential notation manifests itself if one is dealing with resistance laws of the <u>dry friction</u> type, for in that case, the relation cannot be "solved" to express one of the elements v or f as a function of the other.

Generally speaking, a dry friction law is defined by giving a nonempty closed convex subset C of $\mathcal{F}$, containing the origin, and by stating the <u>maximal dissipation principle</u>, i.e. the values of $f \in \mathcal{F}$ which correspond to some given $v \in \mathcal{V}$ are the elements of C which minimize the numerical function $f \mapsto < v,f >$ (usually $- < v,f >$ is called the dissipated power). Such a relation between v and f is immediately found equivalent to

$$- v \in \partial \psi_C(f) . \tag{2.4}$$

In view of (1.3) this is also equivalent to

$$f \in \partial \psi_C^* (-v) \tag{2.5}$$

equivalent to

$$(2.6) \qquad \psi_C^*(-v) + \psi_C(f) + \langle v,f \rangle = 0$$

(where the = sign may be replaced by $\leq$).

Denoting by φ the function $v \mapsto \psi_C^*(-v)$ (it is the support function of the set $-C$) one gives to (2.5) the form

$$-f \in \partial\varphi(v)$$

similar to (2.3) .

Observe that (2.6) is equivalent to

$$(2.7) \qquad \begin{cases} f \in C \\ -\langle v,f \rangle = \varphi(v) . \end{cases}$$

In other words the values of f that the considered relation associates with a given $v \in \mathcal{V}$ are the elements of C for which the dissipated power $-\langle v,f \rangle$ exactly equals $\varphi(v)$; hence φ may be called in the present case the dissipation function of the considered resistance law (for a general discussion of this concept see [21], [24]).

2.c. SYSTEMS WITH LINEAR CONFIGURATION MANIFOLD

Many problems of applied mechanics are treated under the small deviation approximation, i.e. the considered system is assumed to remain "infinitely close" to a given reference configuration. Then all the geometrical and kinematical relations concerning the possible motions are linearized ; thereby the set of the possible configurations of the system is treated as a linear space $\mathcal{U}$, whose the considered reference configuration constitutes the origin. A motion of the system being defined as a mapping $t \mapsto u(t) \in \mathcal{U}$, the velocity at the instant t is the derivative $\dot{u}(t) \in \mathcal{U}$, supposed to exist relatively to some topology on $\mathcal{U}$. Thus, in the present situation one has $\mathcal{V} = \mathcal{U}$, the same velocity space for all configurations, and a single force space $\mathcal{F}$, in duality with $\mathcal{U}$, will be considered.

2.d. THE PRANDTL - REUSS LAW

The classical treatment of plasticity consists in introducing, beside the geometric or "visible" elements depicting the configuration of the considered system, some internal variables or hidden parameters. If the framework of the preceding paragraph, where the configuration manifold is a linear space $\mathcal{U}$, one is naturally induced to interpret also the hidden parameters as defining an element of some linear space. Each possible state of the system is thus described by two components : the <u>visible</u> or <u>exposed</u> component $q \in \mathcal{U}$ and the hidden or <u>plastic</u> component p . The phenomenological representation of <u>strain hardening</u> requires of p to range over a larger space than q (see [28], [29]) ; but as far as <u>perfect plasticity</u> is concerned it suffices to take as p an element of the same space $\mathcal{U}$. The elastic potential is then assumed to depend only on the difference $q - p$, the <u>elastic deviation</u> : this implies that the elastic forces "acting" respectively on the components q and p are two elements of $\mathcal{F}$ with zero sum.

On the other hand, the plastic component p is submitted to a <u>resistance to yielding</u>, which is a resistance law of the dry friction type.

Such is in particular the underlying pattern of the Prandtl - Reuss law of perfect platicity for a continuous medium occupying a domain Ω of the physical spaces. The linear space $\mathcal{U}$ consists of strain tensor fields such as $q : \omega \mapsto \varepsilon(\omega)$, the <u>visible strain</u>. The <u>plastic strain</u> is an element $p : \omega \mapsto \varepsilon_p(\omega)$ of the same space. This latter element is assumed to present only a quasistatic evolution, i.e. at every instant, the elastic force on it equilibrates the resistance to yielding. The linear space $\mathcal{F}$ consists of stress tensor fields such as $s : \omega \mapsto \sigma(\omega)$, the stress in the medium. At every point ω of the medium is given the <u>rigidity set</u> $\gamma(\omega)$ a closed convex subset of E_6 , the six-dimensional linear space of the second order symmetric tensors. The resistance to yielding is an element $\omega \mapsto r(\omega)$ of $\mathcal{F}$, locally related to the "plastic strain velocity" $\dot{\varepsilon}_p(\omega)$ by

(2.8) $$\dot{\varepsilon}_p(\omega) \in \partial\psi_{\gamma(\omega)}(r(\omega))$$

We choose to understand this subdifferential in the sense of the self-duality of E_6 , defined by the natural euclidean scalar product of second order tensors ; hence the sign discrepancy with (2.4) . As the stress s is conceived as "acting" on the visible component, the elastic force acting on the component p is $-s$ and the quasi-equilibrium of this component is finally expressed by

(2.9) $$\dot{\varepsilon}_p(\omega) = \partial\psi_{\gamma(\omega)}(s(\omega))$$

for every ω in Ω .

3. ELASTOPLASTIC ROD : THE PRIMARY FORMULATION

3.a. THE SYSTEM

One considers a rectilinear thin rod occupying the interval $0 \leq x \leq 1$ of the x axis. The elements of this rod are supposed to perform only displacements along the x axis ; let $u(t,x)$ denote the displacement, at the time $t \geq 0$, of the element whose position in some reference state of the rod is $x \in [0,1]$. The values of u are treated as "infinitely small" ; thus the derivative

$$(3.1) \qquad \partial u/\partial x = \varepsilon\,(t,x)$$

constitutes the <u>strain</u> at the time t and at the point x of the rod.

3.b. BOUNDARY CONDITIONS.

The extremity $x = 0$ is maintained fixed, i.e.

$$(3.2) \qquad \forall\, t \geq 0 \; : \qquad u(t,0) = 0 \quad .$$

A given motion is imposed to the extremity $x = 1$, i.e.

$$(3.3) \qquad \forall\, t \geq 0 \quad : \qquad u(t,1) = h(t)$$

where $t \mapsto h(t)$ is a given function.

3.c. LOAD

Distributed external forces parallel to the x axis, depending on time, are exerted on the various elements of the rod. They are at the present stage described by a function $(t,x) \mapsto f(t,x)$, the density of this distribution of forces relatively to the Lebesgue measure on $[0,1]$.

3.d. BEHAVIOR OF THE MATERIAL

According to the Prandtl - Reuss model recalled in Sect. 2, the strain $\varepsilon(t,x)$ is decomposed into the sum of the <u>elastic strain</u> ε_e and of the plastic strain ε_p

$$(3.4) \qquad \varepsilon\,(t,x) = \varepsilon_e(t,x) + \varepsilon_p(t,x)$$

For each (t,x) the value of ε_e is related to the <u>stress</u> or <u>tension</u> s (here reduced to a single scalar component) by a linear law

(3.5) $$\varepsilon_e(t,x) = a(x)\, s(t,x) \quad .$$

The given scalar $a(x) \geq 0$, independent of time, is the <u>elastic compliance</u> of the rod at the point x .

For every $x \in [0,1]$ the local <u>rigidity set</u> is a given interval of $\mathbb{R}$

$$\gamma(x) = [\alpha(x), \beta(x)]$$

whose extremities are the <u>local yield limits</u>. Putting $\partial\varepsilon_p/\partial t = \dot{\varepsilon}_p$, we require as in (2.9) that

(3.6) $$\forall\, t \geq 0 \;,\; \forall\, x \in [0,1] : \dot{\varepsilon}_p\,(t,x) \in \partial\psi_{\gamma(x)}\,(s(t,x)\,)$$

where the subdifferential is relative to the natural self-duality of $\mathbb{R}$.

3.e. INITIAL CONDITIONS

As usual in elastoplastic problems, the <u>initial stress</u> must be given (not arbitrarily, see (3.10) below) :

(3.7) $$\forall\, x \in [0,1] \;:\; s(0,x) = s_0(x)$$

as well as the initial deviation :

(3.8) $$\forall\, x \in [0,1] \;:\; u(0,x) = u_0(x).$$

The latter implies

(3.9) $$\forall\, x \in [0,1] \;:\; \varepsilon(0,x) = \frac{\partial u_0}{\partial x}(x).$$

3.f. QUASI - EQUILIBRIUM

The problem is that of determining the evolution of the rod under conditions (3.1) to (3.9), supposing that this evolution is quasistatic, i.e. the data are subject to such limitations that inertia is negligible in the motion which actually takes place. Thus the dynamical equation reduces to that of pure statics

$$(3.10) \qquad \forall\, t \geq 0 \;, \quad \forall\, x \in [0,1] \;:\; \frac{\partial s}{\partial x} + f = 0 \;.$$

3.g NECESSARY WEAKENING OF THE REQUIREMENTS.

As usual in the primary formulation of mechanical problems, such as they arise from engineering situations, the requirements listed above implicitely involve the smoothness of the considered functions. But in the present example, one easily observesthat, even under very strong regularity assumptions regarding the data, the existence of a solution in terms of smooth functions cannot be expected in general.

Suppose for instance that the given load density f is time-independent, continuous with regard to x and vanishing only at a finite number of points. Then the function

$$x \mapsto F(x) = -\int_0^x f(\xi)\, d\xi$$

attains its extrema at some of these points $x_1, x_2, \ldots, x_n$. The quasi-equilibrium condition (3.10) is equivalent to $s(x,t) = y(t) + F(x)$ where $t \mapsto y(t)$ denotes an unknown function. Suppose that the yield limits α and β are independent of x and that the initial tension $s_0(x) = y(0) + F(x)$ verifies

$$\forall\, x \in [0,1] \quad : \quad \alpha < s_0(x) < \beta \;.$$

This evidently implies the existence of a time interval $[0,t_1[$ during which the evolution, caused by the given continuous motion imposed by (3.3) to the extremity 1 of the rod, takes place in a purely elastic way, i.e. $\dot{\epsilon}_p = 0$ everywhere. Thus $x \mapsto \epsilon_p(x)$ is independent of t and, by easy calculation,

$$y(t) = y(0) + (h(t) - h(0)) \,/ \int_0^1 a(x)\, dx \quad .$$

The function $x \mapsto s(t,x) = y(t) + F(x)$ attains its extrema at some of the points $x_1, x_2, \ldots, x_n$. Supposing for instance that $t \mapsto h(t)$ continuously increases, one finds that this phase of motion ends at the instant where the maximum of $x \mapsto y(t) + F(x)$ attains the yield limit β . Henceforward the condition $\alpha \leq s(x) \leq \beta$ makes that $t \mapsto y(t)$ cannot increase anymore, thus

$$\int_0^1 \epsilon_e(t,x)\, dx = \int_0^1 a(x)\, (y(t) + F(x)\,)\, dx$$

remains constant. The requirement $u(t,1) = h(t)$ can only be met as the result of yielding, localized to some of the points $x_1, x_2, \ldots, x_n$. Consequently the plastic strain cannot be depicted as a function $x \mapsto \varepsilon_p(t,x)$ nor its time-rate as a function $x \mapsto \dot{\varepsilon}_p(t,x)$. One is definitely induced to consider this elements as measures, thus to turn to a weaker formulation of the problem.

4. THE CHOICE OF A PAIR OF SPACES

4.a THE STRAIN AND STRAIN-RATE MEASURES

The elementary treatment sketched in the preceding section suggests that, under reasonable assumptions concerning the data, one may expect a solution of the problem in which the stress in all part of the rod is depicted, for every t, by a continuous numerical function $x \mapsto s(t,x)$. But to describe the strain or its decomposition into the sum of the elastic and plastic terms and also in what concerns the time-rates of these elements, it turned out that the suitable mathematical objects should not be functions but measures.

Therefore we choose as the mathematical framework for all the sequel the following pair of linear spaces :

1° The space $\mathscr{C}([0,1], \mathbb{R})$, abreviatively denoted by $\mathscr{C}$, of the continuous numerical functions defined on the compact interval $[0,1]$.

2° The space $\mathscr{M}([0,1], \mathbb{R})$, abreviatively denoted by $\mathscr{M}$, of the (bounded) scalar measures on the same interval.

$\mathscr{C}$ is a separable nonreflexive Banach space and $\mathscr{M}$ is its dual ; the corresponding bilinear pairing will be denoted by $\ll . , . \gg$

In the elementary setting of Sect.3, the time-rate of strain at the instant t was a numerical function $x \mapsto \dot{\varepsilon}(t,x)$. Similarly to the three-dimensional expression recalled in (2.1), the corresponding power of some stress $x \mapsto \sigma(t,x)$ should be expressed by an integral relative to the Lebesgue measure of $[0,1]$

$$\mathscr{P} = - \int_{[0,1]} \sigma(t,x)\, \dot{\varepsilon}(t,x)\, dx \quad . \tag{4.1}$$

This may equivalently by read as the integral of the function $x \mapsto \sigma(t,x)$ relatively to the measure $\dot{q}$ whose density with regard to the Lebesgue measure is $\dot{\varepsilon}$. The step we take now consists in depicting the time-rate of strain of the rod at some instant t by an element $\dot{q}(t)$ of $\mathcal{M}$ which does not necessarily possess a density with regard to the Lebesgue measure. Then if the stress $x \mapsto \sigma(t,x)$ is a continuous function on $[0,1]$, i.e. an element $\sigma(t)$ of $\mathcal{C}$, (4.1) is to be replaced by

$$(4.2) \qquad \mathcal{P} = - \langle\langle \sigma, \dot{q} \rangle\rangle .$$

By this minus sign the functional analytic pairing $\langle\langle \cdot , \cdot \rangle\rangle$ we shall use in the sequel differs from the mechanical pairing which would be defined, according to Sect.2, between $\mathcal{M}$ considered as the velocity space and $\mathcal{C}$ considered as the force space of our mechanical system.

Similarly, the strain of the rod at each instant t will depicted by an element $q(t)$ of $\mathcal{M}$, without reference to the Lebesgue measure. The connection between q and the displacement u is stated as follows : it is required of the numerical function $x \mapsto u(t,x)$ to have for every t a bounded variation and to admit $q(t)$ as its differential measure.

This classically means that the function $x \mapsto u(t,x)$ possesses at every point $x \in]0,1[$ a left limit and a right limit respectively denoted by $u^-(t,x)$ and $u^+(t,x)$ and that

$$(4.3) \qquad \int_{[0,x[} q(t) = u^-(t,x) - u(t,0)$$

$$(4.4) \qquad \int_{[0,x]} q(t) = u^+(t,x) - u(t,0) .$$

This will still be true for $x = 0$ or $x = 1$ if we agree to write $u^-(t,0) = u(t,0)$ and $u^+(t,1) = u(t,1)$.

Of course we shall require of q to be, in a certain sense, the time integral of $\dot{q}$: see the formulation in Sect.5 below. In addition, some connection will be found between the measure $\dot{p}(t)$ and the derivatives of the functions $t \mapsto u^-(t,x)$ and $t \mapsto u^+(t,x)$.

4.b ELASTICITY LAW

The general pattern of elastoplasticity (cf Parag. 2.d) is now applied by taking $\mathcal{M}$ as the configuration space and $\mathcal{C}$ as the force space. Thus, for every t, the "visible" strain $q(t)$ is decomposed into

(4.5) $$q = e + p \quad ,$$

where $e \in \mathcal{M}$ and $p \in \mathcal{M}$ are respectively called the elastic strain and the plastic strain.

The local linear elasticity relation (3.5) will be replaced by

(4.6) $$e = As$$

where $s \in \mathcal{C}$ is the stress of the rod at the considered instant and A a nonnegative element of $\mathcal{M}$, independent of time, called the compliance measure of the rod ; the right member of (4.6) is to be read as the product of this measure by the continuous function s.

In the elementary setting, the measure A possessed, with regard to the Lebesgue measure of $[0,1]$, a density which was precisely the compliance function $x \mapsto a(x)$. But the Lebesgue measure has no mechanical relevance to the present situation and defining directly the compliance of the rod as a measure appears definitely more convenient. As an illustration, suppose that the measure A presents an atom at some point $x \in]0,1[$, i.e. this measure is the sum of a diffuse measure and of the punctual mass $\alpha > 0$ at the point x. Then (4.6) implies that e is the sum of a diffuse measure and of the punctual mass $\alpha\, s(x)$ at the point x. If on the other hand, the plastic strain p is zero at the considered instant, (4.3) and (4.4) imply

$$u^+(t,x) - u^-(t,x) = \alpha\, s(x) \quad .$$

Such a jump of u means that a gap occurs at the point x of the rod, proportional to the local value of the stress. This may represent some loose elastic connection between the parts $[0,x[$ and $]x,1]$ of the rod ; something like a crack.

4.c PLASTICITY LAW

In the same way as for q , it will be required in Sect.5 of the function $t \mapsto p(t) \in \mathcal{M}$ to be the (weak) integral of some function $t \mapsto \dot{p}(t) \in \mathcal{M}$. The measure $\dot{p}(t)$ is called the time-rate of plastic strain of the rod at the instant t .

Let us denote by C the set of the continuous selections of the multifunction $x \mapsto \gamma(x)$, i.e.

$$C = \{s \in \mathcal{C} \ : \ \forall x \in [0,1], \ \ \alpha(x) \leq s(x) \leq \beta(x) \}$$

This is a closed convex subset of $\mathcal{C}$; Assumptions 2 and 3 to be formulated in Parag.5.c below, will ensure that this set is nonempty and that Rockafellar's result of Parag. 1.e apply. Therefore the writing

(4.7) $$\dot{p}(t) \in \partial\psi_C\,(s(t))$$

in the sense of the duality $\mathcal{C}$, $\mathcal{M}$ will have a local meaning quite similar to the Prandtl - Reuss law (3.6) of the elementary case. The only difference is that the function $x \mapsto \dot{\varepsilon}_p\,(t,x)$ will then be the density of the measure $\dot{p}(t)$ with regard to some nonnegative measure on [0,1] possibly other than the Lebesgue measure.

5. THE MATHEMATICAL PROBLEM

5.a FORMULATION

Let us choose an interval of time $[0,T]$.

We are to determine a mapping $u : [0,T] \times [0,1] \to \mathbb{R}$, three mappings $q,e,p : [0,T] \to \mathcal{M}$, a mapping $s : [0,T] \to \mathcal{C}$ such that :

1° For every $t \in [0,T]$, the function $x \mapsto u(t,x)$ has a bounded variation on [0,1], it agrees with the boundary conditions

(5.1) $$u(t,0) = 0$$

(5.2) $$u(t,1) = h(t)$$

and the measure $q(t)$ is its differential measure according to (4.3) and (4.4) .

2° For every $x \in [0,1]$, the functions $t \mapsto u^-(t,x)$ and $t \mapsto u^+(t,x)$ (i.e. the left and right limits at the point x ; cf. Parag. 4.a : recall that, by convention, $u^- = u$ for $x = 0$ and $u^+ = u$ for $x = 1$) are Lipschitz and they agree with the initial conditions

(5.3) $$u^-(0,x) = u_0^-(x) \quad , \quad u^+(0,x) = u_0^+(x) \quad .$$

3° The function $t \mapsto s(t)$ is Lipschitz in the norm topology of $\mathcal{C}$; it agrees with the initial condition

(5.4) $$s(0) = s_0 \in \mathcal{C}$$

and, for every $t \in [0,T]$ the quasi-equilibrium condition

(5.5) $$\frac{\partial s}{\partial x} + f = 0$$

holds in the elementary sense ; here the given load density $x \mapsto f(t,x)$ is a continuous function on $[0,1]$. More generally the given load could be depictedas a <u>diffuse measure</u> on $[0,1]$; then (5.5) would be replaced by the equality of this measure to the negative of the differential measure of s (in that case s should be a continuous function with bounded variation) : the essential fact is that this quasi-equilibrium condition could be translated into the form (5.15) below.

4° For every $t \in [0,T]$

(5.6) $$q = e + p$$

(5.7) $$e = As$$

with $A \in \mathcal{M}$, a nonnegative given measure.

5° There exists $\dot{p} : [0,T] \to \mathcal{M}$, <u>weakly integrable</u> with values in a ball of $\mathcal{M}$, such that for every $t \in [0,T]$

(5.8) $$p(t) = p(0) + \int_0^t \dot{p}(\tau)d\tau \quad ,$$

with p(0) related to the initial data (see (5.17) below) and that, for almost every t

$$\dot{p} \in \partial\psi_C(s) \quad , \tag{5.9}$$

with C defined as in Parag. 4.c .

6° There exists $\dot{q} : [0,T] \to \mathcal{M}$, weakly integrable with values in a ball of $\mathcal{M}$ such that for every t

$$q(t) = q(0) + \int_0^t \dot{q}(\tau)\, d\tau \quad . \tag{5.10}$$

For almost every t , the measure $\dot{q}(t)$ is the differential measure of a function $x \mapsto \dot{u}(t,x)$ with bounded variation on [0,1], vanishing at x = 0 and such that

$$\frac{d}{dt}\, u^-(t,x) = \dot{u}^-(t,x) \tag{5.11}$$

$$\frac{d}{dt}\, u^+(t,x) = \dot{u}^+(t,x). \tag{5.12}$$

Remark. As $\mathcal{C}$ is separable, (5.8) and (5.10) imply that for almost every t the elements $\dot{p}(t)$ and $\dot{q}(t)$ of $\mathcal{M}$ are the derivatives of the functions $t \mapsto p(t)$ and $t \mapsto q(t)$ in the sense of the weak-star topology $\sigma(\mathcal{M}, \mathcal{C})$.

5.b EQUIVALENT DATA

As we are to focus on the construction of mappings of [0,T] into $\mathcal{C}$ or $\mathcal{M}$, we first represent all the data by elements of these spaces. So to speak, the dual pair of linear spaces $\mathcal{C}$, $\mathcal{M}$ constitute a calculating device and the data must be fed into it under a suitably adapted form.

Let us denote by U the subspace of $\mathcal{M}$ consisting of the measures whose sum is zero ; let us denote by V the subspace of $\mathcal{C}$, isomorphic to $\mathbb{R}$, of the constant numerical functions on [0,1]. Clearly V and U are the orthogonal of each other relatively to the duality $\mathcal{C}$, $\mathcal{M}$.

To the given $h : [0,1] \mapsto \mathbb{R}$ depicting the motion of the extremity 1 of the rod, we associate $g : [0,T] \to V$ by

$$g(t) = h(t) / \int A \tag{5.13}$$

Thus, the existence of u meeting the requirements listed in 1° above is equivalent to

(5.14) $\quad \forall\, t \in [0,T] \; : \qquad q(t) \in U + A\, g(t)$.

On the other hand, the quasi-equilibrium condition (5.5) is equivalent to

(5.15) $\quad \forall\, t \in [0,T] \; : \qquad s(t) \in V + c(t)$

where $c(t) \in \mathcal{C}$ denotes, for each t, a primitive of the function $x \mapsto -f(t,x)$. As an arbitrary constant may be added to this primitive, we suppose it chosen in such a way that

(5.16) $\quad \forall\, t \in [0,T] \; : \qquad A\, c(t) \in U$.

Concerning the initial data, let us denote by $q_0 \in \mathcal{M}$ the differential measure of the given function $x \mapsto u_0(x)$ (function with bounded variation on [0,1]). Thus, in view of (5.4), (5.6), (5.7), condition (5.8) shall be understood with

(5.17) $\qquad p(0) = q_0 - A\, s_0$

and (5.10) with $\qquad q(0) = q_0$.

5.c HYPOTHESES CONCERNING THE DATA

Assumption 1 : The nonnegative measure A does not vanish, i.e. $\int A > 0$.

Assumption 2 : The multifunction $x \mapsto \gamma(x)$ of [0,1] into $\mathbb{R}$ is lower semi continuous (cf Parag. 1.e) ; equivalently the yield limits $x \mapsto \alpha(x)$ and $x \mapsto \beta(x)$ are respectively u.s.c. and l.s.c. numerical functions.

Assumption 3 : For every $t \in [0,T]$, the space of constants V has a nonempty intersection with the interior of the convex subset $C - c(t)$ of $\mathcal{C}$; equivalently

$$\forall\, t \in [0,T] \; : \qquad \sup_x \,(\alpha(x) - c(t,x)\,) < \inf_x \,(\beta(x) - c(t,x)\,) \; .$$

One may call this Assumption the safe load hypothesis because of the following remark : For (5.9) and (5.15) to be satisfied it is obviously necessary that $V + c$ meets C , i.e. $V \cap (C - c) \neq \emptyset$; otherwise the system could not present a quasistatic evolution. The above assumption means that the latter requirement is fullfilled with a certain safety margin. In view of the unavoidable uncertainty in the physical measurement of data, such a margin may be considered as necessary for the problem to be physically well set.

Assumption 4 : The numerical function $t \mapsto h(t)$ is Lipschitz and the function $t \mapsto c(t)$ is Lipschitz in the sense of the norm of $\mathcal{C}$.

This last Assumption is less restrictive than it looks. Intuitively, the evolution of our system, being regulated by some resistance phenomenon of the dry friction type (i.e. the resistance force depends on the oriented direction of the velocity but not on its magnitude), associates the successive configurations of the system to the successive values of the data in a way which does not depend on the timing. This means that the mathematical conditions of the problem are invariant under any absolutely continuous non decreasing change of variable $t \mapsto t'$. Starting from some data $t \mapsto h$ and $t \mapsto c$ which would be only absolutely continuous, one could use such a change of variables to reduce to the Lipschitz case.

5.d AUXILIARY PROBLEM

Let us introduce two new unknowns functions $y : [0,T] \to \mathcal{C}$ and $z : [0,T] \to \mathcal{M}$ by

(5.18) $$y(t) = s(t) - c(t) - g(t)$$

(5.19) $$z(t) = q(t) - A\,c(t) - A\,g(t) \quad .$$

Then, in view of (5.6) and (5.7)

(5.20) $$p(t) = z(t) - A\,y(t)$$

and the requirements of Parag. 5.a take the following equivalent form :

1° For every $t \in [0,T]$:

(5.21) $$y(t) \in V \quad , \qquad z(t) \in U$$

(in other words y is a numerical function) .

2° For $t = 0$:

(5.22) $y(0) = s_0 - c(0) - g(0)$ denoted by $y_0 \in V$

(5.23) $z(0) = q_0 - A\,c(0) - A\,g(0)$ denoted by $z_0 \in U$

3° There exists $\dot{y} \in L^{\infty}(0,T\ ;\ \mathbb{R})$ such that

$$y(t) = y_0 + \int_0^t \dot{y}(\tau)\, d\tau . \tag{5.24}$$

4° There exists $\dot{z} : [0,T] \to U$, weakly integrable with values in a ball of $\mathcal{M}$ such that

$$z(t) = z_0 + \int_0^t \dot{z}(\tau)\, d\tau . \tag{5.25}$$

5° For almost every t , in view of (5.9) and (5.20),

$$\dot{z} - A\,\dot{y} \in \partial\psi_C\,(y + c + g) \tag{5.26}$$

or equivalently

$$\dot{z} - A\,\dot{y} \in \partial\psi_{C - c(t) - g(t)}\,(y) . \tag{5.27}$$

6. DETERMINATION OF s OR y .

6.a. THE PROBLEM FOR y

As usual in the study of elastoplastic systems, determining the stress is the easiest part and it will be found that this preliminary problem possesses a unique solution.

In view of (5.18) it is equivalent to determine the function $t \mapsto y$; this will be done by drawing some consequences of the various conditions formulated in Parag. 5.d.

From $y \in V$ it results (see Parag. 1.b)

$$\partial\psi_V(y) = U .$$

As (5.21) implies $\dot{z} \in U$, (5.27) implies

(6.1) $$- A \dot{y} \in \partial\psi_{C-c-g}(y) + \partial\psi_V(y) \quad .$$

Considering the general inclusion (1.5), this in turn implies that for almost every t ,

(6.2) $$- A \dot{y} \in \partial\psi_{(C-c-g)\cap V}(y).$$

For every $t \in [0,T]$, the intersection

$$(C - c(t) - g(t)) \cap V = I(t)$$

is a convex subset, i.e. an interval, of the one-dimensional space V . Let us identify the elements of V (the constant numerical functions on $[0,1]$) with the real numbers. An element of I(t) is then a real number η such that

$$\forall x \in [0,1] : \alpha(x) - c(t,x) - g(t) \leq \eta \leq \beta(x) - c(t,x) - g(t).$$

In other words

$$I(t) = [\ \sup_x (\alpha(x) - c(t,x) - g(t)),\ \inf_x (\beta(x) - c(t,x) - g(t))\ ]$$

a non empty interval, by virtue of Assumption 3.

The subdifferential in (6.2) must be read in the sense of the duality $\mathscr{C}$, $\mathscr{M}$. But, as (6.2) requires that the function $t \mapsto y$ takes its values in V (otherwise the right member would be empty), we shall be able now to convert (6.2) into an equivalent form, without reference to the imbedding of V in $\mathscr{C}$. In fact, when η, y, $\dot{y}$ are interpreted as real functions on $[0,1]$, (6.2) means

$$\begin{cases} y \in I(t) \\ \forall \eta \in I(t) : \ll \eta - y, - A\dot{y} \gg \leq 0 \end{cases}$$

and the latter inequality is merely

$$- (\eta - y)\, \dot{y} \int A \leq 0 \quad .$$

As the strictly positive factor $\int A$ may be omitted this comes to be equivalent to

(6.3) $$- \dot{y} \in \partial\psi_{I(t)}(y)$$

in the sense of the conventional self-duality of $\mathbb{R}$

6.b. THE SWEEPING PROCESS.

Condition (6.3) is a very special case of the following : Let H be a real Hilbert space and let $\Gamma : [0,T] \to H$ be a given moving closed convex subset of H, i.e. a multifunction of $[0,T]$ into H with nonempty closed convex values. One looks for a moving point, i.e. an absolutely continuous mapping $u : [0,T] \to H$, agreeing with some initial condition $u(0) = u_0 \in \Gamma(0)$ and such that for almost every t in $[0,T]$

$$(6.4) \qquad - \dot{u} \in \partial\psi_{\Gamma(t)}(u)$$

(because H is a reflexive Banach space, the absolutely continuous function u is known to possess a strong derivative $\dot{u}$ for almost every t).

Recall (cf. Parag. 1.b) that $\partial\psi_{\Gamma(t)}(u)$ is the normal outward cone to the set $\Gamma(t)$ at the point u .

In the case where Γ possesses a nonempty interior the meaning of (6.4) may be illustrated as follows : as long as the point u happens to lie in this interior, it remains at rest, for the normal cone reduces at this time to the origin of H . When u is caught up with by the boundary of the moving set, this point can only take a motion in an inward normal direction as if pushed by this boundary, so as to go on belonging to $\Gamma(t)$.

We call this a sweeping process.

Condition (6.4) is a special case of

$$(6.5) \qquad - \dot{u} \in M(t,u)$$

where $x \mapsto M(t,x)$ denotes, for each t, a multifunction of the Hilbert space H into itself which is monotone in the sense of Minty ; monotony trivially implies that, for each initial condition $u(0) = u_0$, (6.5) possesses at most one solution.

The simplest sufficient condition we found for the existence of solutions of (6.4) is the absolute continuity of the multifunction $t \mapsto \Gamma(t)$ in the sense of the Hausdorff distance (see [24]). Such is in particular the case when the multifunction is Lipschitz in the sense of the Hausdorff distance ; then every solution $t \mapsto u(t)$ is also Lipschitz, in the sense of the norm of H , with the same Lipschitz ratio as Γ.

In the present situation $H = \mathbb{R}$ and the Lipschitz property for the multifunction $t \mapsto I(t)$ is an immediate consequence of Assumption 4. We conclude the existence of a unique solution $t \mapsto y(t)$ of (6.3), agreeing with the initial condition (5.22)

This solution is Lipschitz, so that $\dot{y} \in L^\infty(0,T ; \mathbb{R})$.

6.c. THE CATCHING-UP ALGORITHM.

For the numerical solution of (6.4) we proposed (see e.g. [24]) the following algorithm : An increasing sequence is chosen in the interval $[0,T]$:

$$0 = t_0 < t_1 < \dots < t_n = T$$

and a sequence of points of H is constructed by successive projections, i.e.

$$u^0 = u_0$$

$$u^{i+1} = \text{proj}\,(u^i , \Gamma(t_{i+1}))$$

(this denotes the nearest point to u^i in the closed convex set $\Gamma(t_{i+1})$). It is proved that the step function based on this sequence (or also the continuous piecewise linear function of t which interpolates the sequence) converges uniformly to the desired solution when the division of $[0,T]$ is refined in such a way that. $\max_i (t_{i+1} - t_i)$ tends to zero.

In the present one-dimensional case the projection operation is especially simple : u^{i+1} either equals u^i or is one of the two extremities of the interval $I(t_{i+1})$.

A graphical solution would also easily be devised.

7. EXISTENCE PROOF

7.a. THE CONSTRUCTION OF $\dot{p}$ AND p.

By combining the various conditions imposed in section 5 to the unknowns, we obtained condition (6.3), which involves the single unknown function $t \mapsto y$. It was found that, with attention to the initial data, (6.3) possessed a unique solution. We are now to establish that such a procedure properly constitutes an <u>elimination</u>, i.e. (6.3) is not only necessary but also sufficient for the existence of functions $t \mapsto p$, $t \mapsto q$, $t \mapsto s$, $(t,x) \mapsto u$ fullfilling with the considered $t \mapsto y$ all the requirements formulated in Sect. 5.

Henceforth it is supposed that $t \mapsto y$ satisfies (6.3), or equivalently (6.2), for almost every t; then it satisfies also (6.1) by virtue of the <u>addition rule for subdifferentials</u> (cf. Parag. 1.c) and of Assumption 3 which implies that V intersects the interior of $C - c(t) - g(t)$. Now (6.1) means that, for the considered values of t, there exists a non empty set of elements $\dot{z}$ of $\partial\psi_V(y) = U$ such that (5.26) holds. Equivalently there exists a nonempty set $\Gamma(t)$ of elements $\dot{p} = \dot{z} - A\dot{y}$ of $U - A\,\dot{y}(t)$ such that

$$\dot{p} \in \partial\psi_C\,(y + c + g). \tag{7.1}$$

We are to prove that, from the multifunction $t \mapsto \Gamma(t)$, a single-valued function $t \mapsto \dot{p}(t)$ can be <u>selected</u>, which is weakly integrable, so as to permit the construction of $t \mapsto p(t)$ according to (5.8). Then, if we construct $t \mapsto s(t)$ by (5.18), condition (5.9) will be satisfied for almost every t:

Owing to the initial conditions, the fact that $\dot{z}$ takes its values in U implies $z(t) \in U$ for every t if z is the weak integral of $\dot{z}$.

Since by construction $y \in C - c - g$, one has $\psi_C(y + c + g) = 0$; the equivalence of the various relations (1.3) makes that (7.1) may as well be written under the form

$$\psi^*_C(\dot{p}) - \ll y + c + g\,,\,\dot{p} \gg\; \leq\; 0\;.$$

Therefore $\Gamma(t)$ is the intersection of $U - A\,\dot{y}(t)$ with the set

$$\Phi(t) = \{w \in \mathcal{M} : \psi^*_C(w) - \ll y(t) + c(t) + g(t),\, w \gg \leq 0\,\} \tag{7.2}$$

7.b. BOUNDEDNESS OF $\Gamma(t)$

Assumptions 2 and 3 imply the existence of some $\rho > 0$ and of some continuous mapping $k : [0,T] \to \mathscr{C}$ such that, for every $t \in [0,T]$, one has $k(t) \in V + c(t)$ and the closed ball in $\mathscr{C}$ with radius ρ, with center $k(t)$, is contained in C. In fact the interval

$$\left[\sup_x (\alpha(x) - c(t,x)) , \inf_x (\beta(x) - c(t,x)) \right]$$

has a non zero length, depending continuously on $t \in [0,T]$. Let us choose $\rho > 0$ such that 2ρ minorizes this length for every t. The interval

$$\left[\sup_x (\alpha(x) - c(t,x) + \rho) , \inf_x (\beta(x) - c(t,x) - \rho) \right]$$

is nonempty and depends on t in a Lipschitz way (see Sect. 6). Let $t \mapsto u(t) \in \mathbb{R}$ be a solution of the sweeping process for this moving interval ; $u(t)$ may be interpreted as an element of V and $t \mapsto k(t) = u(t) + c(t)$ meets the above requirements.

The inclusion of the ball in C is equivalent to the following inequality between the respective support functions of these closed convex sets

$$(7.3) \qquad \forall w \in \mathcal{M} : \quad \rho \|w\|_{\mathcal{M}} + \langle\langle k(t), w \rangle\rangle \leq \psi_C^*(w) .$$

For any w in $\mathcal{M}$ let us write

$$\langle\langle y + c + g, w \rangle\rangle = \langle\langle k, w \rangle\rangle + \langle\langle y + c + g - k, w + A\dot{y} \rangle\rangle - \langle\langle y + c + g - k, A\dot{y} \rangle\rangle.$$

If $w \in \Gamma(t)$, one has $w + A\dot{y} \in U$, thus the second bracket in the right member vanishes, since $y \in V$, $k - c \in V$, $g \in V$. On the other hand, in view of (5.16),

$$\langle\langle c, A\dot{y} \rangle\rangle = \langle\langle \dot{y}, Ac \rangle\rangle = 0 .$$

Then, by comparing (7.2) with (7.3) one obtains that, for every $w \in \Gamma(t)$,

$$\rho \|w\|_{\mathcal{M}} + \langle\langle y + g - k , A\dot{y} \rangle\rangle \leq 0$$

thus

$$(7.4) \qquad \|w\|_{\mathcal{M}} \leq \frac{1}{\rho} M |\dot{y}|_{\mathbb{R}} \int A \quad ,$$

where M denotes a majorant of the continuous function $t \mapsto \|y(t) + g(t) - k(t)\|_{\mathscr{C}}$

on $[0,T]$.

Recall that $y : [0,T] \to \mathbb{R}$ is Lipschitz i.e. $\dot{y}$ is bounded. Then (7.4) implies the existence of a closed ball B in $\mathcal{M}$ containing the set $\Gamma(t)$ for almost every t.

7.c. MEASURABILITY OF THE MULTIFUNCTION Γ.

As $\mathcal{C}$ is separable, the topology induced by $\sigma(\mathcal{M}, \mathcal{C})$ (the weak-star topology) makes of the preceding ball a compact metrizable topological space B_s. For almost every t the set $\Gamma(t)$ is nonempty and

$$\Gamma(t) = B_s \cap \Phi(t) \cap (U - A\,\dot{y}(t))$$

Let us make use now of the theory of measurable selectors, initiated in [2] and [10]. We are here in the simple case of a multifunction whose values are closed subsets of the metrizable separable complete space B_s (a convenient account of this case may be found in [34]). A necessary and sufficient condition for the multifunction to possess a dense collection of measurable selectors is that its graph, i.e.

$$G = \{ (t,m) \in [0,T] \times B_s \ : m \in \Gamma(t) \}$$

belong to the σ-algebra $\mathcal{L} \otimes \mathcal{B}$ generated by the products of Lebesgue-measurable subsets of $[0,T]$ by Borel subsets of B_s.

Let us establish this property separately for the two multifunctions $t \mapsto B_s \cap \Phi(t)$ and $t \mapsto B_s \cap (U - A\,\dot{y}(t))$, as G is the intersection of their respective graphs.

Concerning the first one, we observe that in the definition (7.2) of $\Phi(t)$, the function $w \mapsto \psi_C^*(w)$ is l.s.c. for $\sigma(\mathcal{M}, \mathcal{C})$; besides, $t \mapsto y + c + g$ is continuous on $[0,T]$ for the norm topology of $\mathcal{C}$; therefore $(t,w) \mapsto \ll y + c + g, w \gg$ is continuous from $[0,T] \times B_s$ into $\mathbb{R}$. Consequently the graph of $t \mapsto B_s \cap \Phi(t)$ is a closed subset of $[0,T] \times B_s$, thus a member of $\mathcal{L} \otimes \mathcal{B}$.

On the other hand, by the definition of U, a couple (t,m) belongs to the graph of $t \mapsto U - A\,\dot{y}(t)$ if and only if

$$\int m + \dot{y}(t) \int A = 0 \ .$$

As $m \mapsto \int m$ is continuous from B_s into $\mathbb{R}$ and as $t \mapsto \dot{y}(t)$ is Lebesgue-measurable from $[0,T]$ into $\mathbb{R}$, one easily sees that this graph belongs to $\mathcal{L} \otimes \mathcal{B}$.

We observe at this stage that our problem will possess in general an infinity of solutions ; let us take as $t \mapsto \dot{p}(t)$ any one of the $\sigma(\mathcal{M}, \mathcal{C})$ - measurable selectors of Γ. In view of $\Gamma(t)$ being contained in B_s for every t, this $\dot{p}$ is an element of the space $L^\infty_{\mathcal{M}_s}$ (cf. [1], Chap.6, § 2 n° 6), the dual of $L^1_{\mathcal{C}}$. Then p is constructed in accordance with (5.8)

$$(7.5) \qquad p(t) = p(0) + \int_0^t \dot{p}(\tau)\, d\tau$$

(the weak integral is in fact an element of $\mathcal{M}$, see e.g. [1], Chap. 6, § 1 n° 4). The function $t \mapsto p(t) \in \mathcal{M}$ is Lipschitz ; it admits for almost every t a $\sigma(\mathcal{M}, \mathcal{C})$ - derivative equal to $\dot{p}(t)$.

7.d. CONSTRUCTION OF e AND $\dot{e}$

Recall that we constructed s according to (5.18), i.e.

$$s(t) = y(t) + c(t) + g(t) .$$

This defines a mapping $s : [0,T] \to \mathcal{C}$ which is Lipschitz relatively to the norm of $\mathcal{C}$, by virtue of Assumption 4.

In order to comply with (5.7) we take now $e(t) = A\, s(t)$. This defines a mapping $e : [0,T] \to \mathcal{M}$ which is Lipschitz relatively to the norm of $\mathcal{M}$. One concludes the existence of $\dot{e} \in L^\infty_{\mathcal{M}_s}$ such that

$$(7.6) \qquad e(t) = e(0) + \int_0^t \dot{e}(\tau)\, d\tau .$$

This existence may be established as follows : The product of some time-independent element a of $\mathcal{C}$ by the characteristic function $\chi_{[t_1,t_2]}$ of some subinterval of $[0,T]$ yields an element u of $L^1_{\mathcal{C}}$. Denoting as before by $\langle\langle \cdot , \cdot \rangle\rangle$ the pairing between $\mathcal{C}$ and $\mathcal{M}$, put

$$(7.7) \qquad \ell(u) = \langle\langle a, e(t_2) - e(t_1) \rangle\rangle$$

If λ is the Lipschitz constant of e, one has

$$|\ell(u)|_{\mathbb{R}} \leq \lambda \, \|a\|_{\mathcal{C}} \, (t_2 - t_1) \leq \lambda \, \|u\|_{L^1_{\mathcal{C}}} .$$

This inequality holds more generally when $u \mapsto \ell(u)$ is extended by linearity to the space of the <u>step-functions</u> from $[0,T]$ into $\mathcal{C}$. As this space is dense in $L^1_{\mathcal{C}}$, one can extend ℓ as an element of the dual of $L^1_{\mathcal{C}}$: let us define $\dot{e}$ as this element of $L^\infty_{\mathcal{M}_s}$. Then (7.6) immediately follows from (7.7), since the weak integral may be characterized by

$$\ll a, \int_0^t \dot{e}(\tau)\, d\tau \gg \; = \; \lll a\, \chi_{[0,t]} , \dot{e} \ggg$$

for every $a \in \mathcal{C}$, where $\lll . , . \ggg$ denotes the pairing between $L^1_{\mathcal{C}}$ and $L^s_{\mathcal{M}_s}$.

7.e. CONSTRUCTION OF u^- and u^+.

Let us take q in accordance with (5.6) ; by addition it comes that

$$\dot{q} = \dot{e} + \dot{p}$$

is an element of $L^\infty_{\mathcal{M}_s}$ and that

$$q(t) = q(0) + \int_0^t \dot{q}(\tau)\, d\tau .$$

For every $t \in [0,T]$ we construct now u according to (4.3), (4.4) and (5.1) :

$$u^-(t,x) = \int_{[0,x[} q(t)$$

$$u^+(t,x) = \int_{[0,x]} q(t)$$

(the very value of u at some discontinuity point is immaterial : one may choose any value between u^- and u^+).

As (5.21) holds, thus also (5.14), one concludes that the boundary condition (5.2) is satisfied.

To verify that the final requirements of Parag. 5.a , are fullfilled is simply a matter of commuting integrations. Some results about weak vector integration (see [1], Chap. 6, § 1) might be used. More elementarily a sequence (φ_n) of elements of

$\mathcal{C}$ may be chosen, with $|\varphi_n| \leq 1$, converging pointwise to the characteristic function of the interval $[0,x[$ (resp. the interval $[0,x]$) ; then one applies the dominated convergence theorem repeatedly.

REFERENCES

1. BOURBAKI, N. Elements de Mathématique, Livre VI, Intégration, Paris, Hermann.

2. CASTAING, C. Sur les multiapplications mesurables. Rev. Franç. Inform. Rech. Operat. 1 (1967), 3-34.

3. ——— Version aléatoire du problème de rafle par un convexe variable, C.R. Acad. Sci. Paris, Sér. A, 277 (1973), 1057-1059.

4. DINCA, G. Sur la monotonie d'après Minty-Browder de l'opérateur de la théorie de la plasticité, C.R. Acad. Sci. Paris, Sér. A, 269 (1969), 535-538.

5. ——— Opérateurs monotones dans la théorie de plasticité, Ann. Scuola Norm. Sup. Pisa, Sci. Fis. Mat. , 24 (1970), 357-399.

6. DUVAUT, G. ; LIONS, J.L. Les inéquations en mécanique et en physique, Dunod, 1972.

7. EKELAND, I. ; TEMAM, R. Analyse convexe et problèmes variationnels, Dunod - Gauthier-Villars, 1974.

8. IOFFE, A.D. ; LEVIN, V.L. Subdifferentials of convex functions, Trudy Mosk. Mat. Ob. , 1972.

9. IOFFE, A.D. ; TIKHOMIROV, V.M. On the minimization of integral functionals, Funkt. Analiz. 3 (1969) 61-70 ; see also:Funct. Anal. Appl. 3 (1969), 218-229.

10. KURATOWSKI, K. ; RYLL-NARDZEWSKI, C. A general theorem on selectors, Bull. Polish Acad. Sci. 13 (1965), 397-411.

11. LANCHON, H. Torsion elastoplastique d'un arbre cylindrique de section simplement ou multiplement connexe, Thèse, Université de Paris VI, 1972.

12. LAURENT, P.J. Approximation et optimisation, Hermann, 1972.

13. LÉNÉ, F. Sur les matériaux élastiques à énergie de déformation non quadratique, J. de Mécanique, 13 (1974), 499-534.

14. LEVIN, V.L. On the subdifferential of convex functionals, Uspekhi Mat. Nauk 25 (1970), 183-184.

15. MICHAEL, E. Continuous selections I, Ann. of Math. 63 (1956), 361-382.

16. MOREAU, J.J. Fonctionnelles sous-différentiables, C.R. Acad. Sci. Paris, Sér.A, 257 (1963), 4117-4119.

17. ———— Principes extrémaux pour le problème de la naissance de la cavitation, J. de Mécanique, 5 (1966), 439-470.

18. ———— One-sided constraints in hydrodynamics, in : (J. Abadie, éditor) Nonlinear programming, North Holland Pub. Co , 1967, p.261-279.

19. ———— Fonctionnelles convexes, Séminaire sur les Equations aux Dérivées Partielles, Collège de France, Paris, 1966-67 (multigraph, 108 p.).

20. ———— La notion de sur-potentiel et les liaisons unilatérales en élastostatique, C.R. Acad. Sci. Paris, Sér. A, 267 (1968), 954-957.

21. ———— Sur les lois de frottement, de plasticité et de viscosité, C.R. Acad. Sci. Paris, Sér. A, 271 (1970), 608-611.

22. ———— Sur l'évolution d'un système élasto-visco-plastique, C.R. Acad. Sci. Paris, Sér. A, 273 (1971), 118-121.

23. ———— Problème d'évolution associé à un convexe mobile d'un espace hilbertien, C.R. Acad. Sci. Paris, Sér.A, 276 (1973), 791-794.

24. ———— On unilateral constraints, friction and plasticity, in : (G. Capriz, G. Stampacchia, editors), New Variational Techniques in Mathematical Physics, Centro Internazionale Matematico Estivo, II Ciclo 1973, Edizioni Cremonese, Roma, 1974, p.175-322.

25. ———— La Convexité en Statique, in : (J.P. Aubin, editor), Analyse Convexe et ses Applications, Lecture Notes in Economics and Mathematical Systems, n°102 (1974), Springer Verlag, p.141-167.

26. NAYROLES, B. Essai de théorie fonctionnelle des structures rigides plastiques parfaites, J. de Mécanique, 9 (1970), 491-506.

27. ———— Quelques applications variationnelles de la théorie des fonctions duales à la Mécanique des solides, J. de Mécanique, 10 (1971), 263-289

28. NGUYEN, Q.S. Matériaux elasto-visco-plastiques et élastoplastiques à potentiel généralisé, C.R. Acad. Sci. Paris, Sér. A, 277 (1973), 915-918.

29. NGUYEN, Q.S. ; HALPHEN, B. Sur les lois de comportement elasto-visco-plastique à potentiel généralisé, C.R. Acad. Sci. Paris, Sér. A, 277 (1973), 319-322.

30. ROCKAFELLAR, R.T. Integrals which are convex functionals, Pacific J. Math. 24 (1968), 525-539.

31. ———— Measurable dependence of convex sets and functions on parameters, J. Math. Anal. Appl. 28 (1969), 4-25.

32. ———— Convex Analysis, Princeton U.P. , 1970.

33. ———— Integrals which are convex functionals, II, Pacific J. Math. 39 (1971), 439-469.

34. ———— Convex integral functionals, in : (E.H. Zarantonello, editor) Contributions to Nonlinear Functional Analysis, Academic Press (1971), 215-236.

THEORY OF MIXED AND HYBRID FINITE-ELEMENT APPROXIMATIONS IN LINEAR ELASTICITY

J. T. ODEN and J. K. LEE

Texas Institute for Computational Mechanics
The University of Texas, Austin

1. Introduction

The complementary and dual variational principles of solid mechanics have always occupied an important place in elasticity theory, not only because they represent something intermediate to the classical extremal principles of potential and complementary energy, but also because they form the basis for a variety of effective methods of approximation. The well-known method of the hypercircle of Prager and Synge [1] is, in fact, rooted in these principles, as are the popular mixed finite-element methods, introduced by Hermann [2] and based on Reissner's principle [3]. In recent years, there has been added to this collection of ideas a new family of special variational methods referred to as hybrid methods. These have been largely developed by Pian and his associates (e.g. [4]) in connection with finite element approximations in solid and continuum mechanics. A summary of these variational concepts, together with some discussion of their mathematical properties, can be found in the book of Oden and Reddy [5].

While the mixed and hybrid finite element methods have proved to be very effective in certain specific numerical experiments, they also frequently meet with disappointing and often unexplained failure. For example, it is not uncommon to construct apparently reasonable approximations of stress and displacement fields in a finite element, only to find that they lead to singular or ill-conditioned stiffness matrices. The mathematical properties of mixed models which are responsible for their delicate behavior were examined in the papers of Oden and Reddy [6,7], Ciarlet and Raviart [8], and Johnson [9] for model second- and fourth-order problems. The theory of hybrid methods is considerably more involved. Certain properties of a class of hybrid methods are discussed in the reports of Raviart [10] and Thomas [11], and a fairly detailed theory of mixed-hybrid methods for the solution of a model second-order problem has been contributed recently by Babuska, Oden, and Lee [12].

In the present paper, we present a theory of mixed and of hybrid finite element approximations of a class of problems in linear plane elasticity. Our treatment of mixed methods generalizes all of those published previously and our method of proof differs considerably from others. Our principal tools are two fundamental theorems due to Babuska [13] (see also [14], [15]), one which generalizes the well-known Lax-Milgram theorem and another which establishes bounds for the approximate problem. After introducing some aspects of the general theory of mixed variational principles in Section 2 of the paper and describing a class of linear elastostatic problems with a collection of pertinent notations in Section 3, we prove, in the fourth section, an existence theorem for the corresponding mixed variational problem. A theory of mixed finite elements is given in Section 5. There we present sufficient conditions for the existence of unique solutions to the approximate problem, convergence criteria, and a-priori error estimates. We are also able to develop, apparently for the first time, L_2-error estimates for mixed finite element approximations. In Section 6 we construct a special "hybrid" variational principle for plane elasticity problems as well as a corresponding existence theorem. We demonstrate in Section 7 that the hybrid principle leads naturally to a theory for "displacement" hybrid finite element approximations in which triangular elements are used, the displacement field is approximated on the interior of the elements, and boundary tractions are approximated on the element boundaries. Here we also give conditions for the existence of approximate solutions as well as a-priori error estimates.

2. General Theory of Mixed Variational Principles

A fairly complete account of the general theory of mixed (dual and complementary) variational principles for linear problems in mathematical physics is given in the book of Oden and Reddy [5]. There it is shown that the functional

$$\begin{aligned} J(u,\varepsilon,\sigma) = \frac{1}{2}\langle E\varepsilon,\varepsilon\rangle_G + \langle\sigma,Tu-\varepsilon\rangle_G - \langle f,u\rangle_H \\ + \langle\delta_0\sigma, bu - g_1\rangle_{\partial G_1} - \langle g_2,\gamma_0 u\rangle_{\partial H_2} \end{aligned} \tag{2.1}$$

assumes a stationary value whenever the triple (u,ε,σ) satisfies the equations

$$\left.\begin{aligned} Tu &= \varepsilon \\ E\varepsilon &= \sigma \\ T^*\sigma &= f \end{aligned}\right\} \text{ in } \Omega \qquad \left.\begin{aligned} bu &= g_1 \text{ on } \partial\Omega_1 \\ \\ b^* &= g_2 \text{ on } \partial\Omega_2 \end{aligned}\right\} \tag{2.2}$$

where H, G, ∂G_1, and ∂H_2 are function spaces defined on the domain $\Omega \subset \mathbb{R}^n$ and boundary segments $\partial\Omega_1$ and $\partial\Omega_2$, H is densely imbedded in G, $\langle\cdot,\cdot\rangle_{(\cdot)}$ denotes the duality pairing on the underlying space $(\cdot)$, γ_0: $H \to \partial H_2$ and δ_0: $G \to \partial G_1$ are trace operators (continuous extensions), T is a continuous linear map of H into G, T* is its formal adjoint of T, b and b* are boundary operators, f, g_1, and g_2 denote data, and E is a canonical isomorphism of G onto its dual G'.

The relationship between the abstract problem (2.2) and the linear elasticity problem shall be made clear in Sections 3 and 4. However, the structure indicated in (2.1) and (2.2) is shared by a multitude of problems in mathematical physics (see [5]). In the case of linear elasticity (when T, T*, H, etc. are defined in (3.14)), $J(u,\varepsilon,\sigma)$ is the Hu-Washizu principle. Since there are then a set of three Euler equations (2.2), a total of seven fundamental variational principles can be derived from (2.1) by imposing various restrictions on $J(u,\varepsilon,\sigma)$.* Of particular interest in the present study is the generalized Riessner functional ($R(u,\sigma)$ obtained from (2.1) by setting $\varepsilon = E^{-1}\sigma$:

$$R(u,\sigma) = \langle\sigma, Tu - \frac{1}{2}E^{-1}\sigma\rangle_G - \langle f,u\rangle_H + \langle\delta_0\sigma, bu - g_1\rangle_{\partial G_1} - \langle g_2,\gamma_0 u\rangle_{\partial H_2} \qquad (2.3)$$

We use (2.3) to construct mixed finite element methods in Section 5.

When T is a differential operator and when Ω is partitioned into E-subdomains Ω_e, we can relax conditions on the admissible spaces G and H by treating them as constraints. We refer to variational principles of this type as <u>hybrid</u>. For example, suppose $\varepsilon = T^{-1}$, $Tu = \varepsilon$, and $\gamma_0 u = 0$ on $\partial\Omega(\partial\Omega_2 = \emptyset)$. Then the corresponding hybrid variational functional would be†

$$I(u,\psi) = \sum_{e=1}^{E}\{\frac{1}{2}\langle ETu,Tu\rangle_{G_e} - \langle f,u\rangle_{H_e} + \langle\psi,\gamma_0 u\rangle_{\partial H_e}\} \qquad (2.4)$$

where ψ is a Lagrange multiplier.

Let $J(\underset{\sim}{\omega})$ be a functional corresponding to either a mixed or a hybrid variational principle, $\underset{\sim}{\omega}$ being an ordered triple or pair of entries (e.g. $\underset{\sim}{\omega} = (u,\varepsilon,\sigma)$ in (2.1) and $\underset{\sim}{\omega} = (u,\psi)$ in (2.4)). Clearly, $\underset{\sim}{\omega}$ is a member of a product space W which is the product of two or more Hilbert spaces (e.g. $W = H \times G \times G'$ in (2.1)) and J is a quadratic functional on W. Then, the functional J assumes a stationary value at the point $\underset{\sim}{\omega}_0 \in W$ whenever

$$B(\underset{\sim}{\omega}_0,\overline{\underset{\sim}{\omega}}) - F(\overline{\underset{\sim}{\omega}}) = 0 \qquad \forall\, \overline{\underset{\sim}{\omega}} \in W \qquad (2.5)$$

where

$$\delta J(\underset{\sim}{\omega},\overline{\underset{\sim}{\omega}}) \equiv B(\underset{\sim}{\omega},\overline{\underset{\sim}{\omega}}) - F(\overline{\underset{\sim}{\omega}}) \qquad (2.6)$$

is the first linear Gateaux differential of J at $\underset{\sim}{\omega}$ in direction $\overline{\underset{\sim}{\omega}}$. In (2.5), $B(\cdot,\cdot)$ is a bilinear form mapping $W \times W$ into $\mathbb{R}$, and $F(\cdot)$ is a linear functional on W. The problem of finding an $\underset{\sim}{\omega}_0 \in W$ such that (2.5) holds for a given F is a <u>variational boundary-value problem</u> associated with the form $B(\cdot,\cdot)$.

The determination of sufficient conditions for the existence of a unique solution to (2.5) plays a critical role in this study. As a basis for establishing such conditions, we call upon a fundamental theorem proved by Babuska [13].

*It is interesting to note that for a given problem of the form (2.2) there are fourteen "fundamental" variational principles, seven associated with (2.2) and seven more corresponding to the dual problem. In elasticity theory, these "dual" principles involve stress functions and the compatibility equations for strains. See [5].

†Many other examples could be cited. See, for example [4] and [12].

Theorem 2.1. Let V_1 and V_2 be real Hilbert spaces and let $B: V_1 \times V_2 \to \mathbb{R}$ denote a bilinear form on $V_1 \times V_2$ which has the following properties:

(i) Continuity. $\exists$ $M > 0$ such that

$$B(u,v) \leq M||u||_{V_1} ||v||_{V_2}, \quad \forall u \in V_1, \quad v \in V_2 \tag{2.7}$$

(ii) Weak Coerciveness. $\exists$ $a_0 > 0$ such that

$$\inf_{||u||_{V_1}=1} \sup_{||v||_{V_2}\leq 1} |B(u,v)| \geq a_0 > 0 \tag{2.8}$$

and

$$\sup_{u \in V_1} B(u,v) > 0, \quad v \neq 0 \tag{2.9}$$

Then there exists a unique $u_0 \in V_1$ such that

$$B(u_0,v) = F(v) \quad \forall v \in V_2, \quad f \in V_2' \tag{2.10}$$

Moreover

$$||u_0||_{V_1} \leq \frac{1}{a_0} ||F||_{V_2'} \tag{2.11}$$

■

Theorem 2.1 represents an important generalization of the Lax-Milgram theorem.

In approximating a given variational problem by constructing proper subspaces we also apply the following approximation theorem, also due to Babuska [13] (see also [14], [15]).

Theorem 2.2. Let V_1^h and V_2^h be finite-dimensional subspaces of real Hilbert spaces V_1 and V_2, respectively, and let the bilinear form $B: V_1 \times V_2 \to \mathbb{R}$ of Theorem 2.1 be such that for $U \in V_k^h$, $V \in V_2^h$, the following hold

$$\inf_{||U||_{V_1}=1} \sup_{||V||_{V_2}\leq 1} |B(U,V)| \geq A_0 > 0, \tag{2.12}$$

$$\sup_{U \in V_1^h} |B(U,V)| > 0, \quad V \neq 0, \quad V \in V_2^h \tag{2.13}$$

In addition, let $F \in V_2'$ be given. Then

(i) there exists a unique $U_0 \in V_1^h$ such that

$$B(U_0,V) = F(V) \quad \forall V \in V_2^h \tag{2.14}$$

with

$$||U_0||_{V_1} \leq \frac{1}{A_0} ||F||_{V_2'} \tag{2.15}$$

(ii) if u_0 is the unique solution of (2.10),

$$||u_0 - U_0||_{V_1} \leq (1 + \frac{M}{A_0}) \inf_{\tilde{U} \in V_1^h} ||u_0 - \tilde{U}||_{V_1} \tag{2.16}$$

■

3. A Weak Boundary-Value Problem in Elastostatics

We shall examine the following two-dimensional* variational boundary-value problem in linear elastostatics: find the displacement vector $\underset{\sim}{u} \in \underset{\sim}{U}(\Omega)$ such that

$$\int_\Omega \mathrm{tr}[\underset{\sim}{E}(\underset{\sim\sim}{\nabla u}) \cdot \underset{\sim\sim}{\nabla v}]dx = \int_\Omega \rho \underset{\sim}{f} \cdot \underset{\sim}{v}\, dx + \int_{\partial\Omega_2} \underset{\sim}{S} \cdot \gamma_0 \underset{\sim}{v}\, ds \qquad \forall\, \underset{\sim}{v} \in \underset{\sim}{U}(\Omega) \tag{3.1}$$

where $\underset{\sim}{U}(\Omega)$ is the space of admissible displacements,

$$\underset{\sim}{U}(\Omega) = \{\underset{\sim}{v}: v \in (H^1(\Omega))^2, \quad \gamma_0 \underset{\sim}{v} = \underset{\sim}{0} \text{ on } \partial\Omega_1\} \tag{3.2}$$

Here and in the sequel we employ the following notations and conventions:

(i) Ω is an open bounded compact domain in $\mathbb{R}^2$ of particles $\underset{\sim}{x} = (x_1,x_2)$ with a Lipschitzian boundary $\partial\Omega$, a differential element in Ω being denoted $dx\ (= dx_1dx_2)$ and an element of $\partial\Omega$ by ds. The boundary $\partial\Omega$ consists of two portions, $\partial\Omega = \overline{\partial\Omega_1} \cup \overline{\partial\Omega_2}$, $\partial\Omega_1 \cap \partial\Omega_2 = \emptyset$, $\mathrm{mes}(\partial\Omega_\alpha) \neq 0$, the displacements being prescribed on $\partial\Omega_1$ and the tractions being prescribed on $\partial\Omega_2$.

(ii) $H^m(\Omega)$, m an integer ≥ 0, is the Sobolev space of order m, and is defined as the closure of the space $C^\infty(\Omega)$ of infinitely smooth functions on Ω in the norm

$$||u||^2_{H^m(\Omega)} = \sum_{|\underset{\sim}{\alpha}| \leq m} \int_\Omega |D^{\underset{\sim}{\alpha}} u|^2\, dx \tag{3.3}$$

where $\underset{\sim}{\alpha} = (\alpha_1,\alpha_2)$, α_i = integer ≥ 0, and $D^{\underset{\sim}{\alpha}} u = \partial^{|\underset{\sim}{\alpha}|} u / \partial x_1^{\alpha_1} \partial x_2^{\alpha_2}$ with $|\underset{\sim}{\alpha}| = \alpha_1 + \alpha_2$. We remark that $H^0(\Omega) = L_2(\Omega)$ and that $H^m_0(\Omega)$ is defined as the completion of the space $C^\infty_0(\Omega)$ of infinitely smooth functions with compact support in Ω.

(iii) If $\underset{\sim}{v}$ is a vector whose components $v_\alpha \in H^m(\Omega)$ and $\underset{\sim}{A}$ is a symmetric second-order tensor whose components $A^{\alpha\beta} \in H^m(\Omega)$, we use the notation

$$\underset{\sim}{H}^m(\Omega) = (H^m(\Omega))^2 \ ; \qquad ||\underset{\sim}{v}||_{\underset{\sim}{H}^m(\Omega)} = \sum_{\alpha=1}^{2} ||v_\alpha||_{H^m(\Omega)}$$
$$\underset{\sim}{H}^m(\Omega) = (H^m(\Omega))^3 \ ; \qquad |||\underset{\sim}{A}|||_{\underset{\sim}{H}^m(\Omega)} = \sum_{\substack{\alpha,\beta=1 \\ \alpha \leq \beta}}^{2} ||A^{\alpha\beta}||_{H^m(\Omega)} \tag{3.4}$$

(iv) ∇ is the material gradient operator $[(\underset{\sim\sim}{\nabla u})_{\alpha\beta} = (\partial u_\alpha/\partial x_\beta);\ \alpha,\beta = 1,2]$.

(v) $\underset{\sim}{E}$ is Hooke's tensor, a fourth order tensor, such that:

(v.1) E is a continuous, invertible, self-adjoint isomorphism from

---

*All of the developments in this section and in Sections 4 and 6 can be trivially extended to three-dimensional problems. However, since our approximation theory developed in Sections 5 and 7 is only valid for functions defined on one- and two-dimensional domains, we limit ourselves to two-dimensional problems at the onset. Thus (3.1) is a weak or variational statement of classical mixed plane stress or plane strain problems in linear elasticity.

$(L_2(\Omega))^3$ onto itself

(v.2) By (v.1) and the Banach Theorem, E^{-1} is continuous, and there exist positive constants μ_0 and μ_1 such that

$$\left.\begin{aligned} \int_\Omega tr[\underset{\sim}{E}(\underset{\sim}{A}) \cdot \underset{\sim}{A}]dx &\geq \mu_0 \int_\Omega tr\underset{\sim}{A}^T\underset{\sim}{A}\, dx = \mu_0 |||\underset{\sim}{A}|||_{\underset{\sim}{L}_2(\Omega)} \\ \int_\Omega tr[\underset{\sim}{E}^{-1}(\underset{\sim}{A}) \cdot \underset{\sim}{A}]dx &\geq \mu_1 \int_\Omega tr\underset{\sim}{A}^T\underset{\sim}{A}\, dx = \mu_1 |||\underset{\sim}{A}|||_{\underset{\sim}{L}_2(\Omega)} \end{aligned}\right\} \quad (3.5)$$

Here tr denotes the trace of a second-order tensor and $\underset{\sim}{A}$ is a symmetric second-order tensor. The parameters μ_0 and μ_1 are defined by constitutive properties of the body.

(v.3) $\underset{\sim}{E}$ has the symmetries,

$$E^{\alpha\beta\lambda\mu} = E^{\beta\alpha\lambda\mu} = E^{\alpha\beta\mu\lambda} = E^{\mu\lambda\alpha\beta} \;;\quad \alpha,\beta,\lambda,\mu = 1,2 \quad (3.6)$$

(vi) ρ is the mass density and $\underset{\sim}{f} \in (L_2(\Omega))^2$ the body force per unit mass (we consider plane deformations of a body $\mathcal{B}$ in the material plane Ω and, thus, $\rho\underset{\sim}{f}$ can be regarded as the force on material volume element $1 \cdot dx$).

(vii) γ_0 is the trace operator extending $\underset{\sim}{u} \in (H^1(\Omega))^2$ continuously to the boundary $\partial\Omega$ and $\underset{\sim}{S} \in (L_2(\partial\Omega))^2$ is a prescribed boundary traction. The trace $\gamma_0 u$ of a function u in $(H^1(\Omega))^2$ belongs to the Hilbert space $(H^{\frac{1}{2}}(\partial\Omega))^2$ where $H^{\frac{1}{2}}(\partial\Omega)$ is furnished with the norm

$$||g||_{H^{\frac{1}{2}}(\partial\Omega)} = \inf_{u \in H^1(\Omega)} \{ ||u||_{H^1(\Omega)} \;;\quad g = \gamma_0 u\} \quad (3.7)$$

(viii) Boundary tractions $\underset{\sim}{n} \cdot \underset{\sim}{E}(\nabla \underset{\sim}{u})$ belong to a special Hilbert space $\underset{\sim}{T}(\partial\Omega)$. To define $\underset{\sim}{T}(\partial\Omega)$, we denote the inner product on $(L_2(\partial\Omega))^2$ by

$$(\underset{\sim}{g},\underset{\sim}{\psi})_{\underset{\sim}{L}_2(\partial\Omega)} = \oint_{\partial\Omega} \underset{\sim}{g} \cdot \underset{\sim}{\psi}\, ds \quad (3.8)$$

Next we denote

$$\underset{\sim}{H}^{\frac{1}{2}}(\partial\Omega) = (H^{\frac{1}{2}}(\partial\Omega))^2 \;;\quad ||\gamma_0\underset{\sim}{u}||^2_{\underset{\sim}{H}^{\frac{1}{2}}(\partial\Omega)} = \sum_{\alpha=1}^{2} ||\gamma_0 u_\alpha||^2_{\underset{\sim}{H}^{\frac{1}{2}}(\partial\Omega)} \quad (3.9)$$

Then $\underset{\sim}{T}(\partial\Omega)$ is defined as the completion of $(L_2(\partial\Omega))^2$ in the norm

$$||\underset{\sim}{\tau}||_{\underset{\sim}{H}^{-\frac{1}{2}}(\partial\Omega)} = \sup_{\underset{\sim}{g} \in \underset{\sim}{H}^{\frac{1}{2}}(\partial\Omega)} \frac{(\underset{\sim}{\tau},\underset{\sim}{g})_{\underset{\sim}{L}_2(\partial\Omega)}}{||\underset{\sim}{g}||_{\underset{\sim}{H}^{\frac{1}{2}}(\partial\Omega)}} \quad (3.10)$$

■

Let Div denote the material divergence, i.e. $(\text{Div}\, \underset{\sim}{v}) = \sum_\alpha \partial v_\alpha/\partial x_\alpha$. Then, if differentiation is interpreted in a distributional sense, (3.1) is equivalent to the weak boundary-value problem,

$$\left.\begin{aligned} \text{Div}[\underset{\sim}{E}(\nabla \underset{\sim}{u})] + \rho\underset{\sim}{f} = \underset{\sim}{0} \quad \text{in } \Omega \\ \gamma_0\underset{\sim}{u} = 0 \;\text{ on } \partial\Omega_1, \quad \underset{\sim}{n} \cdot \underset{\sim}{E}(\nabla \underset{\sim}{u}) = \underset{\sim}{S} \;\text{ on } \partial\Omega_2 \end{aligned}\right\} \quad (3.11)$$

Following standard definitions, we introduce

$$\underset{\sim}{\varepsilon} = \text{strain tensor} = \frac{1}{2}(\underset{\sim\sim}{\nabla u} + \underset{\sim\sim}{\nabla u}^T) = \hat{\underset{\sim\sim}{\nabla u}} \in (L_2(\Omega))^3 \tag{3.12}$$

$$\underset{\sim\sim}{E\varepsilon} = \underset{\sim}{\sigma} = \text{stress tensor} \in (L_2(\Omega))^3 \tag{3.13}$$

Equation (3.1) is then equivalent to the equation of balance of linear monentum for elastic bodies at rest; i.e. the equation of static equilibrium. Clearly, (3.11) is equivalent to the system of equations,

$$\left.\begin{array}{r} \underset{\sim\sim}{\nabla u} = \underset{\sim}{\varepsilon} \\ \underset{\sim}{E}(\underset{\sim}{\varepsilon}) = \underset{\sim}{\sigma} \\ \text{Div}\, \underset{\sim}{\sigma} = -\rho \underset{\sim}{f} \end{array}\right\} \text{ in } \Omega \qquad \left.\begin{array}{r} \gamma_0 \underset{\sim}{u} = \underset{\sim}{0} \text{ on } \partial\Omega_1 \\ \underset{\sim}{n} \cdot \underset{\sim}{\sigma} = \underset{\sim}{S} \text{ on } \partial\Omega_2 \end{array}\right\} \tag{3.14}$$

A great deal of numerical experimentation has indicated that, in certain applications, there are advantages in approximating the system (3.14) (or its equivalent variational statement) instead of (3.11). For example, improved accuracies in stress approximations often result when both $\underset{\sim}{u}$ and $\underset{\sim}{\sigma}$ are approximated simultaneously. This fact has led to the development of a variety of so-called mixed or hybrid variational principles for use as basis for Ritz-Galerkin approximations.

4. A Mixed Variational Principle

In this section, we discuss a mixed variational principle associated with (3.11) and suggested by (2.3). Let the product space M be defined by

$$M = \underset{\sim}{U}(\Omega) \times (L_2(\Omega))^3 \tag{4.1}$$

which is a space of ordered pairs $\underset{\sim}{\phi} = (\underset{\sim}{u}, \underset{\sim}{\sigma})$ consisting of an admissible displacement vector u and a stress tensor $\underset{\sim}{\sigma}$. The space $\underset{\sim}{U}(\Omega)$ is as defined in (3.2) and the space M is provided with the norm

$$||\underset{\sim}{\phi}||_M = [||\underset{\sim}{u}||^2_{\underset{\sim}{H}^1(\Omega)} + |||\underset{\sim}{\sigma}|||^2_{\underset{\sim}{L}_2(\Omega)}]^{\frac{1}{2}} \tag{4.2}$$

in which

$$||\underset{\sim}{u}||^2_{\underset{\sim}{H}^1(\Omega)} = \sum_{\alpha=1}^{2} ||u_\alpha||^2_{H^1(\Omega)} \quad ; \quad |||\underset{\sim}{\sigma}|||^2_{\underset{\sim}{L}_2(\Omega)} = \sum_{\substack{\alpha,\beta=1 \\ \alpha \le \beta}}^{2} ||\sigma^{\alpha\beta}||^2_{L_2(\Omega)} \tag{4.3}$$

We now introduce a bilinear form on $M \times M$ given by

$$B(\underset{\sim}{\phi}, \underset{\sim}{\bar{\phi}}) = \int_\Omega \{ \text{tr}(\underset{\sim}{\sigma} \cdot \underset{\sim\sim}{\nabla \bar{u}}) + \text{tr}[(\underset{\sim\sim}{\nabla u} - \underset{\sim}{E}^{-1}(\underset{\sim}{\sigma})) \cdot \underset{\sim}{\bar{\sigma}}]\} dx \tag{4.4}$$

and a linear functional on M by

$$F(\underset{\sim}{\bar{\phi}}) = \int_\Omega \rho \underset{\sim}{f} \cdot \underset{\sim}{\bar{u}}\, dx + \int_{\partial\Omega_2} \underset{\sim}{S} \cdot \gamma_0 \underset{\sim}{\bar{u}}\, ds \tag{4.5}$$

Obviously the functional $F(\cdot)$ is continuous; indeed,

$$||F||_{M'} \leq C_1(||\rho \underset{\sim}{f}||_{\underset{\sim}{L}_2(\Omega)} + ||\underset{\sim}{S}||_{\underset{\sim}{L}_2(\partial\Omega_2)}) \tag{4.6}$$

where $C_1 > 0$ is a constant.

We now state the <u>mixed variational problem</u> associated with (3.11): Find $\underset{\sim}{\phi} \in M$ such that

$$B(\underset{\sim}{\phi},\overline{\underset{\sim}{\phi}}) = F(\overline{\underset{\sim}{\phi}}) \quad \forall\, \overline{\underset{\sim}{\phi}} \in M \tag{4.7}$$

By integrating (4.4) by parts, it is easily verified that (4.7) is equivalent to

$$\left.\begin{array}{l} \underset{\sim}{E} \cdot \hat{\nabla} \underset{\sim}{u} = \underset{\sim}{\sigma} \\ \\ -\operatorname{Div} \underset{\sim}{\sigma} = \rho \underset{\sim}{f} \end{array}\right\} \text{in } \Omega \qquad \left.\begin{array}{l} \gamma_0 \underset{\sim}{u} = 0 \quad \text{on } \partial\Omega_1 \\ \\ \underset{\sim}{\sigma} \cdot \underset{\sim}{\eta} = \underset{\sim}{S} \quad \text{on } \partial\Omega_2 \end{array}\right\} \tag{4.8}$$

The existence of a unique solution to (4.7) can be shown without much difficulty via Theorem (2.1):

<u>Theorem 4.1.</u> The bilinear form $B(\cdot,\cdot)$ defined by (4.4) satisfies all the hypothesis of Theorem 2.1 when we set $V_1 = V_2 = M$. Hence, there exists a unique element $\phi_0 = (\underset{\sim}{u}^0, \underset{\sim}{\sigma}^0) \in M$ which satisfies (4.7).

<u>Proof</u>: The continuity condition is easy to show by using the Schwarz inequality. As for condition (ii) of Theorem 2.1, we begin by choosing

$$\hat{\underset{\sim}{u}} = 2\underset{\sim}{u} \quad \text{and} \quad \hat{\underset{\sim}{\sigma}} = -\underset{\sim}{\sigma} + \underset{\sim}{E} \cdot \hat{\nabla} \underset{\sim}{u} \tag{4.9}$$

Clearly, $\hat{\phi} = (\hat{\underset{\sim}{u}},\hat{\underset{\sim}{\sigma}}) \in M$ and there exists a constant $C_2 > 0$ such that

$$||\hat{\underset{\sim}{\phi}}||_M \leq C_2 ||\underset{\sim}{\phi}||_M \tag{4.10}$$

Now, by the generalized Korn inequality (cf. pp. 325, [16]), there exists a constant $C_3 > 0$ such that

$$\int_\Omega \operatorname{tr}[(\nabla \underset{\sim}{u})^T \cdot \underset{\sim}{E}(\nabla \underset{\sim}{u})]dx \geq C_3 ||\underset{\sim}{u}||^2_{\underset{\sim}{H}^1(\Omega)} \qquad \forall\, \underset{\sim}{u} \in \underset{\sim}{U}(\Omega) \tag{4.11}$$

Hence, in view of (3.5) and (4.10),

$$B(\underset{\sim}{\phi},\hat{\underset{\sim}{\phi}}) \geq C_4 ||\underset{\sim}{\phi}||_M ||\hat{\underset{\sim}{\phi}}||_M \tag{4.12}$$

where $C_4 = \frac{1}{C_2} \min\{C_3, \mu_1\}$. Thus,

$$\inf_{||\underset{\sim}{\phi}||_M = 1} \; \sup_{||\overline{\underset{\sim}{\phi}}||_M \leq 1} B(\underset{\sim}{\phi},\overline{\underset{\sim}{\phi}}) \geq \inf_{||\underset{\sim}{\phi}|| = 1} B(\underset{\sim}{\phi},\hat{\underset{\sim}{\phi}})/||\hat{\phi}||_M \geq C_4 > 0$$

By taking $a_0 = C_4$ in (2.8), the Theorem is proved due to symmetry of $B(\cdot,\cdot)$. ■

5. A Theory of Mixed Finite Element Approximations

We construct a mixed finite-element approximation of the variational boundary-value problem (4.7) as follows: The domain Ω is partitioned into a collection of E subdomains Ω_e, $1 \leq e \leq E$, such that $\bar{\Omega} = \bigcup_{e=1}^{E} \bar{\Omega}_e$ and $\Omega_e \cap \Omega_f = \emptyset$ for $e \neq f$. For simplicity, we shall assume that Ω is convex polygonal and the partition P of Ω is either a triangulation of Ω or a decomposition of Ω into convex quadrilaterals such that each vertex is either on $\partial\Omega$ or is a vertex or corner of all elements containing that vertex; i.e. Ω is a simple partition into triangles or quadrilaterals.*

Over each element we construct polynomial approximations of the components of the displacement vector $\underset{\sim}{u}$ and the stress tensor $\underset{\sim}{\sigma}$, and in this way we construct families of finite-dimensional subspaces $\underset{\sim}{U}_h(\Omega)$ and $\underset{\sim}{S}_h(\Omega)$, $0 < h \leq 1$, of the spaces $\underset{\sim}{U}(\Omega)$ and $(L_2(\Omega))^3$, respectively, which have the following properties:

$$\bullet\ \underset{\sim}{U}_h(\Omega) = \{\underset{\sim}{U} \in \underset{\sim}{U}(\Omega)\ ;\quad (P_k(\Omega))^2 \subset \underset{\sim}{U}_h(\Omega)\ ,\quad k \geq 1\} \tag{5.1}$$

• For every $\underset{\sim}{u} \in H^m(\Omega) = (H^m(\Omega))^2$, there exists a $\tilde{U} \in \underset{\sim}{U}_h(\Omega)$ such that there is a constant $C_1 > 0$, independent of h, such that

$$||\underset{\sim}{u} - \tilde{\underset{\sim}{U}}||_{H^{\ell}(\Omega)} \leq C_1 h^{\mu} ||\underset{\sim}{u}||_{\underset{\sim}{H}^m(\Omega)} \tag{5.2}$$

where $\ell = 0, 1$, and

$$\mu = \min(k+1-\ell,\ m-1), \quad m \geq 1 \tag{5.3}$$

$$\bullet\ \underset{\sim}{S}_h(\Omega) = \{\underset{\sim}{\Sigma} \in (L_2(\Omega))^3:\ (P_r(\Omega))^3 \subset \underset{\sim}{S}_h(\Omega),\quad r \geq 0\} \tag{5.4}$$

• For every $\underset{\sim}{\sigma} \in \underset{\sim}{H}^s(\Omega) = (H^s(\Omega))^3$, there exists a $\tilde{\Sigma} \in \underset{\sim}{S}_h(\Omega)$ such that there is a constant $C_2 > 0$, independent of h, such that

$$|||\underset{\sim}{\sigma} - \tilde{\underset{\sim}{\Sigma}}|||_{\underset{\sim}{L}_2(\Omega)} \leq C_2 h^{\nu} |||\underset{\sim}{\sigma}|||_{\underset{\sim}{H}^s(\Omega)} \tag{5.5}$$

where

$$\nu = \min(r+1, s), \quad s \geq 0 \tag{5.6}$$

In the above definitions, $P_k(\Omega)$ is the space of polynomials of degree $\leq$ k on Ω, etc., and h is the mesh parameter,

$$h = \max_{1 \leq e \leq E} h_e, \quad h_e = \mathrm{dia}(\Omega_e) \tag{5.7}$$

it being assumed that the mesh refinements are quasiuniform.

The mixed finite-element method consists of seeking a pair

$$\underset{\sim}{\bar{\Phi}} = (\underset{\sim}{U}, \underset{\sim}{\Sigma}) \in M_h = \underset{\sim}{U}_h(\Omega) \times \underset{\sim}{S}_h(\Omega) \subset M \tag{5.8}$$

such that

$$B(\underset{\sim}{\Phi}, \underset{\sim}{\bar{\phi}}) = F(\underset{\sim}{\bar{\Phi}}) \quad \forall\ \underset{\sim}{\bar{\phi}} \in M_h \tag{5.9}$$

*This assumption is made only to simplify the development and is not fundamental to our theory. It is shown in [12] that all of the essential aspects of our theory remain unchanged when Ω is partitioned into an arbitrary collection of nonconvex, curvilinear subdomains with boundaries made up of piecewise smooth arcs.

where $B(\cdot,\cdot)$ is the bilinear form in (4.4) and F is the linear functional in (4.5).

The question of overriding importance at this point is what conditions must be met if (5.9) is to have a unique solution? To resolve this question, we introduce an L_2-orthogonal projector Π_h of the space of strains $\underset{\sim}{\varepsilon}(\Omega) = \{\underset{\sim}{\varepsilon}: \underset{\sim}{\varepsilon} = \underset{\sim\sim}{\nabla U},\ \forall\, \underset{\sim}{U} \in \underset{\sim h}{U}(\Omega)\}$ onto the space $\underset{\sim}{E}^{-1}(\underset{\sim h}{S}(\Omega))$ of strains produced by the approximate stresses $\underset{\sim}{\Sigma}$; i.e., for any $\underset{\sim}{\varepsilon} \in \underset{\sim}{\varepsilon}(\Omega)$,

$$\int_\Omega \mathrm{tr}[\underset{\sim}{E}^{-1}(\underset{\sim}{\Sigma}) \cdot (\Pi_{h}\underset{\sim}{\varepsilon} - \underset{\sim}{\varepsilon})]dx = 0 \quad \forall\, \underset{\sim}{\Sigma} \in \underset{\sim}{S}_h(\Omega) \tag{5.10}$$

Next, we construct the special element,

$$\underset{\sim}{\hat{\Phi}} \in M_h\,, \quad \underset{\sim}{\hat{\Phi}} = (\underset{\sim}{\hat{U}},\underset{\sim}{\hat{\Sigma}})\,, \quad \underset{\sim}{\hat{U}} = 2\underset{\sim}{U}\,; \quad \underset{\sim}{\hat{\Sigma}} = -\underset{\sim}{\Sigma} + \underset{\sim}{E}(\Pi_h\underset{\sim\sim}{\nabla U}) \tag{5.11}$$

where $(\underset{\sim}{U},\underset{\sim}{\Sigma})$ is an arbitrary element of M_h. Then

$$\begin{aligned} B(\underset{\sim}{\Phi}, \underset{\sim}{\hat{\Phi}}) &= \int_\Omega \mathrm{tr}(\underset{\sim}{E}^{-1}\underset{\sim}{\Sigma} \cdot \underset{\sim}{\Sigma})dx + \int_\Omega \mathrm{tr}(\underset{\sim\sim}{\nabla U} \cdot \underset{\sim}{E}(\Pi_h\underset{\sim\sim}{\nabla U}))dx \\ &\geq \mu_1 |||\underset{\sim}{\Sigma}|||^2_{\underset{\sim}{L}_2(\Omega)} + \mu_0 |||\Pi_h\underset{\sim\sim}{\nabla U}|||^2_{\underset{\sim}{L}_2(\Omega)} \end{aligned} \tag{5.12}$$

wherein we have used the orthogonality of Π_h, (5.10), and (3.5).

We now introduce a <u>stability parameter</u> λ defined by

$$\lambda = \inf_{\underset{\sim}{U} \in \underset{\sim h}{U}(\Omega)} \frac{|||\Pi_h\underset{\sim\sim}{\nabla U}|||^2_{\underset{\sim}{L}_2(\Omega)}}{|||\underset{\sim\sim}{\nabla U}|||^2_{\underset{\sim}{L}_2(\Omega)}} \tag{5.13}$$

Then, we see that, from (5.12)

$$B(\underset{\sim}{\Phi},\underset{\sim}{\hat{\Phi}}) \geq \beta||\underset{\sim}{\Phi}||^2_M \tag{5.14}$$

where $\beta = \min\{\mu_1, C_3\lambda\}$ and C_3 is as in (4.11). Noting that $B(\underset{\sim}{\Phi},\underset{\sim}{\Phi})$ is symmetric and continuous on $M_h \times M_h$ and recalling Theorem 2.2, we easily see that by establishing (5.14) we have proved the following theorem:

<u>Theorem 5.1</u>. Let the parameter λ of (5.13) be strictly positive. Then there exists a unique $\underset{\sim}{\Phi}^0 \in M_h$ to the mixed finite element approximation problem (5.9)■

<u>Corollary 5.1</u>. Let the approximate displacement field $\underset{\sim}{U}$ be chosen so that $\underset{\sim}{E} \cdot \underset{\sim\sim}{\nabla U} \in \underset{\sim h}{S}(\Omega)$. Then there exists a unique solution to (5.9)

<u>Proof</u>: In this case $\lambda = 1 > 0$.■

To interpret the parameter λ, we note that the approximation $\underset{\sim}{U}$ and $\underset{\sim}{E}^{-1}(\underset{\sim}{\Sigma})$ are combinations of linearly independent polynomial basis functions of the form

$$U_\alpha = \sum_{i=1}^{K} a^i_\alpha \phi_i(\underset{\sim}{x})\,, \qquad E^{-1}_{\alpha\beta\lambda\mu}\Sigma^{\lambda\mu} = \sum_{\Delta=1}^{R} b^\Delta_{\alpha\beta}\omega_\Delta(\underset{\sim}{x})$$

Denoting

$$\left.\begin{aligned} \underset{\sim}{M} &= M_{i\alpha}^{\Delta} = \int_{\Omega} \frac{\partial \phi_i}{\partial x_\alpha} \omega^{\Delta}\, dx\ , \quad \underset{\sim}{H} = H_{\Delta\Gamma} = \int_{\Omega} \omega_{\Delta}\omega_{\Gamma}\, dx \\ \underset{\sim}{S} &= S_{i\alpha j\beta} = \int_{\Omega} \frac{\partial \phi_i}{\partial x_\alpha}\frac{\partial \phi_j}{\partial x_\beta}\, dx; \quad 1 \leq i,\ j \leq K,\ 1 \leq \Delta \leq R \end{aligned}\right\} \tag{5.14}$$

where ω^{Δ} are the basis functions reciprocal to ω_{Δ}. Then

$$\lambda = \inf_{\underset{\sim}{A}} \frac{\underset{\sim}{A}^T\underset{\sim}{M}^T\underset{\sim}{H}\underset{\sim}{M}\underset{\sim}{A}}{\underset{\sim}{A}^T\underset{\sim}{S}\underset{\sim}{A}} \tag{5.15}$$

Now it is clear that the strict positiveness of the parameter λ is equivalent to the positive definiteness of the <u>stiffness</u> <u>matrix</u> $\underset{\sim}{M}^T\underset{\sim}{H}\underset{\sim}{M}$. It is also interesting to note that whenever $\underset{\sim}{M}$ has rank $\rho(\underset{\sim}{M}) \leq R \leq K$ the stiffness matrix is singular and $\lambda = 0$. As a special case, we have the following:

<u>Theorem 5.3</u>. Let the components of $\underset{\sim}{U}_e$ be complete polynomials of degree k and the components of $\underset{\sim}{\Sigma}_e$ be complete polynomials of degree r. Then there exists a unique solution to the mixed finite-element approximation problem (5.9) if

$$k \leq r+1 \tag{5.16}$$

<u>Proof</u>. If (5.16) holds, the conditions of Corollary 5.1 are clearly met. ∎

<u>Error Estimates</u>. We next address the important question of convergence and error estimates. As noted earlier, our principal tool here is the approximation theorem 2.2 and the interpolation properties (5.2) and (5.5) of the subspaces.

<u>Theorem 5.4</u>. Let the solution $\underset{\sim}{u}^o$ of (3.1) be in $\underset{\sim}{H}^m(\Omega)$, $m \geq 1$, and $\underset{\sim}{\phi}^o \in M_h$ of (5.9). Then there exists a positive constant $C_3 > 0$ such that

$$||\underset{\sim}{e}_u||_{\underset{\sim}{H}^1(\Omega)} + |||\underset{\sim}{e}_\sigma|||_{\underset{\sim}{L}_2(\Omega)} \leq C_3 h^{\alpha} ||\underset{\sim}{u}^o||_{\underset{\sim}{H}^m(\Omega)} \tag{5.17}$$

where $\underset{\sim}{e}_u = \underset{\sim}{u}^o - \underset{\sim}{U}^o$ is the error in the displacement vector, $\underset{\sim}{e}_\sigma = \underset{\sim}{\sigma}^o - \underset{\sim}{\Sigma}^o$ is the error in the stress tensor, and

$$\alpha = \min(k,\ r+1,\ m-1), \quad m \geq 1 \tag{5.18}$$

<u>Proof</u>: Noting that $\underset{\sim}{\phi}^o - \underset{\sim}{\Phi}^o = (\underset{\sim}{e}_u, \underset{\sim}{e}_\sigma)$, we obtain immediately from (2.16), (5.2) and (5.5), the inequality

$$||\underset{\sim}{e}_u||^2_{\underset{\sim}{H}^1(\Omega)} + |||\underset{\sim}{e}_\sigma|||^2_{\underset{\sim}{L}_2(\Omega)} \leq C[h^{2\mu}||\underset{\sim}{u}^o||_{\underset{\sim}{H}^m(\Omega)} + h^{2\nu}|||\underset{\sim}{\nabla}\underset{\sim}{u}^o|||_{\underset{\sim}{H}^{m-1}(\Omega)}]$$

which leads immediately to (5.17). ∎

In certain cases a finer resolution of the error can be obtained, as is indicated in the following theorem:

Theorem 5.5. Let the conditions of Theorem 5.3 hold and let $k \leq r+1$. Then the following L_2-estimates hold:

$$||e_u||_{L_2(\Omega)} \leq C_5 h^{\eta} ||u^0||_{H^m(\Omega)} \tag{5.19}$$

where $\eta = 1 + \min(k, m-1)$ and $m \geq 1$.

Proof: The basic idea here is to apply the technique used by Nitsche [17]. Let $v \in U(\Omega)$ be the solution of the auxiliary problem,

$$- \operatorname{Div}[E(\nabla v)] = e_u \quad \text{in } \Omega \tag{5.20}$$

$$\gamma_0 v = 0 \quad \text{on } \partial\Omega_1 \quad \text{and} \quad n \cdot E(\nabla v) = 0 \quad \text{on } \partial\Omega_2$$

Then we know from the regularity result [18] that there exists a constant $C_1 > 0$ such that

$$||v||_{H^2(\Omega)} \leq C_1 ||e_u||_{L_2(\Omega)} \tag{5.21}$$

Denoting by $\psi = (v, E(\nabla v)) \in M$ and $\phi = (u, \sigma) \in M$, we have

$$B(\psi, \phi) = \int_\Omega tr[E(\nabla v) \cdot \nabla u] dx$$

and

$$B(\psi, \phi) = \int_\Omega e_u \cdot u \, dx \quad \forall u \in U(\Omega) \tag{5.22}$$

Therefore,

$$B(\psi, (e_u, e_\sigma)) = ||e_u||^2_{L_2(\Omega)} \tag{5.23}$$

We also have

$$B((e_u, e_\sigma), (V, \Sigma)) = 0 \quad \forall (V, \Sigma) \in U_h(\Omega) \times S_h(\Omega) \tag{5.24}$$

By choosing $\Sigma = E(\nabla V)$, we see that $\Sigma \in S_h(\Omega)$ because of the assumption $k \leq r+1$, and (5.24) can be written as

$$B((V, E(\nabla V)), (e_u, e_\sigma)) = 0 \quad \forall V \in U_h(\Omega) \tag{5.25}$$

due to the symmetry of $B(\cdot,\cdot)$. Now substracting (5.25) from (5.23) and by using the Schwarz inequality, we obtain for every $V \in U_h(\Omega)$,

$$||e_u||^2_{L_2(\Omega)} \leq C_2 |||\nabla(v - V)|||_{L_2(\Omega)} \, |||\nabla e_u|||_{L_2(\Omega)} \leq C_2 ||v - V||_{H^1(\Omega)} ||e_u||_{H^1(\Omega)}$$

Now by choosing $V = \tilde{U}$ as in (5.2) and using (5.17), we obtain

$$||e_u||^2_{L_2(\Omega)} \leq C_3 h^{\mu} ||v||_{H^2(\Omega)} h^{\xi} ||u||_{H^m(\Omega)} \tag{5.26}$$

where $\mu = \min(k,1) = 1$, $\xi = \min(k, m-1)$, and $m \geq 1$, from which (5.19) is deduced when the regularity result (5.21) is used. ■

6. A Hybrid Variational Principle for Plane Elasticity

We now direct our attention to a considerably more complex type of approximation, the so-called hybrid finite element methods. There are a variety of different hybrid methods, but we shall describe the properties of a representative class of hybrid finite element approximations of the plane elasticity problem: find $\underset{\sim}{u} \in (H_0^1(\Omega))^2$ such that

$$\int_\Omega \mathrm{tr}[\underset{\sim}{E}(\underset{\sim}{\nabla u}) \cdot \underset{\sim}{\nabla v}]dx = \int_\Omega \rho \underset{\sim}{f} \cdot \underset{\sim}{v}\, dx \quad \forall\, \underset{\sim}{v} \in (H_0^1(\Omega))^2 \tag{6.1}$$

This corresponds to the "problem of place" in plane elasticity; a linear elastic body is subjected to body forces and deformed symmetrically with respect to the material plane Ω, the boundary displacements of which are held fixed ($\gamma_{0}\underset{\sim}{u} = \underset{\sim}{0}$). Again, Ω is assumed to be polygonal for simplicity.

We next construct a special variational statement of problem (6.1) that is ideally suited for the development of hybrid finite element approximations,

(i) Ω is viewed as a collection of E <u>triangles</u> Ω_e; vertices of each triangle being either on $\partial\Omega_e$ or shared by two or more triangles. The number of triangular subdomains E depends upon the partition P (i.e., the triangulation of Ω) and we write $E = E(P)$. Again, $\overline{\Omega} = \bigcup_{e=1}^{E} \overline{\Omega}_e$, $\Omega_e \cap \Omega_f = \emptyset$, $e \neq f$, and we denote the diameter of Ω_e by h_e, $1 \leq e \leq E(P)$.

(ii) The boundary $\partial\Omega_e$ of each triangle consists of three vertices $\{x_e^1, x_e^2, x_e^3\}$ and three open line segments Γ_e^i, $i = 1,2,3$, connecting the vertices. The inter-element boundary $\Gamma(P)$ is defined by

$$\Gamma = \Gamma(P) = \bigcup_{\substack{1 \leq e < E \\ i=1,2,3}} \Gamma_e^i \tag{6.2}$$

(iii) Some special Hilbert spaces are now introduced; first

$$\underset{\sim}{H}^1(P) = \{\underset{\sim}{u}: \underset{\sim}{u}_e = \underset{\sim}{u}|_{\Omega_e} \in (H^1(\Omega_e))^2, \quad 1 \leq e \leq E\} \tag{6.3}$$

which is endowed with the norm

$$||\underset{\sim}{u}||_{\underset{\sim}{H}^m(P)} = [\sum_{e=1}^{E} ||\underset{\sim}{u}_e||^2_{\underset{\sim}{H}^1(\Omega_e)}]^{1/2} \tag{6.4}$$

and

$$\underset{\sim}{T}(\Gamma) = \text{completion of } \underset{\sim}{L}_2(\Gamma) = (L_2(\Gamma))^2 \text{ in the norm}$$

$$||\underset{\sim}{\tau}||_{\underset{\sim}{T}(\Gamma)} = [\sum_{e=1}^{P} ||\underset{\sim}{\tau}_e||_{\underset{\sim}{H}^{-1/2}(\partial\Omega_e)}]^{1/2} \tag{6.5}$$

where $\underset{\sim}{H}^{-1/2}(\partial\Omega_e)$ is defined in (3.10). Finally, we introduce the product space

$$H = \underset{\sim}{H}^1(P) \times \underset{\sim}{T}(\Gamma) \tag{6.6}$$

together with the norm

$$||\underset{\sim}{\theta}||_H = [\sum_{e=1}^{E} ||\underset{\sim}{\theta}_e||^2_{H_e}]^{\frac{1}{2}} \tag{6.7}$$

where $\underset{\sim}{\theta}_e = \underset{\sim}{\theta}|_{\overline{\Omega}_e}$ is the pair $(\underset{\sim}{u}_e, \underset{\sim}{\tau}_e)$, $\underset{\sim}{u}_e$ being the local displacement field corresponding to Ω_e and $\underset{\sim}{\tau}_e$ being the local surface tractions acting on the boundary $\partial\Omega_e$, and

$$||\underset{\sim}{\theta}_e||^2_{H_e} = ||\underset{\sim}{u}_e||^2_{\underset{\sim}{H}^1(\Omega_e)} + ||\underset{\sim}{\tau}_e||^2_{\underset{\sim}{H}^{-\frac{1}{2}}(\partial\Omega_e)} \tag{6.8}$$

(iv) A bilinear form $\hat{B}$ on $H \times H$ is introduced, where

$$\hat{B}(\underset{\sim}{\theta},\underset{\sim}{\overline{\theta}}) = \sum_{e=1}^{E} \hat{B}_e(\underset{\sim}{\theta}_e,\underset{\sim}{\overline{\theta}}_e) \tag{6.9}$$

$$\hat{B}_e(\underset{\sim}{\theta}_e,\underset{\sim}{\overline{\theta}}_e) = \int_{\Omega_e} tr[\underset{\sim}{E}(\nabla \underset{\sim}{u}_e) \cdot \nabla \underset{\sim}{\overline{u}}_e]dx + \oint_{\partial\Omega_e} (\underset{\sim}{\tau}_e \cdot \gamma_0 \overline{u}_e + \underset{\sim}{\overline{\tau}}_e \cdot \gamma_0 \underset{\sim}{u}_e)ds \tag{6.10}$$

as well as a linear form on H given by ($\rho\underset{\sim}{f} \in \underset{\sim}{H}^0(\Omega)$)

$$\hat{F}(\underset{\sim}{\theta}) = \sum_{e=1}^{E} \hat{F}_e(\underset{\sim}{\overline{\theta}}_e) = \sum_{e=1}^{E} \int_{\Omega_e} \rho \underset{\sim}{f}_e \cdot \underset{\sim}{\overline{\theta}}_e \, dx \tag{6.11}$$

(v) Finally, we consider the special variational problem: Find $\theta \in H$ such that

$$\hat{B}(\underset{\sim}{\theta},\underset{\sim}{\overline{\theta}}) = \hat{F}(\overline{\theta}) \quad \forall \underset{\sim}{\overline{\theta}} \in H \tag{6.12}$$

The importance of the constructions (i)-(iv) and the special problem (6.12) rests in properties established in the following theorem.

Theorem 6.1. Let $\hat{B}(\cdot,\cdot)$ and $\hat{F}(\cdot)$ be given by (6.9) and (6.11). Then

(i) $\hat{B}(\cdot,\cdot)$ is a continuous bilinear form on $H \times H$

(ii) $\hat{B}(\cdot,\cdot)$ is weakly coercive, in the sense that there exists a constant $a > 0$, independent of the partition P, such that

$$\inf_{||\underset{\sim}{\theta}||_H = 1} \sup_{||\overline{\theta}||_H \leq 1} |\hat{B}(\theta,\overline{\theta})| \geq a > 0 \quad \text{and} \quad \sup_{\underset{\sim}{\theta} \in H} \hat{B}(\underset{\sim}{\theta},\underset{\sim}{\overline{\theta}}) > 0, \; (\underset{\sim}{\overline{\theta}} \neq \underset{\sim}{0}) \tag{6.13}$$

(iii) $\hat{F}$ is a continuous linear functional on H

(iv) there exists a unique solution $\underset{\sim}{\theta}^0$ to (6.12) and

$$||\underset{\sim}{\theta}^0|| \leq \frac{1}{a} ||\rho \underset{\sim}{f}||_{\underset{\sim}{H}^0(\Omega)} \tag{6.14}$$

(v) if $\underset{\sim}{u}^*$ is the solution of (6.1), $\underset{\sim}{\tau}_e^* = \underset{\sim}{n}_e \cdot \underset{\sim}{\underset{\sim}{E}}(\nabla \underset{\sim}{u}_e^*)$, and $\underset{\sim}{\theta}^* \in H$ is such that $\underset{\sim}{\theta}_e^* = (\underset{\sim}{u}_e^*, \underset{\sim}{\tau}_e^*)$, then

$$\underset{\sim}{\theta}^0 = \underset{\sim}{\theta}^* \tag{6.15}$$

i.e. problems (6.1) and (6.12) are equivalent. ■

Sketch of the Proof: The proof of this theorem is lengthy and can be adapted from the proof of a similar theorem in [12]. Parts (i), (iii), and (v) follow easily from standard inequalities and algebraic manipulations. Property (iv) follows from (i), (ii), (iii) and Theorem 2.1.

The proof of (ii) is a bit involved, and its principal feature is the construction of an auxiliary problem,

$$\int_{\Omega_e} \mathrm{tr}[\nabla \underset{\sim}{z} \cdot \nabla \underset{\sim}{u}]dx + \oint_{\partial\Omega_e} \gamma_0 \underset{\sim}{z} \cdot \gamma_0 \underset{\sim}{u}\, ds = \oint_{\partial\Omega_e} \underset{\sim}{\tau}_e \cdot \gamma_0 \underset{\sim}{u}\, ds \qquad \forall\, \underset{\sim}{u} \in \underset{\sim}{H}^1(\Omega_e) \tag{6.16}$$

for a given $\underset{\sim}{\tau}_e = \underset{\sim}{\tau}|_{\Omega_e}$, $\underset{\sim}{\tau} \in \underset{\sim}{T}(\Gamma)$. Clearly, such a $\underset{\sim}{z}$ exists in $H^1(\Omega_e)$. Furthermore, there exist constants $C_1, C_2 > 0$ such that (Cf. [12],[19])

$$C_1 ||\underset{\sim}{\tau}_e||^2_{\underset{\sim}{H}^{-\frac{1}{2}}(\partial\Omega_e)} \leq \oint_{\partial\Omega_e} \underset{\sim}{\tau}_e \cdot \gamma_0 \underset{\sim}{z}\, ds = |\underset{\sim}{z}|^2_{\underset{\sim}{H}^1(\Omega_e)} \leq C_2 ||\underset{\sim}{\tau}_e||^2_{\underset{\sim}{H}^{-\frac{1}{2}}(\partial\Omega_e)} \tag{6.17}$$

where

$$|\underset{\sim}{z}|^2_{\underset{\sim}{H}^1(\Omega_e)} = \oint_{\Omega_e} \mathrm{tr}[\nabla \underset{\sim}{z} \cdot \nabla \underset{\sim}{z}]dx + \oint_{\partial\Omega_e} \gamma_0 \underset{\sim}{z} \cdot \gamma_0 \underset{\sim}{z}\, ds \tag{6.18}$$

is a norm equivalent to $||\cdot||^2_{\underset{\sim}{H}^1(\Omega_e)}$.

Now by choosing

$$\hat{\underset{\sim}{\theta}}_e = (\hat{\underset{\sim}{u}}_e, \hat{\underset{\sim}{\tau}}_e) \in H_e \quad \text{where} \quad \begin{cases} \hat{\underset{\sim}{u}}_e = \underset{\sim}{u}_e + \underset{\sim}{z} \\ \hat{\underset{\sim}{\tau}}_e = -2\underset{\sim}{\tau}_e + \gamma_0 \underset{\sim}{u}_e + \gamma_0 \underset{\sim}{z} \end{cases} \tag{6.19}$$

we see that

$$||\hat{\underset{\sim}{\theta}}_e||_{H_e} \leq C_3 ||\underset{\sim}{\theta}_e||_{H_e} \quad \text{and} \quad B_e(\underset{\sim}{\theta}_e, \hat{\underset{\sim}{\theta}}_e) \geq C_4 ||\underset{\sim}{\theta}_e||_{H_e} \tag{6.20}$$

where $C_3, C_4 > 0$ and independent of P. Upon substituting (6.19) into (6.9), we immediately obtain the first member of (6.13). The second member then follows from the symmetry of $\hat{B}(\cdot,\cdot)$. ■

7. Hybrid Finite Element Approximations

We now pass on to the problem of approximation. The triangular subdomains are now, of course, viewed as finite elements, and over each finite element we introduce

olynomial approximations of $\underset{\sim}{u}_e$ on the interior of the element and polynomial approxi-
ations of the boundary tractions $\underset{\sim}{\tau}_e$ along the sides of each triangle. In this way,
e develop finite-dimensional $\underset{\sim}{U}_h(\Omega)$ and $\underset{\sim}{V}_h(\Omega)$ with the following properties:

$$\underset{\sim}{U}_h(P) = \{\underset{\sim}{U} \quad \underset{\sim}{H}^1(P): \underset{\sim}{U}_e \in \underset{\sim}{P}_k(\Omega_e), \quad k \geq 1, \quad 1 \leq e \leq E\} \tag{7.1}$$

o that for any $u \in (H^m(\Omega))^2$, a constant $M_1 > 0$, independent of h_e, a $\underset{\sim}{U}_e \in \underset{\sim}{U}_h(P)$ can
e found such that

$$||\underset{\sim}{u}_e - \underset{\sim}{U}_e||_{\underset{\sim}{H}^1(\Omega_e)} \leq M_1 h_e^{\mu} ||\underset{\sim}{u}_e||_{\underset{\sim}{H}^m(\Omega_e)} \tag{7.2}$$

here $\mu = \min(k,m-1)$ and $m \geq 1$.

$$\underset{\sim}{V}_h(\Gamma) = \{\underset{\sim}{T} \quad \underset{\sim}{T}(\Gamma): \underset{\sim}{T}|_{\Gamma_e^i} \in \underset{\sim}{P}_t(\Gamma_e^i), \quad t \geq 0, \quad i = 1,2,3, \quad 1 \leq e \leq E\} \tag{7.3}$$

or any $\underset{\sim}{\tau} \quad \underset{\sim}{}(\Gamma)$, there exists a $\tilde{\underset{\sim}{T}} \quad \underset{\sim}{V}_h(\Omega)$ and a constant $M_2 > 0$, not depending on
h_e such that,

$$||\underset{\sim}{\tau}_e - \tilde{\underset{\sim}{T}}_e||_{\underset{\sim}{H}^{-\frac{1}{2}}(\partial\Omega_e)} \leq M_2 h_e^{\nu} ||\underset{\sim}{\tau}_e||_{\hat{\underset{\sim}{H}}^{s-\frac{3}{2}}(\partial\Omega_e)} \tag{7.4}$$

here $\nu = \min(t + \frac{3}{2}, s-1)$, $s \geq 2$, and

$$||\underset{\sim}{\tau}_e||_{\hat{\underset{\sim}{H}}^{s-\frac{3}{2}}(\partial\Omega_e)} = \inf\{||\underset{\sim}{u}||_{\underset{\sim}{H}^s(\Omega_e)} : \underset{\sim}{n} \cdot \nabla \underset{\sim}{u}_e = \underset{\sim}{\tau}_e \text{ on } \Gamma_e^i\}$$

In the above definitions, $\underset{\sim}{P}_k(\Omega_e)$ and $\underset{\sim}{P}_t(\Gamma_e^i)$ are spaces of complete vector poly-
omials of degree k over Ω_e and t over each side of $\partial\Omega_e$, respectively.

We denote by H_h the product space

$$H_h = \underset{\sim}{U}_h(P) \times \underset{\sim}{V}_h(\Gamma) \tag{7.5}$$

ccasionally, we use the notation

$$H_h^e = \underset{\sim}{U}_h^e \times \underset{\sim}{V}_h^e = \underset{\sim}{P}_k(\Omega_e) \times \underset{\sim}{P}_t(\partial\Omega_e) \tag{7.6}$$

The <u>hybrid finite element method consists of seeking a pair</u> $\underset{\sim}{\Theta} \in H_h$ <u>such that</u>

$$\hat{B}(\underset{\sim}{\Theta},\underset{\sim}{\bar{\Theta}}) = \hat{F}(\underset{\sim}{\bar{\Theta}}) \qquad \forall\, \underset{\sim}{\bar{\Theta}} \in H_h \tag{7.7}$$

The fundamental questions, of course, are whether or not (7.7) has a solution and, if so, is it unique? What conditions must be enforced if (7.7) is to have a unique solution for each partition P of Ω, and do the approximate solutions converge in some sense to the solution of (6.1)? In addition, if the hybrid scheme is convergent, in what sense is it convergent, what is the rate-of-convergence, and how does this rate depend upon the subspaces $\underset{\sim}{U}_h(P)$ and $\underset{\sim}{V}_h(\Gamma)$? Finally, a very practical question: how can the local approximations $\underset{\sim}{U}_e$ and $\underset{\sim}{T}_e$ be devised so that we are guaranteed a stable, convergent scheme which is acceptably accurate? All of these questions are essentially answered in a collection of basic theorems that we present below.

We first state a fundamental lemma in hybrid finite element approximations.

Lemma 7.1. Let Ω_e, $1 \leq e \leq E$, be a triangle. Let the local basis functions for displacement vector $\underset{\sim}{U}_e$ and boundary tractions $\underset{\sim}{T}_e$ be complete polynomials of degree k and t, respectively. Then, for a $\underset{\sim}{T} \in V_h(\Gamma)$,

$$\oint_{\partial\Omega_e} \underset{\sim}{T}_e \cdot \gamma_0 \underset{\sim}{U}_e \, ds = 0 \quad \forall\, \underset{\sim}{U}_e \in P_k(\Omega_e) \quad \text{implies that } \underset{\sim}{T}_e = 0 \tag{7.8}$$

if and only if $k \geq t + 1$.

Proof: A complete proof is given in [12] and involves examination of the rank of the matrix representing the integral when local basis functions for $\underset{\sim}{U}_e$ and $\underset{\sim}{T}_e$ are introduced into the integral.■

The condition (7.8) is examined by Raviart [10] and Thomas [11] globally, i.e., by summing the integral over all E rather than locally, and called the "compatibility condition". We prefer to call it a "rank condition" because of its close relation to the rank of the matrix representing the bilinear form.

Corollary 7.1. Let a real parameter μ_e be defined by

$$\mu_e = \mu_e(\underset{\sim}{U}_h^e, \underset{\sim}{V}_h^e) \equiv \inf_{\underset{\sim}{T}_e \in \underset{\sim}{V}_h^e} \left\{ \oint_{\partial\Omega_e} \underset{\sim}{T}_e \cdot \gamma_0(\Pi^1 \underset{\sim}{z}) ds / ||\underset{\sim}{T}_e||_{\underset{\sim}{H}^{-\frac{1}{2}}(\partial\Omega_e)} \right\} \tag{7.9}$$

where $\underset{\sim}{z} \in \underset{\sim}{H}^1(\Omega_e)$ is the solution to (6.16) with $\underset{\sim}{\tau}_e$ replaced by $\underset{\sim}{T}_e$ and Π^1 is an orthogonal projector of $H^1(\Omega_e)$ onto $\underset{\sim}{U}_h^e$ associated with the scalar product

$$\langle \underset{\sim}{u}, \underset{\sim}{v} \rangle = \int_{\Omega_e} \operatorname{tr}[\nabla \underset{\sim}{u} \cdot \nabla \underset{\sim}{v}] dx + \oint_{\partial\Omega_e} \gamma_0 \underset{\sim}{u} \cdot \gamma_0 \underset{\sim}{v} \, ds \tag{7.10}$$

Then $\mu_e > 0$ if and only if $k \geq t+1$. Furthermore,

$$\mu_e \geq C_0 - C_1 h_e \tag{7.11}$$

where $C_0, C_1 > 0$ are constans not depending on h_e.

Proof: The first part is a direct consequence of Lemma 7.1. A proof of the second part is lengthy and can be found in [20].■

Theorem 7.1. Let the bilinear form $\hat{B}(\cdot,\cdot)$ be defined (6.9), and let Lemma 7.1 hold. Then

$$\inf_{||\underset{\sim}{\Theta}||_{H_h} = 1} \; \sup_{||\underset{\sim}{\overline{\Theta}}||_{H_h} \leq 1} \hat{B}(\underset{\sim}{\Theta}, \underset{\sim}{\overline{\Theta}}) \geq A > 0 \tag{7.12}$$

where $A = A_0 + A_1(h)$, $A_0 > 0$ is a constant not depending on h and $A_1(h) \to 0$ as $h \to 0$.

Hence, there exists a unique $\underset{\sim}{\Theta}^0 \in H_h$ which satisfies (7.7). Moreover, the following error estimate holds:

$$||\underset{\sim}{u}^0 - \underset{\sim}{U}^0||_{\underset{\sim}{H}^1(P)} + ||\underset{\sim}{\tau}^0 - \underset{\sim}{T}^0||_{\underset{\sim}{T}(\Gamma)} \leq M_0 h^\alpha \{\sum_{e=1}^{E} ||u_e^o||^2_{\underset{\sim}{H}^m(\Omega_e)}\}^{\frac{1}{2}} \quad (7.13)$$

where $h = \max_{1 \leq e \leq E} h_e$ and

$$\alpha = \min\{k,\ t + \frac{3}{2},\ m-1\}, \quad m \geq 2 \quad (7.14)$$

$$M_0 = C(1 + \frac{M}{A}) = \text{constant} > 0 \quad \forall\ h \quad (7.15)$$

where C and M are constants > 0, independent of h, and A is the constant in (7.12).

Proof: Only a sketch of the proof will be given. For details, see [20]. Let $\underset{\sim}{\Theta}_e = (\underset{\sim}{\hat{U}}_e, \underset{\sim}{\hat{T}}_e) \in H_h^e$ be such that

$$\underset{\sim}{\hat{U}}_e = \underset{\sim}{U}_e + \underset{\sim}{Z} \quad \text{and} \quad \underset{\sim}{\hat{T}}_e = -2\underset{\sim}{T}_e + \Pi^0(\gamma_0 \underset{\sim}{U}_e) \quad (7.16)$$

where $\underset{\sim}{Z} = \Pi^1 \underset{\sim}{z}$ with Π^1 and $\underset{\sim}{z}$ as in Corollary 7.1 and Π^0 is an orthogonal projector of traces of functions in $\underset{\sim}{U}_h^e$ onto $\underset{\sim}{V}_h^e$ associated with the L_2-inner product on the boundary. Then, by substituting (7.16) into (6.10) we obtain, by using Corollary 7.1 and projection properties of Π^1 and Π^0,

$$\hat{B}_e(\underset{\sim}{\Theta}_e, \underset{\sim}{\hat{\Theta}}_e) \geq A_e ||\underset{\sim}{\Theta}_e||_{H_e} \quad (7.17)$$

Here A_e is a function of h_e and μ_e such that

$$A_e = A_e^0 + A_e^1(h_e)$$

in which $A_e^0 > 0$ is a constant not depending on h_e and $A_e^1(h_e) \to 0$ as $h_e \to 0$. Also, from (7.14), we see that there is a constant $C_2 > 0$ not depending on h_e such that

$$||\underset{\sim}{\hat{\Theta}}||^2_{H_e} \leq C_2 ||\underset{\sim}{\Theta}_e||^2_{H_e}$$

Hence, (7.12) is obtained by taking

$$A = \min_{1 \leq e \leq E} \{A_e / C_2\}$$

The existence of a unique solution follows from Theorem 2.2. The error estimate (7.13) is also a direct consequence of Theorem 2.2 and the properties (7.1)-(7.4) of subspaces.■

8. Some Concluding Comments

(i) The existence of a unique solution to the mixed finite element formulation depends on whether or not the parameter λ of (5.13) is positive. When λ = 0, each $\underset{\sim}{\nabla U}$ is orthogonal to the space $\underset{\sim}{E}^{-1}(\underset{\sim}{S}_h(\Omega))$ of approximate strains, and this means that

$$\int_{\Omega} \mathrm{tr}[\underset{\sim}{E}(\nabla \underset{\sim}{U}) \cdot \underset{\sim}{\Sigma}]dx = 0 \qquad \forall\, \underset{\sim}{\Sigma} \in \underset{\sim}{S}_h(\Omega), \Rightarrow \underset{\sim}{U} \neq \underset{\sim}{0}$$

Thus, there is no strain energy produced by the stresses $\underset{\sim}{\Sigma}$ in each element due to the gradients ∇U. Therefore, ∇U must then represent a rigid rotation, and the collection of elements behaves as a mechanism. Similar physical interpretations apply to the hybrid models.

(ii) Conditions for existence of solutions of the mixed formulation are fairly easy to fulfill by reasonable choices of the approximating polynomials, and a significant increase in the accuracy of stress approximations over the convention displacement models can be obtained. For triangular elements on which $\underset{\sim}{U}$ and $\underset{\sim}{\Sigma}$ are complete polynomials of degree k and r, respectively, our L_2-estimates for displacements indicate a convergence rate of 1 order higher than the H^1 estimates as expected.

(iii) The hybrid models are more delicate than the mixed. However, the order-of-accuracy is the same as that of a comparable displacement model obtained using polynomials of the same degree. This accuracy is obtained despite the fact that the displacements are discontinuous across interelement boundaries, interelement continuity requirements being treated as constraints in hybrid approximations.

Acknowledgement. We gratefully acknowledge the support of this work by the U.S. Air Force Office of Scientific Research under Grant No. 74-2660.

REFERENCES

1. Prager, W. and Synge, J. L., "Approximations in Elasticity Based on the Concept of Function Space," Q. Appl. Math., Vol. 5, No. 3, pp. 241-269, 1943.

2. Herrmann, L. R., "A Bending Analysis of Plates," Proc. Conf. Matrix Meth. Struct. Mech., Wright-Patterson AFB, Ohio, AFFDL-TR66-80, 1965.

3. Reissner, E., "On a Variational Theorem in Elasticity," J. Math. Phys., Vol. 29, pp. 90-95, 1950.

4. Pian, T. H. H., and Tong, P., "Basis of Finite Element Methods for Solid Continua," Int. J. Num. Meth. in Eng., Vol. 1, No. 1, pp. 3-28, 1969.

5. Oden, J. T. and Reddy, J. N., Variational Methods in Theoretical Mechanics. Springer-Verlag, Heidelberg, Berlin, and New York, (to appear).

6. Oden, J. T. and Reddy, J. N., "On Mixed Finite Element Approximations," SIAM J. Num. Anal., (to appear).

7. Reddy, J. N. and Oden, J. T., "Mathematical Theory of Mixed Finite Element Approximations," Q. Appl. Math. (to appear).

8. Ciarlet, P. G. and Raviart, P. A., "Mixed Finite Element Method for the Biharmonic Equation," Mathematical Aspects of Finite Elements in Partial Differential Equations, Ed. by C. deBoor, Academic Press, N. Y., pp. 125-145, 1947.

9. Johnson, C., "On the Convergence of a Mixed Finite-Element Method for Plate Bending Problems," Num. Math., Vol. 21, pp. 43-62, 1973.

10. Raviart, P. A., "Hybrid Finite Element Methods for Solving 2nd Order Elliptic Equations," Report ____, Universite Paris VI et Centre National de la Recherche Scientifique, 1974.

11. Thomas, J. M., "Methodes des Elements Finis Hybrides Duaux Pour les Problems Elliptiques du Second-Order," Report 189, Universite Paris VI et Centre National de la Recherche Scientifique, 1975.

12. Babuska, I., Oden, J. T., and Lee, J. K., "Mixed-Hybrid Finite-Element Approximations of Second-Order Elliptic Boundary Value Problems," (to appear).

13. Babuska, I., "Error Bounds for Finite Element Method," Num. Math., Vol. 16, pp. 322-333, 1971.

14. Babuska, I. and Aziz, A. K., "Survey Lectures on the Mathematical Theory of Finite Elements," The Mathematical Foundations of the Finite Element Method with Applications to Partial Differential Equations, Ed. by A. K. Aziz, Academic Press, N. Y., pp. 3-395, 1972.

15. Oden, J. T. and Reddy, J. N., An Introduction to the Mathematical Theory of Finite Elements, Wiley Interscience, New York, (to appear).

16. Hlavacek, I. and Necas, J., "On Inequalities of Korn's Type," Arch. Rat. Mech. Anal., Vol. 36, pp. 305-334, 1970.

17. Nitsche, J., "Ein Kriterium fur die Quasi-Optimalitat des Ritzchen Verfahrens," Num. Math., Vol. 11, pp. 346-348, 1968.

18. Kondratev, V. A., "Boundary Problems for Elliptic Equations with Conical or Angular Points," Trans. Moscow Math. Soc. 1967, pp. 227-313.

19. Babuska, I., "The Finite Element Method with Lagrange Multipliers," Num. Math., Vol. 20, pp. 172-192, 1973.

20. Lee, J. K., "A-Priori Error Estimates of Mixed and Hybrid Finite Element Methods," Ph.D. Dissertation, Div. of Engr. Mech., Univ. of Texas at Austin (forthcoming).

Perturbation of domains in elliptic boundary value problems

By Friedrich Stummel

In a series of papers, the functional analysis for the treatment of general approximations of linear operators and an associated perturbation theory for elliptic sesquilinear forms has been developed (cf. Stummel [9], [1o]). Within this framework, there has further, in [11], been established a perturbation theory for Sobolew spaces $W^{m,p}(G)$ under perturbation of the domain of definition G. The present paper deals with the application of these theories to elliptic boundary value problems under perturbation of domains. Without difficulties, our methods are applied to additional approximations by subspaces and thus, for example, well-known finite element methods.

The first section of this paper gives a brief account of basic concepts and results of functional analysis for the treatment of approximations of inhomogeneous equations and eigenvalue problems defined by sesquilinear forms. Note, that we use fundamental concepts of numerical analysis, in particular, (bi-)stability and consistency in order to establish (bi-)convergence and the validity of associated two-sided error estimates for the solutions of inhomogeneous equations. Important new results are the two fundamental equivalence theorems in Subsection 1.2. The concepts and results of Section 1 may be applied to projection methods, conforming and nonconforming finite element methods, penalty methods and singular perturbations as well as to perturbations of domains and boundary conditions in elliptic boundary value problems (cf. Stummel [12]).

For the sake of brevity, we have to limit the applications in this paper to simple model examples of second order elliptic boundary value problems, defined by the Laplace operator under Dirichlet, Neumann, Robin and Steklov boundary conditions. The well-known inhomogeneous and eigenvalue problems for the vibrating membrane are of the type considered here. The present paper states very general conditions ensuring the continuous dependence of the solutions of these problems upon perturbations or approximations of the given data, the domains of definition and additional approximations by subspaces. Moreover, we show that the resolvent sets and spectra, the eigenvalues and associated eigenspaces depend continuously upon these perturbations. Our methods and results considerably generalize the corresponding previous ones of Babuška in [2], [3], Babuška-Výborný [4], and Nečas in [8]. In addition, our approach clarifies the basic concepts and methods in the treatment of perturbation of domains and enables the application of functional analysis to this class of problems.

Second order elliptic boundary value problems with variable coefficients can be treated in a similar way as the model examples in Section 2. The same is true for higher order elliptic boundary value problems under Dirichlet or homogeneous Neumann boundary conditions. However, note that higher order problems, in general, require convergence conditions on the boundaries, using higher orders of regularity.

1. Basic concepts and results of functional analysis

This section presents the basic definitions and theorems for the study of an important class of approximations of inhomogeneous equations and eigenvalue problems defined by sesquilinear forms. These approximations are specified by a sequence of continuously embedded subspaces E_ι, $\iota = 0,1,2,\ldots$ of a Hilbert space E. The first subsection explains the strong and weak discrete convergence both for sequences of elements in E_ι and for sequences of continuous linear functionals on E_ι, and establishes their basic properties. The second subsection states two fundamental equivalence theorems for the biconvergence of the solutions of sequences of inhomogeneous equations defined in E_ι. The first theorem equivalently characterizes biconvergence by bistability and consistency. A much stronger result is obtained by the second equivalence theorem for uniformly strongly coercive sequences of sequilinear forms. These two theorems and the associated two-sided error estimates have not been stated previously. As we have shown in [12], these theorems follow easily from the corresponding convergence theorems in [9-I], [10]. The third subsection briefly collects basic definitions and theorems concerning the convergence of solutions of resolvent equations, of resolvent sets and spectra, of eigenvalues and associated eigenspaces. These results have already been established in [9-II], [10].

1.1. Preliminaries

Let E be a real or complex Hilbert space with scalar product $(.,.)_E$. Let E_ι be linear subspaces of E which are Hilbert spaces with scalar products $(.,.)_{E_\iota}$ for $\iota = 0,1,2,\ldots$. We shall assume in the following that the associated norms satisfy the conditions

(1) $\quad \|u\|_E = \|u\|_{E_0}, \quad u \in E_0, \qquad \|u\|_E \leq \|u\|_{E_\iota}, \quad u \in E_\iota, \ \iota = 1,2,\ldots$.

Let J_ι be the natural embedding of E_ι into E. In view of (1), J_ι is a continuous linear operator from E_ι to E having norm $\|J_\iota\| \leq 1$. The associated adjoint operator, denoted by $R_\iota = J_\iota^*: E \rightarrow E_\iota$, is defined by the relation

(2) $\quad (J_\iota u,v)_E = (u,R_\iota v)_{E_\iota}, \quad u \in E_\iota, \ v \in E, \ \iota = 0,1,2,\ldots$.

For simplicity of notation, we shall omit writing J_ι in the following. The operators R_ι may be viewed as a sequence of restriction operators (cf. Aubin [1], Stummel [9]). By (1), we have $(.,.)_E = (.,.)_{E_0}$ on E_0 such that R_0 is the orthogonal projection P_0

of E onto E_0 and thus

(3) $$R_0 u = P_0 u_0 = u, \quad u \in E_0.$$

In general, however, the operators R_ι are not projections. Using (1) and the representation (2) for $u = R_\iota v$, one obtains the inequalities

(4) $$\|R_\iota v\|_E \leq \|R_\iota v\|_{E_\iota} \leq \|v\|_E, \quad v \in E, \; \iota = 0,1,2,\ldots .$$

By means of (2), the following, <u>basic identity</u> is easily verified,

(5) $$\|u_\iota - u\|_E^2 + \delta_\iota(u_\iota)^2 = \|u_\iota - R_\iota u\|_{E_\iota}^2 + \rho_\iota(u)^2$$

where the functionals δ_ι, ρ_ι are defined by

(6) $$\delta_\iota(u_\iota)^2 = \|u_\iota\|_{E_\iota}^2 - \|u_\iota\|_E^2, \quad u_\iota \in E_\iota,$$
$$\rho_\iota(u)^2 = \|u\|_E^2 - \|R_\iota u\|_{E_\iota}^2, \quad u \in E,$$

for all $\iota = 0,1,2,\ldots$. By virtue of (1) and (4), we have $\delta_\iota(u_\iota)^2 \geq 0$ and $\rho_\iota(u)^2 \geq 0$ so that the positive square roots $\delta_\iota(u_\iota)$, $\rho_\iota(u)$ are well-defined by (6). The above identity for $u_\iota = R_\iota u$ immediately entails the representation

(7) $$\rho_\iota(u)^2 = \|u - R_\iota u\|_E^2 + \delta_\iota(R_\iota u)^2 = \min_{\varphi_\iota \in E_\iota} \{\|u - \varphi_\iota\|_E^2 + \delta_\iota(\varphi_\iota)^2\}, \; u \in E, \; \iota = 0,1,2,\ldots$$

Note that

(8) $$\delta_0(u) = 0, \; u \in E_0, \quad \rho_0(u) = \|u - P_0 u\|_E = |u, E_0|, \quad u \in E.$$

In our perturbation theory, the space E_0 will be the given or unperturbed space and E_ι, $\iota = 1,2,\ldots$, the sequence of approximating or perturbed spaces. An essential tool in this theory is the notion of discrete convergence. For infinite subsequences $\mathbb{N}' \subset \mathbb{N} = (1,2,3,\ldots)$ and elements $u \in E$, $u_\iota \in E_\iota$, $\iota \in \mathbb{N}'$, the <u>strong discrete convergence</u> s-lim is defined by

$$\text{s-lim}\, u_\iota = u \iff u_\iota \underset{E}{\rightarrow} u, \; \|u_\iota\|_{E_\iota} \rightarrow \|u\|_E,$$

and the <u>weak discrete convergence</u> w-lim by

$$\text{w-lim}\, u_\iota = u \iff u_\iota \underset{E}{\rightharpoonup} u, \quad \|u_\iota\|_{E_\iota} \text{ is bounded,}$$

for $\iota \rightarrow \infty$, $\iota \in \mathbb{N}'$. From the identity (5) it follows immediately that

(9) $$\text{s-lim}\, u_\iota = u \iff \|u_\iota - u\|_E \rightarrow 0, \; \delta_\iota(u_\iota) \rightarrow 0$$
$$\iff \|u_\iota - R_\iota u\|_{E_\iota} \rightarrow 0, \; \rho_\iota(u) \rightarrow 0$$

for $\iota \rightarrow \infty$, $\iota \in \mathbb{N}'$. By virtue of (7), obviously, the statement

(1o) $$\rho_\iota(u) \rightarrow 0 \iff \text{s-lim}\, R_\iota u = u \quad (\iota \rightarrow \infty)$$

is true.

The strong and weak discrete convergence are restrictions of the strong and weak convergence in E. Hence limits of discretely convergent sequences are uniquely defined. Evidently, the weak discrete convergence is linear. Using the second equivalent characterization in (9), one easily proves the relation

$$\text{s-lim } u_\iota = u,\ \text{s-lim } v_\iota = v \Longrightarrow \lim\, (u_\iota, v_\iota)_{E_\iota} = (u,v)_{E},$$

whence it follows that also the strong discrete convergence is linear.

Bounded linear functionals $l \in E'$, $l_\iota \in E_\iota'$ can always be represented by uniquely determined elements $v \in E$, $v_\iota \in E_\iota$ in the form

(11) $\quad l(\varphi) = (\varphi, v)_E,\ \varphi \in E, \quad l_\iota(\varphi) = (\varphi, v_\iota)_{E_\iota},\ \ \varphi \in E_\iota,$

for $\iota = o,1,2,\ldots$. This leads to the following definition of the <u>strong discrete convergence</u> s-lim for sequences of functionals $l_\iota \in E_\iota'$, $\iota \in \mathbb{N}' \subset \mathbb{N}$:

$$\text{s-lim } l_\iota = l_o \Longleftrightarrow \text{s-lim } v_\iota = v_o.$$

The functional l is an <u>extension</u> of l_o if the <u>restriction</u> $l|E_o$ or l_{E_ι} of l to E_o is equal to l_o, that is $l(\varphi_o) = l_o(\varphi_o)$ for all $\varphi_o \in E_o$. Denote by $\hat{l}_o \in E'$ the continuous linear extension of l_o defined by

(12) $\quad \hat{l}_o(\varphi) = (\varphi, v_o)_E = (P_o\varphi, v_o)_E, \quad \varphi \in E.$

By means of the representation

(13) $\quad \| l_\iota - l|E_\iota \|_{E_\iota'} = \sup\limits_{o \neq \varphi \in E_\iota} |(\varphi, v_\iota)_{E_\iota} - (\varphi, R_\iota v)_{E_\iota}| / \|\varphi\|_{E_\iota} = \|v_\iota - R_\iota v\|_{E_\iota}, \quad \iota = 1,2,\ldots,$

and by (9), we then obtain the equivalent characterization of the above discrete convergence by the <u>discretely uniform convergence</u>

(14) $\quad \text{s-lim } l_\iota = l_o \Longleftrightarrow \| l_\iota - \hat{l}_o|E_\iota \|_{E_\iota'} \to o, \quad \rho_\iota(v_o) \to o,$

for $\iota \to \infty$, $\iota \in \mathbb{N}'$.

We shall need the following basic lemma.

(15) <u>If, and only if,</u> $R_\iota \to P_o$ <u>in</u> E $(\iota \to \infty)$, <u>it follows that</u> s-lim $l_{E_\iota} = l_{E_o}$ <u>for each functional</u> $l \in E'$.

<u>Proof.</u> Each $l \in E'$ has the representation (11). From (2) it is seen that $l_{E_\iota}(\varphi) = (\varphi, R_\iota v)_{E_\iota}$ for all $\varphi \in E_\iota$ and $\iota = o,1,2,\ldots$. Hence s-lim $l_{E_\iota} = l_{E_o}$ if and only if s-lim $R_\iota v = P_o v$ for each $v \in E$. The last condition implies $R_\iota v \underset{E}{\longrightarrow} P_o v$, that is, $R_\iota \to P_o$ in E for $\iota \to \infty$. Conversely, under this assumption and using (2), one obtains that also

$$\|R_\iota v\|^2_{E_\iota} = (R_\iota v,v)_E \to (P_o v,v)_E = \|P_o v\|^2_E$$

for $\iota \to \infty$, whence s-lim $R_\iota v = P_o v$. ■

An obvious conclusion from the above is the important theorem.

(16) If $R_\iota \to P_o$ in E $(\iota \to \infty)$, the equivalent characterization

$$\text{s-lim } l_\iota = l_o \iff \lim \| l_\iota - l_{E_\iota} \|_{E'_\iota} = o$$

is true for every sequence $l_\iota \in E'_\iota$, $\iota = o,1,2,\ldots$, and any extension l of l_o such that $l|E_o = l_o$.

The discrete convergence of sequences of functionals can also be characterized equivalently by the concept of continuous convergence:

(17) If $R_\iota \to P_o$ in $E(\iota \to \infty)$, the condition s-lim $l_\iota = l_o$ holds if and only if

$$\text{w-lim } \varphi_\iota = \varphi_o \implies \lim l_\iota(\varphi_\iota) = l_o(\varphi_o)$$

for every weakly discretely convergent sequence of elements $\varphi_\iota \in E_\iota$, $\iota = o,1,2,\ldots$.

This theorem has been proved in [10], 1.(12), using the representation (11), and in [11], 1.3.(3), for reflexive Banach spaces.

1.2. Inhomogeneous equations

Two sequences of continuous sesquilinear forms a_ι and continuous linear functionals l_ι on E_ι for $\iota = o,1,2,\ldots$ define a sequence of inhomogeneous equations

(1) $\qquad a_\iota(\varphi,u_\iota) = l_\iota(\varphi), \quad \varphi \in E_\iota.$

This equation for $\iota = o$ is viewed as the given or unperturbed problem and the equations (1) for $\iota = 1,2,\ldots$ represent the sequence of approximating or perturbed problems. The sesquilinear form a_ι is said to be injective (surjective) if the associated inhomogeneou equation has at most (at least) one solution u_ι in E_ι for each l_ι in E'_ι, the space of continuous linear functionals on E_ι. The sesquilinear form a_ι is bijective if and only if a_ι is both injective and surjective. Problem (1) is said to be properly posed if and only if a_ι is bijective and bicontinuous, that means, the above inhomogeneous equation is uniquely and bicontinuously solvable for all right hand sides $l_\iota \in E'_\iota$.

We next introduce the concept of biconvergence for sequences of sesquilinear forms a_ι on E_ι, $\iota = o,1,2,\ldots$. Note that, for each $u_\iota \in E_\iota$, the sesquilinear form a_ι defines a continuous linear functional $a_\iota(\cdot,u_\iota)$ on E_ι. The sequence (a_ι) is said to be

<u>biconvergent</u> to a_o at the point u_o iff there exists at least some sequence $u_\iota^o \in E_\iota$ such that $s\text{-}\lim u_\iota^o = u_o$ and the <u>biconvergence relation</u>

$$s\text{-}\lim u_\iota = u_o \iff s\text{-}\lim a_\iota(\cdot,u_\iota) = a_o(\cdot,u_o)$$

is true for all $u_\iota \in E_\iota$, $\iota = 1,2,\dots$. To illustrate this concept, let us assume that the sesquilinear forms a_ι are bijective for $\iota = 1,2,\dots$ so that the inhomogeneous equations (1) have uniquely determined solutions u_ι for all right hand sides 1_ι . In the case that (a_ι) is biconvergent to a_o at u_o, the solutions u_ι of these equations for $\iota = 1,2,\dots$ then converge discretely to the solution u_o of (1) for $\iota = o$ if and only if the associated inhomogeneous terms 1_ι converge discretely to 1_o for $\iota \longrightarrow \infty$, that is,

$$s\text{-}\lim u_\iota = u_o \iff s\text{-}\lim 1_\iota = 1_o.$$

In order to establish the basic biconvergence theorems, the concepts of bistability and consistency are needed. The continuous linear functionals $a_\iota(\cdot,u_\iota)$ on E_ι have the norms

$$\|a_\iota(\cdot,u_\iota)\|_{E_\iota'} = \sup_{o \neq \varphi \in E_\iota} |a_\iota(\varphi,u_\iota)| / \|\varphi\|_{E_\iota}$$

for $u_\iota \in E_\iota$, $\iota = o,1,2,\dots$. The sequence (a_ι) is <u>stable</u> iff (a_ι) is uniformly bounded, that is, there exists a positive number α_1 with the property

$$\|a_\iota(\cdot,u_\iota)\|_{E_\iota'} \le \alpha_1 \|u_\iota\|_{E_\iota}, \quad u_\iota \in E_\iota,$$

uniformly for all $\iota = 1,2,\dots$. Analogously, the sequence (a_ι) is said to be <u>inversely stable</u> iff there exists a positive number α_o such that the inequalities

$$\alpha_o \|u_\iota\|_{E_\iota} \le \|a_\iota(\cdot,u_\iota)\|_{E_\iota'}, \quad u_\iota \in E_\iota,$$

hold uniformly for all $\iota = 1,2,\dots$. Finally, we call the sequence (a_ι) <u>bistable</u> iff (a_ι) is both stable and inversely stable, that is, there exist positive numbers α_o, α_1 such that the <u>bistability condition</u>

$$\alpha_o \|u_\iota\|_{E_\iota} \le \|a_\iota(\cdot,u_\iota)\|_{E_\iota'} \le \alpha_1 \|u_\iota\|_{E_\iota}, \quad u_\iota \in E_\iota,$$

is valid for all $\iota = 1,2,\dots$.

For instance, consider a sequence of sesquilinear forms a_ι on E_ι satisfying the inequalities

(2) $$\alpha_o \|u_\iota\|^2_{E_\iota} \le |a_\iota(u_\iota)| \le \alpha_1 \|u_\iota\|^2_{E_\iota}, \quad u_\iota \in E_\iota,$$

uniformly for all $\iota = 1,2,\dots$, with some positive constants α_o, α_1. To simplify notation, we write $a_\iota(u_\iota)$ instead of $a_\iota(u_\iota,u_\iota)$. It is readily seen from the well-known theorem of Lax-Milgram that these sesquilinear forms a_ι are bijective and bicontinuous for each ι. Moreover, this sequence (a_ι) is bistable as one easily shows.

A sequence (a_ι) is said to be <u>consistent</u> at $u_o \in E_o$ iff there exists a discretely convergent sequence $u_\iota^o \in E_\iota$, $\iota = 1,2,\ldots$, with the property

$$\text{s-lim}\, u_\iota^o = u_o, \quad \text{s-lim}\, a_\iota(\cdot, u_\iota^o) = a_o(\cdot, u_o).$$

Using 1.1.(7), (9), one immediately obtains the statement that a stable sequence (a_ι) is consistent at $u_o \in E_o$ if and only if

(3) $$\text{s-lim}\, R_\iota u_o = u_o, \quad \text{s-lim}\, a_\iota(\cdot, R_\iota u_o) = a_o(\cdot, u_o).$$

In this case, we define the associated <u>discretization error</u> of the sequence of inhomogeneous equations (1) at u_o by

$$d_\iota = l_\iota - a_\iota(\cdot, R_\iota u_o), \quad \iota = 1,2,\ldots .$$

Having made these preparations, we are in a position to state the first fundamental equivalence theorem:

(4) <u>Given a sequence of bijective, continuous sesquilinear forms</u> a_ι <u>on</u> E_ι, $\iota = 1,2,\ldots$, <u>bistability and consistency of the sequence</u> (a_ι) <u>at</u> u_o <u>is the necessary and sufficient condition for biconvergence of</u> (a_ι) <u>to</u> a_o <u>at</u> u_o. <u>In both cases,</u>

(i) $$\lim \rho_\iota(u_o) = o, \quad \text{s-lim}\, a_\iota(\cdot, R_\iota u_o) = a_o(\cdot, u_o),$$

<u>and there exist positive constants</u> α_o, α_1 <u>such that for all</u> $l_\iota \in E_\iota'$ <u>and the associated solutions</u> u_ι <u>of</u> (1) <u>the two-sided discretization error estimates</u>

(ii) $$\frac{1}{\alpha_1^2}\|d_\iota\|_{E_\iota'}^2 + \rho_\iota(u_o)^2 \le \|u_\iota - u_o\|_E^2 + \delta_\iota(u_\iota)^2 \le \frac{1}{\alpha_o^2}\|d_\iota\|_{E_\iota'}^2 + \rho_\iota(u_o)^2$$

<u>hold uniformly for all</u> $\iota = 1,2,\ldots$, <u>where</u> $d_\iota = l_\iota - a_\iota(\cdot, R_\iota u_o)$

The above biconvergence theorem can be improved essentially in the case of uniformly strongly coercive sequences of sesquilinear forms. A sesquilinear form a_ι on E_ι is called <u>strongly coercive</u> iff there exist a positive constant γ_ι and a compact sesquilinear form k_ι on E_ι such that

$$\text{Re}\, a_\iota(\varphi) \ge \gamma_\iota \|\varphi\|_{E_\iota}^2 - \text{Re}\, k_\iota(\varphi), \quad \varphi \in E_\iota.$$

The sequence (a_ι) is said to be <u>uniformly strongly coercive</u> iff there exist a positive constant γ and a weakly collectively compact sequence of sesquilinear forms k_ι on E_ι such that the <u>coerciveness inequalities</u>

$$\mathrm{Re}\ a_\iota(\varphi) \geq \gamma \|\varphi\|^2_{E_\iota} - \mathrm{Re}\ k_\iota(\varphi), \quad \varphi \in E_\iota,$$

hold uniformly for all $\iota = 0,1,2,\ldots$. When $R_\iota \to P_0$ in E for $\iota \to \infty$, the sequence (k_ι) is weakly collectively compact iff k_ι is compact for each $\iota = 0,1,2,\ldots$ and the convergence relation

$$\text{w-lim}\ v_\iota = 0 \implies \lim \|k_\iota(v_\iota,\cdot)\|_{E'_\iota} = 0$$

is valid for every sequence $v_\iota \in E_\iota$, $\iota = 1,2,\ldots$.

Using these concepts, we can now formulate the second equivalence theorem, concerning the biconvergence of uniformly strongly coercive sequences of sesquilinear forms.

(5) Let s-lim $R_\iota u_0 = u_0$ for each $u_0 \in E_0$, let (a_ι) be a stable, uniformly strongly coercive sequence of continuous sesquilinear forms and let a_0 be injective. Then the given problem (1) is properly posed, and consistency of the sequence (a_ι) at all points $u_0 \in E_0$ is the necessary and sufficient condition for the validity of the following three statements:

(i) almost all of the approximating problems are properly posed;

(ii) the sequence (a_ι) is biconvergent to a_0 at E_0, that is,

$$\text{s-lim}\ u_\iota = u_0 \iff \text{s-lim}\ l_\iota = l_0$$

for all $u_\iota \in E_\iota$, $a_\iota(\cdot,u_\iota) = l_\iota$ and $\iota = 0,1,2,\ldots$;

(iii) there exist positive constants α_0, α_1, ν such that the associated two-sided discretization error estimates (4ii) hold for all $\iota \geq \nu$.

1.3. Resolvent equations and eigenvalue problems

Given a complex Hilbert space E and two sequences of sesquilinear forms a_ι, b_ι on E_ι for $\iota = 0,1,2,\ldots$, we now consider both the resolvent equations

(1) $$a_\iota(\varphi,u_\iota) - \overline{\zeta}_\iota b_\iota(\varphi,w_\iota) = l_\iota(\varphi_\iota), \quad \varphi \in E_\iota,$$

where l_ι are continuous linear functionals on E_ι and ζ_ι are complex numbers, and the associated eigenvalue problems

(2) $$a_\iota(\varphi,w_\iota) = \overline{\lambda}_\iota b_\iota(\varphi, w_\iota), \quad \varphi \in E_\iota.$$

The resolvent set $\rho(a_\iota,b_\iota)$ of a_ι,b_ι is the set of all those complex numbers ζ_ι for which problem (1) is properly posed. The spectrum $\Sigma(a_\iota,b_\iota)$ of a_ι,b_ι is the complement of $\rho(a_\iota,b_\iota)$ in the complex plane. The spectrum $\Sigma(a_\iota,b_\iota)$ is said to be discrete iff it consists of a countable set of eigenvalues with finite multiplicities, having no finite accumulation point in $\mathbb{C}$. The following theorem establishes conditions such that the resolvent sets $\rho(a_\iota,b_\iota)$ converge with respect to the open limit <u>Lim</u> to $\rho(a_0,b_0)$ and the spectra $\Sigma(a_\iota,b_\iota)$ converge with respect to the closed limit $\overline{\text{Lim}}$ to $\Sigma(a_0,b_0)$ in $\mathbb{C}$. These convergence concepts have been defined by Hausdorff in [7], p. 236. Finally, the pair a_ι,b_ι is said to be <u>strongly definite</u> iff there exist real numbers α_ι,β_ι such that

$$\alpha_\iota \operatorname{Re} a_\iota(\varphi) + \beta_\iota \operatorname{Re} b_\iota(\varphi) > 0, \quad o \neq \varphi \in E_\iota.$$

The following set of assumptions ensures the convergence of the solutions of the resolvent equations, of the resolvent sets and spectra of a_ι,b_ι, the eigenvalues and associated eigenspaces.

(EVP) <u>Let</u> s-lim $R_\iota u_0 = u_0$ <u>for each</u> $u_0 \in E_0$ <u>and</u> lim $\zeta_\iota = \zeta_o$. <u>Let the sequence</u> (a_ι) <u>be stable, consistent and uniformly strongly coercive. Let</u> (b_ι) <u>be stable, consistent and weakly collectively compact. Finally, let</u> a_ι,b_ι <u>be strongly definite for each</u> $\iota = o,1,2,\ldots$.

Under the above assumptions (EVP), the following statements can be proved concerning the solvability of the resolvent equations and the biconvergence of their solutions.

(3) <u>The inhomogeneous equations</u> (1) <u>satisfy the Fredholm alternative for all</u> $\zeta_\iota \in \mathbb{C}$ <u>and the spectra</u> $\Sigma(a_\iota,b_\iota)$ <u>are discrete for all</u> $\iota = o,1,2,\ldots$. <u>If</u> ζ_0 <u>is not an eigenvalue of</u> a_0,b_0, <u>problem</u> (1) <u>for</u> $\iota = o$ <u>and almost all of the approximating problems</u> (1) <u>for</u> $\iota = 1,2,\ldots$ <u>are properly posed, the sequence</u> $(a_\iota - \overline{\zeta}_\iota b_\iota)$ <u>is biconvergent to</u> $a_0 - \overline{\zeta}_0 b_0$ <u>and the two-sided discretization error estimates</u> 1.2.(4ii) <u>hold for almost all</u> ι.

The next theorem establishes the convergence of the resolvent sets and spectra of a_ι,b_ι, of the eigenvalues and associated eigenspaces of problem (2).

(4) <u>Under the assumption</u> (EVP) <u>the following statements are true:</u>

(i) <u>The resolvent sets and spectra of</u> a_ι,b_ι <u>converge in the sense</u>

$$\underline{\text{Lim}}\ \rho(a_\iota,b_\iota) = \rho(a_0,b_0), \quad \overline{\text{Lim}}\ \Sigma(a_\iota,b_\iota) = \Sigma(a_0,b_0).$$

(ii) <u>Let</u> λ_0 <u>be an eigenvalue of</u> a_0,b_0 <u>having the algebraic multiplicity</u> m, <u>let</u> $w^{(1)},\ldots,w^{(m)}$ <u>be a basis of the associated algebraic eigenspace and let</u> U <u>be any compact neighbourhood of</u> λ_0 <u>in</u> $\mathbb{C}$ <u>such that</u> $\{\lambda_0\} = \Sigma(a_0,b_0) \cap U$. <u>Then there are, for almost all</u> $\iota = 1,2,\ldots$, <u>exactly</u> m <u>eigenvalues</u> $\lambda_\iota^{(1)},\ldots,\lambda_\iota^{(m)}$, <u>counted repeatedly according</u>

to their algebraic multiplicities, and linearly independent vectors $w_\iota^{(1)},\dots,w_\iota^{(m)}$ in the sum of the algebraic eigenspace of $\lambda_\iota^{(1)},\dots,\lambda_\iota^{(m)}$ with the properties

$$\Sigma(a_\iota,b_\iota)\cap U = \{\lambda_\iota^{(1)},\dots,\lambda_\iota^{(m)}\}$$

and

$$\lim \lambda_\iota^{(k)} = \lambda_o, \quad \text{s-lim } w_\iota^{(k)} = w^{(k)}, \; k = 1,\dots,m.$$

Finally, let us consider Hermitian sesquilinear forms a_ι, b_ι on E_ι, $\iota = o,1,2,\dots$. In this case, the associated quadratic forms are real, all eigenvalues λ_ι of a_ι, b_ι in (2) are real and the associated algebraic eigenspaces coincide with the geometric eigenspaces, that is, the sets of all solutions w_ι of (2) or $(a_\iota - \lambda_\iota b_\iota)(\cdot, w_\iota) = o$. Under the assumption (EVP), the spectrum of a_o, b_o is discrete. Let μ be some real number in $\rho(a_o,b_o)$. In view of (4i), then $\mu \in \rho(a_\iota,b_\iota)$ for almost all $\iota = 1,2,\dots$. Hence, the eigenvalues of a_ι, b_ι can be arranged in ascending order

$$\dots \le \lambda_\iota^{(-2)} \le \lambda_\iota^{(-1)} < \mu \le \lambda_\iota^{(o)} \le \lambda_\iota^{(1)} \le \lambda_\iota^{(2)} \le \dots$$

for $\iota = o,1,2,\dots$, where each eigenvalue is counted repeatedly according to its multiplicity. Theorem (4) then yields the existence of eigenvalues $\lambda_\iota^{(t)}$ and the convergence of the ordered sequence of these eigenvalues

$$(5) \qquad \lambda_\iota^{(t)} \longrightarrow \lambda_o^{(t)} \quad (\iota \to \infty), \quad t = o, \pm 1, \pm 2, \dots,$$

as far as eigenvalues of a_o, b_o exist.

2. Elliptic boundary value problems of second order under perturbation of domains and approximation by subspaces

We are now in a position to prove that the solutions of inhomogeneous equations and of eigenvalue problems in elliptic boundary value problems depend continuously upon perturbations of the domain of definition. However, this can be shown here only for simple model examples of second order equations. We shall study the inhomogeneous equations $-(\Delta + \zeta_\iota)u_\iota = f_\iota$ in G_ι and the associated eigenvalue equations $-\Delta w_\iota = \lambda_\iota w_\iota$ in G_ι for $\iota = o,1,2,\dots$ under Dirichlet, Neumann, Robin and Steklov boundary conditions. Here $G = G_o$ denotes the given or unperturbed domain and $G_1, G_2, \dots$ a sequence of approximating or perturbed domains of definition. We shall easily be able to apply our methods, simultaneously, to both perturbation of domains and approximation by subspaces and thus to the important class of approximations by finite element methods.

The basic tools in the treatment of these problems are the concepts and results of functional analysis which have been stated in Section 1 as well as our perturbation theory for Sobolev spaces [11]. Subsection 2.1 introduces basic concepts of this perturbation theory. In particular, the natural embedding of the Sobolev spaces $H^m(G_\iota)$ in $L^{m,2}$ and the associated convergence concepts for sequences of functions $u_\iota \in H^m(G_\iota)$, $\iota = o,1,2,\ldots$, are explained. These convergence concepts generalize corresponding concepts of Babuška in [2], [3], Babuška-Výborný [4] and Nečas in [8]. An important role in the following study is played by collectively or discretely compact sequences of natural embeddings of Sobolev spaces under perturbation of domains. The first fundamental results, concerning this concept, have been established by Grigorieff in [6].

The following subsections state those assumptions on the above boundary value problems which ensure the continuous dependence upon perturbations of the inhomogeneous data, the domains of definition and approximations by subspaces. These assumptions first guarantee that the orthogonal projections of $L^{m,2}$ onto the subspaces E_ι converge, that the sequences of associated sesquilinear forms are stable, consistent and uniformly strongly coercive or weakly collectively compact, respectively, and that the sequences of continuous linear functionals, representing the inhomogeneous data, converge. Then the convergence theorems of Section 1 are applied to the boundary value problems considered here and thus the desired convergence results are obtained. For brevity's sake, we cannot, in this paper, explain the fine structure of the error bounds in the discretization error estimates or relate the order of convergence directly to the order of approximation of the boundaries (cf., for example, Blair [5]). Concerning such questions, we refer to our paper [12].

2.1. Subspaces of $L^{m,2}$

Denote by $L^{m,2}$ the Cartesian product $\prod_{|\sigma|\leq m} L^2(\mathbb{R}^n)_\sigma$ of spaces $L^2(\mathbb{R}^n)$ for all indices $\sigma = (\sigma_1,\ldots,\sigma_n)$, of order $|\sigma| = \sigma_1 + \ldots + \sigma_n \leq m$. That is, $L^{m,2}$ is the space of all square integrable functions on $\mathbb{R}^n$ with values $u(x) = (u^\sigma(x))_{|\sigma|\leq m}$ for $x \in \mathbb{R}^n$. This space is a Hilbert space with the scalar product

$$(u,v)_m = \sum_{|\sigma|\leq m} \int_{\mathbb{R}^n} u^\sigma(x)\overline{v^\sigma(x)}dx$$

for $u = (u^\sigma)$, $v = (v^\sigma) \in L^{m,2}$. Let $L_o^{m,2}(G)$ be the subspace of all functions $u \in L^{m,2}$ such that $u = o$ almost everywhere in $\mathbb{R}^n - G$, G being a measurable subset of $\mathbb{R}^n$.

Let us consider a sequence of closed subspaces $E_\iota \subset L^{m,2}$ for $\iota = o,1,2,\ldots$ and let $P_\iota : L^{m,2} \mapsto E_\iota$ be the orthogonal projections of $L^{m,2}$ onto E_ι. We shall use the notation $\lim E_\iota = E_o$ iff $P_\iota \to P_o$ in $L^{m,2}$ for $\iota \to \infty$. As an example, let $G = G_o, G_1, G_2, \ldots$ be a uniformly bounded sequence of measurable subsets in $\mathbb{R}^n$. Then

(1) $\quad \lim L_o^{m,2}(G_\iota) = L_o^{m,2}(G) \iff \lim \text{mes}\,(G \Delta G_\iota) = o,$

where $G \Delta G_\iota$ denotes the symmetric difference of G and G_ι (cf. [11], Theorem 2.1(3)).

Given an open subset $G \subset \mathbb{R}^n$, $H^m(G)$ denotes the well-known Sobolew space of all functions in $L^2(G)$ having generalized derivatives in $L^2(G)$ of all orders less or equal to m. $C_o^\infty(G)$ is the subspace of all test functions with compact support in G, and $H_o^m(G)$ denotes the closure of $C_o^\infty(G)$ in $H^m(G)$. The <u>natural embedding</u> J_G^m of $H^m(G)$ into $L^{m,2}$ is defined by

(2) $$(J_G^m u)(x) = ((D^\sigma u)(x))_{|\sigma| \leq m}, \; x \in G, \quad (J_G^m u)(x) = o, \; x \in \mathbb{R}^n - G,$$

for all $u \in H^m(G)$. Instead of $J_G^m u$, we shall also write $\underline{u}$. J_G^m maps the space $H^m(G)$ isomorphically and isometrically onto a closed subspace ${}_J H^m(G)$ of $L^{m,2}$. Evidently, ${}_J H^m(G)$ is a subspace of $L_o^{m,2}(G)$.

Let G, G_1, G_2,... be a sequence of open sets in $\mathbb{R}^n$. A sequence of functions $u_\iota \in H^m(G_\iota)$, $\iota \in \mathbb{N}' \subset \mathbb{N}$, is said to converge strongly to $u_o \in H^m(G)$ iff the sequence $\underline{u}_\iota = J_{G_\iota}^m u_\iota$ converges strongly to $\underline{u}_o = J_G^m u_o$ in $L^{m,2}$, that is,

$$\text{s-lim}\, u_\iota = u_o \iff \lim \|\underline{u}_\iota - \underline{u}_o\|_m = o.$$

The sequence $u_\iota \in H^m(G_\iota)$, $\iota \in \mathbb{N}' \subset \mathbb{N}$, is said to <u>converge weakly</u> to $u_o \in H^m(G)$ iff the sequence $\underline{u}_\iota$ converges weakly to $\underline{u}_o$ in $L^{m,2}$, that is,

$$\text{w-lim}\, u_\iota = u_o \iff \lim\, (\varphi, \underline{u}_\iota)_m = (\varphi, \underline{u}_o)_m$$

for all $\varphi \in L^{m,2}$. Note that the following representation holds

(3) $$\|\underline{u}_\iota - \underline{u}_o\|_m^2 = \|u_\iota - u_o\|_{m,G \cap G_\iota}^2 + \|u_\iota\|_{m,G_\iota - G}^2 + \|u_o\|_{m,G-G_\iota}^2$$

for $u_\iota \in H^m(G_\iota)$, $\underline{u}_\iota = J_{G_\iota}^m u_\iota$, $\iota = o,1,2,\ldots$, and $G = G_o$. In the case that $G_\iota \subset G$ and $\text{mes}\,(G - G_\iota) \to o$ for $\iota \to \infty$, our strong convergence is, due to (3), characterized by

$$\text{s-lim}\, u_\iota = u_o \iff \lim \|u_\iota - u_o\|_{m,G_\iota} = o.$$

This is Babuška's convergence concept in [2], [3] and Nečas' "convergence presque" in [8], Section 3.6.8. When $G \subset G_\iota$, $\iota = 1,2,\ldots$, we have

$$\text{s-lim}\, u_\iota = u_o \iff \lim \|u_\iota - u_o\|_{m,G} = o, \; \lim \|u_\iota\|_{m,G_\iota - G} = o.$$

Applied to sequences $u_\iota \in H_o^m(G_\iota)$, viewed as subspaces of $H^m(\mathbb{R}^n)$, we obtain

(4) $$\|\underline{u}_\iota - \underline{u}_o\|_m = \|u_\iota - u_o\|_m, \quad \iota = o,1,2,\ldots,$$

such that the strong convergence $\text{s-lim}\, u_\iota = u_o$ is the convergence $\lim u_\iota = u_o$ in $H^m(\mathbb{R}^n)$.

For simplicity of notation, we shall denote functions in $L^{1,2}$ by $(u^0,u^1,\dots,u^n)$. For each $u\in H^1(G)$, for example, thus $\underline{u} = J_G^1 u$ is specified by

$$\underline{u}(x) = (u, \frac{\partial u}{\partial x_1},\dots,\frac{\partial u}{\partial x_n})(x),\ x\in G, \quad \underline{u}(x) = (0,\dots,0),\ x\in \mathbb{R}^n - G.$$

Let the sesquilinear forms a,b on $L^{1,2}$ be defined by

$$(5) \qquad a(\varphi,\psi) = \sum_{k=1}^{n}\int_{\mathbb{R}^n}\varphi^k\overline{\psi}^k\,dx, \quad b(\varphi,\psi) = \int_{\mathbb{R}^n}\varphi^0\overline{\psi}^0\,dx$$

for $\varphi,\psi\in L^{1,2}$. The sesquilinear form b may be viewed as the restriction of the scalar product $(.,.)_0$ to $L^{1,2}$. Given a sequence of closed subspaces $E_\iota\subset L^{1,2}$, we denote by a_{E_ι}, $b_{E_\iota} = (.,.)_0|E_\iota$ the restrictions of the sesquilinear forms $a,b = (.,.)_0$ to E_ι. The following theorem, in particular, establishes the validity of our basic assumption (EVP) in Section 1.3.

(6) Let $\lim E_\iota = E_0$ and let the sequence $(.,.)_0|E_\iota$, $\iota = 0,1,2,\dots$, be weakly collectively compact. Then the sequences (a_{E_ι}), (b_{E_ι}), and $(a_{E_\iota} - \overline{\zeta}_\iota b_{E_\iota})$ for $\zeta_\iota\to\zeta_0(\iota\to\infty)$ are stable and consistent at each $u_0\in E_0$. The sequences (a_{E_ι}), $(a_{E_\iota} - \overline{\zeta}_\iota b_{E_\iota})$ are strongly coercive and the sequence (b_{E_ι}) is weakly collectively compact. Finally, a_{E_ι}, b_{E_ι} is strongly definite for each $\iota = 0,1,2,\dots$.

Proof. Evidently,

$$\|a_{E_\iota}\|\le 1,\ \|b_{E_\iota}\|\le 1,\ \|a_{E_\iota} - \overline{\zeta}_\iota b_{E_\iota}\|\le 1+|\zeta_\iota|, \quad \iota = 0,1,2,\dots,$$

so that these sequences of sesquilinear forms are uniformly bounded and thus stable. Under the above assumption, for each $u_0\in E_0$ there is the sequence $u_\iota = P_\iota u_0\in E_\iota$, $\iota = 1,2,\dots$, converging to u_0 in $L^{1,2}$ for $\iota\to\infty$. Using the estimate

$$|a(\varphi,u_\iota) - a(\varphi,u_0)|\le \|\varphi\|_1\,\|u_\iota - u_0\|_1, \quad \varphi\in L^{1,2},$$

and setting $1 = a(\cdot,u_0)\in(L^{1,2})'$, we obtain $1|E_0 = a_{E_0}(\cdot,u_0)$ and

$$\|a_{E_\iota}(\cdot,u_\iota) - 1|E_\iota\|_{E_\iota'}\to 0 \quad (\iota\to\infty).$$

Hence Theorem 1.1.(6) yields $\text{s-lim}\, a_{E_\iota}(\cdot,u_\iota) = a_{E_0}(\cdot,u_0)$. This proves the consistency of (a_{E_ι}). The consistency of the sequences (b_{E_ι}) and $(a_{E_\iota} - \overline{\zeta}_\iota b_{E_\iota})$ is proved correspondingly. Since $b_{E_\iota} = (.,.)_0|E_\iota$, the sequence (b_{E_ι}) is weakly collectively compact by virtue of our above assumption. Further, the sequences (a_{E_ι}) and $(a_{E_\iota} - \overline{\zeta}_\iota b_{E_\iota})$ are uniformly strongly coercive because

$$a_{E_\iota}(\varphi)\ge\|\varphi\|_1^2 - b_{E_\iota}(\varphi)$$

and

$$\operatorname{Re}(a_{E_\iota} - \overline{\zeta}_\iota b_{E_\iota})(\varphi) \geq \|\varphi\|_1^2 - (1 + \sup|\zeta_\iota|) b_{E_\iota}(\varphi)$$

for all $\varphi \in E_\iota$, $\iota = 0,1,2,\dots$. Finally, a_{E_ι}, b_{E_ι} is strongly definite for each ι, because $\alpha_\iota = \beta_\iota = 1$ yields

$$\alpha_\iota a_{E_\iota}(\varphi) + \beta_\iota b_{E_\iota}(\varphi) = \|\varphi\|_1^2 > 0, \quad 0 \neq \varphi \in E_\iota. \qquad \blacksquare$$

Two sequences of functions $\underline{f}_\iota \in L^{0,2}$, $\underline{g}_\iota \in L^{1,2}$, $\iota = 0,1,2,\dots$ specify a sequence of continuous linear functionals on subspaces E_ι of $L^{1,2}$ by

(7) $$l_\iota(\varphi) = (a - \overline{\zeta}_\iota b)(\varphi, \underline{g}_\iota) - b(\varphi, \underline{f}_\iota), \quad \varphi \in E_\iota .$$

Additionally, let the functional $l \in (L^{1,2})'$ be defined by the equation

$$l(\varphi) = (a - \overline{\zeta}_0 b)(\varphi, \underline{g}_0) - b(\varphi, \underline{f}_0), \quad \varphi \in L^{1,2}.$$

One easily proves the estimate

(8) $$\|l_\iota - l|E_\iota\|_{E_\iota'} \leq \|\underline{f}_\iota - \underline{f}_0\|_0 + |\zeta_\iota - \zeta_0|\,\|\underline{g}_0\|_1 + (1 + \sup|\zeta_\iota|)\|\underline{g}_\iota - \underline{g}_0\|_1,$$

for $\iota = 1,2,\dots$. By virtue of Theorem 1.1.(6), we thus obtain the result:

(9) <u>The convergence statement</u> $\text{s-lim}\, l_\iota = l_0$ <u>is true if</u> $\lim E_\iota = E_0$ <u>and</u>

$$\lim \zeta_\iota = \zeta_0, \quad \text{s-lim}\, f_\iota = f_0, \quad \text{s-lim}\, g_\iota = g_0.$$

<u>Example.</u> Let $f \in L^2$, $g \in H^1(\mathbb{R}^n)$ and choose $f_\iota = f|G_\iota$, $g_\iota = g|G_\iota$, $\iota = 0,1,2,\dots$. Then, by (3),

(10) $$\|\underline{f}_\iota - \underline{f}_0\|_0 = \|f\|_{0,G\Delta G_\iota}, \quad \|\underline{g}_\iota - \underline{g}_0\|_1 = \|g\|_{1,G\Delta G_\iota},$$

where $G\Delta G_\iota$ denotes the symmetric difference of G and G_ι. These terms tend to zero if $\lim \operatorname{mes}(G\Delta G_\iota) = 0$.

2.2. Dirichlet problems

As a first example, let us consider Dirichlet boundary value problems under perturbation of domains. The inhomogeneous equations have the form

(1) $$-(\Delta + \zeta_\iota)v_\iota = f_\iota \text{ in } G_\iota, \quad v_\iota = g_\iota \text{ in } \partial G_\iota,$$

and the associated eigenvalue problems are

(2) $$-\Delta w_\iota = \lambda_\iota w_\iota \text{ in } G_\iota, \quad w_\iota = 0 \text{ in } \partial G_\iota,$$

for $\iota = 0,1,2,\dots$. These problems have to be understood in the generalized sense defined by the equations

(1') $$\int_{G_\iota} \{\nabla\varphi_\iota \cdot \overline{\nabla v_\iota} - \overline{\zeta}_\iota \varphi_\iota \overline{v}_\iota\}\, dx = \int_{G_\iota} \varphi_\iota \overline{f}_\iota\, dx, \quad \varphi_\iota \in \overset{\circ}{H}{}^1(G_\iota);$$
$$g_\iota - v_\iota \in \overset{\circ}{H}{}^1(G_\iota),$$

for $\iota = 0,1,2,\ldots$ and functions $f_\iota \in L^2(G_\iota), g_\iota \in H^1(G_\iota)$. The associated generalized eigenvalue problems have the form

(2') $\int_{G_\iota} \nabla\varphi_\iota \cdot \overline{\nabla w_\iota}\,dx = \lambda_\iota \int_{G_\iota} \varphi_\iota \overline{w_\iota}\,dx, \quad \varphi_\iota \in H^1_o(G_\iota); \quad o \neq w_\iota \in H^1_o(G_\iota),$

for $\iota = 0,1,2,\ldots$. Here ∇u_ι denotes the vector

$$\nabla u_\iota = \left(\frac{\partial u_\iota}{\partial x_1}, \ldots, \frac{\partial u_\iota}{\partial x_n}\right), \quad u_\iota \in H^1(G_\iota).$$

The general theorems in Section 2 may be applied to the above problems. Let the sesquilinear forms a,b on $L^{1,2}$ be defined by

$$a(\varphi,\psi) = \sum_{k=1}^{n} \int_{\mathbb{R}^n} \varphi^k \overline{\psi}^k dx, \quad b(\varphi,\psi) = \int_{\mathbb{R}^n} \varphi^o \overline{\psi}^o dx$$

for $\varphi,\psi \in L^{1,2}$. Then choose

(3)
$$E = L^{1,2}, \qquad E_\iota = {}_JH^1_o(G_\iota),$$
$$a_\iota = a|_JH^1_o(G_\iota), \qquad b_\iota = b|_JH^1_o(G_\iota),$$
$$l_\iota(\varphi) = \int_{G_\iota} \left\{ \sum_{k=1}^{n} \varphi^k_\iota \frac{\partial g_\iota}{\partial x_k} - \overline{\zeta}_\iota \varphi^o_\iota \overline{g}_\iota - \varphi^o_\iota \overline{f}_\iota \right\} dx, \quad \varphi \in L^{1,2},$$

for $\iota = 0,1,2,\ldots$.

In finite element methods, the given problem on the space $H^1_o(G)$ is approximated by a sequence of approximating domains G_ι and subspaces of piecewise polynomial functions in $H^1_o(G_\iota)$. So, more generally than (3), let us consider a sequence of closed subspaces E_ι of $E = L^{1,2}$ and sesquilinear forms a_ι, b_ι, specified by

(4)
$$E_o = {}_JH^1_o(G), \qquad E_\iota \subset {}_JH^1_o(G_\iota)$$
$$a_\iota = a|E_\iota, \qquad b_\iota = b|E_\iota, \qquad \iota = 0,1,2,\ldots,$$

and let l_ι be defined as above in (3). In both cases (3) and (4), the sequence of inhomogeneous equations takes on the form

(1") $(a_\iota - \overline{\zeta}_\iota b_\iota)(\varphi, \underline{u}_\iota) = l_\iota(\varphi), \quad \varphi \in E_\iota; \quad \underline{u}_\iota = g_\iota - \underline{v}_\iota \in E_\iota;$

and the associated eigenvalue problems are

(2") $a_\iota(\varphi, \underline{w}_\iota) = \overline{\lambda}_\iota b_\iota(\varphi, \underline{w}_\iota), \quad \varphi \in E_\iota; \quad o \neq \underline{w}_\iota \in E_\iota,$

for $\iota = 0,1,2,\ldots$. Evidently, the problems (1') and (2') are the same as (1") and (2") for $E_\iota = {}_JH^1_o(G_\iota)$.

The next theorem is basic in the study of Dirichlet problems under perturbation of domains.

(5) <u>Let</u> $G = G_o, G_1, G_2, \ldots$ <u>be a uniformly bounded sequence of open subsets in</u> $\mathbb{R}^n$ <u>having the properties:</u>

(i) <u>For each compact subset</u> $K \subset G$,

$$\lim \operatorname{cap}_m(K - G_\iota) = o;$$

(ii) $\lim \operatorname{mes}(G_\iota - \overline{G}) = o;$

(iii) $\overline{\mathrm{Lim}}\sup(\partial G \cap G_\iota)$ <u>has the segment property.</u>

<u>Then</u> $\lim {}_J H_o^m(G_\iota) = {}_J H_o^m(G)$ <u>for each</u> $m \geq 1$ <u>and the sequence</u> $(.,.)_o | {}_J H_o^m(G_\iota)$, $\iota = o,1,2,...$ <u>is weakly collectively compact.</u>

<u>Proof.</u> We have proved the convergence of the orthogonal projections $P_\iota : L^{m,2} \to {}_J H_o^m(G_\iota)$, $\iota = o,1,2,...$ in [11], Theorem 2.3.(13). The weak collective compactness of the sesquilinear forms $b_{E_\iota} = (.,.)_o$ on $E_\iota = {}_J H_o^m(G_\iota)$, $\iota = o,1,2,...$, follows from the compactness of the natural embeddings of $H_o^m(G_\iota)$ into $L^2(G_\iota)$ and, using the inequality $\|b_{E_\iota}(\underline{v}_\iota,\cdot)\|_{E'_\iota} \leq \|v_\iota\|_o$ for $\underline{v}_\iota = J_{G_\iota}^m v_\iota \in {}_J H_o^m(G_\iota)$, from [11], Theorems 3.1.(1), (2) or Grigorieff [6], p. 77. ■

For definitions and properties of the concepts used here, we refer to Stummel [11]. Let S be any compact subset of $\mathbb{R}^n$ and let $C_1^\infty(S)$ be the set of all those functions $\psi \in C_o^\infty(\mathbb{R}^n)$ which are equal to 1 in some individual neighbourhood of S. By the <u>capacity</u> of S with respect to $H^m(\mathbb{R}^n)$ is meant (cf. Grigorieff [6])

$$\operatorname{cap}_m(S) = \inf_{\psi \in C_1^\infty(S)} \|\psi\|_m^2 .$$

The concepts of open and closed limits inferior and superior have been introduced by Hausdorff [7], VII - 5. It is easily seen, that

(6) $\quad G \subset \underline{\mathrm{Lim}} \inf G_\iota \Longrightarrow$ (5i),

because the first condition holds if and only if for each compact subset $K \subset G$ it follows that $K \subset G_\iota$ for almost all ι. In condition (5ii), mes denotes the Lebesgue measure in $\mathbb{R}^n$. Condition (5iii) is satisfied trivially, for instance, if $G_\iota \subset G$ and thus $G_\iota \cap \partial G = \emptyset$ for all ι, or if the boundary ∂G of G has the segment property.

Babuška-Výborný have studied in [4] the convergence of spectra and eigenspaces of certain generalized Dirichlet eigenvalue problems under perturbation of domains. In approximations of the domain from the interior, that is $G_\iota \subset G$, convergence is shown under the condition $G \subset \underline{\mathrm{Lim}} \inf G_\iota$, using our notation. In approximations of the domain from the exterior, one has $G \subset G_\iota$. In this case, the condition $\complement\overline{G} \subset \underline{\mathrm{Lim}} \inf \complement G_\iota$, being equivalent to $\overline{\mathrm{Lim}} \sup G_\iota \subset \overline{G}$, is assumed in [4] and, additionally, that $H_o^m(G)$ is stable.

So (5ii) generalizes the above convergence condition and (5iii) then ensures the stability.

For the simultaneous treatment of perturbation of domains and approximation by subspaces, under the assumptions of Theorem (5), one has the following obvious corollary to this theorem.

(7) Let $E_o = {}_J\mathring{H}^m(G)$ and let a sequence of closed subspaces $E_\iota \subset {}_J\mathring{H}^m(G_\iota)$, $\iota = 1,2,\dots$, be given such that for each $\varphi \in {}_J C_o^\infty(G)$ there exists a sequence $\varphi_\iota \in E_\iota$ with the property $\|\varphi_\iota - \varphi\|_m \to o$ for $\iota \to \infty$. Then $\lim E_\iota = {}_J\mathring{H}^m(G)$ and the sequence $(.,.)_o|E_\iota$ $\iota = o,1,2,\dots$, is weakly collectively compact.

Due to the above Theorems (5), (7) and Theorem 2.1.(6), the convergence theorems for the solutions of inhomogeneous equations and eigenvalue problems can be applied to the Dirichlet problems (1), (1') or (1") and (2), (2') or (2"). If $\lim \zeta_\iota = \zeta_o$ and $\zeta_\iota \notin]o,\infty[$ for $\iota = o,1,2,\dots$, the sequence $a_{E_\iota} - \overline{\zeta}_\iota b_{E_\iota}$ satisfies the inequalities 1.2.(2) for $\iota = o,1,2,\dots$. Hence these sequilinear forms are bijective, bicontinuous and the sequence $(a_{E_\iota} - \overline{\zeta}_\iota b_{E_\iota})$ is bistable and consistent. Consequently, Theorem 1.2.(4), applied to $a_\iota = a_{E_\iota} - \overline{\zeta}_\iota b_{E_\iota}$, yields the biconvergence of the solutions of (1") and the associated two-sided discretization error estimates.

In the case that ζ_o is not an eigenvalue of (2), (2') for $\iota = o$ and $\lim \zeta_\iota = \zeta_o$, all of the assumptions of Theorem 1.2.(5) are satisfied for $a_\iota = a_{E_\iota} - \overline{\zeta}_\iota b_{E_\iota}$. Thus there exists a natural number ν such that the Dirichlet problems (1") are properly posed for $\iota = o$ and all $\iota \geq \nu$. Moreover, the solutions of these problems satisfy the biconvergence relation

$$\text{s-lim } u_\iota = u_o \iff \text{s-lim } l_\iota = l_o .$$

In view of Theorem 2.1.(9), we have

$$\lim \zeta_\iota = \zeta_o, \ \text{s-lim } f_\iota = f_o, \quad \text{s-lim } g_\iota = g_o \implies \text{s-lim } u_\iota = u_o .$$

In particular, let $f \in L^2(\mathbb{R}^n)$, $g \in H^1(\mathbb{R}^n)$ and $f_\iota = f|G_\iota$, $g_\iota = g|G_\iota$ for $\iota = o,1,2,\dots$, and let mes $(G \Delta G_\iota) \to o$ for $\iota \to \infty$. From 2.1.(1o) it is seen that these inhomogeneous terms converge and hence the corresponding solutions u_ι of (1") and v_ι of (1), (1') converge to u_o and v_o, respectively. The associated two-sided discretization error estimates take on the form

(8) $$\frac{1}{\alpha_1}\|d_\iota\|^2_{E'_\iota} + \rho_\iota(\underline{u}_o)^2 \leq \|u_\iota - u_o\|_1^2 \leq \frac{1}{\alpha_o^2}\|d_\iota\|^2_{E'_\iota} + \rho_\iota(\underline{u}_o)^2, \quad \iota \geq \nu.$$

Note that here $\delta_\iota = o$ and 2.1.(4) is true. From 1.1.(7) it is seen that $\rho_\iota(\underline{u}_o)$ is the shortest distance of $\underline{u}_o$ to E_ι, that is,

(9) $$\rho_\iota(\underline{u}_o) = \|\underline{u}_o - P_\iota \underline{u}_o\|_1 = \min_{\varphi_\iota \in E_\iota} \|\underline{u}_o - \varphi_\iota\|_1 = |\underline{u}_o, E_\iota|$$

for $\iota = 1,2,\ldots$. This term tends to zero because $P_\iota \to P_o$ $(\iota \to \infty)$. The discretization error d_ι has the form $d_\iota = l_\iota - (a_{E_\iota} - \zeta_\iota b_{E_\iota})(\cdot, P_\iota \underline{u}_o)$. We know that the discretization error tends to zero whenever s-lim $l_\iota = l_o$, since s-lim $P_\iota \underline{u}_o = \underline{u}_o$ implies

$$\text{s-lim}\,(a_{E_\iota} - \overline{\zeta}_\iota b_{E_\iota})(\cdot, P_\iota \underline{u}_o) = (a_{E_o} - \overline{\zeta}_o b_{E_o})(\cdot, \underline{u}_o) = l_o$$

and hence s-lim $d_\iota = o$ or $\lim \|d_\iota\|_{E'_\iota} = o$.

Finally, under the assumptions of Theorem (5) or (7) and $\lim \zeta_\iota = \zeta_o$, it follows from Theorem 2.1.(6) that all of the assumptions (EVP) in Section 1.3 are valid for the resolvent equations (1") and the associated eigenvalue problems (2"). Hence Theorems 1.3.(3) and (4), in particular, yield the convergence of the resolvent sets and spectra, the eigenvalues and associated eigenspaces for the Dirichlet problems considered here. Note that a_{E_ι}, b_{E_ι} are Hermitian sesquilinear forms. Hence the eigenvalues are real and converge in order to the eigenvalues of (2), (2') for $\iota = o$, as we have stated in 1.3.(5).

2.3. Neumann Problems

The treatment of Neumann boundary value problems under perturbation of domains and approximation by subspaces is similar to that in the preceding section as far as homogeneous boundary conditions are concerned. Inhomogeneous boundary conditions will be studied in the next section. Consider the equations

(1) $$-(\Delta + \zeta_\iota)u_\iota = f_\iota \text{ in } G_\iota, \quad \frac{\partial u_\iota}{\partial N_\iota} = o \text{ in } \partial G_\iota,$$

and the associated eigenvalue problems

(2) $$-\Delta w_\iota = \lambda_\iota w_\iota \text{ in } G_\iota, \quad \frac{\partial w_\iota}{\partial N_\iota} = o \text{ in } \partial G_\iota,$$

for $\iota = o,1,2,\ldots$, where $\partial/\partial N_\iota$ denotes derivatives in the outward normal direction on ∂G_ι. The above inhomogeneous problems are defined by

(1') $$\int_{G_\iota} \{\nabla \varphi_\iota \cdot \overline{\nabla u_\iota} - \overline{\zeta}_\iota \varphi_\iota \overline{u}_\iota\} dx = \int_{G_\iota} \varphi_\iota \overline{f}_\iota dx, \quad \varphi_\iota \in H^1(G_\iota);\ u_\iota \in H^1(G_\iota),$$

for $\iota = 0,1,2,\ldots$, where $f_\iota \in L^2(G_\iota)$, and the associated generalized eigenvalue problems have the form

(2') $$\int_{G_\iota} \nabla\varphi \cdot \overline{\nabla w_\iota}\,dx = \lambda_\iota \int_{G_\iota} \varphi\overline{w_\iota}\,dx, \qquad \varphi_\iota \in H^1(G_\iota);\ 0 \neq w_\iota \in H^1(G_\iota), \qquad \iota = 0,1,2,\ldots .$$

Let the sesquilinear forms a,b on $L^{1,2}$ be defined as in Section 2.1, 2.2 and set

$$E = L^{1,2}, \qquad E_\iota = {}_JH^1(G_\iota),$$

(3) $$a_\iota = a|{}_JH^1(G_\iota), \qquad b_\iota = b|{}_JH^1(G_\iota),$$

$$l_\iota(\varphi) = \int_{G_\iota} \varphi_\iota^o \overline{f_\iota}\,dx, \qquad \varphi \in L^{1,2}, \qquad \iota = 0,1,2,\ldots .$$

In finite elements, one approximates $H^1(G)$ by finite dimensional subspaces of $H^1(G_\iota)$. In this case, we have a sequence of closed subspaces E_ι and sesquilinear forms a_ι, b_ι specified by

$$E_0 = {}_JH^1(G), \qquad E_\iota \subset {}_JH^1(G_\iota),$$

(4) $$a_\iota = a|E_\iota, \qquad b_\iota = b|E_\iota,$$

$$l_\iota(\varphi) = \int_{G_\iota} \varphi_\iota^o \overline{f_\iota}\,dx, \qquad \varphi \in E_\iota, \qquad \iota = 0,1,2,\ldots .$$

In both cases the inhomogeneous problems have the form 1.3.(1) and the associated eigenvalue problems the form 1.3.(2).

For the treatment of Neumann boundary value problems under perturbation of domains, the following basic theorem has been established in [11]:

(5) <u>Let</u> $G = G_0, G_1, G_2, \ldots$ <u>be a uniformly bounded sequence of open sets in</u> $\mathbb{R}^n$ <u>such that</u> mes $(\partial G) = 0$ <u>and</u>

(i) $\underline{\mathrm{Lim}}\inf G_\iota = G$, (ii) $\overline{\mathrm{Lim}}\sup G_\iota = \overline{G}$,

(iii) $G_0, G_1, G_2, \ldots$ <u>has the uniform segment property.</u>

<u>Then</u> $\lim {}_JH^m(G_\iota) = {}_JH^m(G)$ <u>and the sequence</u> $(.,.)_0|{}_JH^m(G_\iota)$, $\iota = 0,1,2,\ldots$, <u>is weakly collectively compact.</u>

<u>Proof.</u> The convergence of the orthogonal projections $P_\iota : L^{m,2} \mapsto {}_JH^m(G_\iota)$, $\iota = 0,1,2,\ldots$, has been shown in [11], Theorem 2.3.(14). The natural embeddings of $H^m(G_\iota)$ into $L^2(G_\iota)$ are compact so that the sesquilinear forms $b_{E_\iota} = (.,.)_0$ on $E_\iota = {}_JH^m(G_\iota)$ are compact for each $\iota = 0,1,2,\ldots$. Using the inequality

$$\|b_{E_\iota}(\underline{v}_\iota,\cdot)\|_{E_\iota'} \le \|v_\iota\|_0, \qquad \underline{v}_\iota = J^m_{G_\iota} v_\iota \in {}_JH^m(G_\iota),$$

for $\iota = 0,1,2,\ldots$, the weak collective compactness of (b_ι) is finally obtained from

[11], Theorems 3.1.(1), (8) or Grigorieff [6], p. 83.

A subset G in $\mathbb{R}^n$ is said to be an <u>open domain</u> iff $(\overline{G})^o = G$. When G is open, G is an open domain if and only if $(\overline{G})^o \subset G$. Hence an open set G in $\mathbb{R}^n$ is an open domain if and only if no point in the boundary of G is interior to the closure of G. It is easily seen that in this case the above conditions (5i), (5ii) have the following equivalent characterizations:

(6) (i) $G \subset \underline{\mathrm{Lim}}\inf G_\iota$, $\overline{\mathrm{Lim}}\sup G_\iota \subset \overline{G}$;

$\Leftrightarrow$ (ii) $\underline{\mathrm{Lim}}\inf G_\iota = G$, $\overline{\mathrm{Lim}}\sup G_\iota = \overline{G}$;

$\Leftrightarrow$ (iii) $\underline{\mathrm{Lim}}\, G_\iota = G$, $\overline{\mathrm{Lim}}\, G_\iota = \overline{G}$.

These conditions are valid if and only if for all compact subsets $K \subset G$, $K' \subset \complement\overline{G}$ it follows that $K \subset G_\iota$, $K' \subset \complement\overline{G}_\iota$ for almost all ι. It is this convergence concept for sequences of subsets in $\mathbb{R}^n$ which has been used by Babuška in [2], [3]. Note that also the condition $G = (\overline{G})^o$ is assumed in [3], p.16, and [4], p. 176. It is easily seen that conditions (5i), (5ii), (5iii) above imply the corresponding conditions in Theorem 2.2.(5).

The bounded sequence of open sets G_o, G_1, G_2,... has the <u>uniform segment property</u> iff there exist a finite open covering $\mathcal{O}_1,\dots,\mathcal{O}_r$ of $\bigcup_\iota \partial G_\iota$ and an associated system of vectors $a_1,\dots a_r$ in $\mathbb{R}^n$ with the property

$$\overline{G}_\iota \cap \overline{\mathcal{O}}_k + ta_k \subset G_\iota, \quad o < t < 1, \quad k = 1,\dots,r,$$

uniformly for all $\iota = o,1,2,\dots$.

In the study of semihomogeneous Neumann boundary value problems, simultaneously under perturbation of domains and approximation by subspaces, we use the following conditions in addition to the assumptions of Theorem (5).

(7) <u>Let</u> $E_o = {}_J H^m(G)$ <u>and let a sequence of closed subspaces</u> $E_\iota \subset {}_J H^m(G_\iota)$, $\iota = 1,2,\dots$, <u>be given such that, for each</u> $\varphi \in {}_J C_o^\infty(\mathbb{R}^n)$, <u>there exists a sequence of functions</u> $\varphi_\iota \in E_\iota$ <u>such that</u> $\|\varphi_\iota - \varphi|G_\iota\|_{m,G_\iota} \to o$ <u>for</u> $\iota \to \infty$. <u>Then</u> $\lim E_\iota = {}_J H^m(G)$ and the sequence $(.,.)_o|E_\iota$, $\iota = o,1,2,\dots$, <u>is weakly collectively compact.</u>

Theorems (5), (7) in connection with Theorem 2.1.(6) show that the basic assumptions of the convergence Theorems in Section 1.2, 1.3 are satisfied. If $\lim \zeta_\iota = \zeta_o$ and

$\zeta_\iota \notin [o,\infty[$ for $\iota = o,1,2,\ldots$, the sesquilinear forms $a_\iota = a_{E_\iota} - \overline{\zeta}_\iota b_{E_\iota}$ satisfy the inequality 1.2.(2), and thus are bijective, bicontinuous. Moreover, the sequence is bistable and consistent so that the assertions of Theorem 1.2.(4) hold. When $\lim \zeta_\iota = \zeta_o$ and ζ_o is not an eigenvalue of (2), (2'), $a_o = a_{E_o} - \overline{\zeta}_o b_{E_o}$ is injective and all of the assertions of Theorem 1.2.(5) hold. Finally, the condition (EVP) in Section 1.3 is true for the sesquilinear forms a_ι, b_ι and spaces E_ι specified in (3) and (4). Consequently, the fundamental results 1.3.(3), (4), (5) hold for the above Neumann problems (1), (1') and (2), (2') as well as for their approximations by subspaces.

2.4. Robin and Steklov problems

Let us finally study boundary value problems under Robin boundary conditions

(1) $$-(\Delta + \zeta_\iota)u_\iota = f_\iota \text{ in } G_\iota, \quad \frac{\partial u_\iota}{\partial N_\iota} + \eta_\iota u_\iota = g_\iota \text{ in } \partial G_\iota,$$

and the associated eigenvalue problems

(2) $$-\Delta w_\iota = \lambda_\iota w_\iota \text{ in } G_\iota, \quad \frac{\partial w_\iota}{\partial N_\iota} + \eta_\iota w_\iota = o \text{ in } \partial G_\iota,$$

where $\partial/\partial N_\iota$ denotes derivatives in the outward normal direction on ∂G_ι. Note that one obtains Neumann boundary conditions for $\eta_\iota = o$. The generalized form of these problems is

(1') $$\int_{G_\iota}\{\nabla\varphi_\iota \cdot \overline{\nabla u_\iota} - \overline{\zeta}_\iota \varphi_\iota \overline{u_\iota}\}dx + \overline{\eta}_\iota \int_{\partial G_\iota} \varphi_\iota \overline{u}_\iota ds_\iota = \int_{G_\iota}\varphi_\iota \overline{f}_\iota dx + \int_{\partial G_\iota}\varphi_\iota \overline{g}_\iota ds_\iota, \quad \varphi_\iota \in H^1(G_\iota),$$

having solutions $u_\iota \in H^1(G_\iota)$, and

(2') $$\int_{G_\iota}\nabla\varphi_\iota \cdot \overline{\nabla w_\iota} dx + \overline{\eta}_\iota \int_{\partial G_\iota}\varphi_\iota \overline{w}_\iota ds_\iota = \overline{\lambda}_\iota \int_{G_\iota}\varphi_\iota \overline{w}_\iota dx, \quad \varphi_\iota \in H^1(G_\iota),$$

where $o \neq w_\iota \in H^1(G_\iota)$, for $\iota = o,1,2,\ldots$.

More generally, let E_ι be a sequence of closed subspaces in $E = L^{1,2}$ and let a_ι, b_ι be sesquilinear forms such that

(3) $$\begin{aligned} &E_o = {}_J H^1(G), && E_\iota \subset {}_J H^1(G_\iota) \\ &a_\iota = a_{E_\iota} + \overline{\eta}_\iota c_\iota, && b_\iota = b_{E_\iota}, \\ &c_\iota(\varphi,\psi) = \int_{\partial G_\iota} \varphi^o \overline{\psi^o} ds_\iota, && \varphi,\psi \in E_\iota, \\ &l_\iota(\varphi) = \int_{G_\iota}\varphi^o \overline{f}_\iota dx + \int_{\partial G_\iota}\varphi^o \overline{g}_\iota ds_\iota, && \varphi \in E_\iota, \end{aligned}$$

where $f_\iota \in L^2(G_\iota)$, $g_\iota \in L^2(\partial G_\iota)$ for $\iota = o,1,2,\ldots$. Consider the inhomogeneous equations 1.3.(1) and the associated eigenvalue problems 1.3.(2) for the above spaces and sesquilinear forms. These problems represent approximations of the boundary value problems (1) or (1') and (2) or (2') for $\iota = o$ by perturbation of domains and approximation by subspaces.

Let us further consider the sequence of inhomogeneous Steklov problems

(4) $$-\Delta u_\iota = f_\iota \text{ in } G_\iota, \quad \frac{\partial u_\iota}{\partial N_\iota} - \zeta_\iota u_\iota = g_\iota \text{ in } \partial G_\iota,$$

and the corresponding eigenvalue problems

(5) $$-\Delta w_\iota = o \text{ in } G_\iota, \quad \frac{\partial w_\iota}{\partial N_\iota} = \lambda_\iota w_\iota \text{ in } \partial G_\iota,$$

for $\iota = o,1,2,\ldots$. Approximations of these problems on closed subspaces $E_\iota \subset {}_JH^1(G_\iota)$ are represented by 1.3.(1) and 1.3.(2) using the sequilinear forms

(6) $$a_\iota = a|E_\iota, \quad b_\iota = c_\iota, \quad \iota = o,1,2,\ldots,$$

and the same functionals l_ι as in (3).

The following assumptions ensure the basic convergence, stability and compactness properties for this class of problems.

(7) <u>Let</u> $G = G_o, G_1, G_2, \ldots$ <u>be a uniformly bounded sequence of open domains in</u> $\mathbb{R}^n$ <u>such that</u>

(i) $\underline{\text{Lim}}\, G_\iota = G$, (ii) $\overline{\text{Lim}}\, G_\iota = \overline{G}$,

(iii) $G_o, G_1, G_2, \ldots$ <u>has the uniform Lipschitz property and the surface measures</u> s_ι <u>on</u> ∂G_ι <u>satisfy the condition</u>

$$\forall v \in C(\mathbb{R}^n): \quad \lim \int_{\partial G_\iota} v \, ds_\iota = \int_{\partial G} v \, ds.$$

<u>Then</u> $\lim {}_JH^m(G_\iota) = {}_JH^m(G)$ <u>and the sequence</u> $(.,.)_o | {}_JH^m(G_\iota)$, $\iota = o,1,2,\ldots$ <u>is weakly collectively compact. Moreover, the sequence</u> $(.,.)_{\partial G_\iota} | {}_JH^m(G_\iota)$, $\iota = o,1,2,\ldots$, <u>is stable, consistent and weakly collectively compact.</u>

<u>Proof.</u> Under the above assumptions (7i), (7ii), (7iii) the corresponding assumptions of Theorem 2.3.(5) are valid. Therefore, the first assertion above follows from this Theorem. Next, consider the sequence of sesquilinear forms $c_\iota = (.,.)_{\partial G_\iota}$ on $E_\iota = {}_JH^m(G_\iota)$, $\iota = o,1,2,\ldots$. This sequence is stable, because [11], 3.3.(4i), ensures the existence of a positive constant γ such that

$$|c_\iota(\varphi,\psi)| = \Big|\int_{\partial G_\iota} \varphi^o \overline{\psi^o} ds_\iota\Big| \le \gamma \|\varphi\|_m \|\psi\|_m, \quad \varphi,\psi \in {}_JH^m(G_\iota),$$

uniformly for all $\iota = o,1,2,\ldots$. Let $\underline{u}_\iota \in {}_JH^m(G_\iota)$, $\iota = o,1,2,\ldots$, be any discretely convergent sequence and s-lim $\underline{u}_\iota = \underline{u}_o$. By virtue of Theorem 1.1.(17), s-lim $c_\iota(\cdot,\underline{u}_\iota) = c_o(\cdot,\underline{u}_o)$ is equivalent to the convergence relation

$$\text{w-lim}\ \underline{v}_\iota = \underline{v}_o \Longrightarrow \lim \int_{\partial G_\iota} v_\iota \overline{u}_\iota ds_\iota = \int_{\partial G} v_o \overline{u}_o ds$$

for every weakly discretely convergent sequence of functions $\underline{v}_\iota \in {}_J H^m(G_\iota)$, $\iota = 0,1,2,\ldots$. The validity of this relation is seen from [11], 3.3.(13ii). Finally, one has the inequalities

(8) $\quad \|v_\iota\|_{\partial G_\iota} \leq \gamma(\varepsilon \|v_\iota\|_1 + \frac{1}{\varepsilon}\|v_\iota\|_o)$,

uniformly for all $v_\iota \in H^m(G_\iota)$, all $\iota = 0,1,2,\ldots$ and $0 < \varepsilon < 1$. Due to the compactness of the natural embeddings of $H^m(G_\iota)$ into $L^2(G_\iota)$ and the above inequalities, one first obtains that c_ι is compact for each ι. Further (8) and the weak collective compactness of $(.,.)_o | {}_J H^m(G_\iota)$, $\iota = 0,1,2,\ldots$, entail the relation

$$\text{w-lim}\ \underline{v}_\iota = o \Longrightarrow \lim \|v_\iota\|_{\partial G_\iota} = o,$$

for every weakly convergent sequence $\underline{v}_\iota = J^m_{G_\iota} v_\iota \in {}_J H^m(G_\iota)$, $\iota = 0,1,2,\ldots$. Finally

$$\|c_\iota(\underline{v}_\iota;\cdot)\|_{E'_\iota} \leq \gamma \|v_\iota\|_{\partial G_\iota}$$

so that w-lim $\underline{v}_\iota = o$ implies $\lim \|c_\iota(\underline{v}_\iota,\cdot)\|_{E'_\iota} = o$. Consequently, the sequence (c_ι) is weakly collectively compact. ∎

From the definitions in [11], Section 3.2, it is readily seen that open sets G_ι, having the uniform Lipschitz property, are open domains, that is, $(\overline{G}_\iota)^o = G_\iota$. In this case, one has the equivalent characterizations 2.3.(6) of the above conditions (7i) and (7ii). Hence conditions (7i), (7ii), (7iii) are equivalent to condition (G5) in [11]. Note further that under the assumptions of Theorem (7), also the assumptions of Theorems 2.2.(5) and 2.3.(5) are valid.

(9) Let the assumptions of Theorem (7) be valid, let $E_o = {}_J H^1(G)$ and let $E_\iota \subset {}_J H^1(G_\iota)$, $\iota = 1,2,\ldots$, be a sequence of closed subspaces specified as in Theorem 2.3.(7) such that $\lim E_\iota = {}_J H^m(G)$. Then the associated sequences of sesquilinear forms (a_ι), (b_ι) both in (3) and in (6) as well as $(a_\iota - \overline{\zeta}_\iota b_\iota)$ are stable and consistent at each $\underline{u}_o \in {}_J H^1(G)$, provided that $\lim \eta_\iota = \eta_o$, $\lim \zeta_\iota = \zeta_o$. The sequences (a_ι), $(a_\iota - \overline{\zeta}_\iota b_\iota)$ are strongly coercive and the sequence (b_ι) is weakly collectively compact. Finally a_ι, b_ι is strongly definite for each ι.

Proof. The first statement is an immediate consequence of the corresponding statements for the sequences of sequilinear forms (a_{E_ι}), (b_{E_ι}), (c_ι) in Theorem 2.1.(6) and Theorem (7) above. Additionally, these two theorems yield the weak collective compactness of $(b_\iota) = (b_{E_\iota})$ and $(b_\iota) = (c_\iota)$. Note that

$$a(\varphi) \geq \|\varphi\|_1^2 - \|\varphi\|_o^2, \quad \varphi \in E_\iota.$$

Thus the uniform strong coerciveness of the sequences (a_ι), $(a_\iota - \overline{\zeta}_\iota b_\iota)$ is seen from the following representations:

$$\mathrm{Re}\{a_\iota - \overline{\zeta}_\iota b_\iota\} = \|\varphi\|_1^2 - \mathrm{Re}\{(1+\zeta_\iota)b_{E_\iota} - \eta_\iota c_\iota\}(\varphi)$$

when (a_ι), (b_ι) are specified by (3), and

$$\mathrm{Re}\{a_\iota - \overline{\zeta}_\iota b_\iota\} = \|\varphi\|_1^2 - \mathrm{Re}\{b_{E_\iota} + \zeta_\iota c_\iota\}(\varphi)$$

in the case (6). Both $(k_\iota) = ((1+\zeta_\iota)b_{E_\iota} - \eta_\iota c_\iota)$ and $(k_\iota) = (b_{E_\iota} + \zeta_\iota c_\iota)$ are weakly collectively compact for all convergent sequences (ζ_ι), (η_ι) under the assumptions of this theorem. The pairs a_ι, b_ι are strongly definite because

$$(\alpha_\iota \mathrm{Re}\, a_\iota + \beta_\iota \mathrm{Re}\, b_\iota)(\varphi) = \|\varphi\|_o^2 > o, \quad o \neq \varphi \in E_\iota,$$

for $\alpha_\iota = o$, $\beta_\iota = 1$ and a_ι, b_ι as in (3);

$$(\alpha_\iota \mathrm{Re}\, a_\iota + \beta_\iota \mathrm{Re}\, b_\iota)(\varphi) = a(\varphi) + c_\iota(\varphi) > o, \quad o \neq \varphi \in E_\iota,$$

for $\alpha_\iota = \beta_\iota = 1$ and a_ι, b_ι specified by (6). ■

We finally have to establish conditions ensuring the convergence of the sequence of inhomogeneous terms in (1) and (4), that is, of the functionals l_ι specified in (3). Since $G_o, G_1, G_2, \ldots$ has the uniform Lipschitz property, there exists a finite covering of $\bigcup_\iota \partial G_\iota$ by open sets $U_1, \ldots, U_r$ such that, in particular, for each $U \in \{U_1, \ldots, U_r\}$ there exist an orthogonal transformation τ of $\mathbb{R}^n$, open intervals $(a,b) \subset \mathbb{R}^{n-1}$, $(c,d) \subset \mathbb{R}$ and a sequence of uniformly Lipschitz continuous functions $h_\iota : [a,b] \to (c,d)$, $\iota = o,1,2,\ldots$, with the properties $U = \tau\{(a,b) \times (c,d)\}$ and

$$\partial G_\iota \cap \overline{U} = \tau\{(y,z) \mid y \in [a,b],\ h_\iota(y) = z\}.$$

Note that still further properties are required in the definition of the uniform Lipschitz property in [11]. Under the assumptions of Theorem (7) and (9), we have the following statement. Let us remark in passing, that Nečas uses condition (1oi) in the treatment of perturbation of domains in [8], Section 3.6.7.

(1o) <u>The sequence of functionals</u> (l_ι), <u>specified in</u> (3), <u>converges to</u> l_o <u>if</u> $s\text{-}\lim f_\iota = f_o$ <u>in</u> $L^{o,2}$ <u>and, for each neighbourhood</u> $U \in \{U_1, \ldots, U_r\}$,

(i) $$\lim \int_a^b |(g_\iota \circ \tau)(y, h_\iota(y)) - (g_o \circ \tau)(y, h_o(y))|^2 dy = o.$$

<u>In particular, if</u> $\hat{g}_\iota \in H^1(G_\iota)$ <u>and</u> $g_\iota = \hat{g}_\iota | \partial G_\iota$ <u>for</u> $\iota = o,1,2,\ldots$,

(ii) $$s\text{-}\lim f_\iota = f_o, \quad s\text{-}\lim \hat{g}_\iota = \hat{g}_o \implies s\text{-}\lim l_\iota = l_o.$$

<u>Proof.</u> The functionals l_ι may be decomposed in $l_\iota = l_\iota^o + l_\iota^1$ where

$$l_\iota^o(\varphi) = \int_{G_\iota} \varphi \bar{f}_\iota^o dx, \quad l_\iota^1(\varphi) = \int_{\partial G_\iota} \varphi_\iota \bar{g}_\iota^o ds_\iota, \ \varphi \in E_\iota,$$

$\iota = 0,1,2,\dots$. Due to Theorem 2.1.(9) we have s-lim $l_\iota^o = l_o^o$. Using Theorem 1.1.(7), the convergence of (l_ι^1) is equivalent to the relation

$$\text{w-lim } \varphi_\iota = \varphi_0 \implies \lim \int_{\partial G_\iota} \varphi_\iota^0 \bar{g}_\iota ds_\iota = \int_{\partial G_0} \varphi_0^0 \bar{g}_0 ds,$$

for every weakly convergent sequence of functions $\varphi_\iota \in E_\iota$, $\iota = 0,1,2,\dots$. Hence

$$\text{s-lim } l_\iota^1 = l_o^1 \iff \text{s-lim } g_\iota = g_0$$

in the sense of [11], Section 3.3. The sufficient condition (loi) has been established in [11], Theorem 3.3.(17). Additionally, [11], Theorem 3.3.(13) shows that

$$\text{s-lim } \hat{g}_\iota = \hat{g}_0 \implies \text{s-lim } g_\iota = g_0$$

for all $\hat{g}_\iota \in H^1(G_\iota)$ and $g_\iota = \hat{g}_\iota | \partial G_\iota$, $\iota = 0,1,2,\dots$, which entails (loii). ■

We are now in a position to apply the fundamental convergence theorems of Section 1.2, 1.3 to the Robin and Steklov boundary value problems (1), (1'), (2), (2') and (4), (5) as well as to its approximations by subspaces $E_\iota \subset {}_J H^1(G_\iota)$, $\iota = 1,2,\dots$. We shall assume in the following that lim $\zeta_\iota = \zeta_o$, lim $\eta_\iota = \eta_o$. For example, if $\zeta_\iota \in]-\infty,o]$ and $\eta_\iota \in]o,\infty[$, or if $\zeta_\iota \in]-\infty,o[$ and $\eta_\iota \in [o,\infty[$ for $\iota = 0,1,2,\dots$, the sesquilinear forms $a_\iota - \bar{\zeta}_\iota b_\iota$ in the Robin problem (1), (3) satisfy the assumption 1.2.(2) such that Theorem 1.2.(4) may be applied. In the Steklov problem (4), (6) and $\zeta_\iota \notin [o,\infty[$ for $\iota = 0,1,2,\dots$, condition 1.2.(2) holds so that Theorem 1.2.(4) becomes applicable.

Let ζ_0 be not an eigenvalue of (2) or (5) for $\iota = o$. Then Theorem 1.2.(5) shows that the given boundary value problem and almost all of the approximating problem are properly posed, the biconvergence relation is valid and the associated two-sided discretization error estimates hold. Using the representation 2.1.(3), these estimates now have the form

$$(11) \quad \frac{1}{\alpha_1^2}\|d_\iota\|^2_{E'_\iota} + \rho_\iota(\underline{u}_0)^2 \le \|u_\iota - u_o\|^2_{1,G\cap G_\iota} + \|u_\iota\|^2_{1,G_\iota - G} + \|u_o\|^2_{1,G-G_\iota} \le \frac{1}{\alpha_0^2}\|d_\iota\|^2_{E'_\iota} + \rho_\iota(\underline{u}_0)^2$$

for $\iota = 1,2,\dots$. In view of 1.1.(7), $\rho_\iota(\underline{u}_0)$ is the shortest distance of $\underline{u}_0$ to E_ι in $L^{1,2}$, just as in 2.2.(9). Since $P_\iota \to P_0$ for $\iota \to \infty$, we have

$$\rho_\iota(\underline{u}_0) = \|\underline{u}_0 - P_\iota \underline{u}_0\|_1 = |\underline{u}_0, E_\iota| \to o \quad (\iota \to \infty).$$

Moreover, using Theorem (9),

$$\text{s-lim}\,(a_\iota - \bar{\zeta}_\iota b_\iota)(\cdot, P_\iota \underline{u}_o) = (a_o - \bar{\zeta}_o b_o)(\cdot, \underline{u}_o) = 1_o.$$

Therefore, under the assumptions of Theorem (1o), s-lim $1_\iota = 1_o$ and thus

$$\|d_\iota\|_{E'_\iota} = \|1_\iota - (a_\iota - \bar{\zeta}'_\iota b_\iota)(\cdot, P_\iota \underline{u}_o)\|_{E'_\iota} \to o \quad (\iota \to \infty).$$

The assumptions (EVP) in Section 1.3 are valid, both for the above Robin and Steklov problems. Therefore, Theorem 1.3.(3), additionally, establishes the Fredholm property for the inhomogeneous equations (1) and (4) as well as the discreteness of the spectra of (2) and (5) and their approximations by subspaces. Moreover, Theorem 1.3.(4) ensures the continuous dependence upon perturbation of domains and approximation by subspaces for the resolvent sets and spectra, the eigenvalues and eigenspaces of the above Robin and Steklov model problems. When η_ι is real, the sesquilinear forms a_ι, b_ι are Hermitian. Thus, 1.3.(5), additionally, ensures the convergence of the ordered sequences of eigenvalues.

References

1. Aubin, J.P.: Approximation des espaces de distribution et des opérateurs différentiels. Bull. Soc. Math. France Mém. 12, 1-139 (1967).

2. Babuška, I.: Stabilität des Definitionsgebietes mit Rücksicht auf grundlegende Probleme der Theorie der partiellen Differentialgleichungen auch im Zusammenhang mit der Elastizitätstheorie. Czechoslovak Math. J. 11, 76-1o5, 165-2o3 (1961).

3. —— The theory of small changes in the domain of existence in the theory of partial differential equations and its applications. Proc. Conference Diff. Equations, Prague 1962, 13-26. Prague: Czechoslovak Academy of Sciences 1963.

4. Babuška, I., and Výborný, R.: Continuous dependence of eigenvalues on the domain. Czechoslovak Math. J. 15, 169-178 (1965).

5. Blair, J.J.: Bounds for the change in the solutions of second order elliptic PDE's when the boundary is perturbed. SIAM J. Appl. Math. 24, 277-285 (1973).

6. Grigorieff, R.D.: Diskret kompakte Einbettungen in Sobolewschen Räumen. Math. Ann. 197, 71-85 (1972).

7. Hausdorff, F.: Grundzüge der Mengenlehre. New York: Chelsea 1965.

8. Nečas, J.: Les méthodes directes en théorie des équations elliptiques. Paris: Masson 1967.

9. Stummel, F.: Diskrete Konvergenz linearer Operatoren. I. Math. Ann. 19o, 45-92 (197o). II. Math. Z. 12o, 231-264 (1971). III. Proc. Oberwolfach Conference on Linear Operators and Approximation 1971. Int. Series of Numerical Mathematics 2o, 196-216. Basel: Birkhäuser 1972.

1o. Stummel, F.: Singular perturbations of elliptic sesquilinear forms. Proc. Conference on Differential Equations, Dundee, March 1972. Lecture Notes in Mathematics 28o, 155-18o. Berlin-Heidelberg-New York: Springer 1972.

11. —— Perturbation theory for Sobolev spaces. Proc. Royal Soc. Edinburgh 73A, 5-49 (1974/75).

12. —— Perturbation theory for elliptic sesquilinear forms and boundary value problems in mathematical physics. Lecture Notes, March 1975, Institute for Fluid Dynamics and Applied Mathematics, College Park, University of Maryland. To appear.

Prof. Dr. F. Stummel
Department of Mathematics
Johann Wolfgang Goethe-Universität

D-6ooo Frankfurt am Main
Robert-Mayer-Strasse 1o
West Germany

FROST PROPAGATION IN WET POROUS MEDIA

J. Aguirre-Puente* M. Frémond**

Résumé : Un milieu poreux saturé d'eau gèle lorsqu'il est soumis à l'action du froid. Le front de gel qui sépare la partie non gelée de la partie gelée est une surface libre. L'expérience montre qu'il apparaît une dépression sur le front de gel. L'eau est alors aspirée vers ce front et gèle en l'atteignant.

Le problème est un problème de Stefan couplé liant les équations de la diffusion de la chaleur et de l'eau. L'équation de la conservation de l'énergie couple les équations sur le front de gel. Les équations d'évolution obtenues sont résolues en introduisant une nouvelle inconnue, l'indice de gel, et en utilisant les techniques des inéquations variationnelles. On présente enfin un exemple numérique.

Summary : A water saturated porous medium freezes when it is chilled. The frost line which separates the frozen part and the unfrozen part is a free surface. Experiments show that a depression appears on the frost line. Water is thus sucked in through the unfrozen part. It freezes when it reaches the frost line.

The problem is a coupled Stefan problem linking the heat and water equations of diffusion. The energy conservation law couples the equations on the frost line. The equations are solved using a new unknown the freezing index and the methods of variationnal inequalities. A numerical example is given.

---

* Laboratoire d'Aérothermique du CNRS, 4ter route des Gardes - 92190 MEUDON-BELLEVUE.

** Laboratoire Central des Ponts et Chaussées, 58 Bld Lefebvre 75732 PARIS CEDEX 15.

FROST PROPAGATION IN WET POROUS MEDIA

J. Aguirre-Puente* M. Frémond**

I. INTRODUCTION THE PHYSICAL PROBLEM.

A water saturated porous medium freezes when it is chilled. It occupies an open part Ω of $\mathbb{R}^n$ (n = 1, 2 or 3). The frost line which separates the frozen and unfrozen parts is a free surface: i.e., it is an a priori unknown surface. Experiences show that a depression appears on the frost line. Water is thus sucked in through the unfrozen part. This sucked in water freezes when it reaches the frost line. This accumulation of ice induced by the frost results in an heaving of the structure.

This phenomenon is important for road maintenance in cold weather. It is known that ice accumulation and frost heaving results in a decrease of bearing capacity during thaw.

The physical experiments have allowed the comprehension of the phenomenon and the construction of a mathematical model [1]. The unknowns of the problem are the temperature $\theta(x,t)$ and the head of water $h(x,t)$ at any point x of Ω and any time t of the $[0,T]$ period during which this phenomenon is being investigated. The data are the initial state of the medium and the external actions which determine the boundary conditions. The hydraulic and thermal phenomenons described by the classical diffusion equations are, as already said, coupled. The coupling occurs on the frost line according to the energy conservation law.

This problem is solved in two steps. First we introduce a new unknown: the freezing index, roughly a heat quantity, which is important from the technical and physical point of view. If there is no hydraulic phenomenon this step solves again the thermal problem in a different manner, which is however equivalent to previous results [3, 4, 5, 6, 9, 11, 12, 15]. The advantage of this presentation is to use the freezing index and a formulation in terms of variational inequalities. The second step allows the calculation of the head of water. The problem is entirely solved by the knowledge of the freezing index and of the head of water allowing to compute the frost heaving.

Computer programs are used by the Laboratoire Central des Ponts et Chaussées to study the freezing of pavements and to protect them against its harmful consequences.

The equations are established in §II where the freezing index is introduced. The variational formulation is also given in §II. The main results are given in §III: they concern the existence and the uniqueness of solutions. The §IV and V contain the proof pattern. Numerical results are presented in §VI.

---

* Laboratoire d'Aérothermique du CNRS, 4ter route des Gardes - 92190 MEUDON-BELLEVUE
** Laboratoire Central des Ponts et Chaussées, 58 Bld Lefebvre 75732 PARIS CEDEX 15.

II. THE EQUATIONS.

We assume that the temperature of water circulating in the unfrozen part is equal to the temperature of the porous medium skeleton. We assume also that the speed of this water is low, so we can neglect the heat transmitted by convection comparatively to the heat transmitted by conduction. Of course, we take into account the latent heat of fusion of this circulating water.

At any time t, Ω is divided into two parts, the unfrozen part $\Omega_1(t)$ and the frozen part $\Omega_2(t)$ separated by the frost line with local equation $t = \gamma(x)$ (fig.1).

The equation are derived by using the conservation and behavioral laws [10]. The behavioral laws for the diffusion of water and heat are the Darcy and Fourier laws. Let θ_i be the restriction of θ to $\Omega_i(t)$. The energy conservation law gives then:

$$- \text{ in } \Omega_1(t) : b_1 \frac{\partial \theta_1}{\partial t} - \Delta\theta_1 = 0, \tag{1}$$

$$- \text{ in } \Omega_2(t) : b_2 \frac{\partial \theta_2}{\partial t} - \Delta\theta_2 = 0, \tag{2}$$

where $b_i = C_i/\lambda_i$; C_i and λ_i are the heat capacities by unit volume and the thermal conductivities of the frozen and unfrozen parts.

The mass conservation law gives :

$$- \text{ in } \Omega_1(t) : \frac{\nu}{\varepsilon k} \frac{\partial h}{\partial t} - \Delta h = 0, \tag{3}$$

where ν is a coefficient representing the compressibility of the porous medium under head variations, ε is the porosity and k the hydraulic conductivity. To simplify, we assume that $b_1 = \nu/\varepsilon k$. This assumption is not restrictive because the usual numerical values are close. It has also been shown that those two terms are not important for the phenomenon.

On the frost line, mass and energy conservation laws give :

$$- \ell = (\lambda_1 \text{grad}\theta_1 - \lambda_2 \text{grad}\theta_2 + \varepsilon k \ell' \text{grad}h)\cdot \text{grad}\gamma, \tag{4}$$

where ℓ and ℓ' are the latent heat of fusion per unit volume of the water-saturated porous medium and of the water. We also have :

$$\theta_1(x,\gamma(t)) = \theta_2(x,\gamma(x)) = 0,\ h(x,\gamma(x)) = -d, \tag{5}$$

where $-d(d \geq 0)$ is a constant which measures the depression due to the frost. Moreover, experiments have shown that :

$$\text{in } \Omega_1(t),\ h(x,t) \geq -d, \tag{6}$$

The head which is a priori defined in $\Omega_1(t)$ only, is extented to $\Omega_2(t)$ by $h = -d$.

We assume that the boundary of Ω is divided into four parts Γ_j (mes $\Gamma_j \neq 0$), where :

- on Γ_1 which is in general unfrozen, the temperature $\overline{\theta}_1(x,t) \geq 0$ and the head $\overline{h}_1(x,t) \geq -d$ are given :

$$\theta = \overline{\theta}_1,\ h = \overline{h}_1, \tag{7}$$

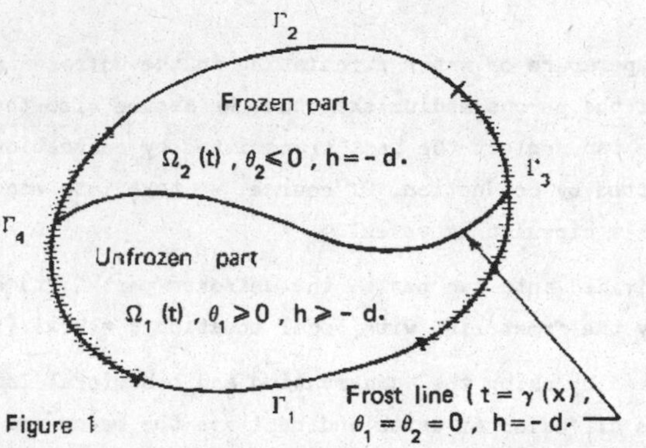

Figure 1

- on Γ_2 which is in general frozen, the temperature $\overline{\theta}_2(x,t)$ and the head $\overline{h}_2 = -d$ are given :

$$\theta = \overline{\theta}_2, \; h = -d, \tag{8}$$

- on Γ_3, the flows are proportional to the differences between inside and outside ($\frac{\partial}{\partial n}$ is the outer normal derivative),

$$\frac{\partial h}{\partial n} + \alpha(h+d) = 0, \; \lambda_i \frac{\partial \theta}{\partial n} + \alpha(\lambda_i \theta - s) = 0 \text{ on } \Gamma_3 \cap \overline{\Omega}_i(t), \tag{9}$$

where the constant $\alpha > 0$ and the function $s(x,t)$ are given.

- on Γ_4 which is water-tight the heat flow $\omega(x,t)$ is given :

$$\frac{\partial h}{\partial n} = 0, \lambda_i \frac{\partial \theta}{\partial n} = \omega \text{ on } \Gamma_4 \cap \overline{\Omega}_i(t), \tag{10}$$

The initial temperature and head $(\theta_0(x), h_0(x))$ are given as well as the parts $\Omega_1(0)$ and $\Omega_2(0)$; they verify :

$$\theta_0|\Omega_1(0) \geq 0, \; h_0|\Omega_1(0) \geq -d, \quad \theta_0|\Omega_2(0) \leq 0, \; h_0|\Omega_2(0) = -d, \tag{11}$$

a/ Change of unknown : the freezing index. To solve the equations (1) to (11), we define the new unknown : the freezing index

$$\hat{u}(x,t) = \int_0^t \{\lambda_1 \theta^+(x,\tau) - \lambda_2 \theta^-(x,\tau) + \delta[h(x,\tau)+d]\} d\tau,$$

where $\delta = \varepsilon k \ell'$ and $\phi^+ = \sup\{0,\phi\}$, $\phi^- = \sup\{0,-\phi\}$. The function $\hat{u}$ and similar ones are widely used in freezing problems [7,8,14,16] . To find equations verified by $\hat{u}$, let us first consider, at a time t, a point $x \in \Omega_1(0) \cap \Omega_2(t)$. We assume that $\theta(x,\tau)$ changes sign an odd number of times : $\gamma_j(x)$ ($j = 1, \ldots, 2n+1$) between 0,t. We have :

$$\operatorname{grad}\hat{u}(x,t) = \int_0^{\gamma_1(x)} \lambda_1 \operatorname{grad}\theta_1(x,\tau) + \delta \operatorname{grad}h(x,\tau) d\tau + \int_{\gamma_1(x)}^{\gamma_2(x)} \lambda_2 \operatorname{grad}\theta_2(x,\tau) d\tau$$

$$+ \ldots + \int_{\gamma_{2n}(x)}^{\gamma_{2n+1}(x)} \lambda_1 \operatorname{grad}\theta_1(x,\tau) + \delta \operatorname{grad}h(x,\tau) d\tau + \int_{\gamma_{2n+1}(x)}^{t} \lambda_2 \operatorname{grad}\theta_2(x,\tau) d\tau,$$

because θ and h are continuous across the frost line. We have then :

$$\Delta\hat{u}(x,t) = \int_0^{\gamma_1(x)} \lambda_1\Delta\theta_1(x,\tau)+\delta\Delta h(x,\tau)d\tau + \int_{\gamma_1(x)}^{\gamma_2(x)} \lambda_2\Delta\theta_2(x,\tau)d\tau + \ldots$$

$$+ \int_{\gamma_{2n}(x)}^{\gamma_{2n+1}(x)} \Delta\theta_1(x,\tau)+\delta\Delta h(x,\tau)d\tau + \int_{\gamma_{2n+1}(x)}^{t} \lambda_2\Delta\theta_2(x,\tau)d\tau + \big(\lambda_1 \mathrm{grad}\phi_1(x,\gamma_1(x))$$

$$+ \delta\mathrm{grad}h(x,\gamma_1(x)) - \lambda_2\mathrm{grad}\theta_2(x,\gamma_1(x))\big)\cdot\mathrm{grad}\gamma_1(x) + \ldots + \big(\lambda_1\mathrm{grad}\theta_1(x,\gamma_{2n+1}(x)) +$$

$$\delta\mathrm{grad}h(x,\gamma_{2n+1}(x)) - \lambda_2\mathrm{grad}\theta_2(x,\gamma_{2n+1}(x))\big)\cdot\mathrm{grad}\gamma_{2n+1}(x).$$

Making use of the relations (1) to (4), we obtain :

$$\Delta\hat{u}(x,t) = \int_0^{\gamma_1(x)} \{C_1 \frac{\partial\theta_1}{\partial t}(x,\tau)d\tau + \nu\ell'\frac{\partial}{\partial t}(h(x,\tau)+d)\}d\tau + \int_{\gamma_1(x)}^{\gamma_2(x)} C_2 \frac{\partial\theta_2}{\partial t}(x,\tau)d\tau + \ldots$$

$$+ \int_{\gamma_{2n}(x)}^{\gamma_{2n+1}(x)} \{C_1 \frac{\partial\theta_1}{\partial t}(x,\tau) + \nu\ell'\frac{\partial}{\partial t}(h(x,\tau)+d)\}d\tau + \int_{\gamma_{2n}(x)}^{t} \{C_2\frac{\partial\theta_2}{\partial t}(x,\tau)\}d\tau$$

$$- \ell + \underbrace{(+\ell-\ell)}_{2n \text{ times}} = - b_1\big(\lambda_1\theta_0(x) + \delta(h_0(x)+d)\big) + b_2\big(\lambda_2\theta(x,t)\big) - \ell .$$

We eventualy obtain : $\lambda_2(\theta_2(x,t)) = \frac{\partial\hat{u}}{\partial t} < 0$ and $b_2 \frac{\partial\hat{u}}{\partial t} - \Delta\hat{u} = b_1(\lambda_1\theta_0+\delta(h_0+d)) + \ell$ in $\Omega_1(0) \cap \Omega_2(t)$.

We obtain in the same way :

$$b_1 \frac{\partial\hat{u}}{\partial t} - \Delta\hat{u} = b_1\big(\lambda_1\theta_0+\delta(h_0+d)\big) \text{ and } \frac{\partial\hat{u}}{\partial t} > 0, \text{ in } \Omega_1(0) \cap \Omega_1(t),$$

$$b_2 \frac{\partial\hat{u}}{\partial t} - \Delta\hat{u} = b_2\lambda_2\theta_0 \text{ and } \frac{\partial\hat{u}}{\partial t} < 0, \text{ in } \Omega_2(0) \cap \Omega_2(t),$$

$$b_1 \frac{\partial\hat{u}}{\partial t} - \Delta\hat{u} = b_2\lambda_2\theta_0 - \ell \text{ and } \frac{\partial\hat{u}}{\partial t} > 0 \text{ in } \Omega_2(0) \cap \Omega_1(t), \tag{12}$$

Let $\sum_j = \Gamma_j \times]0,T[$. The boundary conditions are :

$$\hat{u} = f \text{ on } \textstyle\sum_1 \cup \sum_2,\ \frac{\partial\hat{u}}{\partial n} + \alpha(\hat{u}-g) = 0 \text{ on } \sum_3,\ \frac{\partial\hat{u}}{\partial n} = k \text{ on } \sum_4 \tag{13}$$

where $f(x,t) = \int_0^t \lambda_i\overline{\theta}_i(x,\tau) + \delta[\overline{h}_i(x,\tau)+d]d\tau$ on Γ_i, $i = 1,2$.

$$g(x,t) = \int_0^t s(x,\tau)d\tau \text{ and } k(x,t) = \int_0^t \omega(x,\tau)d\tau.$$

The initial condition is indeed : $\hat{u}(x,0) = 0$, (14)

Once $\hat{u}$ has been determined by the equations (12) to (14), we seek h such that :

$$- d < h < \psi = \frac{1}{\delta}\left(\frac{\partial\hat{u}}{\partial t}\right)^+ - d, \tag{15}$$

according to relation (6) and definition of $\hat{u}$,

$$\forall v < \psi,\ (v-h)\ \left(b_1 \frac{\partial h}{\partial t} - \Delta h\right) > 0, \tag{16}$$

It is indeed easy to check that $b_1 \frac{\partial h}{\partial t} - \Delta h$ is a non-negative distribution carried by the frost line. This relation means also that the frost line is a water sink. The equations (15), (16) and the boundary and initial conditions (7) to (11) allow to find h.

b/ The equations. Prior to giving a variational formulation, we perform a translation on u and h to have homogeneous Dirichlet boundary conditions on $\Sigma_1 \cup \Sigma_2$. Let $\hat{u} = u+\tilde{u}$, $h = p+\tilde{h}$ where $\tilde{u}$ and $\tilde{h}$ verify :

$$\tilde{u} = f \text{ on } \Sigma_1 \cup \Sigma_2,\ \tilde{u}(x,0) = 0,\ \frac{\partial \tilde{u}}{\partial t}(x,0) = \lambda_1\theta_0^+(x) - \lambda_2\theta_0^-(x) + \delta\big(h_0(x)+d\big) ;$$

$$\tilde{h} = h_1 \text{ on } \Sigma_1,\ \tilde{h} = -d \text{ on } \Sigma_2,\ \frac{\partial h}{\partial n} + \alpha(\tilde{h}+d) = 0 \text{ on } \Sigma_3,\ \frac{\partial \tilde{h}}{\partial n} = 0 \text{ on } \Sigma_4,$$

$$\tilde{h}(x,0) = h_0(x),\ b_1 \frac{\partial \tilde{h}}{\partial t} - \Delta\tilde{h} = 0 \text{ in } Q = \Omega\times\,]0,T[\,.$$

The equations verified by u and p are :

$$\begin{aligned}
&b_1 \frac{\partial u}{\partial t} - \Delta u = b_1 \frac{\partial \tilde{u}}{\partial t}(x,0) - \Big(b_1 \frac{\partial \tilde{u}}{\partial t} - \Delta\tilde{u}\Big),\ \frac{\partial u}{\partial t} > - \frac{\partial \tilde{u}}{\partial t},\ \text{in } \Omega_1(0)\cap\Omega_1(t),\\
&b_2 \frac{\partial u}{\partial t} - \Delta u = b_1 \frac{\partial u}{\partial t}(x,0) - \Big(b_2 \frac{\partial u}{\partial t} - \Delta\tilde{u}\Big)+\ell,\ \frac{\partial u}{\partial t} < - \frac{\partial \tilde{u}}{\partial t},\ \text{in } \Omega_1(0)\cap\Omega_2(t),\\
&b_1 \frac{\partial u}{\partial t} - \Delta u = b_2 \frac{\partial u}{\partial t}(x,0) - \Big(b_1 \frac{\partial u}{\partial t} - \Delta\tilde{u}\Big)-\ell,\ \frac{\partial u}{\partial t} > - \frac{\partial u}{\partial t},\ \text{in } \Omega_2(0)\cap\Omega_1(t),\\
&b_2 \frac{\partial u}{\partial t} - \Delta u = b_2 \frac{\partial u}{\partial t}(x,0) - \Big(b_2 \frac{\partial u}{\partial t} - \Delta\tilde{u}\Big),\ \frac{\partial u}{\partial t} < - \frac{\partial \tilde{u}}{\partial t},\ \text{in } \Omega_2(0)\cap\Omega_2(t).\\
&u = 0 \text{ on } \Sigma_1\cup\Sigma_2,\ \frac{\partial u}{\partial n} = -\frac{\partial \tilde{u}}{\partial n} + k \text{ on } \Sigma_4,\ \frac{\partial u}{\partial n} + \alpha(u-g) = -\Big(\frac{\partial \tilde{u}}{\partial n} + \alpha\tilde{u}\Big) \text{ on } \Sigma_3,\\
&u(x,0) = 0,
\end{aligned} \tag{17}$$

$$\begin{aligned}
&-d-\tilde{h} = \psi_1 < p < \psi_2 = \psi - \tilde{h},\\
&\forall v < \psi_2,\ (v-p)\Big(b_1 \frac{\partial p}{\partial t} - \Delta p\Big) > 0 \text{ in } Q,\\
&p = 0 \text{ on } \Sigma_1\cup\Sigma_2,\ \frac{\partial p}{\partial n} + \alpha p = 0 \text{ on } \Sigma_3,\ \frac{\partial p}{\partial n} = 0 \text{ on } \Sigma_4,\ p(x,0) = 0,
\end{aligned} \tag{18}$$

c/ Variational formulation. We now introduce :

$$V = \{v \mid v \in H^1(\Omega);\ v = 0 \text{ on } \Gamma_1\cup\Gamma_2\},$$

with the norm $\|v\|^2 = \int_\Omega v^2\, d\Omega + \int_\Omega \text{grad}v.\text{grad}v\, d\Omega$; and $\forall v, w \in V$:

$$a(v,w) = \int_\Omega \text{grad}v.\text{grad}w\, d\Omega + \alpha\int_{\Gamma_3} v\, w\, d\Gamma,$$

$$(v,w) = \int_\Omega v\, w\, d\Omega,\ |v| = \sqrt{(v,v)},$$

$$L(t,v) = -a\big(\tilde{u}(t),v\big) + \alpha\int_{\Gamma_3} g(t)v\, d\Gamma + \int_{\Gamma_4} k(t)v\, d\Gamma + \Big(\beta\frac{\partial \tilde{u}}{\partial t}(0),v\Big)$$

$$\Phi(v) = \ell\int_{\Omega_1(0)} v^-\, d\Omega + \ell\int_{\Omega_2(0)} v^+\, d\Omega.$$

We define also :

$$\forall y \in \mathbb{R},\ \beta(y) = b_1y^+ - b_2y^-,$$

and

$$W = \Big\{ v \mid v \in L^2(0,T;V);\ \frac{\partial v}{\partial t} \in L^2(0,T;V');\ v(0) = 0\Big\}$$

where V' is the dual space of V,

$$K(t) = \{v \mid v \in V;\ v \leq \psi_2(t)\},$$
$$\mathcal{K} = \{v \mid v \in W;\ v(t) \leq \psi_2(t)\}.$$

One can show that the equations (17) and (18) are formally equivalent to find the functions u and p which verify :

$$\begin{cases} u(0) = 0, \\ \forall t \in [0,T],\ \ \forall v \in V,\ \left(\beta\left(\frac{du}{dt} + \frac{d\tilde{u}}{dt}\right), v - \frac{du}{dt}\right) + a\left(u, v - \frac{du}{dt}\right) + \phi\left(v + \frac{d\tilde{u}}{dt}\right) \\ - \phi\left(\frac{du}{dt} + \frac{d\tilde{u}}{dt}\right) \geq L\left(t, v - \frac{d\tilde{u}}{dt}\right), \end{cases} \tag{20}$$

$$\begin{cases} p(0) = 0,\ p(t) \in K(t) \text{ a.e. in } t, \\ \forall v \in \mathcal{K},\ \int_0^T \{b_1\left(\frac{dv}{dt}, v-p\right) + a(p, v-p)\}d\tau \geq \frac{b_1}{2}\,|v(T)-p(T)|^2, \end{cases} \tag{21}$$

Note.- The condition $p \geq \psi_1$ is not part of the definition of $\mathcal{K}$ and K. We will see later on that the solution of problem (21) verifies it.

III. THE MAIN RESULTS.

The following theorems define the condition of existence and uniqueness of the solutions of equations (20), (21) which are then solutions of the problem.

THEOREM I. *If* $\tilde{u}$, $\frac{d\tilde{u}}{dt}$, $\frac{d^2\tilde{u}}{dt^2} \in L^2(0,T;H^1(\Omega))$, $\frac{d^3\tilde{u}}{dt^3} \in L^2(Q)$ *and if* $\frac{d\tilde{u}}{dt}(0)|\Omega_1(0) \geq 0$, $\frac{d\tilde{u}}{dt}(0)|\Omega_2(0) \leq 0$ *(i.e. if* θ_0 *and* h_0 *verify the relations* (11)), *if* k, $\frac{dk}{dt}, \frac{d^2k}{dt^2} \in L^2(\sum_4)$ *and if* g, $\frac{dg}{dt}$, $\frac{d^2g}{dt^2} \in L^2(\sum_3)$, *there exists a unique solution* u *of problem* (20) *which verifies :*

$$u \in C([0,T];V),\ \frac{du}{dt} \in L^\infty(0,T;V) \cap C([0,T];L^2(\Omega));\ \frac{d^2u}{dt^2} \in L^2(Q).$$

The freezing index $\hat{u} = u + \tilde{u}$ *is also unique.*

THEOREM II. *If the hypotheses of theorem* I *are verified and if* $\tilde{h} \in L^2(0,T;H^1(\Omega))$, $\frac{d\tilde{h}}{dt} \in L^2(Q)$, *there exists a unique solution* p *of problem* (21) *which verifies* $p \in L^\infty(0,T;V) \cap C([0,T];L^2(\Omega))$, $p \leq 0$, $p \geq \psi_1$. *The head of water is also unique.*

IV. PROOF OUTLINE OF THEOREM I.

We solve first a regularized problem in a finite dimension space. We obtain next a priori estimates which allow to conclude.

a/ Regularized problem. Let $\eta > 0$ and

$$\phi_\eta(y) = \begin{cases} 0 \text{ if } y \leq 0, \\ -\frac{y^4}{2\eta^3} + \frac{y^3}{\eta^2}, \text{ if } 0 \leq y \leq \eta, \\ y - \frac{\eta}{2}, \text{ if } y \geq \eta, \end{cases}$$

$$\beta_\eta(y) = (b_1 - b_2)\phi_\eta(y) + b_2 y$$

The function ϕ_η is convex and ϕ_η et β_η are C^2-functions. We define also :

$$\Phi_\eta(v) = \ell\int_{\Omega_1(0)} \phi_\eta(-v)d\Omega + \ell\int_{\Omega_2(0)} \phi_\eta(v)d\Omega.$$

Let $[w_1, \ldots, w_n]$ be a basis of $V_n \subset V$ such that $\bigcup_n V_n$ is dense in V. Let $D^1\Phi_\eta(v)$ and $D^2\Phi_\eta$ the first derivatives of Φ_η. We seek $u_n = \sum_{i=1}^{n} y_i(t)w_i$, solution of the ordinary differential equations :

$$\begin{cases} u_n(0) = 0, \\ \left(\beta_\eta\left(\frac{du_n}{dt} + \frac{d\tilde{u}}{dt}\right), w_j\right) + a(u_n, w_j) + \left(D^1\Phi_\eta\left(\frac{du_n}{dt} + \frac{d\tilde{u}}{dt}\right), w_j\right) = L(t, w_j),\ j=1,\ldots,n \end{cases} \quad (22)$$

One can observe that $\frac{du_n}{dt}(0) = 0$ for $D^1\Phi_\eta\left(\frac{d\tilde{u}}{dt}(0)\right) = 0$. The function $\frac{du_n}{dt}(t)$ is then solution of the ordinary differential equations obtained by differentiation of (22):

$$\begin{cases} \frac{du_n}{dt}(0) = 0, \\ \left(\frac{d\beta_\eta}{dy}\left(\frac{du_n}{dt} + \frac{d\tilde{u}}{dt}\right)\left(\frac{d^2u_n}{dt^2} + \frac{d^2\tilde{u}}{dt^2}\right) + D^2\Phi\left(\frac{du_n}{dt} + \frac{d\tilde{u}}{dt}\right)\left(\frac{d^2u_n}{dt^2} + \frac{d^2\tilde{u}}{dt^2}\right), w_j\right) \\ + a\left(\frac{du_n}{dt}, w_j\right) = \frac{\partial L}{\partial t}(t, w_j),\ j = 1, \ldots, n, \end{cases} \quad (23)$$

One can show by using the properties of Φ_η and β_η that these differential equations have solutions on $(0,t^n)$. The following a priori estimates will show that $t^n = T$.

b/ <u>A priori estimates</u>. From relation (23) we obtain :

$$\left(\frac{d\beta_\eta}{dy}\left(\frac{du_n}{dt} + \frac{d\tilde{u}}{dt}\right)\left(\frac{d^2u_n}{dt^2} + \frac{d^2\tilde{u}}{dt^2}\right) + D^2\Phi_\eta\left(\frac{du_n}{dt} + \frac{d\tilde{u}}{dt}\right)\left(\frac{d^2u_n}{dt^2} + \frac{d^2\tilde{u}}{dt^2}\right), \frac{d^2u_n}{dt^2} + \frac{d^2\tilde{u}}{dt^2}\right) + a\left(\frac{du_n}{dt}, \frac{d^2u_n}{dt^2}\right)$$

$$= \left(\frac{d}{dt}\left[\beta_\eta\left(\frac{du_n}{dt} + \frac{d\tilde{u}}{dt}\right) + D^1\Phi_\eta\left(\frac{du_n}{dt} + \frac{d\tilde{u}}{dt}\right)\right], \frac{d^2\tilde{u}}{dt^2}\right) + \frac{d}{dt}\left[\frac{\partial L}{\partial t}\left(t, \frac{du_n}{dt}\right)\right] - \frac{\partial^2 L}{\partial t^2}\left(t, \frac{du_n}{dt}\right)$$

We integrate this relation from 0 to t $(t < t^n)$ and use the following relations:

$$\forall v, w \in W,\ \left(\frac{d\beta_\eta}{dy}(w)v, v\right) \geq C|v|^2,\ |\beta_\eta(v)| \leq C|v|,\ \left(D^2\Phi_\eta(w)v, v\right) \geq 0,\ |D^1\Phi_\eta(w)v| \leq C|v|,$$

$$C\|v\|^2 \leq a(v,v),\quad |a(w,v)| \leq C\|w\|\,\|v\|,$$

where C is now and later on a strictly positive constant independant of n and η. After some computation we obtain :

$$C\int_0^t \left|\frac{d^2u_n}{dt}\right|^2 d\tau + C\left\|\frac{du_n}{dt}(t)\right\|^2 \leq C + C\left\|\frac{du_n}{dt}(t)\right\| + \int_0^t \xi(\tau)\left\|\frac{du_n}{dt}(\tau)\right\| d\tau$$

where $\xi \in L^2(0,T)$.

It follows that $t^n = T$ and that :

$$C\left\|\frac{du_n}{dt}(t)\right\|^2 \leq C + \int_0^t \xi(\tau)\left\|\frac{du_n}{dt}(\tau)\right\| d\tau$$

We now use Granwall's lemma :

$$\forall t \in [0,T], \left\| \frac{du_n}{dt}(t) \right\| < C, \tag{24}$$

It follows that :

$$\left\| \frac{d^2u_n}{dt^2} \right\|_{L^2(Q)} < C, \quad \forall t \in [0,T], \| u_n(t) \| < C, \tag{25}$$

c/ Limiting process - Uniqueness. We now let $n \to \infty$ and $\eta \to 0$. From the estimates (23), (25) we obtain there exists a function u and a subsequence (u_n) such that :

$u_n \to u$ in $L^2(0,T;V)$, $L^\infty(0,T;V)$ strong,

$\frac{du_n}{dt} \to \frac{du}{dt}$ in $L^2(0,T;V)$ weak, $L^\infty(0,T;V)$ weak*,

$\frac{d^2u_n}{dt^2} \to \frac{d^2u}{dt^2}$ in $L^2(Q)$ weak.

We know [13] that we can choose $u \in C([0,T],V)$ and $\frac{du}{dt} \in C([0,T];L^2(\Omega))$. By letting $n \to \infty$ and $\eta \to 0$ in the relation (23), we obtain that u is a solution of the problem (20). The proof of the uniqueness is straightforward.

V. PROOF OUTLINE OF THEOREM II.

The proof uses the penalization of the projection on the convex set K(t) which depends on time.

a/ Penalized problem. Let $\eta > 0$; according to [13] there exists $p_\eta \in W$ such that

$$\forall v \in V, \left(\frac{dp_\eta}{dt}, v\right) + a(p_\eta v) + \frac{1}{\eta}((p_\eta - \psi_2)^+, v) = 0 \tag{26}$$

By letting $v = -(p_\eta - \psi_1)^-$ and then p_η^+ in relation (26), one can successively show that $(p_\eta - \psi_1)^- = 0$ and that $p_\eta^+ = 0$. We have then :

$$p_\eta > \psi_1 \text{ and } p_\eta < 0, \tag{27}$$

b/ A priori estimates. From the properties of $\hat{u}$ and $\tilde{h}$ we have : $\psi_2^- \in L^2(0,T;V)$ and $\frac{d\psi_2^-}{dt} \in L^2(Q)$. By letting $v = p_\eta + \psi_2^-$ in relation (26), we obtain :

$$t \in [0,T], |p_\eta(t)| < C, \|p_\eta\|_{L^2(0,T;V)} < C$$

and

$$\frac{1}{\eta} \int_0^T |(p_\eta - \psi_2)^+|^2 d\tau < C, \tag{28}$$

c/ Limiting process. We now let $\eta \to 0$. From the estimates (28) we obtain that there exists a function p and a subsequence (p_η) such that :

$p_\eta \to p$ in $L^2(0,T;V)$ weak, in $L^\infty(0,T;L^2(\Omega))$ weak*,

$p \in K(t)$ a.e. in t.

We now have for any $v \in \mathcal{K}$:

$$\int_0^T \{b_1(\frac{dv}{dt},v-p_\eta) + a(p_\eta,v-p_\eta)\}d\tau = \int_0^T \{b_1(\frac{dv}{dt} - \frac{dp_\eta}{dt},v-p_\eta) - \ldots$$

$$- \frac{1}{\eta}((p_\eta-\psi_2)^+,v-p_\eta)\}d\tau \geq \frac{b_1}{2}|v(T)-p_\eta(T)|^2, \qquad (29)$$

for $-\frac{1}{\eta}\int_0^T((p_\eta-\psi_2)^+,v-p_\eta)d\tau \geq 0$ because of $v \in \mathcal{K}$.

By letting $\eta \to 0$ in relation (29), we obtain that p is a solution of problem (21). The relation (27) shows that $p \leq 0$ and $p \geq \psi_1$. The uniqueness and the continuity of p can be shown by using a method introduced in [13].

Note :

1/ The relation $p \leq 0$ shows that $h \leq \tilde{h}$ and means that the head is lower in the case of freezing than in the case of no freezing. It means also that in a neighbourhood of Γ_1 where $\psi_2 \geq 0$ the medium is not frozen.

2/ It is also possible to consider the quasi-static situation ($b_1 = b_2 = 0$). Similar results can be obtained. In this case the main term only, i.e. the latent heat of fusion, is left in the energy of the materials.

VI. NUMERICAL EXAMPLES.

For application to road problems, a one dimensional program has been developped. The abscisse stands for the depth inside a multilayered pavement. The computer drawn figures (2) and (3) represents the surface temperature of a road v. time and the position of the frost line v. time. Developpments concerning applications will be found in [2,16].

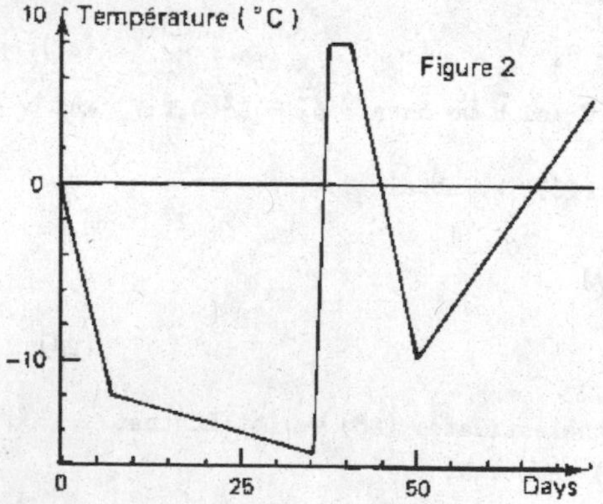

Figure 2

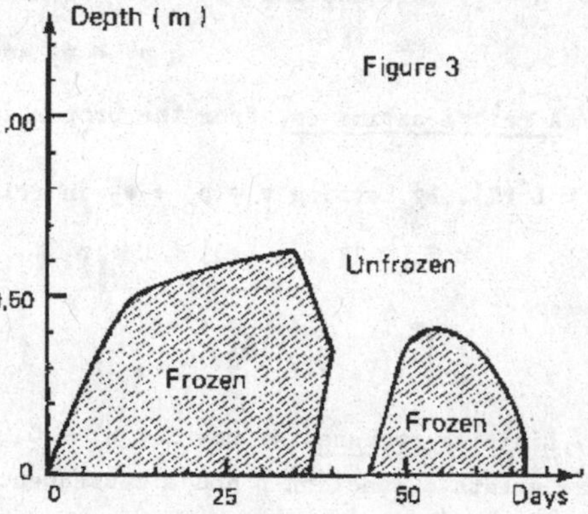

Figure 3

R E F E R E N C E S

[1] J. AGUIRRE-PUENTE, M. FREMOND : Congélation d'un milieu poreux de texture fine, saturé d'eau, considérée comme un couplage de plusieurs phénomènes élémentaires. XIVème Congrès International du Froid. Moscou. Septembre 1975.

[2] L. CANIARD, A. DUPAS, M. FREMOND, M. LEVY : Comportement thermique d'une structure routière soumise à un cycle de gel-dégel. Simulations expérimentale et numérique. VIème Congrès International de la Fondation Française d'Etudes Nordiques. Les problèmes posés par la gélifraction. Le Havre. Avril 1975.

[3] H. BREZIS : On some degenerate nonlinear parabolic equations. Proc. Symp. Pure Math., 18 (pt. 1), Amer. Math. Soc., Providence, R.I., 1970, pp. 28-38.

[4] H. BREZIS : Communication personnelle.

[5] J.R. CANNON, C.D. HILL : Existence, uniqueness, stability and monotone dependance in a Stefan problem for the heat equation, J. Math. Mech. 17, p.1,1967.

[6] J.R. CANNON, M. PRIMICERIO : A two phase Stefan problem with temperature boundary conditions, Ann. Mat. Pura Appl. 88 (IV), p. 177, 1971.

[7] G. DUVAUT : Comptes Rendus à l'Académie des Sciences de Paris, 276, Série A, p. 1461, 1973.

[8] M. FREMOND : Frost propagation in porous media. International Conference on Computational Methods in Nonlinear Mechanics - Austin, 1974.

[9] A. FRIEDMAN : The Stefan problem in several space variables, Transaction American Mathematical Society 132, p. 51, 1968.

[10] P. GERMAIN: Cours de Mécanique des Milieux Continus. Masson - Paris 1973.

[11] S.L. KAMENOMOSTSKAJA: On Stefan's problem. Mat. Sb. 53, 489, 1961.

[12] O.A. LADYZENSKAYA, V. SOLLOMIKOV, N. URALCEVA : Linear and quasilinear equations of parabolic type. English transl. Transl. Math. Monographs. Vol. 23, Amer. Math. Soc. 1968.

[13] J.L. LIONS : Quelques méthodes de résolution des problèmes aux limites non linéaires. Dunod. 1969.

[14] L.L. LIONS : Introduction to some aspects of free surface problems. I.R.I.A. Paris 1975.

[15] O. OLEINIK : On Stefan-type free boundary problems for parabolic equations. Ist. Naz. Acta. Mat. p. 388, Ed. Cremonese. 1963.

[16] A. PHILIPPE, J. AGUIRRE-PUENTE, H. BERTOUILLE, M. FREMOND : La propagation du gel dans les chaussées et sa simulation à la station de gel expérimentale de Caen. Bulletin de liaison des Laboratoires des Ponts et Chaussées. Supplément au n°68 - 1973.

[17] A. DAMLAMIAN, Thesis, Havard University 1974 - Thèse Paris, to appear.

VISCOUS FLUID FLOW IN CHEMICALLY REACTING AND DIFFUSING SYSTEMS

R. AMIEL – Département de Mathématiques – Université de Nice
G. GEYMONAT – Istituto Matematico – Politecnico – Torino

1 – Introduction and statement of the problem.

We are interested to study the functional analysis approach to the flow of a newtonian or not-newtonian fluid with two chemically reacting components and with heat and mass transfer; the governing equations which determine the fluid velocity and temperature and the chemical species concentrations are based on conservation laws involving diffusion, convection, chemical reactions and external sources. Such type of flow are usual in chemical reactor theory [1], [2], [9].

More precisely we study the steady state in a smooth region $\Omega \subset \mathbb{R}^2$; in this case we can introduce the stream function Ψ and in an extended Boussinesq approximation we obtain that the unknown Ψ, T (absolute temperature), c (concentration of diffusing substance) will satisfy the following system of nonlinear dimensionless equations :

$$(1.1)\quad \frac{1}{Re}\,\mathrm{Trace}\Big(D(2\mu\, D\Psi)\Big) + \frac{\partial}{\partial y}\left(\frac{\partial\Psi}{\partial x}\,\Delta\Psi\right) - \frac{\partial}{\partial x}\left(\frac{\partial\Psi}{\partial y}\,\Delta\Psi\right) + S_1\left(c,T,\frac{\partial\Psi}{\partial x},\frac{\partial\Psi}{\partial y}\right) = f_1(x,y)$$

$$(1.2)\quad -(Pe_T)^{-1}\mathrm{div}\,(k\ \mathrm{grad}\ T) + \frac{\partial}{\partial y}\left(\frac{\partial\Psi}{\partial x}\,T\right) - \frac{\partial}{\partial x}\left(\frac{\partial\Psi}{\partial y}\,T\right) + S_2\left(c,T,\frac{\partial\Psi}{\partial x},\frac{\partial\Psi}{\partial y}\right) = f_2(x,y)$$

$$(1.3)\quad -(Pe_D)^{-1}\mathrm{div}\,(\rho\delta\ \mathrm{grad}\ c) + \frac{\partial}{\partial y}\left(\frac{\partial\Psi}{\partial x}\,c\right) - \frac{\partial}{\partial x}\left(\frac{\partial\Psi}{\partial y}\,c\right) + S_3\left(c,T,\frac{\partial\Psi}{\partial x},\frac{\partial\Psi}{\partial y}\right) = f_3(x,y)$$

where :

$$D = \begin{bmatrix} \partial^2/\partial x\partial y & \tfrac{1}{2}(\partial^2/\partial y^2 - \partial^2/\partial x^2) \\ \tfrac{1}{2}(\partial^2/\partial y^2 - \partial^2/\partial x^2) & -\partial^2/\partial x\partial y \end{bmatrix}$$ is the differential matrix operator of strain rate .

Re, Pe_T, Pe_D : Reynolds number, Péclet numbers of heat and diffusion

$\mu\left(c,T,\frac{\partial\Psi}{\partial x},\frac{\partial\Psi}{\partial y},D\Psi\right)$: viscosity

$\delta\left(c,T,\frac{\partial\Psi}{\partial x},\frac{\partial\Psi}{\partial y}\right)$, $k\left(c,T,\frac{\partial\Psi}{\partial x},\frac{\partial\Psi}{\partial y}\right)$: diffusivity and thermal conductivity

$\rho(c,T)$: mass density of the mixture

$S_1\left(c,T,\frac{\partial\Psi}{\partial x},\frac{\partial\Psi}{\partial y}\right)$: action of external forces (e.g. gravity, Coriolis type force,...)

$S_2\left(c,T,\frac{\partial\Psi}{\partial x},\frac{\partial\Psi}{\partial y}\right)$, $S_3\left(c,T,\frac{\partial\Psi}{\partial x},\frac{\partial\Psi}{\partial y}\right)$: thermal and mass reaction rates

$f_1(x,y)$, $f_2(x,y)$, $f_3(x,y)$: eventual distributed sources

For example the functions considered can be of the following type :

(1.4) $\mu\left(c,T,\frac{\partial\Psi}{\partial x},\frac{\partial\Psi}{\partial y},D\Psi\right) = \mu_1\left(c,T,\frac{\partial\Psi}{\partial x},\frac{\partial\Psi}{\partial y}\right) + \beta\left[(\text{Trace }(D\Psi)^2)^{\alpha/2} + \gamma\right]^{-\delta} \quad \alpha,\delta > 0, \alpha\delta \le 1$

where μ_1 and $\mu_2 = \mu - \mu_1$ are the newtonian and non-newtonian parts of viscosity;

μ_1 can be given for instance by the Andrade law ;

(1.5) $\rho^{-1}(c,T) = c\left(1+\alpha_1(T-T^*)\right)\rho_1^{-1} + (1-c)\left(1+\beta_1(T-T^*)\right)\rho_2^{-1}$ where ρ_1, ρ_2 are the references mass densities of each component at the reference temperature T^*.

(1.6) $S_1(c,T) = g\ \partial\rho/\partial x$ where g is the gravity coefficient and the x-axis is the vertical direction ;

(1.7) $S_2(c,T) = \Delta H\ \rho\ c\ \exp(-E/T)$, $S_3(c,T) = \rho\ c\ \exp(-E/T)$ (Arrhenius form)

(1.8) $S_i(c,T) = -d_i\,(1-c)\exp[T/(1+\gamma T)] - h_i$, $d_i,\gamma,h_2 > 0$, $h_3 = 0, i = 2,3$ [3]

The system (1.1),(1.2),(1.3) must be complemented with the boundary conditions on $\partial\Omega = \Gamma_0 \cup \Gamma_1$ with n outward normal and meas $\Gamma_0 > 0$:

(1.9) $\Psi|_\Gamma, \frac{\partial\Psi}{\partial n}|_\Gamma, T|_{\Gamma_0}, c|_{\Gamma_0}$ are given functions with $0 \le c|_{\Gamma_0} \le 1$ and $T|_{\Gamma_0} \ge 0$.

(1.10) $k(Pe_T)^{-1}\ \partial T/\partial n + h_T(x,y)(T - g_T(x,y)) = 0$ on Γ_1

(1.11) $\rho\delta(Pe_D)^{-1}\ \partial c/\partial n + h_c(x,y)(c - g_c(x,y)) = 0$ on Γ_1 .

where h_T and h_c are transfer functions and g_T, g_c are the surroundings temperature and concentration.

Moreover the compatibility condition (total mass conservation)

(1.12) $\int_\Gamma \rho(c,T)\left(\cos(n,x)\ \partial\Psi/\partial y - \cos(n,y)\ \partial\Psi/\partial x\right)d\Gamma = 0$

will be <u>a priori</u> satisfied assuming on Γ_1 Ψ constant, and by a suitable choice of the boundary data (1.9) on Γ_0 .

The physically significant restrictions :

(1.13) $T \ge 0$ and $0 \le c \le 1$

will be <u>a posteriori</u> satisfied by a maximum principle (see §.6).

Our purpose is to prove that this problem has a solution, in particular for the examples previously stated; more general results and complete proofs will be given in [O].

2 - Functional analysis approach

Let $V = H^2(\Omega) \times H^1(\Omega) \times H^1(\Omega)$ and $V_o = H_o^2(\Omega) \times H^1(\Omega\,;\,\Gamma_o) \times H^1(\Omega\,;\,\Gamma_o)$ where $H^1(\Omega\,;\,\Gamma_o) = \left\{\varphi \in H^1(\Omega)\ ;\ \varphi|_{\Gamma_o} = o\right\}$ equipped with usual norms (equivalent onto V_o) where $u = (\Psi, T, c)$ is the generic element

$$\|u\|^2 = \|\Psi\|^2_{H^2(\Omega)} + \|T\|^2_{H^1(\Omega)} + \|c\|^2_{H^1(\Omega)}$$

$$\|u\|_o^2 = \|\Delta\Psi\|^2_{L^2(\Omega)} + \int_\Omega |\mathrm{grad}\ T|^2 d\,\Omega + \int_\Omega |\mathrm{grad}\ c|^2 d\,\Omega$$

The problem (1.1) - (1.3) , (1.9) - (1.12) can be written in a symbolic way as:

(2.1) $$\mathcal{A}(u) \equiv \mathcal{D}(u) + \mathcal{B}(u) + \mathcal{S}(u) = f$$

where $\mathcal{D}(u)$ represents the diffusion (or conduction) terms and $\mathcal{B}(u)$ represents the convection terms.

Let R_o the set of $u_o = (\Psi_o, T_o, c_o) \in V$ verifying (1.9) and (1.12) ; for any fixed $u_o \in R_o$ we can study the weak problem : Let $f \in V_o'$ be given ; find $\tilde{u} \in V_o$ such that

(2.2) $$\left(\mathcal{A}(\tilde{u} + u_o),\ w\right) = (f\ ,\ w) \qquad \forall\, w \in V_o\ .$$

A wellknown theorem of LERAY-LIONS [6] states that if $\mathcal{A}$ is pseudo-monotone and coercive then the weak problem has at least one solution.

We shall now give sufficients conditions on $\mathcal{D}$, $\mathcal{B}$ and $\mathcal{S}$ to ensure the pseudo-monotonicity and coercivity of $\mathcal{A}$.

Conditions on $\mathcal{D}$:

D1 There exists a form $a(u;v,w)$ defined on $V \times V \times V_o$ and linear with respect to the 2nd and 3rd variables such that

$$\left(\mathcal{D}(u)\,,w\right) = a(u;u,w) \quad \forall\, u \in V, \quad \forall\, w \in V_o$$

D2 $|a(u;v,w)| \leq M\,\|v\|\,\|w\|_o \quad \forall\, u,\ v \in V\ , \quad \forall\, w \in V_o$

D3 $\exists\, A > o$ such that $a(v+z_o; v,v) \geq A\,\|v\|_o^2 \quad \forall\, v \in V_o, \quad \forall\, z_o \in R_o$

D4 (pseudo-monotonicity) : if $w_j \to w$ weakly in V_o and $\limsup \left(\mathcal{D}(w_j+z_o), w_j - w\right) \leq o$ then $\liminf \left(\mathcal{D}(w_j + z_o)\,, w_j - v\right) \geq \left(\mathcal{D}(w + z_o), w-v\right) \quad \forall\, v \in V_o\,, \forall\, z_o \in R_o$

Conditions on $\mathcal{B}$

B1 There exists a trilinear form $b(u,v,w)$ defined on $V \times V \times V_o$ such that

$$\left(\mathcal{B}(u),w\right) = b(u,u,w) \quad \forall\, u \in V,\ \forall\, w \in V_o$$

B2 $b(v, w+z_o, v) = 0 \quad \forall\, v,w \in V_o,\ \forall\, z_o \in R_o$

B3 $|b(u,v,w)| \le B\, \|u\|\, \|v\|\, \|w\|_o \quad \forall\, u,v \in V,\ \forall\, w \in V_o$

B4 If $w_j \to w$ weakly in V_o then $b(v+z_o, w_j+z_o, w_j - v) \to b(v+z_o, w+z_o, w-v)$

$$\forall\, v \in V_o,\ \forall\, z_o \in R_o$$

B5 $\forall\, \varepsilon > 0$ there exists $z_o^\varepsilon \in R_o$ such that $|b(z_o^\varepsilon, v, v)| \le \varepsilon \|v\|_o^2 \quad \forall\, v \in V_o$

Conditions on $\mathcal{S}$

S1 $\mathcal{S}$ has at most a linear growth , i.e. there exist $L_1, L_2 \ge 0$ and $0 \le \alpha \le 1$ such that $\left|\left(\mathcal{S}(v+z_o), w\right)\right| \le L_1 \|w\|_o \left(\|v\|_o^\alpha + \|z_o\|^\alpha + L_2\right) \quad \forall\, v,w \in V_o,\ \forall\, z_o \in R_o$

S2 If $w_j \to w$ weakly in V_o then $\left(\mathcal{S}(w_j+z_o), w_j - v\right) \to \left(\mathcal{S}(w+z_o), w-v\right) \quad \forall\, v \in V_o, \forall\, z_o \in R_o$

We can state without proof the following existence theorem.

THEOREM 1 Let be verified the previous conditions on $\mathcal{A}$, $\mathcal{B}$, $\mathcal{S}$; then $\forall\, f \in V_o'$ there exist $u_o \in R_o$ and $\tilde{u} \in V_o$ satisfying (2.2) if $0 \le \alpha < 1$ or if $\alpha = 1$ and $A > L_1$.

In order to use theorem 1 for the problem considered in §1 we need some natural assumptions on the non linear functions μ , S_1 , k , S_2 , ρ , δ , S_3 . Indeed such functions are defined only for arguments belonging to special intervals depending on the physics of the model and obtained by theoretical laws or experimental data. However from a mathematical point of view we need that such functions shall be defined in all $\mathbb{R}^N$; for this we shall extend the functions from the original range in a good way. In § 3-5 we introduce some qualitative hypothesis on such non linear functions in order to satisfy the conditions on $\mathcal{A}$, $\mathcal{B}$, $\mathcal{S}$. In § 7 we go back to the examples of § 1.

3 - Verification of the conditions on $\mathcal{B}$.

D1 For $u = (\Psi,T,c), v = (\eta,S,e) \in V$ and $w = (\varphi,R,d) \in V_0$ and with the notations of § 1 via a Green formula we have :

$$(3.1)\quad \left(\mathcal{B}(u),w\right) = \frac{1}{Re}\int_\Omega \text{Trace}\left([2\mu\, D\Psi][D\varphi]d\,\Omega + \frac{1}{Pe_T}\int_\Omega k \text{ grad } T \text{ grad } R \; d\,\Omega\right.$$

$$+\int_{\Gamma_1} h_T\, T\, R\, d\Gamma + \frac{1}{Pe_D}\int_\Omega \rho\delta \text{ grad } c \text{ grad } d\; d\Omega + \int_{\Gamma_1} h_c\, c\, d\, d\Gamma .$$

We suppose that $\mu = \mu_1 + \mu_2$ with μ_1 newtonian and μ_2 not-newtonian; then we can define the forms

$$(3.2)\quad a_1(u;v,w) = \frac{1}{Re}\int_\Omega \text{Trace}[2\mu_1(u)\, D\eta][D\varphi]d\,\Omega + \frac{1}{Pe_T}\int_\Omega k(u)\text{grad } S \text{ grad } R\; d\,\Omega +$$

$$+\int_{\Gamma_1} h_T\, S R\, d\Gamma + \frac{1}{Pe_D}\int_\Omega \rho(u)\,\delta(u) \text{ grad } e \text{ grad } d\; d\,\Omega + \int_{\Gamma_1} h_c\, e\, d\, d\Gamma$$

$$(3.3)\quad a_2(u;v,w) = \frac{1}{Re}\int_\Omega \text{Trace}[2\mu_2(u)\, D\eta][D\varphi]d\,\Omega \; .$$

Such forms are well-defined if $\mu_i(u)$, $k(u)$, $\rho(u)\delta(u) \in L^\infty(\Omega)$ for all $u \in V$ and this condition is verified under the hypothesis :

(3.4) $\mu_1(\xi_1,\xi_2,\xi_3,\xi_4)$, $\mu_2(\xi_1,\xi_2,\xi_3,\xi_4,\xi_5)$, $k(\xi_1,\xi_2,\xi_3,\xi_4)$, $\rho(\xi_1,\xi_2)$, $\delta(\xi_1,\xi_2,\xi_3,\xi_4)$ are Caratheodory functions and $\mu_1,\mu_2,k,\rho\delta$ are uniformly bounded for all $\xi_i \in \mathbb{R}$.

D2 It is verified with

$$(3.5)\quad M = \sup\left\{ \frac{M_1}{Re}\, |\mu|_{L^\infty} \; ; \; \frac{C_1}{Pe_T}|k|_{L^\infty} + C_2|h_T|_{L^\infty(\Gamma_1)} \; ; \; \frac{C_1}{Pe_D}|\rho\delta|_{L^\infty} + C_2|h_c|_{L^\infty(\Gamma_1)}\right\}$$

where M_1, C_1, C_2 are positive constants depending only from Ω and Γ_1.

D3 It is verified with

(3.6) $$A = \inf \left\{ \frac{1}{Re} \inf \mu \; ; \; \frac{1}{Pe_T} \inf k + \gamma \inf_{\Gamma_1} h_T \; ; \; \frac{1}{Pe_D} \inf(\rho\delta) + \gamma \inf_{\Gamma_1} h_c \right\}$$

where $\gamma > 0$ depends from Γ_1. Obviously

(3.7) if $h_c, h_T \geq 0$ and $\mu \geq \mu_0 > 0$, $k \geq k_0 > 0$, $\rho\delta \geq \delta_0 > 0$ then $A > 0$.

D4 LEMMA. The conditions D4.1 - D4.6 imply D4.

D4.1 $a_1(v+z_0; w,w) \geq 0 \quad \forall\, v,w \in V_0, \quad \forall\, z_0 \in R_0$

D4.2 $a_2(v+z_0; v+z_0; v-w) \geq a_2(w+z_0; w+z_0, v-w) \quad \forall\, v,w \in V, \quad \forall\, z_0 \in R_0$

D4.3 $w_j \to w$ weakly in $V_0 \Rightarrow a_1(w_j+z_0; v+z_0, w_j-w) \to 0 \quad \forall\, v,w \in V, \quad \forall\, z_0 \in R_0$

D4.4 $w_j \to w$ weakly in $V_0 \Rightarrow a_2(v+z_0; v+z_0, w_j-w) \to 0 \quad \forall\, v,w \in V, \quad \forall\, z_0 \in R_0$

D4.5 $w_j \to w$ weakly in V_0 and $\lim a_1(w_j+z_0; w_j+z_0, w_j-w) = 0 \Rightarrow$
$\lim a_1(w_j+z_0; w_j+z_0, v) = a_1(w+z_0; w+z_0, v) \quad \forall\, v \in V_0, \quad \forall\, z_0 \in R_0$

D4.6 $\lim_{\lambda \to 0} a_2(v+\lambda w+z_0; v+\lambda w+z_0, z) = a_2(v+z_0; v+z_0, z) \quad \forall\, v,w,z \in V_0, \quad \forall\, z_0 \in R_0$

D4.1 Follows from (3.7) and the hypothesis $\mu_1 \geq 0$.

D4.2 It is verified only for particular μ_2; for instance if μ_2 is function only of $\left(\mathrm{Trace}[D\Psi]^2\right)^{\frac{1}{2}}$ it suffies that $\lambda \to \lambda\, \mu_2(\lambda)$ is not decreasing from $\mathbb{R}_+$ to $\mathbb{R}$, like in the pseudoplastic case $\mu_2(\lambda) = \beta(\lambda^{\alpha} + \gamma)^{-\delta}$ of (1.4).

D4.3 If $w_j \to w$ weakly in V_0 then $w_j \to w$ strongly in $W_0^{1,p}(\Omega) \times L^q(\Omega) \times L^r(\Omega)$ for all $p,q,r \geq 1$; from (3.4) we can then apply a well known theorem of KRASNOSEL'SKII-VAINBERG [5], [10], to obtain the desired result.

D4.4 Obvious

D4.5 The proof is the same that [6] lemma 3.3.

D4.6 Follows from (3.4) using the KRASNOSEL'SKII-VAINBERG theorem for $[\mu_2\, D\Psi]$.

4 - Verification of the condition $\mathcal{B}$

B1 With notations of §3 via a Green formula we obtain

$$(4.1)\quad \big(\mathcal{B}(u),w\big) = b_1(\Psi,\Psi,\varphi) + b_2(T,\Psi,R) + b_3(c,\Psi,d)$$

where

$$(4.2)\quad b_1(\Psi,\eta,\varphi) = \int_\Omega \left(\frac{\partial\Psi}{\partial y}\frac{\partial\varphi}{\partial x} - \frac{\partial\Psi}{\partial x}\frac{\partial\varphi}{\partial y}\right)\Delta\eta \, d\Omega$$

is the usual trilinear form of NAVIER-STOKES equations,

$$(4.3)\quad b_2(T,\eta,R) = \tfrac{1}{2}\int_\Omega \left\{\frac{\partial\eta}{\partial x}\left(\frac{\partial T}{\partial y}R - \frac{\partial R}{\partial y}T\right) + \frac{\partial\eta}{\partial y}\left(\frac{\partial R}{\partial x}T - \frac{\partial T}{\partial x}R\right)\right\} d\Omega$$

$$+ \tfrac{1}{2}\int_{\Gamma_1}\left(\frac{\partial\eta}{\partial x}\cos(n,y) - \frac{\partial\eta}{\partial y}\cos(n,x)\right)T\,R\,d\Gamma$$

and analougsly for $b_3(c,\eta,d)$.

B2 The compatibility condition (1.12) implies that also the boundary integrals in b_2 and b_3 are $= 0$.

B3 , B4 are proved as for the classical NAVIER-STOKES equations [8] .

B5 is proved as for the classical NAVIER-STOKES equations [8] because the boundary integrals in b_2, b_3 are $= 0$ for $\eta \in H_0^2(\Omega)$.

5 - Verification of the conditions on $\mathcal{S}$

S1 For $w = (\varphi,R,d) \in V_0$ we have

$$\big(\mathcal{S}(u),w\big) = \big(S_1(u),\varphi\big) + \big(S_2(u),R\big) + \big(S_3(u),d\big)$$

with $H^{-2}(\Omega) \ni S_1(u) = \sum_{|\lambda|\le 2} D^\lambda S_{1\lambda}\left(c,T,\frac{\partial\Psi}{\partial x},\frac{\partial\Psi}{\partial y}\right)$ where $D^\lambda = \partial^{\lambda_1+\lambda_2}/\partial x^{\lambda_1}\partial y^{\lambda_2}$

and $S_2(u) = S_2\left(c,T,\frac{\partial\Psi}{\partial x},\frac{\partial\Psi}{\partial y}\right)$, $S_3(u) = S_3\left(c,T,\frac{\partial\Psi}{\partial x},\frac{\partial\Psi}{\partial y}\right) \in L^2(\Omega)$.

To obtain the desired estimate we shall suppose :

(5.1) $S_{1\lambda}(\xi_1,\xi_2,\xi_3,\xi_4)$ and $S_i(\xi_1,\xi_2,\xi_3,\xi_4)$ are Carathéodory functions with at most a linear growth .

S2 It follows from the KRASNOSEL'SKII-VAINBERG theorem .

6 - Maximum principle

In this § the order relation between the functions must be intended in the sense of $H^1(\Omega)$; for more details on this order relation see [7] , [4] .

THEOREM 2 Let the hypothesis of theorem 1 satisfied. If moreover

(6.1) $f_2 \geq 0 \quad , \quad g_T|_{\Gamma_1} \geq 0$

(6.2) $S_2(\xi_1,\xi_2,\xi_3,\xi_4)\min(0,\xi_2) \geq 0 \qquad \forall\, \xi_i \in \mathbb{R} \quad i = 1,\dots,4$

then $T \geq 0$

THEOREM 3 Let the hypothesis of theorem 1 satisfied. If moreover

(6.3). $f_3 = 0 \quad , \qquad 0 \leq g_c|_{\Gamma_1} \leq 1$

(6.4) $S_3(\xi_1,\xi_2,\xi_3,\xi_4)\max(0,\xi_1 - 1) \geq 0 \qquad \forall\, \xi_i \in \mathbb{R}, \quad i = 1,\dots,4$

(6.5) $S_3(\xi_1,\xi_2,\xi_3,\xi_4)\min(0,\xi_1) \geq 0 \qquad \forall\, \xi_i \in \mathbb{R} , \quad i = 1,\dots,4$

then $0 \leq c \leq 1$

Proof of theorem 2 Let $u = \tilde{u} + u_0 = (\Psi,T,c) \in V$ verifying (2.2) , i.e. such that for all $w \in V_0$ $\big(\mathcal{G}(u),w\big) = (f,w)$. Taking $w = (0,R,0)$ with $R \in H^1(\Omega\,;\,\Gamma_0)$ we have .

$$\frac{1}{Pe_T}\int_\Omega k \operatorname{grad} T \operatorname{grad} R\, d\Omega + b_2(T,\Psi,R) + \int_\Omega S_2\, R\, d\Omega + \int_{\Gamma_1} h_T(T - g_T)R\, d\Gamma =$$
$$= \int_\Omega f_2\, R\, d\Omega \quad .$$

Let $T_1 = \min(0,T)$, then $T_1 \in H^1(\Omega\,;\,\Gamma_0)$ because $T \geq 0$ on Γ_0 ; so we can take $R = T_1$ and from $b_2(T,\Psi,T_1) = 0$ we have

$$0 = \frac{1}{Pe_T}\int_\Omega k\,|\operatorname{grad} T_1|^2\, d\Omega + \int_\Omega S_2\, T_1\, d\Omega + \int_{\Gamma_1} h_T\, T_1^2\, d\Gamma - \int_{\Gamma_1} h_T\, g_T\, T_1\, d\Gamma -$$
$$- \int_\Omega f_2\, T_1\, d\Omega \quad .$$

From (6.1) , (6.2) and D3 it follows $T_1 = o$. Q.E.D.

In the same way we can prove the theorem 3 and the following results .

<u>THEOREM 4</u> <u>Let the hypothesis of theorem 1 satisfied. If, moreover</u>

(6.6) $f_2 \geq o$, $g_T\Big|_{\Gamma_1} \geq T_*$, $T\Big|_{\Gamma_0} \geq T_* > o$

(6.7) $S_2(\xi_1,\xi_2,\xi_3,\xi_4)\min(o,\xi_2 - T_*) \geq o \quad \forall\ \xi_i \in \mathbb{R}$, $i = 1,\ldots,4$

<u>then</u> $T \geq T_*$.

<u>THEOREM 5</u> <u>Let the hypothesis of theorem 1 satisfied. If , moreover</u>

(6.8) $f_2 \leq o$, $g_T\Big|_{\Gamma_1} \leq T^*$, $T\Big|_{\Gamma_0} \leq T^*$

(6.9) $S_2(\xi_1,\xi_2,\xi_3,\xi_4)\max(o,\xi_2 - T^*) \geq o \quad \forall\ \xi_i \in \mathbb{R}$, $i = 1,\ldots,4$

<u>then</u> $T \leq T^*$.

7- <u>Examples</u>

We shall exhibit some examples to illustrate how the hypothesis are verified in practice.

7.1 <u>Newtonian viscosity</u> $\mu_1(c,T) = A(c)\exp(K/T)$ (Andrade Law) where $K > o$ is a constant and $A(c) \geq A_o > o$ is a polynomial (usually obtained by interpolation of experimental data) ; this law is valid when $o < T_* \leq T \leq T^* < +\infty$ and obviously $o \leq c \leq 1$.

We can extend this function to all $\mathbb{R}^2$ in agreement with (3.4),(3.7) in the following way :

$$\mu_1(\xi_1,\xi_2) = \mu_1(\xi_1,T^*) \quad \text{if} \quad o \leq \xi_1 \leq 1 \quad \text{and} \quad \xi_2 \geq T^*$$

$$\mu_1(\xi_1,\xi_2) = \mu_1(\xi_1,T_*) \quad \text{if} \quad o \leq \xi_1 \leq 1 \quad \text{and} \quad \xi_2 \leq T_*$$

$$\mu_1(\xi_1,\xi_2) = \mu_1(o\ ,\xi_2) \quad \text{if} \quad \xi_1 \leq o \quad \text{and} \quad \xi_2 \in \mathbb{R}$$

$$\mu_1(\xi_1,\xi_2) = \mu_1(1\ ,\xi_2) \quad \text{if} \quad \xi_1 \geq 1 \quad \text{and} \quad \xi_2 \in \mathbb{R}$$

7.2 <u>Mass density</u> If $\rho(c,T)$ is given by (1.5) we can extend $\rho(\xi_1,\xi_2)$ to all $\mathbb{R}^2$ with the same procedure as μ_1 ; in this way ρ verifies (3.4) and (3.7) .

7.3 Action of external forces $S_1(c,T)$ defined by (1.6), with ρ given by (1.5) and extended to $\mathbb{R}^2$ as in 7.2, verifies the conditions S1 with $\alpha = 0$ and S2.

7.4 Thermal and mass reaction rates If $S_i(c,T)$, $i = 2,3$ are defined by (1.7), with ρ as in 7.2, then we can extend $S_i(\xi_1, \xi_2)$ to $\xi_2 < o$ putting $S_i(\xi_1, \xi_2) = o$ and for the other values we use the same technique as in 7.1. In this way we obtain $S_i(\xi_1, \xi_2)$ verifying (5.1),(6.2),(6.4) and (6.5) and so the physically significant conditions : $T \geq o$ and $o \leq c \leq 1$ are true.
Moreover the condition S1 is verified with $\alpha = o$.
Let us also remark that, as is physically reasonable, if the reaction is exothermic ($\Delta H < o$) then (6.7) is true for each T_* and if the reaction is endothermic ($\Delta H > o$) then (6.9) is true for each T^*.
Similar results are true for $S_i(c,T)$, $i = 2,3$ given by (1.8).

Concluding remark In all the examples given here (newtonian or pseudoplastic viscosity) and for the usual reaction rates (1.7) and (1.8) we have obtained the existence of a steady state verifying (1.13) for all Reynolds and Péclet numbers, all eventual distributed sources and all natural boundary conditions verifying the compatibility condition (1.12), when the "resistance" for chemical reaction and natural convection become greater than the "resistance" for transport.

Aknowledgement On the physics of this subject we had some useful discussions with the researchers of A.R.S. - Milano (Italy).

REFERENCES

[0] R.AMIEL - G.GEYMONAT : to appear.

[1] R.ARIS : Elementary chemical reactor analysis. Prentice Hall, 1969.

[2] G.R.GAVALAS : Non linear differential equations of chemically reacting systems. Springer Tracts in Natural Philosophy vol 17, 1968.

[3] V.HLAVACEK - H.HOFMANN : Modeling of chemical reactors.XVI. Steady state axial heat and mass transfer in tubular reactors, Chem.Eng.Sci.25(1970),173-185.

[4] M.JEAN : Un cadre abstrait pour l'espace vectoriel topologique ordonné $W^{1,p}(\Omega)$,Séminaire d'analyse convexe, Montpellier 1975.

[5] M.A.KRASNOSEL'SKII : Topological methods in the theory of nonlinear integral equations, Pergamon Press, 1964.

[6] J.LERAY - J.L.LIONS : Quelques résultats de Višik sur les problèmes elliptiques non linéaires par les méthodes de Minty-Browder; Bull.Soc.Math. France,93,1965, p.97 à 107.

[7] H.LEWY - G.STAMPACCHIA : On the regularity of the solution of a variational inequality, Com.Pure.Appl.Math., XXII,153-188,(1969)

[8] J.L.LIONS : Quelques méthodes de résolution des problèmes aux limites linéaires, Gauthier-Villars, 1969.

[9] R.SALA, F.VALZ-GRIS, L.ZANDERIGHI : A fluid-dynamic study of a continuous polymerisation reactor, Chem.Eng.Sci, 29(1974) p.2205-2212.

[10] M.M.VAINBERG : Variational methods for the study of nonlinear operators, Holden-Day,1964.

LOCAL INVERTIBILITY CONDITIONS FOR GEOMETRICALLY EXACT NONLINEAR ROD AND SHELL THEORIES

Stuart S. Antman and Russell C. Browne
Department of Mathematics
University of Maryland
College Park, Maryland 20742, USA

1. Introduction. Many engineering structures suffer only small strains while undergoing large displacements. For this reason it is often advocated that the exact strain-displacement relations or "approximate" but nonlinear strain-displacement relations be used in conjunction with constitutive relations that give an appropriate stress as a linear function of one of the common strains. (The stress may also depend in some arbitrary manner on other kinematic variables.) Because such models lead to certain analytic simplifications for special problems, these assumptions have been uncritically accepted by analysts. In this paper we first discuss the defects of such models. We then briefly describe the mathematical structure of geometrically exact, nonlinearly elastic rod and shell theories, which do not suffer from such defects. These theories, however, provide a number of serious technical problems of analysis. We finally indicate how to overcome these difficulties.

Let $\underset{\sim}{r}(\underset{\sim}{X})$ represent the position of particle $\underset{\sim}{X}$ in some deformed configuration. Let $\partial\underset{\sim}{r}/\partial\underset{\sim}{X}$ denote the gradient of $\underset{\sim}{X} \to \underset{\sim}{r}(\underset{\sim}{X})$ and let $(\partial\underset{\sim}{r}/\partial\underset{\sim}{X})^*$ denote the transpose of this gradient. Set $\underset{\sim}{C} \equiv (\partial\underset{\sim}{r}/\partial\underset{\sim}{X})^* \cdot (\partial\underset{\sim}{r}/\partial\underset{\sim}{X})$ and $2\underset{\sim}{E} \equiv \underset{\sim}{C} - \underset{\sim}{I}$, where $\underset{\sim}{I}$ is the identity tensor. ($\underset{\sim}{C}$ is the Green deformation tensor.)

Perhaps the most common stress-strain law proposed for an isotropic, nonlinearly elastic body is

$$\underset{\sim}{T}(\underset{\sim}{X}) = \lambda(\underset{\sim}{X})\,\mathrm{tr}\,\underset{\sim}{E}(\underset{\sim}{X})\underset{\sim}{I} + 2\mu(\underset{\sim}{X})\underset{\sim}{E}(\underset{\sim}{X}), \tag{1.1}$$

where $\underset{\sim}{T}$ is the second Piola-Kirchhoff stress tensor, tr is the trace, and λ and μ are the usual Lamé scalars of linear elasticity. To understand the nature of some long-ignored problems caused by (1.1), consider the deformation $\underset{\sim}{X} \mapsto \underset{\sim}{r}(\underset{\sim}{X})$ given in Cartesian coordinates by

$$r_1(\underset{\sim}{X}) = \alpha X_1,\ r_2(\underset{\sim}{X}) = X_2,\ r_3(\underset{\sim}{X}) = X_3, \qquad 0 \le X_1, X_2, X_3 \le 1, \tag{1.2}$$

where α is <u>any</u> real number. When $\alpha = 0$, (1.2) represents the squashing of a unit cube into a unit square. When $\alpha < 0$, (1.2) represents the transformation of this cube into a rectangular block of opposite orientation: "The cube is turned inside out." When (1.1) is used in a nontrivial problem, such singularities may occur throughout a body and be undetectable by either analytic or convenient numerical techniques.

There are other serious difficulties attending the use of (1.1). An elastic body under a compressive loading may undergo buckling through a bifurcation process. When a degenerate stress-strain law such as (1.1) is used, the distribution and number of eigenvalues of the linearizarion of the governing equations about a trivial solution may be very strange.

The very choice of a particular stress-strain law of the type (1.1) may itself help obscure underlying physical processes. Let $\beta \underset{\sim}{E}_{(\beta)} = \underset{\sim}{C}^{\beta/2} - \underset{\sim}{I}$ with $\beta > 0$. When $\underset{\sim}{E}$ of (1.1) is replaced by any one of the tensors $\{\underset{\sim}{E}_{(\beta)}, \beta > 0\}$, the response is unaffected for small strains but differs considerably for large strains. These differences become critical when the prescribed forces depend on $\underset{\sim}{r}$, for then the coercivity of a problem and therefore the very question of existence may depend upon the strain measure used. (Hydrostatic pressures and centrifugal forces are of this kind. (Cf. [2,3,7].)

2. <u>The nature of geometrically exact rod and shell theories</u>. The common ingredients of virtually all geometrically exact rod and shell theories are i) their equations of motion relate a finite number of stress averages and their derivatives to a finite number of acceleration terms, ii) their constitutive relations relate these stress averages to a finite number of kinematic variables, iii) there are a finite number of governing equations with those for rods having but a single independent space-like variable and those for shells having but two space-like variables. A natural interpretation of these theories is that they describe families of three-dimensional bodies that are constrained to undergo only certain kinds of deformations. Such theories are termed <u>exact</u> when no approximations are used to modify the kinematic variables. E.g., if an angle θ is a kinematic variable, then the approximations $\sin\theta \cong \theta$ or $\sin\theta = \theta - \theta^3/6$ are prohibited in an exact theory. The accurary of such approximations for small values of θ is counterbalanced by their errors for large θ. Moreover, the cubic nonlinearity need not be any easier to handle than the sin itself. Many technical theories of rods and shells such as those of von Kármán type make use of approximations like these. Such theories may be of some use in the study of small deformations and local bifurcation processes, but they have no value for the study of large deformations. (The beautiful global analyses of [13,14,15] for von Kármán plates are important for their mathematical content.)

The nature of rod and shell theories inheriting their structure from nonlinear three-dimensional theories is examined in [3,7]. Their development is remarkably simple. In particular, the configuration of a rod or axisymmetrically deformed axisymmetric shell is defined by a function

(2.1) $$\underline{u} : [S_1, S_2] \to \mathbb{R}^n .$$

The requirement that the deformation $\underset{\sim}{r}$ be locally invertible and orientation-preserving embodied in the inequality

(2.2) $$\det (\partial \underset{\sim}{r} / \partial \underset{\sim}{X}) > 0$$

has a one-dimensional analog that there is a region $G \subset \mathbb{R}^n \times \mathbb{R}^n \times [S_1, S_2]$ with certain geometrical properties listed below such that locally invertible and orientation-preserving configurations must satisfy

(2.3) $$(\underline{u}(S), \underline{u}'(S), S) \in G \quad \forall S \in [S_1, S_2].$$

The requirement that the three-dimensional body be an elastic material satisfying the strong ellipticity condition leads immediately to the requirement that the elliptic differential operator of the one-dimensional theory be semi-monotone. Finally, for static problems of hyperelasticity, one can construct a sequence of one-dimensional problems that have solutions that generate three-dimensional fields converging weakly to solutions of the three-dimensional theory. (Cf. [7,8,12].)

3. <u>The invertibility problem</u>. To prevent the absurdities attending the use of (1.1) of the sort that arise in deformations like (1.2), we require (2.2) or its one-dimensional analog (2.3) to hold at least almost everywhere. This leads to some serious technical difficulties. Here we show how to treat these problems in the context of variational problems.

By studying the construction of G appearing in (2.3), we are led to assume that G has the following properties

(3.1) There exists a Lipschitz continuous mapping $\underline{a}, S \mapsto h(\underline{a}, S)$ from $\mathbb{R}^n \times [S_1, S_2]$ to $\mathbb{R}$ such that

$$H \equiv \{(\underline{a}, S) \in \mathbb{R}^n \times [S_1, S_2] : h(\underline{a}, S) > 0\}$$

is an unbounded, open, proper subset of $\mathbb{R}^n \times [S_1, S_2]$ and G is an unbounded, open, proper subset of $H \times \mathbb{R}^n$.

(3.2) ∂G is Lipschitz continuous and $\partial G \cap \partial(H \times \mathbb{R}^n) \neq \emptyset$.

(3.3) $G(\underline{a}, S) \equiv \{\underline{b} : (\underline{a}, \underline{b}, S) \in G\}$ is an unbounded, open, convex subset of $\mathbb{R}^n$.

Let

(3.4) $$G(S) = \{(\underline{a}, \underline{b}) : (\underline{a}, \underline{b}, S) \in G\}.$$

Note that G can be a very wild-looking set. Only in the most trivial of problems can the Lipschitz continuity of $\partial G \cap \partial(H \times \mathbb{R}^n)$ and of $\partial G \backslash \partial(H \times \mathbb{R}^n)$ be strengthened to differentiability [7]. Condition (3.3) immediately implies that there is a Lipschitz continuous mapping $\underline{a}$, $S \mapsto \underline{e}(\underline{a},S) \in$ unit ball of $\mathbb{R}^n$ such that if $\underline{b} \in G(\underline{a},S)$, then $\underline{b} + \beta\underline{e} \in G(\underline{a},S)$ for all $\beta \geq 0$. Thus

(3.5) $$\underline{b} = [\underline{I} - \underline{e} \otimes \underline{e}]\underline{b} + \beta\underline{e}, \qquad \beta = \underline{b}\cdot\underline{e},$$

where $\underline{I} - \underline{e} \otimes \underline{e}$ is the projection of $\mathbb{R}^n$ onto the orthogonal complement of $\underline{e}$. Moreover, there is a Lipschitz continuous mapping $\underline{a},\underline{c},S \mapsto g(\underline{a},\underline{c},S) \in \mathbb{R}^n$ such that points of G satisfy

(3.6) $$\beta > g(\underline{a},[\underline{I} - \underline{e} \otimes \underline{e}]\underline{b}, S).$$

We consider only conservative problems for which the one-dimensional body possesses a strain energy function $\Psi : G \to \mathbb{R}$, with $\Psi(\underline{a},\underline{b},S) \to \infty$ as $(\underline{a},\underline{b}) \to \partial\{(\underline{a},\underline{b}) : (\underline{a},\underline{b},S) \in G\}$ and as $|\underline{a}| + |\underline{b}| \to \infty$ with $\underline{a}$ and $\underline{b}$ subject to some additional requirements ensuring that this growth is not due to a rigid displacement. We further require that $\Psi(\underline{a},\cdot,S)$ be strictly convex. (This is the one dimensional analog of the strong Legendre-Hadamard condition of the three-dimensional theory of nonlinear hyperelasticity.) Set

(3.7) $$U[\underline{w}] = \int_{S_1}^{S_2} \Psi(\underline{w}(S),\underline{w}'(S),S)dS, \qquad E \equiv \{\underline{w} : U[\underline{w}] < \infty\}.$$

We assume that there is a reflexive Banach space W (modeled after but possibly much more complicated than W_p^1 with $p > 1$) whose elements are absolutely continuous on every compact subset of (S_1,S_2). Let V denote the potential functional of the applied loads. Then under mild and physically reasonable conditions that we do not spell out, $U + V$ has a minimizer $\underline{u}$ on a suitable manifold in E [8]. Our conditions on Ψ imply that $\underline{u}$ violates invertibility, i.e., $(\underline{u}(S),\underline{u}'(S)) \in \partial G(S)$, only for S in a set of measure zero.

To obtain a regularity theory including the requirement that $\underline{u}$ be everywhere invertible, we add the informal assumptions that $\Psi \in C^2(G)$, that components of $\Psi_{\underline{a}}$ and $\Psi_{\underline{b}}$ "parallel" to $\partial G(S)$ are "well-behaved", and that V is "well-behaved". We outline the principal steps of the rather intricate development given in [8]. For each $\varepsilon > 0, \delta > 0$, we set

(3.8) $$P_{\varepsilon,\delta} = \{S : 0 \leq h(\underline{u}(S),S) \leq \varepsilon\} \cup [S_1,S_1 + \delta] \cup [S_2 - \delta,S_2],$$

(3.9) $$Q_\varepsilon = \{S : 0 \leq \underline{u}'(S)\cdot\underline{e}(\underline{u}(S),S) - g(\underline{u}(S),[\underline{I} - \underline{e}(\underline{u}(S),S) \otimes \underline{e}(\underline{u}(S),S)]\underline{u}'(S),S) \leq \varepsilon\}.$$

Cf. (3.5),(3.6). The properties of W and h ensure that $[S_1,S_2]\setminus P_{\varepsilon,\delta}$ is open. Let $(A_\varepsilon,B_\varepsilon)$ be a component open interval of this set. To proceed it is necessary to construct a rich collection of variations $\underline{v}$ of $\underline{a}$ that lie in G.

Let $\eta_\varepsilon > 0$ and $\underline{y}'_\varepsilon$ be given. Consider the boundary value problem

(3.10a) $\underline{v}' = \underline{u}' + t\underline{y}'_\varepsilon$ for $S \in (A_\varepsilon, B_\varepsilon)\backslash Q_\varepsilon$,

(3.10b) $\underline{v}' = [\underline{I} - \underline{e}(\underline{v},S) \otimes \underline{e}(\underline{v},S)](\underline{u}'+t\underline{y}'_\varepsilon)$

$$+ [\underline{u}'\cdot\underline{e}(\underline{u},S) - g(\underline{u},[\underline{I} - \underline{e}(\underline{u},S) \otimes \underline{e}(\underline{u},S)]\underline{u}',S) + t\eta_\varepsilon$$

$$+ g(\underline{v},[\underline{I} - \underline{e}(\underline{v},S) \otimes \underline{e}(\underline{v},S)](\underline{u}'+t\underline{y}'_\varepsilon),S)]\underline{e}(\underline{v},S), S\in(A_\varepsilon,B_\varepsilon)\cap Q_\varepsilon,$$

(3.10c) $\underline{v}(A_\varepsilon) = \underline{u}(A_\varepsilon), \quad \underline{v}(B_\varepsilon) = \underline{u}(B_\varepsilon)$.

By using the perturbation theory for initial value problems in the sense of Carathéodory, the Gronwall inequality, and the contraction mapping principle, one can show that the set of function $\{\underline{y}_\varepsilon\}$ for which this boundary value problem has a solution in E for small enough t is dense in $\mathring{W}(A_0,B_0)$. ($\mathring{W}(A_0,B_0)$ consists of the elements of W with compact support in (A_0,B_0).) If $\underline{v}$ satisfies (3.10) and if V is Gâteaux differentiable, then the mean value theorem implies that

(3.11)
$$0 \le t^{-1}\int_A^B [\Psi(\underline{v},\underline{v}',S) - \Psi(\underline{u},\underline{u}',S)]dS + \ldots$$
$$\le \int_{(A_\varepsilon,B_\varepsilon)\cap Q_\varepsilon} [\text{integrable function}]dS + \int_{(A_\varepsilon,B_\varepsilon)\backslash Q_\varepsilon} [\Psi_{\underline{a}}\cdot\underline{y}_\varepsilon + \Psi_{\underline{b}}\cdot\underline{y}'_\varepsilon]dS + \ldots .$$

Here the dots represent differences for V. The last integral has arguments depending on t. The first inequality in (3.11) is an immediate consequence of the definition of a minimum. The second inequality is a consequence of the decomposition of differences of Ψ on $(A_\varepsilon,B_\varepsilon)\cap Q_\varepsilon$ into a badly behaved but negative term and a nicely behaved difference "parallel" to ∂G. We may take the lim sup of (3.11) as $t \to 0$ by using the Lebesgue Dominated Convergence Theorem. The density result ultimately yields

(3.12) $$\int_{A_0}^{B_0} \Psi_{\underline{a}}(\underline{u},\underline{u}',S)\cdot\underline{x} + \Psi_{\underline{b}}(\underline{u},\underline{u}',S)\cdot\underline{x}' + \ldots = 0 \quad \forall\underline{x} \in \mathring{W}(A_0,B_0)$$

to which we may apply a version of the Fundamental Lemma of the Calculus of Variations to conclude ultimately that $\underline{u}$ is invertible and classically regular on compact subsets of (A_0,B_0). We use a similar development across the ends A_0 and B_0 to show that $\Psi_{\underline{a}}$ is integrable on (A_0,B_0). A similar regularity theory then shows that $\underline{u}$ is Lipschitz continuous on $[A_0,B_0]$. Finally we suppose that for bounded $\underline{b}$, $\Psi_{\underline{a}}$

approaches ∞ as $h(\underline{a},S) \to 0$ at a rate greater than const/dist$((\underline{a},S),\partial H)$. This implies that the termini $A_0, B_0 \in (S_1,S_2)$ by an argument analogous to the result that if z is uniformly Lipschitz on $[A,B]$ and if $\int_A^B [z(s)]^{-1} dS < \infty$, then z cannot vanish on $[A,B]$. The treatment of boundary conditions follows along these lines.

The results form the essential groundwork for a number global and local analyses of concrete problems of buckling, necking, and shear instabilities for nonlinearly elastic rods and shells in [1,4,5,6,9, 10,11]. Applications of such methods to problems of nonlinear viscoelasticity are under study. Here the growth of the stress near $\partial G(S)$ causes special difficulties because it prevents the use of standard results for operators of Carathéodory-Nemytskii type that play a central role in the analysis of simpler models.

In conclusion, we observe that the work described in this paper indicates how functional analysis is particularly well suited for handling whole classes of nonlinear problems.

Acknowledgment. The research of the first author was supported by National Science Foundation Grant MPS73-08587A02.

References

1. S. S. Antman, The Shape of Buckled Nonlinearly Elastic Rings, Z.A.M.P. 21(1970), 422-438.

2. _____, Existence and Nonuniqueness of Axisymmetric Equilibrium States of Nonlinearly Elastic Shells, Arch. Ratl. Mech. Anal. 40 (1971), 329-371.

3. _____, The Theory of Rods, Handbuch der Physik Vol. VIa/2, Springer-Verlag, 1972, 641-703.

4. _____, Nonuniqueness of Equilibrium States for Bars in Tension, J. Math. Anal. Appl. 44(1973), 333-349.

5. _____, Qualitative Theory of the Ordinary Differential Equations of Nonlinear Elasticity, in Mechanics Today, 1972, edited by S. Nemat-Nasser, Pergamon Press, 1974, 58-101.

6. _____, Monotonicity and Invertibility Conditions in One-Dimensional Nonlinear Elasticity, Symposium on Nonlinear Elasticity, Mathematics Research Center, Univ. Wisconsin, edited by R. W. Dickey, Academic Press, 1973, 57-92.

7. _____, Boundary Value Problems of One-Dimensional Nonlinear Elasticity I: Foundations of the Theories of Nonlinearly Elastic Rods and Shells, Arch. Rational Mech. Anal., to appear.

8. _____, Boundary Value Problems of One-Dimensional Nonlinear Elasticity II: Existence and Regularity Theory for Conservative Problems, Arch. Rational Mech. Anal., to appear.

9. _____ & E. Carbone, to appear.

10. _____ & K. B. Jordan, Qualitative Aspects of the Spatial Deformation of Nonlinearly Elastic Rods, Proc. Roy. Soc. Edinburgh, to appear.

11. _____ & G. Rosenfeld, in preparation.

12. J. Ball, to appear.

13. M. S. Berger, On von Kármán's Equations and the Buckling of a Thin Elastic Plate, I. Comm. Pure Appl. Math. 20(1967), 687-719.

14. M. S. Berger and P. Fife, On von Kármán's Equations and the Buckling of a Thin Elastic Place, II, Comm. Pure Appl. Math. 21(1968), 227-241.

15. J. H. Wolkowisky, Existence of Buckled States of Circular Plates, Comm. Pure Appl. Math. 20(1967), 549-560.

SOME APPLICATIONS OF FUNCTIONAL ANALYSIS IN THE MATHEMATICAL THEORY OF STRUCTURES

E.R. Arantes Oliveira
Technical University of Lisbon

Instituto Superior Técnico
Av. Rovisco Pais, Lisbon/Portugal

1. A modern view of the Theory of Structures

Let us start by explaining what the Theory of Structures really means to the author as a part of Solid Mechanics (see [1] , [2] , [3]).

Solid Mechanics comprehends different models conceived for the equilibrium and deformation analysis of solids, namely three-dimensional models, two-dimensional models, one-dimensional models and discrete models.

The methods of solution of particular problems within the frame of each model do not fall within the scope of the Theory of Structures but of other theories, like the Theory of Elasticity, the Theory of Shells, the Theory of Rods, and so on.

The formal analogies between models, together with the generation of models from other models, fall however within the scope of what the author calls the Mathematical Theory of Structures, which therefore may be formulated as consisting of three parts:

- a generic model,
- rules for generating models from other models,
- a justification for such rules.

Only elastic structures under static equilibrium will be considered in the present paper. Physical linearity will not be required.

2. The generic model

The generic model consists in three groups of equations — force-stress, strain-displacement and stress-strain equations — involving four kinds of magnitudes — stresses, strains, forces and displacements. Such equations are supplemented by force and displacement boundary conditions (external and eventually internal) and supposed such that the work principle holds.

A couple of a stress and a strain field related by the stress-strain equations form a structural field. A structural field is called a compatible field with respect to a system of incompatibilities (a set of prescribed initial strains and displacement boundary conditions), or is said to compatibilize such system of incompatibilities, if it satisfies the strain-displacement conditions and the displacement boundary conditions. A structural field is called an equilibrated field with respect to a system of external forces (a set of prescribed body forces and force boundary conditions), or is said to equilibrate such system of external forces, if it satisfies the force-stress equations and the force boundary conditions. A field which is simultaneonsly a compatible and an equilibrated one is an exact solution (to the structural equations and boundary conditions) with respect to given systems of incompatibilities and external forces.

Let X be the set of all the structural fields associated to a given elastic structure.

Set X can be made a vector space by defining the operations of addition and multiplication by a scalar. Such definitions can be made in different ways among which the most natural ones are schematized below:

$$\begin{array}{lll|lll} x_1 & \text{corresponds to} & \underset{\sim}{e}_1 & x_1 & \text{corresponds to} & \underset{\sim}{s}_1 \\ x_2 & \text{"} & \underset{\sim}{e}_2 & x_2 & \text{"} & \underset{\sim}{s}_2 \\ x_1+x_2 & \text{"} & \underset{\sim}{e}_1+\underset{\sim}{e}_2 & x_1+x_2 & \text{"} & \underset{\sim}{s}_1+\underset{\sim}{s}_2 \end{array} \qquad (2.1)$$

where $\underset{\sim}{e}$ and $\underset{\sim}{s}$ represent the strain and stress vectors.

X can still be made a Banach space by associating a norm to each of its elements. Such norm can also be defined in several ways as, for instance,

$$\|x\| = \sqrt{\int_\Delta \underset{\sim}{e}\cdot\underset{\sim}{e}\, d\Delta} \qquad \Bigg| \qquad \|x\| = \sqrt{\int_\Delta \underset{\sim}{s}\cdot\underset{\sim}{s}\, d\Delta} \qquad (2.2)$$

where Δ denotes the domain corresponding to the structure.

The set of all the fields in X which |compatibilize / equilibrate| a given system of |incompatibilities / external forces| is called an |isocompatible / isoequilibrated| subset of X.

The set of all the |isocompatible / isoequilibrated| subsets of X is assumed also a Banach space and denoted by $\left|\begin{matrix}\mathcal{J}\\ \mathcal{L}\end{matrix}\right|$. A system of |incompatibilities / external forces| corresponds thus to each element of $\left|\begin{matrix}\mathcal{J}\\ \mathcal{L}\end{matrix}\right|$.

As a unique element of $\left|\begin{matrix}\mathcal{J}\\ \mathcal{L}\end{matrix}\right|$ corresponds to each element of X, a function $\left|\begin{matrix}\iota\\ \varepsilon\end{matrix}\right|$, assumed continuous, with domain X and range $\left|\begin{matrix}\mathcal{J}\\ \mathcal{L}\end{matrix}\right|$, can be considered which associates to each element of X the corresponding element of $\left|\begin{matrix}\mathcal{J}\\ \mathcal{L}\end{matrix}\right|$. We write therefore

$$\iota(x) = I \qquad \Bigg| \qquad \varepsilon(x) = E \qquad (2.3)$$

where $\left|\begin{matrix}I\in\mathcal{J}\\ E\in\mathcal{L}\end{matrix}\right|$.

Equations (2.3) left and right are respectively called compatibility and equilibrium equations.

Assuming that the intersection of each isocompatible and each isoequilibrated subset of X contains one and no more than one element, a one-to-one correspondence can be established between the elements of X and the elements of the cartesian product $\mathcal{J}\times\mathcal{L}$.

Such cartesian product will be called the space $\mathcal{A}$ of the external actions while X may be called the space of responses.

Although ι and ε have no inverse, a function χ can still be considered with domain $\mathcal{A}$ and range X, which associates to each pair (I, E) of an isocompatible and an isoequilibrated subset the corresponding unique intersection, x. We write

$$x = \chi(I, \epsilon) \tag{2.4}$$

and assume χ a continuous function.

With the help of the work principle, the total potential and complementary energy theorems can be proved. The total $\left|\begin{matrix}\text{potential}\\ \text{complementary}\end{matrix}\right|$ energy theorem states that the exact solution makes the total $\left|\begin{matrix}\text{potential}\\ \text{complementary}\end{matrix}\right|$ energy stationary on the set of the $\left|\begin{matrix}\text{compatible}\\ \text{equilibrated}\end{matrix}\right|$ fields. Such theorems become minimum theorems if stability is admitted.

The total $\left|\begin{matrix}\text{potential}\\ \text{complementary}\end{matrix}\right|$ energy being a continuous functional, $\left|\begin{matrix}T_E(x)\\ T_I^*(x)\end{matrix}\right|$, on X (index $\left|\begin{matrix}E\\ I\end{matrix}\right|$ refers to the system of $\left|\begin{matrix}\text{external forces}\\ \text{imcompatibilities}\end{matrix}\right|$ which the functional corresponds to), the minimum total $\left|\begin{matrix}\text{potential}\\ \text{complementary}\end{matrix}\right|$ energy theorem may be enounced by stating that $\chi(I,E)$ minimizes $\left|\begin{matrix}T_E(x') & \text{on} & I\\ T_I^*(x) & \text{on} & E\end{matrix}\right|$.

The distance between two structural fields, i.e., two elements of X, could of course be defined as the norm (defined as above) of their difference. As it will become clear later, however, a more convenient definition consists in making

$$d^2 s = \tfrac{1}{2}\int_\Delta \delta \underset{\sim}{s} \cdot \delta \underset{\sim}{e}\, d\Delta \tag{2.5}$$

for two elements such that the norm of their difference is small, and then defining the distance between two arbitrary points x_1 and x_2 as the smallest obtained by adding the infinitesimal distances along all possible continuous* pathes in X between x_1 and x_2, i.e., by making

$$d(x_1, x_2) = \min \int_{x_1}^{x_2} ds \tag{2.6}$$

where ds denotes the distance, defined through (2.5), between two points very near each other.

As it may be proved that

$$\delta^2 T_E = \int_\Delta \delta \underset{\sim}{s} \cdot \delta \underset{\sim}{e}\, d\Delta \quad \text{on } I \quad \Bigg| \quad d^2 T_I^* = \int_\Delta d\underset{\sim}{s} \cdot d\underset{\sim}{e}\, d\Delta \quad \text{on } E \tag{2.7}$$

at $x = \chi(I,E)$ and as, at the same point, $\delta T_E = \delta T_I^* = 0$ on $\left|\begin{matrix}I\\ E\end{matrix}\right|$, there follows, considering (2.5), that, if $\left|\begin{matrix}h \in I_0\\ h \in E_0\end{matrix}\right|$ and $\|h\|$ is very small,

$$d^2(x+h, x) = T_E(x+h) - T_E(x) \quad \Bigg| \quad d^2(x+h, x) = T_I^*(x+h) - T_I^*(x) \tag{2.8}$$

(*) Continuity gained of course a meaning as soon as the norm was defined.

where E_o and I_o denote the zero elements of $\mathcal{L}$ and $\mathcal{J}$.

A more general scheme may be introduced [4] which covers any situation in which an extremum principle is known to exist. Such scheme simply considers space X and a family Φ of continuous functionals φ which are assumed to admit a proper minimizer s on each subset C belonging to a certain class of subsets of X, called constrained subsets of X. s is assumed a critical point of φ on C, i.e., at point s grad φ vanishes on C.

The constrained subsets are assumed homeomorphic to a certain linear subspace of X and their union is assumed to coincide with X.

The set of all the minimizers corresponding to all the constrained subsets of X is called a minimizing subset of X. Each minimizing subset corresponds to a certain functional $\varphi \in \Phi$, and the union of the minimizing subsets corresponding to all the functionals of family Φ is also assumed to coincide with X.

The intersection of each constrained and each minimizing subset of X is assumed to contain one and no more than one element.

Elements belonging to the same $\left|\begin{array}{l}\text{constrained}\\ \text{minimizing}\end{array}\right|$ subset of X will be called $\left|\begin{array}{l}\text{iso-constrained}\\ \text{iso-minimizing}\end{array}\right|$ in X.

Reference to such general scheme will be made, for sake of commodity, along the next Sections.

3. The generation rules

In what concerns the rules for generating models from other models, their importance in the Theory of Structures can be realized if it is considered that discretization, i.e., the generation of discrete models from continuous models is the typical method of the structural analyst. As, on the other hand, the generation of a discrete model is quite analogous to, for instance, the generation of a two-dimensional model from a three-dimensional one, there is no reason why the study of discretization should not be included within a general theory of the generation of models, which comprehends the description of general generation rules and their justification with the help of convergence analyses.

The procedures for generating new models in the Theory of Structures are essentially two dual ones: the potential and the complementary energy methods [1, 2, 3].

The $\left|\begin{array}{l}\text{potential}\\ \text{complementary}\end{array}\right|$ energy method starts by defining the generated $\left|\begin{array}{l}\text{strains and displacements}\\ \text{stresses and tractions}\end{array}\right|$ in terms of the generating ones. Generated $\left|\begin{array}{l}\text{strain-displacement}\\ \text{force-stress}\end{array}\right|$ equations and $\left|\begin{array}{l}\text{displacement}\\ \text{force}\end{array}\right|$ boundary conditions are simultaneously introduced which must satisfy the condition of being exact in the frame of the generating model. Such equations introduce a correspondence between the generated and the generating fields which may be represented by the equation

$$y = \mathcal{B}_j\, x \tag{3.1}$$

where B_1 is a linear operator whose domain is the space X of the generating fields and whose range is the space Y of the generated ones.

As the generating and generated models are analogous to each other, constrained and minimizing subsets can be considered in Y as well as in X. The fact that the generated $\left|\begin{matrix}\text{compatibility}\\ \text{equilibrium}\end{matrix}\right|$ equations are exact in the generating model implies that the B_1-images of isoconstrained elements in X are isoconstrained in Y.

The work principle, or some variational principle resulting from the work principle, is subsequently used for deriving the generated $\left|\begin{matrix}\text{equilibrium}\\ \text{compatibility}\end{matrix}\right|$ equations.

An assumption is then introduced, like the Kirchhoff's assumption of the theory of shells, or the Bernouilli's assumption of the theory of rods, or the definition of the fields allowed within the elements in the finite element technique, which establishes a correspondence between the elements of Y and elements of X. A new linear operator, B_2, is thus introduced with domain Y and range X', X' being a subspace of X (the space of the allowed fields), and we write

$$x' = B_2(y) \tag{3.2}$$

The $\left|\begin{matrix}\text{potential}\\ \text{complementary}\end{matrix}\right|$ energy method postulates the invariance of the total $\left|\begin{matrix}\text{potential}\\ \text{complementary}\end{matrix}\right|$ energy. This means that

$$T[B_2(y)] = V(y) \quad \Bigg| \quad T^*[B_2(y)] = V^*(y) \tag{3.3}$$

where $\left|\begin{matrix}T \text{ and } V\\ T^* \text{and } V^*\end{matrix}\right|$ denote the total $\left|\begin{matrix}\text{potential}\\ \text{complementary}\end{matrix}\right|$ energies in the generating and in the generated models, and permits to express the generalized elastic coefficients and external force defining magnitudes in terms of the generating ones [1].

According to what was written above and namely to equation (3.3), the linear operator

$$B = B_2 B_1 \tag{3.4}$$

with domain X and range X', which we call the interpolation operator, enjoys the following properties:

i) the B-image of any element belonging to X' coincides with the element itself, i.e.,

$$B(x') = x' \qquad \text{if } x' \in X' \tag{3.5}$$

ii) Constrained and minimizing subsets meeting the same requirements as those in X can be defined in X', with respect to the same family of functionals, ϕ.

iii) The B-images of isoconstrained elements in X are isoconstrained in X', although not necessarily in X.

If any two isoconstrained elements in X' are isoconstrained in X, operator B is said to be conforming.

The constrained subset C' which contains the B-images of the elements of a given constrained subset C of X is said to correspond to C. A given minimizing subset D' of X' is said to correspond to a certain minimizing subset D of X if they both correspond to the same functional φ.

A second operator, A, also with domain X and range X', can still be considered which makes the intersection of each constrained and each minimizing subset of X correspond to the intersection of the corresponding constrained and minimizing subsets of X'.

Operator A, called the approximation operator, is assumed:

i) bounded and continuous;

ii) such that the A-image of any element belonging to X' coincides with the element itself, i.e.,

$$A(x) = x \qquad \text{if } x \in X' \tag{3.6}$$

The A-image x'_a of an element $x \in X$ is called the approximation of x in X'.

We remark that the A-images of any two isoconstrained elements in X are isoconstrained elements in X', and that the A-images of any two isominimizing elements in X are isominimizing elements in X'.

Consideration of operators A and B permits to discuss the conversion of a given variational problem into a new one, eventually simpler than the first and capable of providing an approximate solution to it.

If the interpolation operator B is chosen in such a way that X' is N-dimensional, then, the variational problem becomes particularly easy to solve, as functional φ becomes a function of N variables, and the element which minimizes the functional can be determined by just solving the system of N equations to N unknowns which results from equating to zero the derivatives of the function with respect to each variable. The problem is said to have been discretized. Of course, the equations will be linear only if functional φ is a quadratic function of the variables.

If operator B is conforming and $C' \subset C$, this discretization procedure is nothing else than the classical Ritz method. In the case of the finite element method, however, operator B is generally non-conforming.

4. Justification by convergence

The evolution of the modern Theory of Structures was deeply influenced by the finite element method, and it was namely in connexion with the finite element method that the role of convergence in the theory was fully appreciated.

The importance of convergence (to the exact solution) was truly realized by the finite element analysts not earlier than 1965.[5] Before 1965, indeed, too much stress was put upon conformity and monotonic convergence, so that the search for conforming elements was then a popular topic of research.

Later, if was realized that conformity is not strictly necessary and that it can even be inconvenient. The main theoretical reason for a given type of element being accepted is thus really its capacity for generating sequences of approximate solutions tending to the exact solutions of a sufficiently wide class of problems.

It cannot be forgotten however that, if conformity is violated, the finite element method, regarded as a variational technique or, more precisely, as a technique which provides an approximate solution to a variational problem consisting in minimizing a certain functional φ on a certain space C , by just minimizing φ on a space C' not generally contained in C , becomes distinct from Ritz's method, which supposes that C' is contained in C . The convergence theorems of the last [6] could not therefore be used with the former and a new convergence theory had to be built.

The generation of discrete models by the finite element technique being thus justified by convergence, it is natural that convergence considerations be made also for justifying the generation methods even if they are used for the generation of continuous models like the theory of shells. And this explain how the Theory of Structures can give answers to problem, like the one of the foundations of the theory of shells, which have been seen, until quite recently, exclusively from other points of view.

The statement that the generation methods are justified by convergence requires some explanation, however, because, although it is clear what convergence means in the case of the finite element method, it seems less clear what it means in the case of the generation of the theory of shells, for instance.

Let us consider then the case of a shell and assume that its thickness is made smaller and smaller. What the convergence analysis is required to prove is that the two-dimensional (generated) solution becomes more and more near the corresponding three-dimensional (exact) solution as the shell becomes thinner and thinner, provided the relative values of the bending and membrane elastic coefficients (generated elastic coefficients) are not changed when the thickness tends to zero.

If such coefficients really do not change, then, the two-dimensional (generated) solution does not change also and what must be appreciated is the convergence of a sequence of generating (three-dimensional) solutions towards a generated one, and not, like in the finite element case, the convergence of a sequence of generated (approximate) solutions towards a generating (exact) one.

There remains to remark that the condition of unchangeability of the two-dimensional stiffness coefficients cannot be satisfied by an ordinary shell, in which the bending moments result merely from the ordinary stresses distributed in the thickness t , because, the bending stiffness coefficients being then proportional to t^3 and the membrane coefficients simply to t , the shell becomes more and more a membrane when the thickness tends to zero. But it can be satisfied if the shell is a generalized one, i.e., if non-vanishing couple-stresses are admitted to exist (see [7])

Convergence analyses can be based in any case on an approximation theorem which states that the distance between an arbitrary element s in X , which minimizes φ on C , and its approximation in X', s'_a, satisfies the inequality

$$d(s,s'_a) \leq d(s_a, s'_a) + \sqrt{|\delta_s \varphi| + |\delta_a \varphi|} \tag{4.1}$$

where

$$\delta_s \varphi = \varphi(s') - \varphi(s) \tag{4.2}$$

$$\delta_a \varphi = \varphi(s'_a) - \varphi(s_a) \tag{4.3}$$

and s_a is isoconstrained with s and such that

$$s'_a = B(s_a) \tag{4.4}$$

Indeed, as s minimizes φ on C,

$$\varphi(s) \leq \varphi(s_a) \tag{4.5}$$

On the other hand, as s'_a is the approximation of s in X', s'_a minimizes φ on C', so that

$$\varphi(s'_a) \leq \varphi(s') \tag{4.6}$$

Introducing (4.2) and (4.3) into (4.6), there results

$$\varphi(s_a) + \delta_s \varphi - \delta_a \varphi \leq \varphi(s) \tag{4.7}$$

Combination of (4.7) and (4.5) yields

$$\varphi(s_a) + \delta_s \varphi - \delta_a \varphi \leq \varphi(s) \leq \varphi(s_a) \tag{4.8}$$

Therefore,

$$\varphi(s_a) - \varphi(s) \leq |\delta_s \varphi| + |\delta_a \varphi| \tag{4.9}$$

On the other hand, (2.8) leads to

$$d^2(s, s_a) = \varphi(s_a) - \varphi(s) \tag{4.10}$$

so that inequality (4.9) can be transformed into

$$d(s,s_a) \leq \sqrt{|\delta_s \varphi| + |\delta_a \varphi|} \tag{4.11}$$

or, by virtue of the triangular inequality, into (4.1).

Let us apply the theorem to the case of Ritz's method in which $C' \subset C$. This means that s' and s'_a are isoconstrained with s and s_a. Then, as s'_a belongs to C, s_a can be identified with s'_a. (4.1) becomes therefore

$$d(s,s'_a) \leqslant \sqrt{|\delta_s \varphi|} \tag{4.12}$$

On the otherhand, as s is isoconstrained with s', (2.8) leads to

$$d^2(s,s') = \varphi(s') - \varphi(s) = \delta_s \varphi \tag{4.13}$$

so that (4.12) yields

$$d(s,s'_a) \leqslant d(s,s') \tag{4.14}$$

The approximation error is thus bounded by the interpolation error. In other words, as far as Ritz's method is concerned, completeness implies convergence.

Inequality (4.1) can be simplified if it is considered that $\left|\begin{matrix}\delta T \\ \delta T^*\end{matrix}\right|$ is the sum of two terms, one the order of the $\left|\begin{matrix}\text{strain} \\ \text{stress}\end{matrix}\right|$ variation modulus $\left|\begin{matrix}|\delta \underset{\sim}{e}| \\ |\delta \underset{\sim}{s}|\end{matrix}\right|$ and the other of the order of the $\left|\begin{matrix}\text{displacement} \\ \text{traction}\end{matrix}\right|$ variation modulus $\left|\begin{matrix}\delta \underset{\sim}{u} \\ \delta \underset{\sim}{p}\end{matrix}\right|$, i.e.,

$$\delta T = O(|\delta \underset{\sim}{e}|) + O(|\delta \underset{\sim}{u}|) \qquad \Big| \qquad \delta T^* = O(|\delta \underset{\sim}{s}|) + O(|\delta \underset{\sim}{p}|) \tag{4.15}$$

On the other hand, by virtue of (2.5),

$$d(s,s'_a) = O(|\delta_a \underset{\sim}{e}|) \qquad \Big| \qquad d(s,s'_a) = O(|\delta_a \underset{\sim}{s}|) \tag{4.16}$$

so that the first term in (4.1) can be neglected and (4.1) be transformed into

$$d^2(s,s'_a) \leq |\delta\varphi| + |\delta_a \varphi| \tag{4.17}$$

Any convergence or accuracy analysis[7,8] is thus reduced, according to the present theory, to two essential steps: a) the determination of a field s_a isoconstrained with s and interpolated by the approximate solution, b) the evaluation of the order of $\delta\varphi$, i.e. of the variation of the functional associated with the interpolation error, both for the exact and for the approximate solutions. As this evaluation is not always easy to make in connexion with the approximate solution, resorting to especial techniques like the so-called patch-test, becomes sometimes necessary[8].

5. Conclusions

It would not be fair finishing this paper on the applications of Functional Analysis to the Theory of Structures without referring Prager[9] and Synge's[10] pioneer work, as well as Mikhlin's book[6]. A more recent book by Oden[11] should also not be omitted.

In all such papers no attempt was made however to presenting the Theory of Structures, as a scheme of Mathematical Physics general enough to become a theoretical support not only to discretization but to the general problem of the generation of models from other models.

The theory of convergence contained in the scheme presented by the author can itself still be used, outside structural analysis, for every case in which a minimum principle is known to exist. (see [12]).

ACKNOWLEDGMENT

The present paper was financially supported by Instituto de Alta Cultura through the research project TLE/4.

REFERENCES

1 - OLIVEIRA, E.R.A. - "The Convergence Theorems and their Role in the Theory of Structures", in the Proc. IUTAM Symp. on High Speed Computing of Elastic Structures (1970), Liège, 1971.

2 - OLIVEIRA, E.R.A. - "Mathematical Theory of Linear Structures", in "Lectures on Finite Element Methods in Continuum Mechanics", ed. by J.T. Oden and E. R.A. Oliveira, Un. of Alabama in Huntsville Press, 1973.

3 - OLIVEIRA, E.R.A. - Notes of Lectures on the Foundations of the Theory of Structures, delivered at LNEC (Lisbon) and CISM (Udine) in 1972.

4 - OLIVEIRA, E.R.A. - "A Theory of Variational Methods with Application to Finite Elements", IST, Lisbon, 1972.

5 - BAZELEY, G.R.; CHEUNG, Y.K.; IRONS, B.M.; ZIENKIEWICZ, O.C. -"Triangular Elements in Plate Bending. Conforming and Non-Conforming Solutions",Proc.of the 1 st. Conf. on Matrix Meth. in Struct. Mech., Wright-Patterson AFB,Ohio, 1965.

6 - MIKHLIN, S.G. - "Variational Methods in Mathematical Physics", Pergamon Press, Oxford, 1964.

7 - OLIVEIRA, E.R.A. - "The Role of Convergence in the Theory of Shells", Int.J.Sol. Struct., 10, pp. 531-553, 1974.

8 - OLIVEIRA, E.R.A. - "Results on the Convergence of the Finite Element Method in Structural and Non-Structural Cases", Proc. of the Conf. on Finite Element Methods in Engineering, Sydney, 1974.

9 - PRAGER, W.; SYNGE, J.L. - "Approximations in Elasticity Based on the Concept of Function Spaces", Quart. Appl. Math., 5, pp. 241-269, 1947.

10 - SYNGE, J.L. - "The Hypercircle Method in Mathematical Physics", Cambridge Un.Press. Cambridge, 1957.

11 - ODEN, J.T. - "Finite Elements of Non-Linear Continua", Mc Graw-Hill, New York, 1972.

12 - OLIVEIRA, E.R.A. - "Convergence of Finite Element Solutions in Viscous Flow Problems", Swansea, 1974.

FUNCTIONAL ANALYSIS APPLIED TO THE OPTIMISATION OF A TEMPERATURE PROFILE

D.J. BALL
Department of Electrical, Electronic and Control Engineering,
Sunderland Polytechnic, England.

and J.R. HEWIT
Department of Mechanical Engineering, University of Newcastle
upon Tyne, England.

1.- Introduction -

In the testing of a material to determine its behaviour under mechanical and thermal stress, it is usual to submit a standard specimen of the material to a predetermined program of stress and record the strain induced. The imposition of mechanical and thermal stress can be viewed separately and only the latter is considered here. The basic problem to be studied is that of controlling the temperature profile within a specimen under test in an optimal manner. This clearly represents a distributed parameter optimal control problem.

This paper seeks to show how concepts and notation borrowed from functional analysis are used to extend a well-known method of finite dimensional optimisation to this practical problem defined in infinite dimensional space. Since the geometrical interpretation is closely analoguous to the geometry of three dimensional Euclidean space, the engineer is able to retain an insight into the method of solution. The application of functional analysis thus provides a unifying influence on the various classes of optimisation problem. It also permits a conciseness and clarity of presentation which would otherwise be unobtainable. Throughout the paper points of mathematical rigour are kept to a minimum as these are only of value to the engineer insofar as they assist in the solution of the practical problem.

The paper takes the following form. After an introduction to the problems of material testing, a mathematical model of the process is developed in the form of a linear integral equation. Next, the system Green's function is evaluated using a numerical least squares procedure. The optimum forcing function required to yield the desired temperature profile is then determined. The constrained optimisation involved is performed using a sequential unconstrained minimisation procedure in Hilbert space. Finally, the computed results are compared with those obtained experimentally using an actual test rig. Close agreement between these results validate the approach described in the paper.

2.- Materials Testing -

For mechanical (tensile or compressive) loading of a specimen, it is an easy matter to suddenly change the load from one constant value to another. High frequency transients due to the longitudinal wave equation may be neglected. However for thermal loading, the problem is complicated by the method used to heat (or cool) the specimen in a controlled manner and by the distributed nature of the specimen itself. Quite simply it is not possible to cause the temperature at each point in the specimen to change suddenly by a predetermined amount.

The experimental rig employed does not correspond precisely to a practical creep testing furnace because of the high temperatures usually required in such testing. This would necessitate the use of exotic materials and sophisticated design far beyond that which is necessary to simply validate the technique proposed in this paper. It is hoped in the future, however, to obtain results on a practical system.

Fig. 1 shows the arrangement used to control the specimen temperature profile on the test rig. The following important features are evident :

(i) The furnace temperature is controlled by a closed loop system.

(ii) The specimen temperature profile is related to the furnace temperature in an open loop manner.

(iii) There is no provision for controlled cooling.

The problem is then posed as follows. Find the desired furnace temperature (used as the set point in Fig. 1) so that the actual temperature profile within the specimen (monitored by eight thermocouples spaced longitudinally) is as close as possible (according to some criterion to be defined) to the given desired profile. This is essentially a distributed parameter control problem. The control variable is a scalar function of time whereas the output is a function of time and a spatial co - ordinate.

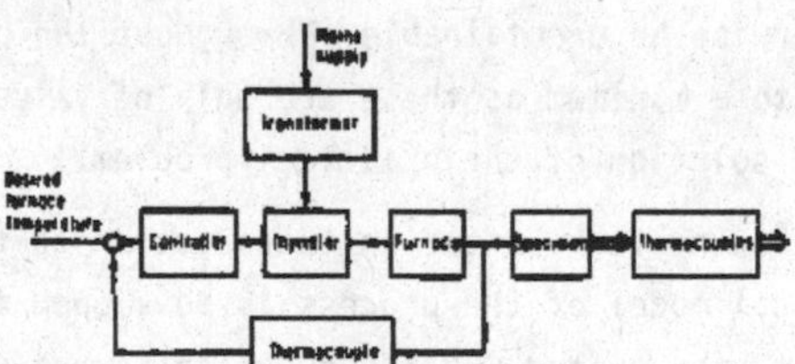

Fig. 1: General arrangement of control scheme

3.- Mathematical Model -

The one-dimensional linear mathematical model of the physical system takes the form :

$$u(x,t) = u_0(x,t) + \int_0^t g(x,t-\tau)f(\tau)d\tau \qquad (1)$$

where u(x,t) is the temperature distribution within the specimen, x being a spatial co-ordinate measured along its axis and t being time. f(t) is the furnace temperature neglecting small spatial variations, g(x,t) is the system Green's function and $u_0(x,t)$ represents the effect of initial conditions. This model neglects nonlinearities inherent in the system and radial temperature differences in the specimen. The validity of these assumptions is apparent from the ability of the model to represent the system behaviour over its normal operating range.

The model can be rearranged to give

$$u(x,t) = u_0(x,t) + \int_0^t f(t-\eta)g(x,\eta)d\eta \qquad (2)$$

and since the forcing function $f(t-\eta)$ can be taken as zero for all times $t \leqslant \eta$, the above is rewritten as

$$u = u_0 + Lg \qquad (3)$$

where L is a linear integral operator defined by

$$L\nu = \int_T f(t-\eta)\nu(x,\eta)d\eta \qquad (4)$$

T being some time interval $[0,t_f]$.

4.- Estimation of Green's Function -

The technique employed is a least squares estimation developed from the work of HSIEH {1} on systems governed by ordinary differential equations. The problem involves the determination of the Green's function such that the error between measured and computed states of the system is minimised over a fixed time interval. This is equivalent to the minimisation of a quadratic functional of the form

$$J(g) = \int_T\int_X [u(x,t) - m(x,t)]^2 \, dx\, dt \qquad (5)$$

where m(x,t) is the measured state of the system over $X \times T$.

Now consider a real square integrable Hilbert space H_1 defined on $X \times T$ with the inner product of two elements given by

$$\langle\nu,\omega\rangle_{H_1} = \int_T\int_X \nu(x,t)\omega(x,t)\, dx\, dt \qquad (6)$$

and the norm of an element by

$$||\nu||_{H_1} = \langle\nu,\nu\rangle_{H_1}^{1/2} \qquad (7)$$

Then substituting for u from (3) into (5) yields

$$J(g) = ||u_0 + Lg - m||^2_{H_1} \qquad (8)$$

Let L^* denote the adjoint of the operator L so that

$$L^*\nu = \int_T f(t - \eta)\nu(x,t)\, dt \tag{9}$$

Thus expanding (8) and rearranging

$$J(g) = \langle g, L^*Lg\rangle_{H_1} - 2\langle g, L^*(m - u_0)\rangle_{H_1} + ||m - u_0||^2_{H_1} \tag{10}$$

The problem is therefore formulated as a quadratic minimisation in H_1.

Next consider some element of H_1 displaced from the minimum $\hat{g}$ by a distance ε in a direction z. Evaluating $J(\hat{g} + \varepsilon z)$ and rearranging gives

$$J(\hat{g} + \varepsilon z) = J(\hat{g}) + 2\varepsilon\langle z, L^*L\hat{g} - L^*(m - u_0)\rangle_{H_1} + \varepsilon^2\langle z, L^*Lz\rangle_{H_1} \tag{11}$$

The condition to ensure a minimum for arbitrary z is then clearly

$$L^*L\hat{g} = L^*(m - u_0) \tag{12}$$

In addition, for practical systems, it is unlikely that zero will be an eigenvalue of the operator L . Hence

$$\langle z, L^*Lz\rangle_{H_1} = ||Lz||^2_{H_1} > 0 \qquad z \neq 0 \tag{13}$$

so that L^*L is a positive-definite self-adjoint operator and the final term in (11) is positive for all non-zero scalars ε. It follows therefore that

$$J(\hat{g} + \varepsilon z) > J(\hat{g}) \qquad \varepsilon, z \neq 0 \tag{14}$$

and hence (12) gives the condition for a unique minimum of J(g).

This condition is however based on measured values of the forcing and state variables and so an analytic solution is inappropriate. An iterative method using the conjugate gradient minimisation procedure is therefore employed. Although this technique was originally developed for parameter minimisation {2}, using functional analytic concepts the algorithm together with its associated convergence theorems can easily be generalised to deal with minimisations in function space {3}. The algorithm proceeds from some initial estimate g_0 as follows

(i) Set $z_0 = - r_0$ and $k = 0$

(ii) Choose ε_k to minimise $J(g_k + \varepsilon_k z_k)$

(iii) Set $g_{k+1} = g_k + \varepsilon_k z_k$ (15)

(iv) Set $z_{k+1} = - r_{k+1} + \beta_k z_k$ where $\beta_k = \dfrac{||r_{k+1}||^2_{H_1}}{||r_k||^2_{H_1}}$

(v) Set $k = k+1$

(vi) Go to (ii) unless the convergence limit is reached;

where r_k is the direction of the gradient of the functional which is given from (11) as

$$r_k = L^*Lg_k - L^*(m - u_o) \tag{16}$$

Also by differentiating (11) with respect to ε and equating to zero, the scalar ε_k which minimises $J(g_k + \varepsilon_k z_k)$ is given by

$$\varepsilon_k = - \frac{\langle z_k, r_k \rangle}{\langle z_k, L^* L z_k \rangle} \tag{17}$$

It can be shown {4} that the directions of search are conjugate with respect to the second derivative of the functional such that

$$\langle z_j, L^* L z_k \rangle = 0 \qquad j \neq k \tag{18}$$

and hence that the directions $\{z_k\}$ form an infinite sequence of linearly independant functions. The extension of the technique for infinite dimensional minimisation problems therefore means that finite quadratic convergence can no longer be guaranteed. However, the technique has been found to be the most efficient iterative procedure available for minimising the majority of quadratic and nonquadratic functionals.

A test signal involving a step change in furnace temperature from ambient to 200°C for one hour was applied to the experimental rig and the resultant specimen temperature profiles recorded. The integral square error between these measured profiles and those computed from the estimated Green's function was then minimised using the technique described above. The system Green's function obtained is shown in Fig. 2.

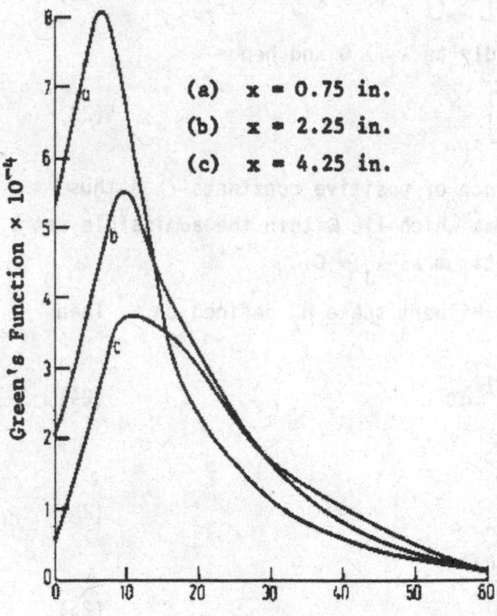

Fig. 2: System Green's function

5.- Temperature Profile Optimisation-

It has already been emphasised that the furnace cannot be cooled in a controlled manner. It has a natural cooling rate determined by the system configuration. Nevertheless, since the heat losses are small, this is relatively slow. This is approximately equivalent to a constraint of the form

$$f(t) \geqslant 0 \tag{19}$$

where f(t) denotes the rate of change of furnace temperature with time.

The problem is to determine the forcing function which minimises

$$J(f) = \int_X \left[u(x,t_f) - d(x,t_f)\right]^2 dx \tag{20}$$

within the admissible set D defined by (19), $d(x,t_f)$ being the desired profile at the final time t_f. The method employed is based on the use of penalty functions which have been studied extensively by FIACCO and McCORMICK {5} for finite dimensional problems. These reduce the optimisation to a series of unconstrained minimisations in Hilbert space and have been used for constrained problems involving lumped parameter systems by LASDON et al {6}.

The performance criterion is augmented by a further term to give

$$J_1(f,\lambda_j) = J(f) + \lambda_j \int_T \dot{f}^{-1}\, dt \qquad \lambda_j > 0 \tag{21}$$

This ensures that the criterion tends to infinity as the rate of change of the forcing function approaches zero. The penalty term is always positive provided that the constraint (19) is not violated. It therefore follows that

$$\min_{f\varepsilon D}\left[J_1(f,\lambda_j)\right] > \min_{f\varepsilon D}\left[J_1(f,\lambda_{j+1})\right] > \min_{f\varepsilon D}\left[J(f)\right] \qquad \lambda_j > \lambda_{j+1} > 0 \tag{22}$$

In addition, the penalty term decreases rapidly as $\lambda_j \to 0$ and hence

$$\lim_{\lambda_j \to 0} \left(\min_{f\varepsilon D}\left[J_1(f,\lambda_j)\right] \right) = \min_{f\varepsilon D}\left[J(f)\right] \tag{23}$$

Minimising $J_1(f,\lambda_j)$ for the decreasing sequence of positive constants $\{\lambda_j\}$ thus yields a series of artificially created minima which lie within the admissible set D and converge toward the local constrained optimum as $\lambda_j \to 0$.

Now consider a real square integrable Hilbert space H_2 defined on X. Then (21) can be written as

$$J_1(f,\lambda_j) = ||u - d||^2_{H_2} + \lambda_j \int_T \dot{f}^{-1}\, dt \tag{24}$$

In addition from (1)

$$u = u_o + Hf \tag{25}$$

where H is a linear integral operator such that

$$H\nu = \int_T g(x,t_f - \tau)\nu(\tau)\, d\tau \tag{26}$$

Substituting for u from (25) into (24) and rearranging

$$J_1(f,\lambda_j) = \langle f, H^*Hf\rangle_{H_3} - 2\langle f, H^*(d-u_o)\rangle_{H_3} + ||d-u_o||^2_{H_2} + \lambda_j \int_T \dot{f}^{-1}\,dt \qquad (27)$$

H_3 being the real square integrable Hilbert space defined on T.

The minimum of $J_1(f,\lambda_j)$ is found by equating its first differential given by

$$\frac{\partial}{\partial\varepsilon}\Big[J_1(f+\varepsilon z,\lambda_j)\Big]\Big|_{\varepsilon=0} = 2\langle z, H^*Hf - H^*(d-u_o)\rangle_{H_3} - \lambda_j\langle \dot{z}, \dot{f}^{-2}\rangle_{H_3} \qquad (28)$$

to zero. Integrating the final term by parts

$$\langle \dot{z}, \dot{f}^{-2}\rangle_{H_3} = \Big[z\dot{f}^{-2}\Big]_0^{t_f} + 2\langle z, \ddot{f}\dot{f}^{-3}\rangle_{H_3} \qquad (29)$$

For the required minimum denoted by $\hat{f}$, the differential (28) must be zero for all functions z. Thus

$$H^*H\hat{f} - \lambda_j\ddot{\hat{f}}\dot{\hat{f}}^{-3} = H^*(d-u_o) \qquad \hat{f}\ \varepsilon\ D \qquad (30)$$

with the boundary conditions

$$\dot{\hat{f}}^{-2} = 0 \qquad t = 0, t_f \qquad (31)$$

The sequential solution of the nonlinear equation (30) using the conjugate gradient method as $\lambda_j \to 0$ will give the optimal forcing function. In this instance the direction of the gradient of the functional is given by

$$r_k = H^*Hf_k - \lambda_j\ddot{f}_k\dot{f}_k^{-3} - H^*(d-u_o) \qquad (32)$$

and the scalars $\{\varepsilon_k\}$ are found by a univariate search technique {7}.

The boundary condition (31) implies that $\dot{\hat{f}}$ must be infinite at the initial and final times. Although this effect may be extremely localised it follows that $\dot{\hat{f}}^{-1}$ and hence the penalty term is not convex throughout T. The minimisation therefore may not yield a unique global minimum.

The optimal control was computed which takes the temperature of the specimen from ambient to a linear profile varying from 150°C to 100°C along the specimen over a control interval of 30 minutes. Various initial estimates were tried to ensure the attainment of a global minimum. The computed optimum solution indicated the forcing function shown in Fig. 3. When this function was applied, the final specimen temperature profile was as shown in Fig.4. The computed and measured profiles are extremely close which suggests that the model adequately represents the system. The proximity of these profiles to the desired temperature distribution is obvious.

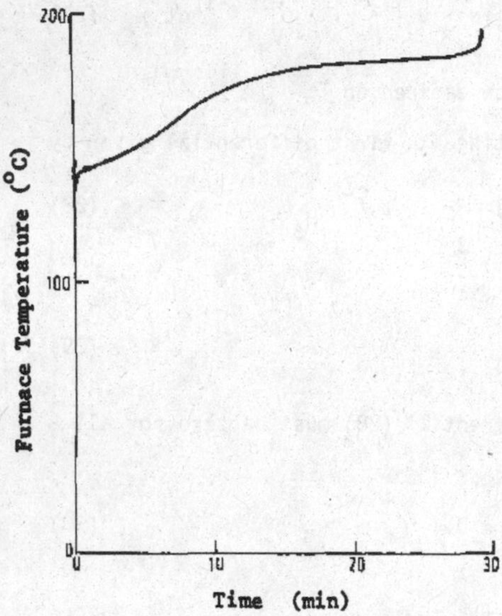

Fig. 3: Optimal desired furnace temperature

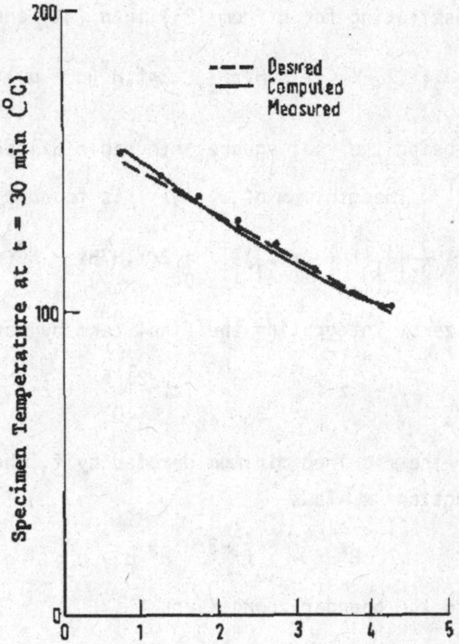

Fig. 4: Optimal specimen temperature profile

6.- Conclusions -

In this paper the application of the concepts of functional analysis to the solution of an optimal heating problem in the field of material testing is considered. In essence these concepts provide the framework for extending the method of conjugate gradient to perform the infinite dimensional minimisations associated with the evaluation of the systems Green's function and of the optimum forcing function. Emphasis is placed on documenting experimental experience of the application of the technique proposed to this one particular practical problem and no attempt is made to develop generalised theory.

The basic limitation of the analysis presented is the assumption of system linearity. This is supported in this application by the close correlation between the computed and measured optimal temperature profiles. However, this assumption may become more suspect at higher temperatures than are considered in this paper where radiation heat transfer predominates. Also at these higher temperatures, the constraint (19) will be modified such that the rate of change of furnace temperature must be less than a negative function of the temperature representing the system natural cooling rate.

The ideas of functional analysis can also be used in developing solutions for the problem of heating specimens to a desired temperature profile in minimum time {8} and for the problem where the system is represented by a partial differential rather than an integral equation {9}.

7.- References -

{1} H.S.HSIEH : Least squares estimation of linear and nonlinear weighting function matrices. Inform. Control 7, 84-115 (1964).

{2} R.FLETCHER and C.M.REEVES : Function minimisation by conjugate gradients. Computer J. 7, 149-154 (1964).

{3} L.S.LASDON, S.K.MITTER and A.D.WAREN : The conjugate gradient method for optimal control problems. IEEE Trans. Auto. Control AC-12, 132-138 (1967).

{4} E.POLAK : Computational methods in optimisation. Academic Press (1971).

{5} A.V.FIACCO and G.P.McCORMICK : The sequential unconstrained minimisation technique for nonlinear programming - a primal dual method. Management Sci. 10, 360-366 (1964).

{6} L.S.LASDON, A.D.WAREN and R.K.RICE : An interior penalty method for inequality constrained optimal control problems. IEEE Trans. Auto. Control AC-12, 388-395 (1967).

{7} M.J.BOX, D.DAVIES and W.H.SWANN : Nonlinear optimisation techniques. Oliver and Boyd (1969).

{8} D.J.BALL and J.R.HEWIT : The time optimal control of a class of distributed systems. J.IMA. 10, 193-201 (1972).

{9} D.J.BALL and J.R.HEWIT : An alternative approach to a distributed parameter optimal control problem. Int. J. Systems Sci. 5, 309-316 (1974).

GLOBAL FREE BOUNDARY PROBLEMS AND THE CALCULUS OF VARIATIONS IN THE LARGE

M.S. BERGER AND L.E. FRAENKEL

1. Introduction

An interesting gap in the methods of classical mathematical physics concerns the treatment of global free boundary problems (i.e. free boundary problems in which a good first approximation to the desired solution cannot be obtained a priori). Moreover, we limit our discussion to those problems that can be obtained as Euler - Lagrange equations of some variational problems. Such problems include figures of equilibrium of a rotating fluid mass, shapes of vortex cores in steady separated ideal fluid flow and the shape of electrified fluid drops, to mention only a few. These problems have a considerable history, often associated with the great figures of mathematical physics, but nonetheless the mathematical results concerning them remain incomplete. Possible reasons for this state of affairs are sketched later but for the moment we suggest the difficulty inherent in studying saddle points of a functional defined over an infinite dimensional admissible class of functions, as being particularly crucial.

2. Relevance to Contemporary Physics

Of course, the mathematical difficulties in dealing with such classic global problems are bound to recur in a more virulent form in contemporary physics. In fact, the theory of rotating fluid masses with a surface tension and a uniform electric charge arose in nuclear physics in the Bohr-Wheeler "liquid drop" model of nuclear fission. This model and its subsequent refinements have proven basic in understanding the equilibrium configurations of an idealized nucleus.

3. Mathematical formulations and Associated Difficulties

Generally speaking, the mathematical formulation of the free boundary problems we are describing requires (i) a smooth function U, (defined on some given unbounded domain D), and (ii) a bounded subset A of D satisfying the following properties:

(a) U satisfies a nonlinear elliptic second order partial differential equation:

$$(1) \qquad CU = \begin{cases} -f(x,U) & \text{in } A \\ 0 & \text{in } D-A \end{cases}$$

Here f represents a "potential" distribution and so can be considered smooth, nondecreasing in U, with at most linear growth in U

(b) On the boundary of A, bA, the function U satisfies $U + q(x) = 0$, whereas across bA, U and its gradient and required to be continuous.

(c) Moreover U is required to have a specific asymptotic behaviour on the unbounded portion of the boundary of D, bD. The boundary bA is then the desired free boundary. (See Figure 1).

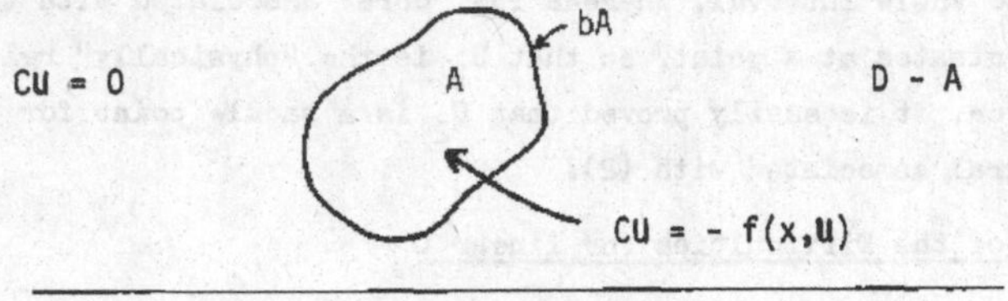

Figure 1 : Notation

From a mathematical point of view the difficulties inherent in finding a complete solution to this class of problem are rather formidable. These difficulties include:

1. the unknown function U and the domain A are simultaneous unknowns.
2. the infinite number of degrees of freedom of the problem.
3. the desired solution is generally a saddle point for the associated variational integral that does not satisfy standard "compactness" restrictions.
4. the discontinuity of the derivatives of U and the function f across bA.
5. the possibility of multiple solutions for the problem with the domain A of differing topological types.
6. the crucial occurrence of parameters in the problem giving rise to "branches" of solutions joining solutions of differing topological type.

4. A One Dimensional Example

The following example is instructive:

(2) $$U_{xx} = \begin{cases} -w & \text{in } A \qquad w > 0, \text{ a constant} \\ 0 & \text{in } [0,1] - A \end{cases}$$

$$U(0) = 0 \quad U(1) = -1 \qquad U = 0 \text{ on } \partial A. \quad U(x) \in C^1 \text{ across } \partial A.$$

In this case it turns out that the only possible domains A have the form $A = (0,k)$ where k satisfies the cubic equation,

$k = \frac{1}{2} w k^2 (1-k)$. This equation has 3 roots $k_o = 0$

$$k_1 = \tfrac{1}{2}\left(1 + \sqrt{1 - 8/w}\right) \qquad k_2 = \tfrac{1}{2}\left(1 - \sqrt{1 - 8/w}\right)$$

The solution corresponding to $k = o$ we call trivial.

Then for $w < 8$, (2) has only one trivial solution, whereas for $w > 8$, (2) has two nontrivial solutions. Now if we let w become infinite (or equivalently consider an expansion of the domain (0,1) to a half-line) we observe that K_1 tends to unity whereas k_2 tends to zero. Thus the "core" of the solution U_1 tends to smear out over the whole interval, whereas the "core" associated with the solution U_2 concentrates at a point, so that U_2 is the "physically" relevant solution. Moreover, it is easily proved that U_2 is a saddle point for the variational integral associated with (2).

5. Resolution of the Difficulties for linear C

Here we list the main ideas needed to resolve the free boundary problem associated with (1) in case C is a linear self-adjoint positive elliptic operator:

1) Conversion of (1) to a boundary value problem on an infinite domain

We extend the function $f(x,U)$ to $\tilde{f}(x,U)$ by setting $\tilde{f}(x,U) = 0$ for $U < o$. Then the maximum principle for C implies, that if we consider the boundary value problem

(3) $CU = -\tilde{f}(x,U)$ in D together with the boundary condition (IC)

We can recover the core A and the free boundary bA by letting A be the set where U is positive and bA be the level set $U = 0$. The price to be paid for such a simplification is that the new $\tilde{f}$ is often discontinuous but this can be overcome by an additional approximation step, as below.

2). Use of Hilbert space techniques for the generalized solutions of (3)

It is easy to formulate a variational integral I(U) whose Euler-Lagrange equation coincides with (3). Then, by considering, the critical points of I(U), defined on an appropriate Hilbert space H, we can find "generalised" solutions of (3) by utilizing the modern techniques of the calculus of variations in the large. The fact that the generalized solutions of (3) are actually solutions in the classical sense, can then be established by the L_2 regularity theory for linear elliptic equations. Moreover in discussing convergence, crucial use can be made of "weak" convergence in H.

(3) Isoperimetric Variational Problem and Approximation Procedure

Modern approaches to study saddle points of functionals I(U) require certain compactness criteria for their implementation. Because the domain D is unbounded these criterial fail to hold for I(U). To overcome this, we approximate D by a nested sequence of bounded domains D (a,b) (for real numbers (a,b) so that the boundary value problem (3) is approximated by the analogous problem for (3) with D replaced with D (a,b). This regains compactness for the approximate problem. Moreover it is important to assert that the solutions of the approximate problem for D(a,b), once obtained, converge to a true nontrivial solution of D. Also it is important to investigate precise properties of bA; this requires even more precise information about the solution for D. To this end it is useful to characterize the saddle point as the solution to a isoperimetric variational problem of the form(P_1) Minimize $J(U) = \int_D F(U)$ subject to the constraint $\|u\|^2$=const.R say, where the norm represents the norm obtained from the Dirichlet integral associated with the operator C. The price to be paid from this transition to an isoperimetric variational problem is that instead of (3) we now solve the problem:

$$(4) \qquad Cu = -\lambda \tilde{f}(x,u) \quad \text{in} \quad D$$

where λ is a Lagrange multiplier with definite bounds. For many physical problems, λ has a definite physical interpretation to that the transition from (3) to (4) is quite satisfactory. Moreover the isoperimetric characterization enables one to link these solutions with classical explicit solutions and thus link solutions of differing topological type. See (6) below.

(4) Symmetrization

Once the desired critical point U of I(U) is characterized by an isoperimetric variational problem P_1, more precise information concerning U (as described (3) above) and its approximations can be found by classical symmetrization techniques, provided P_1 possesses certain symmetry restrictions.

(5) A priori bounds for solutions of (P_1)

These bounds can be obtained via (4) and specific properties of the Green's function for C relative to D. Once obtained, such bounds enable us to take limits of approximate solutions to (P_1) on D(a,b) as D(a,b) tends to D, and thus to assert the existence of nontrivial solutions to(P_1) on D.

6. Steady Vortex Motion in an Ideal Fluid

All the above considerations apply with great success to the study of steady separated ideal fluid flow in two or three dimensions. In the case of three space, it is desirable to find a family of steady vortex rings of varying cross section linking the two classical explicit solutions: Helmholtz's Singular Vortex Ring and Hill's spherical vortex. This can, of course, be accomplished utilizing the special geometry of the situation. Since an axisymmetric solution is desired, we can choose D to the $\frac{1}{2}$ plane Π , U can be chosen to be the Stokes stream function and (as in Reference [1])

$$C(U) = r\left(\frac{1}{r} U_r\right)_r + U_{zz}$$

The asymptotic condition at infinity is specified by demanding that the desired ring more relative to the fluid at infinity with constant velocity in the z direction. The Hilbert space H associated with the operator C is defined by the inner product.

$$(U,V) = \iint \frac{1}{r} \; (U_r V_r + U_z V_z) \; drdz$$

Using the techniques described above we prove the existence of a one parameter family of vortex rings V_R $(0 < R < \infty)$ characterized as a solution to isoperimetric problem (P_1). This result holds for a large class of vorticity distributions f. Moreover the variational characterization of V_R can be used to prove that for convex f the associated vortex core has only one component in Π so that the solutions V_R do actually represent vortex rings physically. Moreover as $R \to o$ we conjecture that the vortex rings V_R tend to a singular vortex ring, analogous to Helmholtz's. In current investigations we have taken the first steps in establishing this fact and studying its general mathematical significance. The situation represents a new type of bifurcation phenomena, which we call "bifurcation from Green's function".

7. Historical Remarks

Kelvin seems to have been the first to have studied both the vortex ring and rotating fluid problem simultaneously. He also attempted to find variational characterizations of vortex rings and, in fact, noted that vortex rings were saddle points of an associated energy functional. Poincaré and Liapunov focussed attention on the "bifurcation" problems of rotating fluids and this work was carried on with great vigour in an attempt to study the "fission" properties of planets. In a large number of papers, L. Lichtenstein studied local aspects of both the vortex ring and rotating fluid problem. However in all these works, very little mathematical work was achieved in studying global aspects of these problems. In an interesting paper of 1918, Carleman used symmetrization techniques to show that spheres were the only free boundaries for problems of the form (4) under "static" conditions. Finally, we note there is a remarkable similarity in the known free boundaries in the vortex ring and rotating fluid problems. Thus, for example, the rotating fluid problem also has ringed shaped figures as free boundaries and these are conjectured to evolve by a continuation process from the well-known Maclaurin ellipsoids.

BIBLIOGRAPHY

1. L.E. Fraenkel and M.S. Berger. A Global Theory of Steady Vortex Rings in an Ideal Fluid. Acta Math 132 pp 13-51 1974.

APPENDIX

APPLICATION OF THE CALCULUS OF VARIATIONS IN THE LARGE TO DETERMINE EQUILIBRIUM STATES IN NON LINEAR ELASTICITY by M.S. Berger.

The modern theory of the calculus of variations in the large is particularly useful in studying the structure of global equilibrium states in non-linear elasticity. In functional analytic language, such states can often be found by finding the solutions of a nonlinear operator equation,

(1) $A(u,\lambda) = f$, λ a real parameter.

Moreover A is often a gradient mapping i,e. it is the Frechet derivative of a smooth real valued functional $I(u,\lambda)$ defined on an appropriate Hilbert space X. For the von Karman equations X is a closed subspace of the Sobolev space $W_{2,2}$.

We shall suppose that A is a gradient mapping defined on a Hilbert space X and moreover that A is smooth. The key compactness property of A that allows the application of modern methods to (1) is the so-called "properness" of A. A mapping N is called proper if the inverse image of a compact set (relative to N) is again compact in the domain of N. The operator $A(u,\lambda)$ associated with the von Karman equations satisfies this property for fixed parameter values λ . Under this assumption the following facts about the solutions of (1) for fixed λ , emerge:

(i) the singular values of A, Σ form a closed subset of X and on each component of $(X-\Sigma)$, the number of solutions of (1) is fixed.

(ii) assuming the kernel of the Frechet derivative of A, is finite dimensional and that $I(u,\lambda) - (f,u)$ is bounded from below, we can prove that the singular values Σ of A are nowhere dense, and for $f \notin \Sigma$, the following Morse inequalities for the solutions of (1) hold:

$$M_0 \geqslant 1$$

$$M_1 - M_0 \geqslant 1$$

$$M_2 - M_1 + M_0 \geqslant 1 \quad \text{etc.}$$

here M_i denotes the number of solutions of (1) with Morse Index i.

These inequalities can be used to predict the existence of new equilibrium states in non linear elasticity once some rough data about known equilibrium

states is obtained. They also ensure the existence of a solution of (1) corresponding to the absolute minimum of I.

The behaviour of the equilibrium states obtained in this way, <u>as λ varies</u> is a crucial unresolved problem for the theory and its application.

PROOF OF EXISTENCE AND UNIQUENESS OF TIDAL WAVES WITH GENERAL VORTICITY DISTRIBUTIONS

K.Beyer & E.Zeidler
Leipzig

A plane dynamic tidal model is considered here within the framework of the channel theory, in which the sea current is characterized by a prescribed vorticity distribution.

Several tidal models are discussed in the handbook article [2] by Defant and in the monograph [5] by Zeidler. Non-vortical tidal waves were shown to exist by Beckert [1] who used topological methods; a constructive proof of existence is given in [5]. In [3] and [4] Maruhn considers tidal waves of homogeneous and non-homogeneous liquids in a static tidal model.

The authors proceed from the dynamic tidal model used in [1]. It is assumed that the sea current, in an equatorial channel , covers the followigg domain in a plan having the rectangular coordinates (x,y)

$$G = \{ x=r \cos \alpha,\ y=r \sin \alpha,\ R < r < r_1(\alpha),\ -\pi\leq\sigma\leq\pi \}$$

(R = radius of the earth). The moon is assumed to be in $x=x_M$, $y=0$. In order to make allowance for spring tides and neap-tides, two sun positions are considered:

$$x=x_S,\ y=0 \quad \text{(new moon)}, \quad x=0,\ y=y_S \quad \text{(half-moon)}.$$

When the monthly and yearly movements of the earth, moon and sun are neglected, then the (x,y)-system is an inertial system. The earth is assumed to rotate in this system with the angular velocity ω. The current with the density μ is assumed to be incompressible. Then the following differential equations apply to the velocity vector (u,v) of the current

$$(1) \quad u_x + v_y = 0, \quad u_y - v_x = cf(\phi)/R$$

with the stream fuction ϕ : $u = \phi_y$, $v = -\phi_x$.

According to [5] the boundary conditions of our problem are as follows:

(2) $\psi = 0$

$$\mu k_E \frac{r}{R} + \mu U_M + \mu U_S + p_o + \frac{\mu}{2}(u^2 + v^2) = \text{const.}$$

along $r = r_1(\alpha)$ and

(3) $\psi = -\psi_o$

along $r = R$ (U_M potential of the moon, U_S potential of the sun, p_o pressure).

A determination is to be made for a current area G and for the velocity distribution (u,v) which in G is the solution of (1) and satisfies the boundary conditions of (2),(3).

By way of quasiconformal mapping of the current area bounded by the unknown sea area the free boundary value problem (1)-(3) is reduced to a non-linear eigenvalue problem (R) for an integro-differential equation over a circular ring

$$E_{q1} = \{ \zeta = \rho \exp i\sigma;\ q \le \rho \le 1,\ -\pi \le \sigma \le \pi \}.$$

(R) Proceeding from $q \in (0,1)$ and the vorticity distribution f

$$f \in C_\beta[q,1],\ \exp \tau_o(\rho) := \int_q^\rho \rho f(\rho)/q^2 \, d\rho + 1 > 0.$$

the determination is for two functions ϑ, $\tau \in C_\beta^{(1)}[E_{q1}]$ which are characterized by

a) $\vartheta(\rho, -\sigma) = -\vartheta(\rho, \sigma)$, $\tau(\rho, -\sigma) = \tau(\rho, \sigma)$

b) $\oint \tau(q, \sigma)\, d\sigma = 0$

c) $\vartheta_\rho - \rho^{-1} \tau_\sigma = A(\vartheta, \tau, \tau_\sigma)$

$$\tau_\rho + \rho^{-1} \vartheta_\sigma + 2\tau_o'\left(\tau + \int_q^\rho \rho^{-1} \tau \, d\rho\right) = B(\vartheta, \tau, \vartheta_\sigma)$$

d) $\vartheta(q, \sigma) = 0$

$$T\tau(1, \sigma) - \lambda T \int_0^\sigma \vartheta(1, \sigma)\, d\sigma = \Gamma(\vartheta, \tau; \nu_M, \nu_S, \lambda),$$

where

$$A = O_2(\vartheta, \tau, \tau_\sigma), \quad B = O_2(\vartheta, \tau, \vartheta_\sigma),$$

$$\Gamma = (\nu_M \pm \nu_S)(\cos 2\sigma + O_1(\vartheta, \tau) + (2+\lambda) O_2(\vartheta, \tau));$$

$$T\tau\,(\rho, \sigma) = \tau(\rho, \sigma) - (2\pi)^{-1} \oint \tau(\rho, \sigma)\, d\sigma.$$

The symbols O_1, O_2 designate quantities of first/second order in the related variables. The numbers ν_M, ν_S result from the disturbing effects of the moon and sun which have been assumed to be small.

The solutions of the boundary value problem (R) are studied in the spaces $C_\beta^{(1)}[E_{q1}]$ of Hölder continuous differentiable functions. It is shown that the linearized problem is of index zero and has positive and simple eigenvalues. In the non-homogeneous case (weak interference from moon and sun) the solutions to the non-linear problem are demonstrated with Banach's fixed point theorem. In order to prove the existence and uniqueness of tidal waves in the absence of perturbing bodies (natural oscillations) a functional analysis variant of the bifurcation theory according to Lyapunov-Schmidt is used.

For noncritical angular velocities $\omega = \omega_j$ of the earth in the presence of perturbing bodies there exist tidal waves; in the absence of perturbing bodies there exist non-trivial tidal waves near ω_j. The solutions are in the vicinity of a radially symmetric current and may be developed to give power series which are absolutely convergent with respect to the norm of $C_\beta^{(1)}$ according to the small parameters ν_M, ν_S, s.

A more detailed demonstration is to appear in Archive for Rational Mechanics and Analysis.

References

[1] Beckert, H.: Existenzbeweis zur Kanaltheorie der Gezeiten. Arch. Rat. Mech. Anal. 9 (1965), 379-396.

[2] Defant, A.: Flutwellen und Gezeiten. Handbuch der Physik XLVIII, 846-927, Springer Verlag Berlin-Göttingen-Heidelberg 1957.

[3] Maruhn, K.: Einige Bemerkungen zur Theorie der Gezeiten. Math. Ann. 56 (1966), 111-116.

[4] Maruhn, K.: Über ein Modell aus der Theorie der Gezeiten. Beiträge zur Analysis, S.9-16, Akademie-Verlag Berlin 1968.

[5] Zeidler, E.: Beiträge zur Theorie und Praxis freier Randwertaufgaben, 220 S., Akademie-Verlag Berlin 1971.

A CRITICAL APPRAISAL OF CERTAIN CONTEMPORARY SHIP MODEL TESTING TECHNIQUES

R.E.D. Bishop and W.G. Price
University College London
Mechanical Engineering Dept
London WC1E 7JE, England

Summary

A linear functional representation of hydrodynamic actions on ships during small departures from steady motion ahead provides insight into the nature and known shortcomings of conventional 'slow motion derivatives'. The forms of the functionals may be found by means of a planar motion mechanism (PMM). If PMM results are interpreted directly in terms of derivatives, the results obtained may be substantially in error.

Introduction

The experimental techniques used to find the fluid forces and moments that act on a ship when it departs from a steady reference motion require that measurements be made on models under either steady or unsteady conditions. In general, the hydrodynamic actions are described mathematically in terms of 'slow motion derivatives', under the assumption of 'quasi-steady flow'. (That is to say the generalised fluid forces at any instant are supposed to be dependent upon the generalised motions at that same instant.) In unsteady testing, this is open to question, and it has been suggested(2),(3),(6) that functional representation may provide a better alternative.

Constrained model tests

The object of constrained model tests is to measure the appropriate derivatives needed for predictions of ship manoeuvring characteristics. Usually, for a surface ship, the motions of interest are:-

Sway: velocity, $v(t)$; acceleration, $\dot{v}(t)$; force derivatives Y_v, $Y_{\dot{v}}$;
moment derivatives N_v, $N_{\dot{v}}$.

Yaw: angular velocity, $r(t)$; angular acceleration, $\dot{r}(t)$; force derivatives Y_r, $Y_{\dot{r}}$; moment derivatives N_r, $N_{\dot{r}}$.

We shall mainly consider, by way of example, the sway motions and derivatives.

The simplest test is the oblique tow test, in which the ship model is towed at a constant velocity and at various drift angles to the direction of motion and the sway force, Y, and yaw moment, N, are measured. The gradient of the curve at zero drift angle in a graph of Y plotted against drift angle gives an experimental value of the derivative Y_v. Likewise, the slope at zero drift angle in the moment curve gives N_v. No comparable experimental technique has been derived by which the acceleration derivatives $Y_{\dot{v}}$ and $N_{\dot{v}}$ can be measured.

The derivatives Y_r and N_r are nowadays determined by means of a 'rotating arm'. A model is towed in a circular path. Only the angular velocity derivatives can be determined by this technique and there still remains the problem of obtaining the acceleration derivatives $Y_{\dot{r}}$ and $N_{\dot{r}}$.

The planar motion mechanism (PMM) was developed[11],[16] with a main objective of determining acceleration derivatives. This mechanism is mounted on the carriage of a towing tank and imposes an oscillatory motion (e.g. of pure sway or pure yaw) on the model as it is towed along the tank at constant speed. Both the velocity and acceleration vary harmonically, and the corresponding forces and moments can be measured, whence the derivatives may be determined.

<u>Derivative representation of hydrodynamic actions</u>

Bryan[9] originally proposed the use of derivatives as a means of representing fluid actions; he assumed the perturbations of those actions to be linearly proportional to the velocity of the perturbed motion. Later, a linear acceleration term was introduced, as discussed by Babister[1]. Thus, if a model, travelling with a steady reference motion, acquires a sway velocity perturbation v(t) and a sway acceleration perturbation $\dot{v}(t)$, then the linear representation of the sway fluid force arising from the perturbed motion is expressed as

$$\Delta Y(v,\dot{v}) = Y_v v + Y_{\dot{v}}\dot{v}$$

where Y_v and $Y_{\dot{v}}$ are the derivatives of sway velocity and acceleration respectively. A similar representation may be developed for other hydrodynamic actions.

In a steady oblique towing test, in which the model has constant forward speed U and is inclined at a small drift angle β to the direction of travel,

$$v(t) = U \sin \beta \doteqdot U\beta = v_0 \ ; \quad \dot{v}(t) = 0 \ .$$

The perturbed sway force becomes

$$\Delta Y = Y_v v_0 ,$$

and if ΔY is plotted against either v_0 or β, the slope of the curve at the origin gives the derivative Y_v.

In practice, both in oblique tow and rotating arm experiments, it is found that the fluid forces and moments do not produce straight lines when plotted against sway velocity or yaw angular velocity. This has resulted in modifications of the previous theory by the introduction of extra terms. Unfortunately, no consensus of opinion exists among experimenters as to what terms should be included and what should not. For example, in the representation of sway force, terms such as $Y_{vvv}v^3$ or $Y_{v|v|}v|v|$ are sometimes included and evaluated by some process of curve fitting. In effect the interpretation of the experimental data then depends on the curve fitting technique employed. Thus the same experimental results may be fitted by the alternative expressions

$$\Delta Y = Y_v v + Y_{v|v|}v|v| ,$$

$$\Delta Y = Y_v v + Y_{vvv}v^3 ;$$

but this results, in general, in the arrival at differing values of the derivative Y_v. The linear derivative - or, more correctly here, 'coefficient' - has lost its uniqueness and, if it is used in a linear stability analysis, the conclusions reached may well depend on which of the two representations was chosen.

The results obtained from a PMM test raise other difficulties. For a pure sinusoidal sway motion $y(t) = y_0 \sin \omega t$,

$$v(t) = y_0 \omega \cos \omega t = v_0 \cos \omega t ,$$

$$\dot{v}(t) = -y_0 \omega^2 \sin \omega t = \dot{v}_0 \sin \omega t .$$

The corresponding sinusoidal fluid force may be measured and expressed as

$$\Delta Y = \Delta Y_{IN} \sin \omega t + \Delta Y_{QUAD} \cos \omega t .$$

But

$$\Delta Y = Y_v v + Y_{\dot{v}}\dot{v} = Y_v v_0 \cos \omega t + Y_{\dot{v}}\dot{v}_0 \sin \omega t ,$$

and so

$$\Delta Y_{IN} = Y_{\dot{v}}\dot{v}_0 \quad ; \quad \Delta Y_{QUAD} = Y_v v_0 .$$

Thus, in theory, the derivatives of velocity *and acceleration* may be determined from ΔY_{IN} and ΔY_{QUAD}. In practice, however, as we have already noted the simple straight-line experimental plot is not obtained. This has led some experimenters(10),

(14) to define and obtain 'non-linear derivatives'.

Comparisons of experimental data for the same ship form, but obtained by various experimenters, revealed great disparities(12). While some discrepancies were undoubtedly due to the use of models of varying quality made to different scales, the information presented clearly illustrates the great difficulty attending the analysis and interpretation of experimental data.

Functional representation of hydrodynamic actions

To avoid the assumption of 'quasi-steady flow', a linear representation may be adopted.(2) Thus, when specifying the perturbed sway force, it was suggested that

$$\Delta Y[v(t)] = \int_{-\infty}^{\infty} y_v(\tau)v(t-\tau)d\tau$$

$$= Y_{\dot{v}}(\infty)\dot{v}(t) + Y_v(\infty)v(t) + \int_{-\infty}^{\infty} y_v^*(\tau)v(t-\tau)d\tau .$$

Here the function $y_v(\tau)$ is the variation of the sway force deviation ΔY caused by a unit 'impulse' of sway velocity, and it is such that

$$y_v(\tau) = 0 \qquad \tau < 0 ,$$

$$y_v(\tau)v(t-\tau) = 0 \qquad \tau > t .$$

The response function $y_v^*(\tau)$ is a well behaved function having a Fourier transform so that there exists a function $Y_v(\omega)$ such that

$$y_v^*(\tau) = \frac{1}{2\pi}\int_{-\infty}^{\infty} Y_v(\omega)e^{i\omega\tau}d\omega ,$$

and

$$Y_v(\omega) = \int_{-\infty}^{\infty} y_v^*(\tau)e^{-i\omega\tau}d\tau$$

$$= \int_{-\infty}^{\infty} y_v^*(\tau)\cos\omega\tau d\tau - i\int_{-\infty}^{\infty} y_v^*(\tau)\sin\omega\tau d\tau$$

$$= Y_v^R(\omega) + iY_v^I(\omega) .$$

If $y_v^*(\tau)$ is known, then, it is possible to deduce $Y_v(\omega)$, and *vice versa*. In practice, the real quantities $Y_v^R(\omega)$ and $Y_v^I(\omega)$ can be measured in an oscillatory test on a model. These integral relationships between the frequency and time domains form the basis of the theoretical derivations of the Hilbert transform and Kramers-Kronig relations.[17]

Alternatively,[4] we may write

$$\Delta Y[\dot{v}(t)] = \int_{-\infty}^{\infty} y_{\dot{v}}(\tau)\dot{v}(t-\tau)d\tau$$

$$= Y_{\dot{v}}(\infty)\dot{v}(t) + Y_v(o)v(t) + \int_{-\infty}^{\infty} y^*_{\dot{v}}(\tau)\dot{v}(t-\tau)d\tau \quad ,$$

where the indicial response function $y_{\dot{v}}(\tau)$ is the variation in the perturbed sway force ΔY caused by an indicial sway velocity, and is such that

$$y_{\dot{v}}(\tau) = 0 \qquad \tau < 0 \quad ,$$

$$y_{\dot{v}}(\tau)\dot{v}(t-\tau) = 0 \qquad \tau > t \quad ,$$

The well behaved function $y^*_{\dot{v}}(\tau)$ has a Fourier transform $Y_{\dot{v}}(\omega)$ such that

$$y^*_{\dot{v}}(\tau) = \frac{1}{2\pi}\int_{-\infty}^{\infty} Y_{\dot{v}}(\omega)e^{i\omega\tau}d\omega \ .$$

The coefficients $Y_{\dot{v}}(\infty)$, $Y_v(\infty)$ and $Y_v(o)$ are all constants associated with the sway force at either $\omega = \infty$ or $\omega = o$.

Functional interpretation of hydrodynamic actions

In a steady tow test, in which the sway velocity $v(t)$ is constant, the sway force is

$$\Delta Y[v(t)] = Y_v v_0 = Y_v(o)v_0 = \left\{Y_v(\infty) + \int_{-\infty}^{\infty} y^*_{\dot{v}}(\tau)d\tau\right\}v_0 \ .$$

A simple relationship exists between the derivative Y_v, the coefficients $Y_v(o)$, $Y_v(\infty)$ and the function $y^*_v(\tau)$.

In an oscillatory test, on the other hand,[4] if

$$\Delta Y[v(t)] = \Delta Y_{IN}\sin\omega t + \Delta Y_{QUAD}\cos\omega t \ ,$$

then

$$\Delta Y_{IN} = \{Y_{\dot{v}}(\infty) + Y^I_v(\omega)/\omega\}\dot{v}_o \ ,$$

$$\Delta Y_{QUAD} = \{Y_v(\infty) + Y^R_v(\omega)\}v_o \ .$$

That is, $Y_v^R(\omega)$ and $Y_v^I(\omega)$ are related to the measured quadrature and in-phase components of the sway force and the impulse response function $y_v^*(\tau)$ can therefore be determined from the Fourier transform relationship.

In the majority of studies of the motions of surface ships, the sway velocity $v(t)$ can be relied upon to be analytically well behaved and to have derivatives of all orders. In this event, it has been shown(8) that for a pure oscillatory sway motion the in-phase and quadrature components of the perturbed sway force may be expressed as

$$\Delta Y_{IN} = [\{Y_{\dot{v}}(\infty) - \int_{-\infty}^{\infty} \tau y_v^*(\tau)d\tau\} + \frac{\omega^2}{3!}\int_{-\infty}^{\infty} \tau^3 y_v^*(\tau)d\tau - \ldots]\dot{v}_0 ,$$

$$\Delta Y_{QUAD} = [\{Y_v(\infty) + \int_{-\infty}^{\infty} y_v^*(\tau)d\tau\} - \frac{\omega^2}{2!}\int_{-\infty}^{\infty} \tau^2 y_v^*(\tau)d\tau + \ldots]v_0 .$$

Thus the coefficients occurring in this expansion may be determined from the impulse response function. The measurements made with the PMM (i.e. ΔY_{IN} and ΔY_{QUAD}) are therefore expressed in terms of the impulse response function $y_v^*(\tau)$ rather than with some arbitrarily chosen curve fitting technique.

Application of functional analysis

In a conventional PMM test undertaken in a towing tank, the lowest frequency of oscillation from which worthwhile results can be obtained may be of the order of 1 rad sec^{-1}. During the course of the test the in-phase and quadrature components of the sway force are recorded. The main drawbacks are that

1) only a very few oscillations can be completed because the model has to be towed at the scaled operational speed in a towing tank of limited length;

2) at low frequencies the oscillatory forces are of very small amplitude and so a large number of oscillations are required to obtain confidence in the data produced;

3) such is the dependence of the experimental data on frequency, extrapolation back to $\omega = 0$ is difficult.

In this last connection, figs. 1 and 2 show some typical results and it will be seen that the 'straight-line analysis' (according to which ΔY_{IN} and ΔY_{QUAD} are constants independent of frequency) is far from appropriate.

From the foregoing theory we see that only ΔY_{QUAD}, or $Y_v^R(\omega)$, need be measured.

This quantity poses no difficulty at zero frequency since its value at this point is the derivative Y_v that is determined from the oblique tow test. In effect, a curve like that of fig. 2 contains all the necessary experimental information for the determination of the impulse response function. By applying the Fourier transform technique as previously described the impulse response function $y_v^*(t)$ was determined, as shown in fig. 3. Again by transforming this function into the in-phase plane we obtain $Y_v^I(\omega)$ *by calculation*, so that a comparison can be made with the measured data. This is illustrated in fig. 1. The advantage of this technique lies in the fact that extrapolation to zero frequency is built into the analysis and is based on data obtained over a wide range of driving frequency.

Alternatively the in-phase data may be transformed into the quadrature plane. This is done by means of the indicial response function $y_{\dot{v}}^*(t)$, illustrated in fig. 4. The resulting comparison is made in fig. 2.

The range of frequency employed by Glansdorp[14] is far greater than that usually covered, the maximum frequency rarely exceeding 4 rad sec^{-1} in practice. Thus in the conventional analysis, a curve fitting technique is sometimes used in the approximate range $1 \leqslant \omega \leqslant 4$ rad sec^{-1} to help with the extrapolation. In this way, some polynomial expression is arrived at and the experimental results are referred to as being 'non-linear with frequency'. But by using a *linear* functional analysis good agreement has been found between experimental data and theory, as shown in figs. 1 and 2.

In a PMM towing tank test for which a large number of driving frequencies are used, the time required for experimentation becomes very large. To overcome this drawback, it has been proposed[7] that such tests could be done satisfactorily in a circulating water channel. The model is then oscillated about a fixed position while the water moves at the scaled operational speed of the ship. Preliminary tests of this sort were reported and it was observed that

1. the time taken for tests at a large number of frequencies is not excessive;
2. any number of oscillations can be performed at a given frequency, there being now no limit set by the length of a tank;
3. a sufficient number of oscillations can be performed at very low frequency to give much greater confidence in the data.

Further PMM tests have been undertaken in a circulating water channel.[13] Various types of model were used, ranging from a tanker form to a fine form. The experiments concentrated on effects at very low frequency - in the range $0.07 < \omega \leqslant 1.5$

rad sec^{-1}. Preliminary findings indicate that the in-phase and quadrature curves can be extrapolated back into the low frequency range by means of the functional analysis method. That is to say the curves rise when $\omega \to 0$ in the low frequency range, as illustrated in figs. 1 and 2. Thus, if one had only measured a few data points at low frequency (as is usual in a conventional PMM test) and then applied the derivative analysis of fitting a straight line extended to $\omega = 0$, it is unlikely that the value obtained for the derivative would be at all accurate. Moreover this would be true of both the in-plane and the quadrature results.

In addition, both the functional analysis and the results of Gill and Price(13) indicate that the derivative obtained from the oblique tow test and the PMM oscillatory tests produce the same value. This has not been so when conventional analytical methods are used.

Linear functional analysis has been found to offer an improved method of finding values of derivatives. It has also proved a useful means of determining the stability of a ship.(5),(8) It seems probable that a less restricted form of functional analysis would provide a rational approach to certain other problems of ship hydrodynamics - notably in the analysis of violent manoeuvres.

Notation

N	Yaw fluid moment with perturbations expressible as $\Delta N(t)$, $\Delta N[\ ]$
N_r	Slow motion yaw moment derivative with respect to yaw angular velocity
$N_{\dot r}$	Slow motion yaw moment derivative with respect to yaw angular acceleration
N_v	Slow motion yaw moment derivative with respect to sway velocity
$N_{\dot v}$	Slow motion yaw moment derivative with respect to sway acceleration
$r(t)$	Yaw angular velocity
$\dot r(t)$	Yaw angular acceleration
t	Time
U	Steady forward reference velocity
$v(t)$	Sway velocity
$\dot v(t)$	Sway acceleration
v_o $(=y_o\omega)$	Amplitude of harmonic sway velocity
$\dot v_o$ $(=-y_o\omega^2)$	Amplitude of harmonic sway acceleration
Y	Sway fluid force with perturbations expressible as $\Delta Y(t)$, $\Delta Y[\ ]$
Y_r	Slow motion sway force derivative with respect to yaw angular velocity
$Y_{\dot r}$	Slow motion sway force derivative with respect to yaw angular acceleration

Y_v	Slow motion sway force derivative with respect to sway velocity
$Y_{\dot{v}}$	Slow motion sway force derivative with respect to sway acceleration
$Y_v(\omega)$	Fourier transform of $y_v^*(t)$ $(= Y_v^R(\omega) + iY_v^I(\omega))$
$Y_v(o)$	Value of $-\Delta Y_{QUAD}/v_o$ at zero frequency
$Y_v(\infty)$	Value of $-\Delta Y_{QUAD}/v_o$ at infinite frequency
$Y_{\dot{v}}(\infty)$	Value of $-\Delta Y_{IN}/\dot{v}_o$ at infinite frequency
$Y_{v\|v\|}$, Y_{vvv}	Hydrodynamic coefficients
$y(t)$	Horizontal displacement to starboard (say)
y_o	Amplitude of harmonic displacement to starboard
$y_v(t)$	Variation of ΔY following unit step displacement to starboard
$y_v^*(t)$	Modified form of $y_v(t)$
$y_{\dot{v}}(t)$	Variation of ΔY following unit step variation of $v(t)$
$y_{\dot{v}}^*(t)$	Modified form of $y_{\dot{v}}(t)$
ΔY_{IN}	Amplitude of harmonic component of ΔY in phase with $y(t)$
ΔY_{QUAD}	Amplitude of harmonic component of ΔY in quadrature with $y(t)$
β	Drift angle
τ	Time variable
ω	Circular frequency

References

1. Babister, A.W. Aircraft stability and control. Pergamon Press, 1961.
2. Bishop, R.E.D., Burcher, R.K. and Price, W.G. The uses of functional analysis in ship dynamics. Proc.R.Soc.Lond. A.332, 23-35, 1973.
3. Bishop, R.E.D., Burcher, R.K. and Price, W.G. Functional analysis in oscillatory ship model testing. Proc.R.Soc.Lond. A.332, 37-49, 1973.
4. Bishop, R.E.D., Burcher, R.K. and Price, W.G. On the linear representation of fluid forces and moments in unsteady flow. Jour. Sound Vib. 29, 113-128 (The Fifth Fairey Lecture), 1973.
5. Bishop, R.E.D., Burcher, R.K. and Price, W.G. Directional stability analysis of a ship allowing for time history effects on the flow. Proc.R.Soc.Lond. A.335, 341-354, 1973.
6. Bishop, R.E.D., Burcher, R.K. and Price, W.G. A linear analysis of planar motion mechanism data. Journal of Ship Research, 18, 242-251, 1974.
7. Bishop, R.E.D., Burcher, R.K. and Price, W.G. The determination of ship manoeuvring characteristics from model tests. Trans.Roy.Inst.Nav.Arch. 117, 1975.
8. Bishop, R.E.D., Parkinson, A.G. and Price, W.G. Modified stability analysis of a surface ship. Fourth Ship Control Systems Symp, The Hague (to appear).

9. Bryan, G.H. Stability in Aviation. London: Macmillan, 1911.

10. Chislett, M.S. and Strøm-Tejsen, J. A model testing technique and method of analysis for the prediction of steering and manoeuvring of surface ships. HyA Report No. Hy-7, 1966.

11. Gertler, M. The DTMB planar motion mechanism system. Symp. of Towing Tank Facilities, Zagreb, Yugoslavia, paper 6, 1959.

12. Gertler, M. 10th International Towing Tank Conference, Rome, 1969.

13. Gill, A. and Price, W.G. Planar motion mechanism testing in a circulating water channel. 1975 (to appear)

14. Glansdorp, C.C. Some notes on oscillator techniques. Shipbuilding Laboratory of the Technological University, Delft, Report No. 231, 1969.

15. Glansdorp, C.C. Horizontal high frequency PMM - tests with a Mariner model. Shipbuilding Laboratory of the Technological University, Delft, Report No. 381-M, 1973.

16. Goodman, A. Experimental techniques and methods of analysis used in submerged body research. Proceedings of Third Symposium on Naval Hydrodynamics, Scheveningen, 379-449, 1960.

17. Kotik, J. and Mangulis, V. On the Kramers-Kronig relations for ship motions. International Shipbuilding Progress, 97, 3-10, 1962.

Fig. 1 The dots represent experimental data obtained by Glansdorp (15) and the full line is calculated from $y^*_{\dot v}(t)$ of fig. 3.

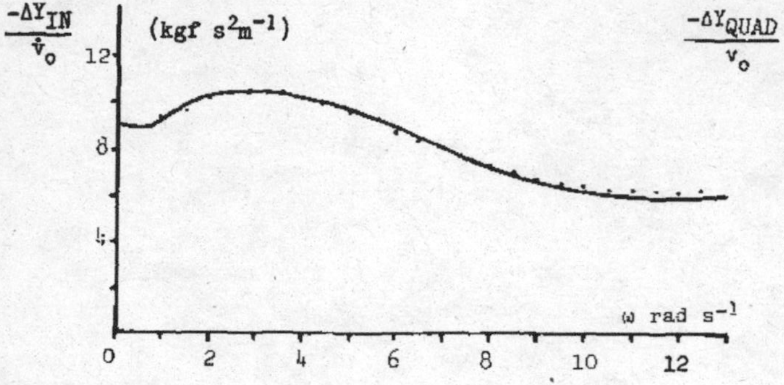

Fig. 2 The dots represent experimental data obtained by Glansdorp (15) from which the curve of fig. 3 was computed. The full line was calculated from fig. 4.

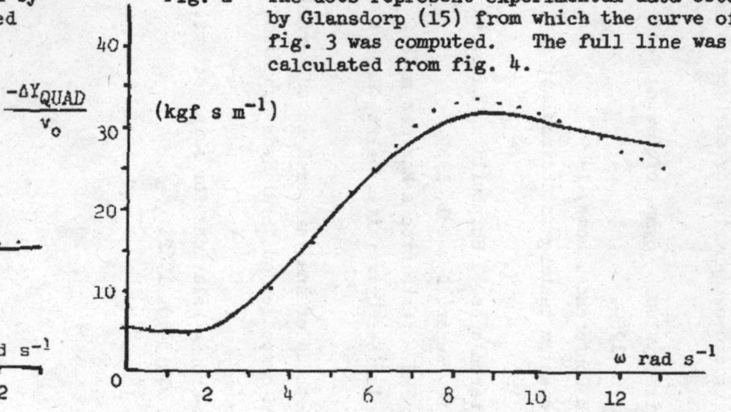

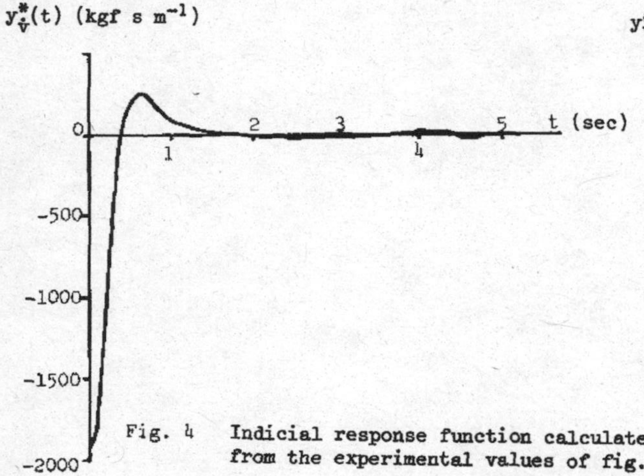

Fig. 4 Indicial response function calculated from the experimental values of fig. 1.

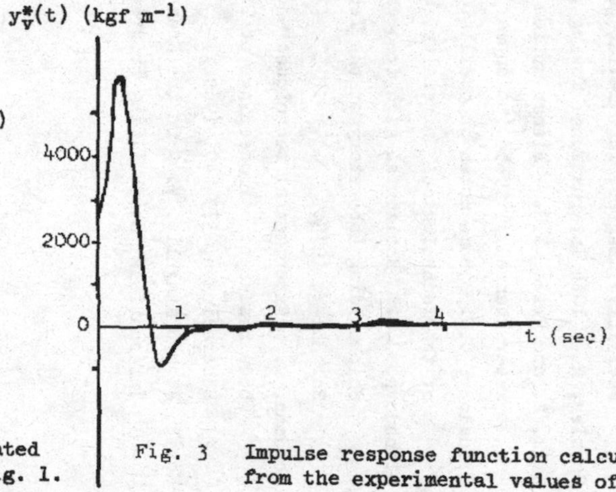

Fig. 3 Impulse response function calculated from the experimental values of fig. 2.

SOLITARY-WAVE SOLUTIONS FOR SOME MODEL EQUATIONS FOR WAVES IN NONLINEAR DISPERSIVE MEDIA

J.L. Bona†   D.K. Bose*   T.B. Benjamin

Fluid Mechanics Research Institute,
University of Essex, Colchester, UK.

I. Introduction

Solitary waves were first observed and described by Scott Russell in the early 1840's. He witnessed the generation and evolution of a single-crested steadily propagating wave of elevation when a barge came to an abrupt halt in a canal in Scotland. The observation of this phenomenon in nature inspired him to conduct a sequence of careful laboratory experiments on the generation and properties of such waves.

The existence of a solitary wave, as reported by Scott Russell in 1844, could not be explained in terms of the then current theories for surface waves. Boussinesq (1871) was able to give an approximate explanation by means of a system of nonlinear model equations which now bear his name. Rayleigh (1876) also gave an approximate expression for the solitary wave. The matter was elucidated further by Korteweg and de Vries (1895), who derived a model equation for the uni-directional propagation of long surface waves in a uniform rectangular channel. Their equation has an exact solution in the form of a steadily-translating single-crested wave of elevation. For surface waves in a channel, Friedrichs and Hyers (1954) extended the approximation of Boussinesq and Rayleigh and proved existence of small-amplitude solitary-wave solutions of the full equations of motion (the Euler equations with nonlinear boundary conditions at the free surface).

In the late 1950's and in the 1960's several other physical systems were shown to manifest solitary waves (e.g. rotating fluids, bubbly liquids, crystalline lattices and density stratified fluids). In the late 1960's a formalism relating to the Korteweg-de Vries equation was developed by Gardner, Greene, Kruskal and Miura (1967, 1974) which indicates that, for a large class of initial wave profiles, solitary waves play an important role in the solution of the pure initial-value problem. Experimental evidence (Hammack 1973 and Hammack and Segur 1974)

†Also Department of Mathematics, Univeristy of Chicago.

*Present address: Institut für Mathematik, Ruhr-Universität Bochum, 463 BOCHUM, West Germany.

for water waves in a channel are more or less consistent with the general conclusion described by this formalism, that the long-term evolution of an initial wave profile of elevation leads to a sequence of solitary waves. (Firm conclusions on this point appear, however, somewhat premature in light of the evidence so far available.)

It is now generally understood (cf. Benjamin, Bona and Mahony 1972, section 2) that equations of the Korteweg-de Vries type will arise in first-order one-dimensional approximations for uni-directional propagation of waves whenever the dominant physical consideration is, on the one hand, a balance between small nonlinear effects that tend to steepen the wave profile and, on the other, smooth dispersive effects which tend to spread the profile. At this level of approximation many physical systems lead to the Korteweg-de Vries equation or to the alternative model proposed by Benjamin et. al. (1972). However, for long-wave models in which the dispersion relation is not a C^2 function near the origin, a generalized version of these equations obtains, namely

$$u_t + u_x + uu_x + Hu_t = 0, \tag{1}$$

or, with allowance for other forms of nonlinearity,

$$u_t + f(u)_x + Hu_t = 0. \tag{2}$$

Here, $u = u(x,t)$, where x and t are real variables representing space and time respectively and H is a convolution operator determined by the dispersion relation. More precisely,

$$\widehat{Hu}(k) = \alpha(k)\,\hat{u}(k). \tag{3}$$

where the circumflex denotes Fourier transform. Such models were first considered by Benjamin (1967). In the case $\alpha(k) = k^2$, the equation studied by Benjamin et. al. (1972) is recovered from (1).

In two instances, for the class of internal waves treated by Benjamin (1967) and for a model suggested by Whitham (1967) to simulate the peaking and breaking of surface gravity waves, explicit solutions of (1) representing waves propagating steadily without change of form have been given. In various other theoretical discussions (cf. Leibovich 1970, Leibovich and Randall 1972, Smith 1972, and Pritchard 1969, 1970) solitary-wave solutions of models in the form (1) have been assumed to exist. Such waves have been produced in the laboratory by Pritchard (1969) who observed them travelling along the vortex of a rotating fluid, a system for which the symbol $\alpha(k) = k^2(1 + K_0(|k|))$ in suitably scaled variables (here K_0 is the zeroth order modified Bessel function). Hence a significant question naturally arises: under what conditions do equations of the form (1) or (2) have solitary-wave solutions? If such a model does not have solitary-wave solutions, a potentially important aspect of the physical situation is lost and the model may be judged to be inadequate. The question can be formulated for (1) or (2) with respect not only to the solitary-wave problem on the infinite domain $\mathbb{R}$ but also for a spatially periodic version of the same problem. (For the original equation of Korteweg and de Vries, solutions representing permanent periodic wave trains were called cnoidal waves on account of their representation in terms of the Jacobian elliptic function cn.)

Sufficient conditions for the existence of solitary-wave solutions of (1) have been derived by Bona and Bose (1974), using an extension of the positive operator methods of Krasnosel'skii. A feature of the analysis, which recurs below, is that there are two trivial solutions to the problem and the object is to establish a third, non-trivial solution. Here a different approach to the problem is considered. The question for periodic waves is settled first by means of a variational argument, and then it is shown that as the wavelength becomes large, the periodic wave-train tends to a solitary wave in an appropriate metric.

One possible advantage of developing the proof as outlined below is that stability to perturbations periodic of the same period of the waves in question may be inferred. For it is shown that solutions to (1) or (2) that are the counterparts of cnoidal waves realize a maximum of a certain functional, subject to constraints on another functional. Such a situation has been exploited by Benjamin (1972) and Bona (1975) in a proof of the stability of the solitary-wave solution of the Korteweg-de Vries equation. Of course it is a long way from the variational principle to a complete proof of stability, but nevertheless such an approach seems to be the best technique available in this type of problem.

In section II the periodic problem is discussed. Section III is devoted to the solitary-wave problem. To keep the technical details at a minimum, attention is restricted to equation (1).

II. The problem of steady periodic waves

The question of existence of solutions $u(x,t) = \phi(x-\lambda t)$ to (1) such that ϕ is a periodic function of period 2ℓ is now considered. A solution of this problem has already been sketched by Benjamin (1974). Benjamin's method and results along with a few extensions, will be briefly recalled.

When the desired form of solution is substituted into (1) and the equation is then integrated once, there results the 'ordinary' pseudo-differential equation

$$\lambda(\phi + H\phi) = \phi + \tfrac{1}{2}\phi^2. \tag{4}$$

By inversion of the operator I + H in (4), a Hammerstein integral equation is obtained.

$$\lambda\phi = K*(\phi + \tfrac{1}{2}\phi^2) \tag{5}$$

where $\hat{K}(k) = \{1 + \alpha(k)\}^{-1}$, α being the symbol of H, and * denotes convolution over the entire real axis. Henceforth it is assumed, in keeping with most of the applications in view, that $\alpha(k)$ is even, non-negative, increasing on $\mathbb{R}^+$, that $\alpha(0) = 0$ and that $\{1 + \alpha(k)\}^{-1} = O(|k|^{-\beta})$ with $\beta > 1$ as $|k| \to +\infty$.

Equation (5) is appropriate for steady-wave solutions of arbitrary period 2ℓ and also for solitary-wave solutions of (1) (since the Green function obtained by inverting I + H subject to periodic boundary conditions of period 2ℓ is simply the periodic function $K_\ell(x) = \sum_{m=-\infty}^{\infty} K(x + 2m\ell)$). Subsequently it will be desirable to fix ℓ and view (5) only on the fundamental period interval $[-\ell,\ell]$. If g is a periodic function of period 2ℓ, then $K*g(x) = \int_{-\infty}^{\infty} K(x-y)g(y)dy = \int_{-\ell}^{\ell} K_\ell(x-y)g(y)dy$, hence (5) is equivalent to

$$\lambda\phi = K_\ell*(\phi + \tfrac{1}{2}\phi^2) \tag{6}$$

where the convolution is now understood to be over the interval $[-\ell,\ell]$.

Following Benjamin (1973), the operator K is now split into positive square roots. More precisely, let M be defined by $\hat{M}(k) = \{1 + \alpha(k)\}^{-\frac{1}{2}}$. Then if $B\psi = M*\psi$ (convolution over $\mathbb{R}$) and $B_\ell\psi = M_\ell*\psi$ (convolution over $[-\ell,\ell]$) where M_ℓ is defined from M as K_ℓ is from K, (5) and (6) become respectively

$$\lambda\phi = B^2(\phi + \tfrac{1}{2}\phi^2) \quad \text{and} \quad \lambda\phi = B_\ell^2(\phi + \tfrac{1}{2}\phi^2) \tag{7}$$

where B^2 means the operator B applied twice. In (7) make the substitution $\psi = B_\ell\phi$. The corresponding equation for periodic steady waves becomes

$$\lambda\psi = B_\ell(B_\ell\psi + \tfrac{1}{2}(B_\ell\psi)^2), \tag{8}$$

and similarly for the equation for solitary waves. The functionals

$$V(u) = \int_{-\ell}^{\ell} u^2(x)dx \quad \text{and} \quad W(u) = \int_{-\ell}^{\ell} \{(B_\ell u)^2 + \tfrac{1}{3}(B_\ell u)^3\}dx$$

are well-defined on the Hilbert space $\mathcal{H}$ of periodic square-integrable function on $\mathbb{R}$ with period 2ℓ. Moreover, both these functionals possess a gradient at any point of $\mathcal{H}$ and (8) is identical with

$$\lambda G_V(\psi) = G_W(\psi) \tag{9}$$

where the operators G_V and G_W are the respective gradients of V and W. Because of the assumed growth condition on α, B_ℓ maps L_2 continuously into the Sobolev space $H^{\beta/2}$. Since $\beta > 1$, both L_2 and L_3 are compactly imbedded in $H^{\beta/2}$ and it follows that W is a weakly continuous functional on $\mathcal{H}$.

Hence if the constrained maximization problem

$$\text{maximize } W(u), \text{ subject to } V(u) \leq R^2, \tag{10}$$

is posed, standard results insure that this problem has a solution, say ψ_ℓ. It is straightforward to check that ψ_ℓ cannot lie in the interior of the ball $\{u : V(u) \leq R^2\}$, and hence the usual theory (Vainberg, 1964, chapter IV) implies the existence of a constant λ_ℓ such that

$$\lambda_\ell G_V(\psi_\ell) = G_W(\psi_\ell). \tag{11}$$

There are two trivial solutions of (11). One is the function identically zero, which is excluded since it lies in the interior of the ball $\{u : V(u) \leq R^2\}$. The other, representing a so-called conjugate flow (cf. Benjamin 1971), is the constant function $\psi_0 \equiv R/\sqrt{2\ell}$, corresponding to which $\lambda_\ell = 1 + \frac{1}{2}\psi_0$. By considering the second derivative of W at ψ_0, Benjamin showed that for ℓ larger than a certain critical value ℓ_c, dependent only on α, ψ_0 does not achieve a maximum of W on $\{u : V(u) \leq R^2\}$. Hence ψ_ℓ is a non-constant 2ℓ-periodic solution of (8) and accordingly $\phi_\ell = B_\ell\psi_\ell$ is a non-constant 2ℓ-periodic solution of (6).

Various additional properties of ψ_ℓ, and so of ϕ_ℓ, may be established by use of the extremal property of ψ_ℓ. On the additional assumption that the kernel M of the operator B is a non-negative even function, non-increasing on $[0,\infty)$, so that similar properties accrue for M_ℓ on the period $[-\ell,\ell]$, it may be inferred that the following conditions can be satisfied by a maximizing function.

(a) $\psi_\ell \geq 0$ and ψ_ℓ may be normalized against translations in x so that it is even and monotone non-increasing on $[0,\ell]$. (Other possible solutions of (8) whose fundamental period is a fraction of 2ℓ, which will realize only a conditional stationary value of W, will not have this property.) Because of the assumptions made concerning the operator B, it follows that $\phi_\ell = B_\ell \psi_\ell$ may be chosen with the same properties.

(b) The 'Lagrange multiplier' λ_ℓ satisfies $1 < \mu_0 \leq \lambda_\ell \leq \mu_1 < +\infty$ for all $\ell \geq \ell_c$. The constants μ_0 and μ_1 depend only on α and not on ℓ. (These bounds on λ_ℓ are obtained by evaluating W for particular functions in the ball $\{u : V(u) \leq R^2\}$.)

Finally a standard 'bootstrap' argument shows that ψ_ℓ must be an H^∞ function on $[-\ell,\ell]$ (i.e. an $L_2(-\ell,\ell)$ function with derivatives of all orders which are also in $L_2(-\ell,\ell)$).

III. Existence of solitary waves

The facts outlined in section II will be used to show that, as the period of the steady periodic waves tends to infinity, the wave profile converges, in a sense to be described below, to a solitary-wave solution of (1). That is, there is a finite constant $\lambda > 1$ and a non-negative even C^∞ function ϕ, defined on $\mathbb{R}$, which is monotone decreasing to 0 as $x \to +\infty$ and satisfies equation (5).

Let $\mathcal{C}(\mathbb{R})$ denote the class of continuous real-valued functions defined on $\mathbb{R}$. $\mathcal{C}(\mathbb{R})$ is given the structure of a Fréchet space by introducing the semi-norms

$$p_j(u) = \sup_{-j\leq x\leq j} |u(x)| . \tag{11}$$

The corresponding metric may be taken to be, for example,

$$d(u,v) = \sum_{j=0}^{\infty} \frac{1}{2^j} \frac{p_j(u-v)}{1+p_j(u-v)} . \tag{12}$$

Thus the statement $u_n \to u$ in the metric d means that $\{u_n\}$ converges to u pointwise, and uniformly on compact subsets of $\mathbb{R}$. The notation B_r will be used for the ball $\{u\in\mathcal{C}(\mathbb{R}) : d(u,0) \leq r\}$. Note that $B_1 = \mathcal{C}(\mathbb{R})$.

Now $\mathcal{C}(\mathbb{R})$ has two properties of particular importance in the present context. First, the periodic permanent-wave solutions of (6) and solitary-wave solutions of (5) are all members of $\mathcal{C}(\mathbb{R})$. Secondly, the operation of convolution with the kernel K is a compact mapping of certain convex subsets of $\mathcal{C}(\mathbb{R})$ which will be defined below. For $\ell > 0$, let

$C_\ell = \{u\in\mathcal{C}(\mathbb{R}) :$ u is non-negative, 2ℓ-periodic, even and monotone non-increasing on $[0,\ell]\}$, and let

$$C = \{u \in \mathcal{C}(\mathbb{R}) : u \text{ is non-negative, even and monotone non-increasing on } [0,\infty)\}.$$

The sets C_ℓ and C are closed and convex in $\mathcal{C}(\mathbb{R})$. In fact, they are closed cones. Let $Au = K*(u + \frac{1}{2}u^2)$, where the convolution is over the entirety of $\mathbb{R}$. Of course, A cannot be defined on the whole of $\mathcal{C}(\mathbb{R})$, which includes functions unbounded at infinity, but, since $K \in L^1(\mathbb{R})$, it can be considered as a mapping of C or of C_ℓ for any $\ell > 0$. As a mapping of these cones, A has some useful properties summarized in a preparatory lemma.

LEMMA 1. A is a continuous map of C into itself and, for each $\ell > 0$, a continuous map of C_ℓ into itself. Moreover, for fixed $\ell > 0$ and r in (0,1), $A(C_\ell \cap B_r)$ (respectively $A(C \cap B_r)$) is a relatively compact subset of C_ℓ (respectively C).

A relationship between the cones C_ℓ and C is now needed. Define mappings $r_\ell : C \to C_\ell$ and $s_\ell : C_\ell \to C$ as follows. For $u \in C$ and $v \in C_\ell$,

$$(r_\ell u)(x) = \begin{cases} u(x) & \text{for } -\ell \le x \le \ell, \\ u(n\ell - |x|) & \text{for } (n-1)\ell \le |x| \le (n+1)\ell, \\ & \quad n = 2,3,\cdots. \end{cases}$$

$$(s_\ell v)(x) = \begin{cases} v(x) & \text{for } -\ell \le x \le \ell, \\ v(\ell) & \text{for } \ell \le |x|. \end{cases}$$

These maps are pictured in the accompanying sketch.

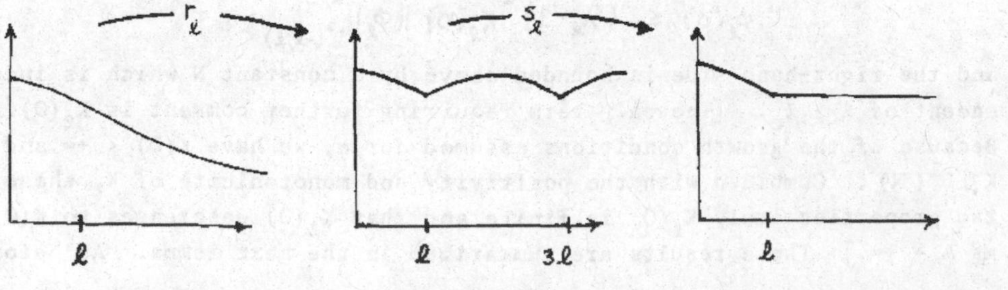

LEMMA 2. Let $\ell > 0$. Then $r_\ell : C \to C_\ell$ and $s_\ell : C_\ell \to C$ are continuous with respect to the relative topology induced by $\bar{C}(\mathbb{R})$. The composition $r_\ell \circ s_\ell$ equals id_{C_ℓ}, the identity mapping of C_ℓ. If γ is a positive constant and $f \in C_\ell$, then $s_\ell(\gamma f) = \gamma s_\ell(f)$.

In topological language, the mappings r_ℓ and s_ℓ are an r-domination of the cone C_ℓ by the cone C (cf. Granas 1972).

Let A_ℓ denote the restriction of A to C_ℓ. If $u \in C_\ell$, then $A_\ell u = K*(u + \frac{1}{2}u^2) = K_\ell*(u + \frac{1}{2}u^2)$, where the first convolution is over $\mathbb{R}$ and the second over $[-\ell,\ell]$. The composition $s_\ell A_\ell r_\ell$ maps C to itself. Moreover, $s_\ell A_\ell r_\ell \to A$ on C as $\ell \to +\infty$. More precisely, we have:

LEMMA 3. Let $r \in (0,1)$ and $\varepsilon > 0$ be given. There exists an $\ell_o = \ell_o(\varepsilon, r)$ such that if $\ell \geq \ell_o$,

$$d(Au, s_\ell A_\ell r_\ell u) < \epsilon$$

for all u in $C \cap B_r$.

A few additional pieces of information are needed concerning the periodic permanent-wave solutions ϕ_ℓ determined in section II. Fix the parameter R, namely the $L^2(-\ell,\ell)$ norm of ϕ_ℓ, and let $\ell \geq \ell_c$. Then ϕ_ℓ is a member of C_ℓ and

$$\lambda_\ell \phi_\ell = K_\ell * (\phi_\ell + \tfrac{1}{2}\phi_\ell^2) = A_\ell \phi_\ell . \tag{13}$$

When this relation is evaluated at 0, and account is taken of the facts that

$$1 = \hat{K}(0) = \int_{-\infty}^{\infty} K(y)dy = \int_{-\ell}^{\ell} K_\ell(y)dy$$

and that $0 \leq \phi_\ell(x) \leq \phi_\ell(0)$ for all x, there appears the lower bound $\lambda_\ell - 1 \leq \phi_\ell(0)$. It is also easily confirmed that

$$\phi_\ell(0) \leq (\lambda_\ell - 1)^{-1} K_\ell(0) \, \|\phi_\ell\|^2_{L^2(-\ell,\ell)},$$

and the right-hand side is bounded above by a constant N which is independent of $\ell \geq \ell_c$. [The only term requiring further comment is $K_\ell(0)$. Because of the growth conditions assumed for α, we have $K(0) < +\infty$ and $K \in L^1(\mathbb{R})$. Combined with the positivity and monotonicity of K, these two properties imply $K_\ell(0)$ is finite and that $K_\ell(0)$ descreases to $K(0)$ as $\ell \to +\infty$.] These results are summarized in the next lemma. As before,

ϕ_ℓ denotes a periodic steady-wave solution of (6) of fundamental period 2ℓ and with $\|\phi_\ell\|_{L^2(-\ell,\ell)} = R$, determined as in section II.

LEMMA 4. There are constants μ_0 and N, independent of $\ell \geq \ell_c$, such that

$$\mu_0 \leq \phi_\ell(0) \leq N \tag{14}$$

for all $\ell \geq \ell_c$.

Armed with these facts, we are ready to consider the existence problem for solitary waves. Define

$$\rho_\ell(x) = \begin{cases} \phi_\ell(x) & \text{for } -\ell \leq x \leq \ell, \\ 0 & \text{otherwise.} \end{cases} \tag{15}$$

Then $\rho_\ell \in L^2(\mathbb{R})$ for $\ell \geq \ell_c$ and $\|\rho_\ell\|_{L^2(\mathbb{R})} = R$. Here is the main result.

THEOREM. There is a non-constant $H^\infty(\mathbb{R})$ function ϕ in the cone C and a finite constant $\lambda > 1$ such that $\lambda\phi = A\phi$. This function is the limit, uniformly on compact subsets of $\mathbb{R}$, of a sequence $\{\phi_{\ell_m}\}_{m=1}^\infty$ of periodic functions satisfying (6) for $\ell_m \to +\infty$. Moreover, the associated cut-off functions $\{\rho_{\ell_m}\}_{m=1}^\infty$ defined in (15) converge to ϕ in $L^2(\mathbb{R})$ and so $\|\phi\|_{L^2(\mathbb{R})} = R$.

Proof. The conclusion (14) of lemma 4 may be interpreted to mean that $\mu_0 \leq p_1(\phi_\ell) \leq N$ for all $\ell \geq \ell_c$. Hence also $\mu_0 \leq p_1(s_\ell\phi_\ell) \leq N$. Referring to the definition (12) of the metric on $C(\mathbb{R})$, it is concluded that there are constants δ and Δ with $0 < \delta < \Delta < 1$ such that $\delta < d(s_\ell\phi_\ell, 0) < \Delta$, provided $\ell \geq \ell_c$. In particular, $s_\ell\phi_\ell \in C \cap B_\Delta$ for $\ell \geq \ell_c$.

Let $\epsilon > 0$ be given. Lemma 3 implies that there is an ℓ_0 such that if $\ell \geq \ell_0$,

$$\epsilon > d(Au, s_\ell A_\ell r_\ell u) \quad \text{for all } u \text{ in } C \cap B_\Delta .$$

Now, $s_\ell A_\ell r_\ell(s_\ell\phi_\ell) = s_\ell A_\ell \phi_\ell = s_\ell(\lambda_\ell\phi_\ell) = \lambda_\ell(s_\ell\phi_\ell)$. Thus for $\ell \geq \ell_1 = \max(\ell_0, \ell_c)$,

$$\epsilon > d(As_\ell\phi_\ell, s_\ell A_\ell r_\ell s_\ell\phi_\ell) = d(As_\ell\phi_\ell, \lambda_\ell s_\ell\phi_\ell). \tag{16}$$

Since ε is arbitrary, it can be concluded that $As_\ell\phi_\ell - \lambda_\ell s_\ell\phi_\ell \to 0$ with respect to the metric d as $\ell \to +\infty$.

Lemma 1 implies that $A(C \cap B_\Delta)$ is a relatively compact subset of C. A subsequence $\{\ell_m\}_{m=1}^\infty$ of wavelengths can therefore be found, with $\ell_m < \ell_{m+1}$, $\ell_m \to +\infty$ as $m \to +\infty$, together with an element ψ in C, such that if $\psi_m = s_{\ell_m}\phi_{\ell_m}$, then $A\psi_m \to \psi$ in the metric d. Since $1 < \mu_0 \leq \lambda_\ell \leq \mu_1$, it may be assumed there exists a λ such that $\{\lambda_m\}_{m=1}^\infty$, with $\lambda_m = \lambda_{\ell_m}$, has $\lambda_m \to \lambda$ as $m \to +\infty$. Obviously $1 < \mu_0 \leq \lambda \leq \mu_1$. In consequence of the conclusion (16), $A\psi_m - \lambda_m\psi_m \to 0$ with respect to d as $m \to +\infty$. Hence $\lambda_m\psi_m \to \psi$, with respect to d as $m \to +\infty$, or, since $\lambda_m \to \lambda$ in $\mathbb{R}$, $\psi_m \to \lambda^{-1}\psi = \phi$, say. As A is continuous on C, $A\psi_m \to A\phi$ for the metric d. But $A\psi_m \to \psi = \lambda\phi$. Hence $A\phi = \lambda\phi$. Note that ϕ is non-zero since $\delta \leq d(\psi_m, 0)$ for all m. Further, the convergence of the sequence $\{\psi_m\}$, with $\psi_m = s_m\phi_m$, to ϕ uniformly on compact subsets of $\mathbb{R}$ implies that the sequence $\{\phi_m\}$ also converges to ϕ uniformly on compact subsets of $\mathbb{R}$, for s_m alters ϕ_m only outside the fundamental period $[-\ell_m, \ell_m]$ of ϕ_m. Thus it is proved that there exists a non-zero solution ϕ of (5) in C which is the limit of a sequence of periodic steady-wave solutions of (6).

Let $\rho_m = \rho_{\ell_m}$ be the cut-off functions associated with ϕ_m as in (15). View $\{\rho_m\}$ as a sequence in $L^2(\mathbb{R})$. Then $\|\rho_m\|_{L^2(\mathbb{R})} = R$ for all m. Moreover, $\rho_m \to \phi$ uniformly on compact subsets of $\mathbb{R}$, hence certainly pointwise. Fatou's lemma implies that $\phi \in L^2(\mathbb{R})$ and that $\rho_m \to \phi$ in $L^2(\mathbb{R})$. It follows that $\|\phi\|_{L^2(\mathbb{R})} = R$. This incidentally shows that ϕ is not the trivial (conjugate flow) solution $\psi_0(x) \equiv 2(\lambda-1)$. Since ϕ is not the zero-function or the constant function ψ_0, ϕ cannot be a constant function.

Finally, since $\phi \in L^2(\mathbb{R})$ and ϕ is bounded, it follows that $\phi^2 \in L^2(\mathbb{R})$. Therefore $A\phi = K*(\phi+\frac{1}{2}\phi^2) \in H^1(\mathbb{R})$, whence $\phi \in H^1(\mathbb{R})$. Continuing this argument shows that $\phi \in H^\infty(\mathbb{R})$. This concludes the proof.

A computation using the Fourier transform shows that ϕ is a solution of the pseudo-differential equation (4), and hence ϕ provides a permanent-wave solution u_s of the evolution equation (1) by setting $u_s(x,t) = \phi(x-\lambda t)$.

The approach to the problem presented here is to specify the total 'energy' of the wave in question (the L^2 norm of the wave) and to then determine a wave-speed for the resulting solution. The approach followed earlier by Bona and Bose (1974) was to specify the wave speed. The view taken here seems to be the right one from the experimental standpoint. The possibility of establishing a stability result by means of the variational method is also inviting. Note, incidentally, that in the special case where the symbol α of H is homogeneous of degree $\sigma > 0$, a change of variables of the form $\psi(x) = a\phi(bx)$, where a and b are positive and satisfy $a(\lambda-1)+1 = a\lambda b^\sigma$, converts the solution ϕ of (5) to a solution ψ of (5) with λ replaced by $a(\lambda-1)+1$.

In conclusion, it deserves remark that the approach presented here can be carried over to certain two-dimensional problems, notably internal waves in heterogeneous fluid flows along a channel and rotating flows down a pipe. The details are naturally different, but the main outline and general conclusions are the same.

REFERENCES

T.B. Benjamin, 1967 'Internal waves of permanent form in fluids of great depth', J. Fluid Mech. 29, p.559.

-------------, 1971 'A unified theory of conjugate flows', Philos. Trans. Roy. Soc. London Ser. A. 269, p.587.

-------------, 1972 'The stability of solitary waves', Proc. Roy. Soc. London Ser A. 328, p.153.

-------------, 1974 'Lectures on nonlinear wave motion', Nonlinear Wave Motion (Proceedings of the Summer Seminar, Potsdam, New York, 1972), p.3, Ammerican Math. Society, Lectures in Applied Mathematics Vol 15.

T.B. Benjamin, J.L. Bona and J.J. Mahony, 1972 'Model equations for long waves in nonlinear dispersive systems', Philos. Trans. Roy. Soc. London Ser. A. 272, p.47.

J.L. Bona, 1975 'On the stability theory of solitary waves', Proc. Roy. Soc. London Ser. A. 344, p.363.

J.L. Bona and D.K. Bose, 1974 'Fixed point theorems for Fréchet spaces and the existence of solitary waves', Nonlinear Wave Motion (Proceedings of the Summer Seminar, Potsdam, New York, 1972) p.175, American Math. Society, Lectures in Applied Mathematics Vol 15.

J. Boussinesq, 1871 'Théorie de l'intumescence liquide appelée onde solitaire ou de translation se propageant dans un canal rectangulaire', Comptes Rendus 72, 755.

K.O. Friedricks and D.H. Hyers, 1954 'The existence of solitary waves', Comm. Pure Appl. Math. 7, p.517.

C.S. Gardner, J.M. Greene, M.D. Kruskal, and R.M. Miura, 1967 'Method for solving the Korteweg-de Vries equation', Phys. Rev. Letters 19, p.1095.

-------------, 1974 'Korteweg-de Vries equation and generalizations',VI. Methods for exact solution, Comm. Pure Appl. Math 27, p.27.

J.L. Hammack, 1973 'A note on tsunamis: their generation and propagation in an ocean of uniform depth', J.Fluid Mech. 60, p.769.

J.L. Hammack and H. Segur, 1974 'The Korteweg-de Vries equation and water waves. Part 2. Comparison with experiments', J. Fluid Mech. 65, p.289.

D.J. Korteweg and G. De Vries, 1895 'On the change of form of long waves advancing in a rectangular canal, and on a new type of long stationary waves', Philos. Mag (5) 39, p.422.

S. Leibovich, 1970 'Weakly nonlinear waves in rotating fluids', J. Fluid Mech. 42, p.803.

S. Leibovich and J.D. Randall, 1972 'Solitary waves in concentrated vortices', J. Fluid Mech. 51, p.625.

W.G. Pritchard, 1969 'The motion generated by a body moving along the axis of a uniformly rotating fluid', J. Fluid Mech. 39, p.443.

-------------, 1970 'Solitary waves in rotating fluid', J. Fluid Mech. 42, p.61.

Lord Rayleigh, 1876 'On waves', Philos Mag (5) 1, p.257.

J. Scott Russell, 1844 'Report on waves', Rep. Fourteenth Meeting of the British Assoc., John Murray, London, p.311.

R. Smith, 1972 'Nonlinear Kelvin and continental-shelf waves', J. Fluid Mech. 57, p.393.

M.M. Vainberg, 1964 Variational methods for the study of nonlinear operators, Holden-Day, San Francisco, Calif.

HILBERTIAN UNILATERAL PROBLEMS IN VISCOELASTICITY

R.Bouc, G.Geymonat, M.Jean, B.Nayroles
Laboratoire de Mécanique et d'Acoustique
31, ch. Joseph Aiguier - 13274 Marseille Cédex 2, France

I.- INTRODUCTION.

The study of the cracks propagation due to the fatigue in a viscoelastic medium leads, for a given configuration of cracks, to search for the mechanic response of the medium submitted to time periodic forces or stresses, of mechanical or thermal nature. This problem bears unilateral constraints since the edges of a crack can part but cannot interpenetrate each other. The hypothesis of contact without friction being unrealistic in physics, the crack is supposed to lie in a plane of symmetry for the mechanical problem, which allows us to formulate conditions of contact mathematically identical to those of a contact without friction.

A typical boundary value problem is considered for an extended Maxwell type viscoelastic material for a plate with plane stresses assumption and quasi-static equilibrium.

The three following problems are considered :

1°) The Cauchy problem (see theorem IV.1).

2°) Asymptotic stability of the solutions (see theorem IV.2).

3°) Periodic problem : all the data and the "stiffness" of the constitutive equation are time T-periodic (see § IV.2).

A preliminary algebraic study, based on the virtual work method and some classical considerations of functional analysis reduce the problems to a standard system of the type of a complementarity system (see § III).

The periodic problem leads to a weakly coupled system of variational inequalities with a non unique solution ; an essential step of the proof is the existence of a greatest lower bound in $H^1(\Omega)^2$ for the family $\{u(t)\}_{t \in [0,T]}$ where $u \in H^1([0,T] ; H^1(\Omega)^2)$ (with suitable properties of continuity for the mapping $u \mapsto \inf \{u(t)\}$ see [3]), which also allows us to caracterize the indetermined part of the solution (theorem IV.4).

For details in a more general situation, see [1].

II.- A TYPICAL BOUNDARY VALUE PROBLEM.

Let us consider a plane medium for which classical hypothesis of infinitesimal displacements and plane stresses are assumed. For the reasons mentioned in the

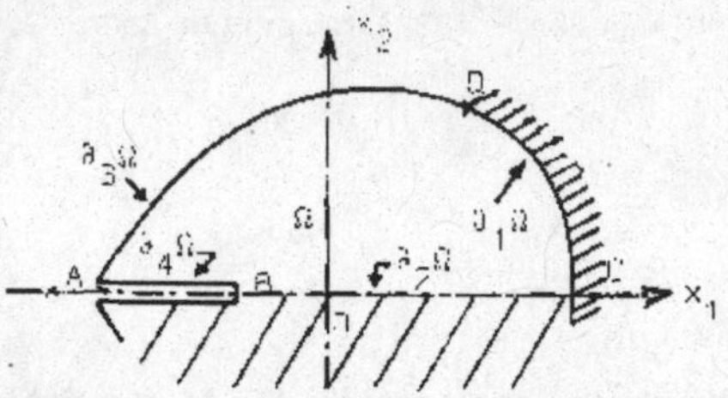

Fig 1

introduction, we suppose the problem to be symmetric with respect to the axis Ox_1, so that we need only to consider the part located in the positive x_2 half plane. We denote Ω the open set limiting the medium, $\bar{\Omega}$ its closure, $\partial\Omega$ its boundary.

$\mathcal{J}$ will denote an interval of time depending on the different problems. For instance $\mathcal{J}$ will be $[o, +\infty[$ for the Cauchy problem. All the considered fields, given or unknown, will be functions of time t and of the position $x = (x_1, x_2)$; for example the condition

$$\forall\ (t,x) \in \mathcal{J} \times \partial_1\Omega \qquad u(t,x) = u^o(t,x)$$

will be written

$$u = u^o \quad \text{on} \quad \partial_1\Omega$$

The unknown fields of displacements and stresses will be respectively denoted u and s.

A field Θ of temperature and a field χ of thermal dilatation symmetric tensor is given on $\mathcal{J} \times \Omega$. The product $\chi\Theta$ is then the field of deformation corresponding to a non stressed state for each element of surface when considered alone. If we set

$$(1) \qquad e_{ij} = \frac{1}{2}\left(\frac{\partial u_j}{\partial x_i} + \frac{\partial u_i}{\partial x_j}\right) - \chi_{ij}\Theta$$

the tensor e_{ij} represents the deformation with respect to the non stressed state.

As for now we shall write the constitutive law under the symbolic form :

$$s = \mathcal{K}(e)$$

which represents a functional relation between the stress field and the strain one.

$\mathcal{K}$ will be precised in paragraph III.

Equilibrium equations are classicaly written

(2) $$s_{ij,j} = - f_i$$

where f denotes a given field of forces per unit area on $\mathcal{J} \times \Omega$.

The boundary is divided into four disjoint open arcs. We prescribe

(3) $$u = u^0 \quad \text{on} \quad \partial_1\Omega \qquad (\partial_1\Omega = CD)$$

where u^0 is a given field of displacement on $\mathcal{J} \times \partial_1\Omega$.

The symmetry hypothesis is written under the form

(3') $$\left.\begin{array}{l} u_2 = 0 \\ s_{12} = 0 \end{array}\right\} \text{ on } \partial_2\Omega \qquad (\partial_2\Omega = \text{arc } BC).$$

(4) $$s_{ij}n_j = h_i \quad \text{on} \quad \partial_3\Omega \qquad (\partial_3\Omega = \text{arc } AD)$$

where h is a given field of forces per unit length on $\partial_3\Omega$. The arc AB = $\partial_4\Omega$ is the upper edge of the crack. It is submitted to contact forces from the lower edge and they are vertical due to the symmetry. Let us suppose that these forces are represented by a field of forces (0,g) per unit lenght on $\mathcal{J} \times \partial_4\Omega$; then the equilibrium equation on $\partial_4\Omega$ are written

(5) $$\left.\begin{array}{l} s_{12} = 0 \\ - s_{22} = g \end{array}\right\} \text{ on } \partial_4\Omega$$

Let us note that g is unknown. But we have to add the following unilateral constraints. First the edges of the crack can be parted but cannot interpenetrate each other, in other words

(6) $$u_2 \geq 0 \quad \text{on} \quad \partial_4\Omega$$

Then g is a compression force

(7) $$g \geq 0 \quad \text{on} \quad \partial_4\Omega$$

At last this contact force vanishes when the edges are parted

(8) $$u_2 g = 0 \quad \text{on} \quad \partial_4\Omega$$

These five later relations are identical to those that would have been written in the case of a contact without friction.

III.- THE STANDARD PROBLEM.

III.1.- Functional spaces and virtual work.

The system of relations written in the above paragraph will be studied from the point of view of the duality method or method of virtual work. We refer to [1] for technical details. We suppose first that the data u^0, met in (1), can be "extended to $\bar{\Omega}$", the extension being still denoted u^0, so that for every $t \in \mathcal{J}$, $u^0(t) \in H^1(\Omega)^2$. Let $v = u - u_0$. Then we take the conditions $u = u_0$ on $\partial_1\Omega$, $u_2 = 0$ on $\partial_2\Omega$ into account in writing $v(t) \in V$ where V is the closed subspace of $H^1(\Omega)^2$ defined by

$$V = \{w \in H^1(\Omega)^2 \mid w = 0 \text{ on } \partial_1\Omega \quad w_2 = 0 \text{ on } \partial_2\Omega\}$$

It is suitable to represent a symmetric tensor by a vector the components of which are the independent components of the tensor. For instance we shall write $a = (a_{11}, a_{22}, \sqrt{2}\, a_{12})$, instead of the tensor (a_{ij}). For every $t \in \mathcal{J}$, $x \mapsto e(t,x)$, will be in $E = L^2(\Omega)^3$, $x \mapsto s(t,x)$ will be in $S = L^2(\Omega)^3$. E and S are placed in duality by the usual scalar product, denoted $\varepsilon.\sigma$. The quantity $-\varepsilon.\sigma$ represents the virtual work of the stress σ with respect to the strain ε. Let D' be the mapping

$$u \in H^1(\Omega)^2 \mapsto D'u = \left(\frac{\partial u_1}{\partial x_1}, \frac{\partial u_2}{\partial x_2}, \frac{1}{\sqrt{2}}\left(\frac{\partial u_2}{\partial x_1} + \frac{\partial u_1}{\partial x_2}\right)\right) \in E$$

We denote D the restriction of D' to V. Thanks to Korn's inequality and to the fact that meas $\partial_1\Omega > 0$, D is a one-to-one bicontinuous mapping from V onto DV and DV is closed in E, [2] . We put $e^0(t) = D'u^0(t) - \chi(t)\Theta(t)$. We suppose that $e^0(t) \in E$. Then the equation (1) can be written $\varepsilon = e - e^0$, $\varepsilon(t) = Dv(t)$. At every time t the pair of force fields $(f(t),h(t))$ can be identified with an element $\varphi(t)$ of Φ the topological dual of V, in the following way (the natural duality between V and Φ is denoted by a point) : $\varphi(t)$ is the continuous linear form $w \in V \mapsto w.\varphi(t) \in \mathbb{R}$, where $w.\varphi(t)$ is the virtual work of the pair $(f(t),h(t))$ for the virtual displacement w. ${}^tD$ will denote the transpose of D for the duality between V and Φ, E and S.

Let L be the linear mapping which transforms each $w \in V$ into the "restriction" to $\partial_4\Omega$ of the trace of w_2 on $\partial\Omega$, which is in $H^{1/2}(\partial_4\Omega)$, $w \in V \mapsto Lw = \mathcal{R}_{\partial_4\Omega}(\gamma w_2)$. Let $B = L(V)$. B is a subspace of $L^2(\partial_4\Omega)$. Properties of B are given in [1] . In particular it is proved that B can be provided with a topology of Hilbert space with continuous injection into $L^2(\partial_4\Omega)$, and for which L is continuous. G, the dual of B, is an Hilbert space of distributions defined on $\partial_4\Omega$. The duality between B and G will be denoted by a point. ${}^tL$ will be the transpose of L for the dualities between B and

G, V and Φ. ${}^tLG$ is closed in Φ. The contact force g(t) can be represented as an element of G in the following way $g(t) : b \in B \mapsto b.g(t)$ and b.g(t) is the virtual work of g(t) for the virtual boundary displacement b.

B may be provided with the order relation induced by the one on $L^2(\partial_4\Omega)$. It is proved that the cone K of positive elements of B, is closed in B and is a lattice [3] . As positive cone in G one classicaly chooses $-K^0$. Due the fact that DV and ${}^tLG$ are closed respectively in E and Φ.K and K^0 have mutually polar images by L^{-1} and ${}^tL$, $\Gamma = L^{-1}(K) \subset V$ and $\Gamma^0 = {}^tL(K^0) \subset \Phi$, and also Γ and Γ^0 have mutually polar images by D and ${}^tD^{-1}$, $C = D(\Gamma) \subset E$ and $C^0 = {}^tD^{-1}(\Gamma^0) \subset S$. The unilateral constraints (6), (7), (8), are written as follows: $b(t) = Lv(t) \in K$, $g(t) \in -K^0$, $b(t).g(t) = 0$.

Let s^0 be such that ${}^tDs^0(t) = \varphi(t)$ and $\sigma = s - s^0 \in S$. The equation (1) to (8) can be rewritten as follows :

$$\varepsilon(t) = Dv(t),\ {}^tD\sigma(t) = \psi(t),\ b(t) = Lv(t),\ \psi(t) = {}^tLg(t),\ b(t).g(t) = 0,$$

$\sigma(t) + s^0(t) = \mathcal{H}(\varepsilon + e^0)(t)$ where the unknown functions v, ψ, ε, σ, b, g take their values in the following functional spaces, $v(t) \in V$, $\varepsilon(t) \in E$, $\psi(t) \in \Phi$, $\sigma(t) \in S$, $b(t) \in K \subset B$, $g(t) \in -K^0 \subset G$. The equation ${}^tD\sigma(t) = \psi(t)$ is a weak formulation of **equations (2)(3')(4)(5),which derives from the virtual work principle**

$$\forall\, w \in V \qquad Dw.s(t) = w.[\varphi(t) + \psi(t)]$$

The situation is summed up in the following diagram:

$$\begin{array}{ccccccc}
K & \subset B & & & G & \supset K^0 & \\
 & \uparrow & L & {}^tL & \downarrow & & \\
L^{-1}(K) = \Gamma & \subset V & & & \Phi & \supset \Gamma^0 = {}^tL\,K^0 & \\
 & \downarrow & D & {}^tD & \uparrow & & \\
D\Gamma = C & \subset E & & & S & \supset C^0 = {}^tD^{-1}(\Gamma^0) &
\end{array}$$

Some unknown variables can be eliminated from the previous equations in order to keep only a pair of variables, for instance the pair (ε,σ). Then one can write the general

Standard system.

$\mathcal{J}$ is an interval of $\mathbb{R}$. $E = S = L^2(\Omega)^3$. e^0, s^0 are given functions defined on $\mathcal{J}$ with values respectively in E and S. $\mathcal{H}$ is a mapping transforming a function ε defined on $\mathcal{J}$ with values in E into a function σ defined on $\mathcal{J}$ with values in S. $C \subset E$, $C^0 \subset S$ are a pair of mutually polar closed convex cones with vertex at the origin. A pair $(\varepsilon,\sigma) \in E \times S$ is said to satisfy the standard system if

$$\begin{cases} \forall\, t \in \mathcal{J}\ ,\ \varepsilon(t) \in C,\ -\sigma(t) \in C^{o},\ \varepsilon(t)\ .\ \sigma(t) = 0 \\ \sigma(t) + s^{o}(t) = \mathcal{K}(\varepsilon + e^{o})(t) \end{cases}$$

III.2.- $\mathcal{K}$, the constitutive law.

We consider in this paper a constitutive law of linear viscoelastic type, locally represented by the model in figure 2. This model allows to take into account the flow phenomenom, i.e. the deformation increases indefinitely when the medium is submitted to a constant stress. The equations are the following at time t and at the point $x \in \Omega$

$$(9)\qquad \begin{cases} s(t,x) = K(t,x)\ [e(t,x) - \xi(t,x)] \\ s(t,x) = V(t,x)\ \dot{\xi}(t,x) \end{cases}$$

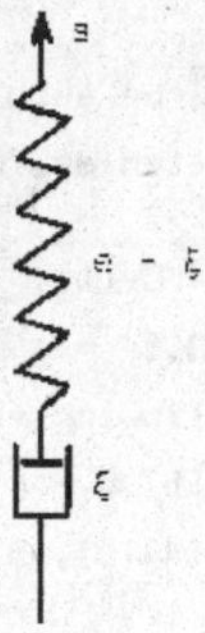

Fig.2

The dot denotes differentiation with respect to time t. The "strains" and "stress", $e(t,x)$, $\xi(t,x)$, $s(t,x)$ are three dimensional vectors. K and V are mappings defined on $\mathbb{R} \times \Omega$ with values in the space of 3 x 3 matrices, measurable on Ω, locally measurable on $\mathbb{R}$. $K(t,x)$ is the stiffness matrix, $V(t,x)$ is the viscosity matrix. These two matrices satisfy

$$(10)\qquad \begin{cases} K(t,x) \text{ and } V(t,x) \text{ are symmetric.} \\ \text{There exist constants } \underline{k},\ \bar{k},\ \underline{v},\ \bar{v}\ ;\ 0 < \underline{k} \le \bar{k}\ ;\ 0 < \underline{v} \le \bar{v} \\ \text{a.e } (t,x) \in \mathbb{R} \times \Omega\ ,\ \forall\, a \in \mathbb{R}^{3}\ ,\ \underline{k}|a|^{2} \le a.K(t,x)a \le \bar{k}|a|^{2} \\ \text{a.e } (t,x) \in \mathbb{R} \times \Omega\ ,\ \forall\, a \in \mathbb{R}^{3}\ ,\ \underline{v}|a|^{2} \le a.V(t,x)a \le \bar{v}|a|^{2} \end{cases}$$

(The dot denotes the scalar product in $\mathbb{R}^{n}$, $|a| = a.a^{1/2}$). These conditions ensure that $K(t,x)$ and $V(t,x)$ have inverses. The study of system (9) is classical. Let $t \mapsto A(t,\tau,x)$ the absolutely continuous mapping defined on $\mathbb{R}$ with values in the space of 3 x 3 matrices and satisfying for every $t \in \mathbb{R}$, $\tau \in \mathbb{R}$,

$$\frac{\partial A}{\partial t}(t,\tau,x) + V^{-1}(t,x)\, K(t,x)\, A(t,\tau,x) = 0 \qquad A(\tau,\tau,x) = I$$

(I identy matrix). One has

$$\text{a.e } x \in \Omega\ \ |A(t,\tau,x)| \le \exp(-(t - \tau)\underline{k}/\bar{v})$$

so that if we put

$$G(t,\tau,x) = K(t,x)\ A(t,\tau,x) \quad , \quad F(t,\tau,x) = \frac{\partial G}{\partial \tau}(t,\tau,x)$$

we first have

(11) $$\text{a.e } (t,\tau,x) \quad |F(t,\tau,x)| \leq \bar{k}^2\ \underline{v}^{-1} \exp(-(t-\tau)\underline{k}/\bar{v})$$

and then

(12) $$s(t,x) = G(t,t,x)\ e(t,x) - \int_{-\infty}^{t} F(t,\tau,x)\ e(\tau,x)d\tau$$

which can also be written

(13) $$s(t,x) = \int_{-\infty}^{t} F(t,\tau,x)\ (e(t,x) - e(\tau,x))d\tau$$

every time the integrals can be defined. Of course we can also derive the inverse form

(14) $$e(t,x) = \bar{\xi}(\tau,x) + K(t,x)^{-1}\ s(t,x) + \int_{\tau}^{t} V(\theta,x)^{-1}\ s(\theta,x)d\theta$$

In the following we may need the additional hypothesis (15), (16), (17) :

(15) $$\exists\ \ell > 0 \text{ a.e } (t_1,t_2,x) \ , \ |K(t_1,x) - K(t_2,x)| \leq \ell |t_1 - t_2|$$

(16) $$\exists\ T > 0 \text{ a.e } (t,x) \quad . \quad K(t+T,x) = K(t,x)$$

(17) the matrix $V(t,x)$ is <u>time independent.</u>

With (16) (17) we have $F(t+T,\tau+T,x) = F(t,\tau,x)$ and with (15) (17) we have :

(18) $$\exists\ k_1,k_2 > 0, \text{ a.e } (t,\tau,x) \ , \ \forall\ h \geq 0 \ , \ |F(t+h,\tau+h,x) - F(t,\tau,x)| \leq hk_1 \exp[-k_2(t-\tau)]$$

IV.- SOLUTION OF THE STANDARD SYSTEM.

The system

$$\forall\ t \in \mathcal{J}, \quad \varepsilon(t) \in C \quad , \quad -\sigma(t) \in C^0 \quad , \quad \varepsilon(t) \ . \ \sigma(t) = 0$$

is equivalent to the inequation

$$\forall t \in \mathcal{J} \ , \quad \forall \eta \ \in C \qquad (\eta - \varepsilon(t)) \ . \ \sigma(t) \geq 0$$

which can also be written

$$\forall t \in \mathcal{J} \ , \quad - \sigma(t) \in \partial\Psi_C(\varepsilon(t))$$

where $\partial\Psi_C$ is the subdifferential cone of the indicatrix function of the closed convex cone C, at the point $\varepsilon(t)$. Let us consider the Hilbert space $\tilde{E} = L^2(\mathcal{J}, E)$ with the scalar product

$$(e^1, e^2) \ \in \ \tilde{E} \mapsto < \ e^1, e^2 > \ = \int_{\mathcal{J}} e^1(t) \ . \ e^2(t)dt$$

the norm being denoted $|||\ \ |||$. Let

$$\tilde{C} = \{e \in \tilde{E} / \text{a.e } t \in \mathcal{J}, \ e(t) \in C\}$$

The following proposition is fundamental.

PROPOSITION IV.1.

Let $\mathcal{J}$ be bounded and $e, \sigma \in \tilde{E}$. The following properties are equivalent :

i) a.e $t \in \mathcal{J}, - \sigma(t) \in \partial\Psi_C(\varepsilon(t))$ *(for the autoduality of* E*).*

ii) $- \sigma \in \partial\Psi_{\tilde{C}}(\varepsilon)$ *(for the autoduality of* $\tilde{E}$*).*

Furthermore $\tilde{C}$ *is a non empty closed convex cone of* $\tilde{E}$.

IV.1.- The Cauchy problem.

$\mathcal{K}$ is given under the form (12). Let $\hat{e}_{t_0} :]-\infty, t_0[\ \to E$ be given, (the history of e till the initial time t_0). We suppose that the integral $\int_{-\infty}^{t_0} F(t,\tau) \ \hat{e}_{t_0}(\tau)d\tau$ is defined. For $t \geq t_0$ we put

$$- s^0(t) + \ \mathcal{K}(\varepsilon + e^0)(t) = \ \mathcal{K}_{t_0}(\varepsilon)(t) + \sigma^0_{t_0}(t)$$

$$\mathcal{K}_{t_0}(\varepsilon)(t) = G(t,t)\varepsilon(t) - \int_{t_0}^{t} F(t,\tau)\varepsilon(\tau)d\tau$$

$$\sigma^0_{t_0}(t) = - s^0(t) + \mathcal{K}_{t_0}(e^0)(t) - \int_{-\infty}^{t_0} F(t,\tau)\hat{e}_{t_0}(\tau)d\tau$$

From (10),(11), the following proposition can easily be proved.

PROPOSITION IV.2.

Let $\mathcal{J} = [t_0, t_0 + t^*[$ *be an interval.* $\mathcal{K}_{t_0}$ *so defined is a mapping from* $\tilde{E}$ *in itself satisfying*

$$\forall e \in \tilde{E} \quad |||\mathcal{K}_{t_0}(e)||| \leq \lambda\, |||e|||$$

$$< e, \mathcal{K}_{t_0}(e) > \; C\, |||e|||^2$$

with

$$\lambda = (\bar{k}^2 \underline{v}^{-1} \bar{v}\, \underline{k}^{-1}) + (\bar{k}^2 \underline{v}^{-1} \bar{v}\, \underline{k}^{-1})(1 - \exp(-t^*/\bar{v}\underline{k}^{-1}))$$

$$C = \underline{k} - (\bar{k}^2 \underline{v}^{-1} \bar{v}\, \underline{k}^{-1})(1 - \exp(-t^*/\bar{v}\, \underline{k}^{-1}))$$

and there exists a strictly positive t^*, *small enough so that* $C > 0$.

Theorem IV.1 follows easily from Proposition IV.2. (Cf. [4]).

THEOREM IV.1. (Existence and unicity).

Let $\hat{e}, e^0, s^0$, *be given mappings with values in* E *defined almost everywhere and locally square integrable respectively on* $\mathbb{R}^-, \mathbb{R}, \mathbb{R}$. *Furthermore we suppose that the integral* $\int_{-\infty}^{0} F(t,\tau)\, \hat{e}(\tau)^* d\tau$ *is defined. Then there exists a unique pair of functions* (ε, σ) *with values in* E, *defined almost everywhere and locally square integrable on* $\mathbb{R}$ *such that*

$$\text{a.e } t \in \mathbb{R} \qquad \sigma(t) + s^0(t) = \mathcal{K}(\varepsilon + e^0)(t)$$

$$\text{a.e } t \in \mathbb{R}^- \qquad \varepsilon(t) = \hat{e}(t) - e^0(t)$$

$$\text{a.e } t \in \mathbb{R}^+ \qquad -\sigma(t) \in \partial\Psi_C(\varepsilon(t))$$

THEOREM IV.2.

We assume (18) is satisfied. Then the solution $(\varepsilon^1, \sigma^1)$ *of the Cauchy problem for data* $\hat{e}^1, e^0, s^0$ *(satisfying the hypothesis of theorem IV.1) is asymptotically stable "in average", which means that if* $(\varepsilon^2, \sigma^2)$ *is the solution of the Cauchy problem for the datas* $\hat{e}^2, e^0, s^0$, *then for every positive* Θ, *the integral* $\int_{\bar{t}}^{\bar{t}+\Theta} \|\sigma^1(t) - \sigma^2(t)\|^2 dt$ *vanishes when* $\bar{t} \in \mathbb{R}$ *tends to* $+\infty$.

Proofs are given in [1].

---

*In fact, as we deal with the differential from (9) the history is reduced to $\xi(0)$ (cf. (14)).

IV.2.- The periodic problem.

In this paragraph we suppose that the hypothesis (15)(16)(17) about $\dot{K}(t,x)$ $V(t,x)$ are satisfied; the expressions (13) and (14) of the constitutive law show that if the stress s is T-periodic, the deformation has the form $e(t) = at + p(t)$, where a is time constant and p is T-periodic. Conversely if e takes this form then the corresponding s is T-periodic.

Notations.

$L^2_{\#}$ $((0,T),E)$ or more briefly $L^2_{\#}$ will denote the space of (classes of) functions defined almost everywhere and T-periodic on $\mathbb{R}$, with values in E, and square integrable. This is an Hilbert space for the scalar product $\langle p_1,p_2\rangle = \int_0^T p_1(t).p_2(t)dt$. The norm will be $|||.|||$. In the following we will use the decomposition : $g \in L^2_{\#}$, $g = \tilde{g} + \bar{g}$ where $\bar{g}$ is the function with value $\frac{1}{T}\int_0^T g(t)dt$. We shall use the same notations for a time constant function and its value. $\mathcal{K}_0$ and $\mathcal{K}_1$ are the continous linear mappings

$$\mathcal{K}_0 : E \to L^2_{\#} \;, \alpha \mapsto \mathcal{K}_0(\alpha) \;, \quad \mathcal{K}_0(\alpha)(t) = \left(\int_{-\infty}^t F(t,\tau)(t-\tau)d\tau\right)\alpha$$

$$\mathcal{K}_1 : L^2_{\#} \to L^2_{\#} \;, q \mapsto \mathcal{K}_1(q) \;. \quad \mathcal{K}_1(q)(t) = \int_{-\infty}^t F(t,\tau)[q(t) - q(\tau)]d\tau$$

The periodic problem is

PROBLEM 1.

Let e^0, s^0, *be given with* $e^0(t) = a^0 t + p^0(t)$, $a^0 \in E$, $p^0 \in L^2_{\#}$, $s^0 \in L^2_{\#}$. *Find* $(t_0,a,p,\sigma) \in \mathbb{R} \times E \times L^2_{\#} \times L^2_{\#}$ *such that*

$$\text{a.e } t \in \mathbb{R} \qquad \sigma(t) + s^0(t) = \mathcal{K}_0(a^0 + a)(t) + \mathcal{K}_1(p^0 + p)(t)$$

$$\text{a.e } t \in [t_0, +\infty[\qquad -\sigma(t) \in \partial\Psi_c(at + p(t))$$

With the methods of convex analysis it can be proved that this problem can be reduced to

PROBLEM 2.

The data are as in problem 1. Find $(a,p,\sigma) \in E \times L^2_{\#} \times L^2_{\#}$ *such that*

i) a.e $t \in R$ $\qquad \sigma(t) + s^0(t) = \mathcal{K}_0(a + a^0)(t) + \mathcal{K}_1(p^0 + p)(t)$

ii) $\qquad -\bar{\sigma} \in \partial\Psi_c(a)$

iii) a.e $t \in R$ $\qquad -\sigma(t) \in \partial\Psi_c(p(t))$

Comment :

Connection between problems 1 and 2 can be more easily understood if they are expressed in terms of the duality between B, the space of traces on $\partial_4\Omega$, and G, the space of associated forces. For problem 1 unilateral constraints are written

a) a.e. $t \in [t_o, +\infty[$: $\quad - g(t) \in \partial \Psi_K(b(t))$

where b(t) should be of the form

$b(t) = \beta t + q(t)$

For problem 2 :

b) $\quad -\bar{g} \in \partial\Psi_K(\beta)$

c) $\quad$ a.e. $t \in \mathbb{R}$: $- g(t) \in \partial\Psi_K(q(t))$

The constitutive equation is implicitely determined as the general solution of a boundary value problem on Ω, a data of which is, for instance, b :

d) $\quad g = \mathcal{G}(b)$

where $\mathcal{G}$ is a affine functional of $L^2_{\#}(0,T),G)$, which only depends of $\tilde{b}$:

e) $\quad \tilde{b}_1 = \tilde{b}_2 \Rightarrow \mathcal{G}(b_1) = \mathcal{G}(b_2)$

Equation b) implies that

a.e. $x \in \partial_4\Omega$: $\quad \bar{g}(x)\beta(x) = 0$

and, furthermore , since g is non negative :

$\bar{g}(x) = 0 \Leftrightarrow$ p.p. $t \in \mathbb{R}$: $g(x,t) = 0$

Hence we see that the force g(x,t) is zero at (almost) every point x where $\beta(x)$ is positive. This is a mechanically obvious result : for large t there is no contact at such x points.

Then it is easy to understand why q may be choosen as a positive function. Let (b,g) a solution of equations a) and d), and b_o a constant field belonging to B such that

f) $\quad b_o\bar{g} = 0$

g) $\quad$ p.p. $t \in \mathbb{R}$ $\quad q(t) + b_o \in K$

Then $(b + b^o, g)$ is another solution of a) and d). Let us choose :

h) $\quad b_o(x) = - \inf_{t \in R} q(x,t)$

which obviously satisfies g). We have :

$$\bar{g}(x) \neq 0 \Rightarrow \bar{g}(x) > 0 \Rightarrow \beta(x) = 0$$

Hence a) implies that $b_o(x)$ is zero at almost every point x where $\bar{g}(x)$ is not zero, and b_o satisfies #).Finally $(b + b^o, g)$ is a solution of a) and d). Thus we always may ask q to be a positive function, at least if h) defines a function b_o that belongs to B.

We understand why we shall be led to study the existence of such an infimum, which will be ensured by the proposition IV.4.

Proposition IV.3.

Under the hypotheses (16) and (17) the linear mappings $\mathcal{K}_o$ *and* $\mathcal{K}_1$ *are continuous. Furthermore*

(19) $$\forall \alpha \in E \qquad \overline{\mathcal{K}_o(\alpha)} = V\alpha$$

(20) $$\forall p \in L^2_{\#} \qquad \mathcal{K}_1(p) = \mathcal{K}_1(\tilde{p}) \quad , \quad \overline{\mathcal{K}_1(p)} = 0$$

(21) $$\exists c > o \quad \forall p \in L^2_{\#} \qquad < p, \ \mathcal{K}_1(p)> \ \geq \ c \ |||\tilde{p}|||^2$$

Outline of the proof.

(19) and (20) can easily be proved from the definitions of $\mathcal{K}_o$ and $\mathcal{K}_1$ and can also be checked on (9). (20) implies $< p, \ \mathcal{K}_1(p) > = < \tilde{p}, \ \mathcal{K}_1(\tilde{p}) >$. In fact $\mathcal{K}_1$ maps onto itself the subspace of $L^2_{\#}$ of functions with mean value zeros. Let $\tilde{\sigma} = \mathcal{K}_1(\tilde{p})$. One has

$$\mathcal{K}_1^{-1}(\tilde{\sigma})(t) = K^{-1}(t)\ \tilde{\sigma}(t) - \frac{1}{T}\int_0^t K^{-1}(\tau)\ \tilde{\sigma}(\tau)d\tau + \int^t V\tilde{\sigma}(\tau)d\tau$$

where $t \mapsto \int^t V\tilde{\sigma}(\tau)d\tau$ is the T-periodic primitive of $V\tilde{\sigma}$ with mean value 0. There exists $C_1 > 0$ such that $|||\ \mathcal{K}_1^{-1}(\tilde{\sigma})||| \leq C_1|||\tilde{\sigma}|||$. As $V^{-1}(x)$ is symmetric

$$\tilde{\sigma}(t).\ \mathcal{K}_1^{-1}(\tilde{\sigma})(t) = \tilde{\sigma}(t).K^{-1}(t)\tilde{\sigma}(t) + \frac{1}{2}\frac{d}{dt}\left[\int^t \tilde{\sigma}(\tau)d\tau . V^{-1}\int^t \tilde{\sigma}(\tau)d\tau\right] +$$

$$- \tilde{\sigma}(t)\ .\ \frac{1}{T}\int_0^T K^{-1}(\tau)\tilde{\sigma}(\tau)d\tau$$

so that

$$<p, \mathcal{K}_1(p)> \ = \ <\tilde{\sigma}\ ,\ \mathcal{K}_1^{-1}\tilde{\sigma}> \ \geq \underline{k}^{-1}|||\tilde{\sigma}|||^2 \geq \underline{k}^{-1}\ C_1^{-1}|||\tilde{p}|||^2 = c|||\tilde{p}|||^2$$

It results from propositions IV.1 and IV.3 that the problem 2 is equivalent to

Problem 3

Let $\tilde{C} = \{q \in L^2_{\#} / \text{a.e } t \in \mathbb{R},\ q(t) \in C\}$. $\tilde{C}$ *is a non empty closed convex cone of* $L^2_{\#}$. *Find* $a \in C$, $p \in \tilde{C}$ *such that*

(22) $\quad \forall \nu \in C \quad (\nu - a) \cdot (\bar{\sigma}^0 + Va) \geq 0$

(23) $\quad \forall \eta \in \tilde{C} \quad \langle \eta - p,\ \sigma^0 + \mathcal{K}_0(a) + \mathcal{K}_1(p) \rangle \geq 0$

where $\sigma^0 = -s^0 + \mathcal{K}_0(a^0) + \mathcal{K}_1(p^0) \in L^2_{\#}$.

The inequalities (22) and (23) are weakly coupled. Since V is coercive, the inequality (22) admits a unique solution a^*. Let $\mathcal{K}' : L^2_{\#} \to L^2_{\#}$ $p \mapsto \mathcal{K}'(p) = \sigma^0 + \mathcal{K}_0(a^*) + \mathcal{K}_1(p)$. We have now to solve

(23') $\quad$ Find $p \in \tilde{C}$, $\forall \eta \in \tilde{C}$, $\langle \eta - p,\ \mathcal{K}'(p) \rangle \geq 0$

$\mathcal{K}'$ is a continuous affine monotone operator. Unfortunately $\mathcal{K}'$ is coercive only on the subspace of $L^2_{\#}$ of functions with mean value zero. We shall use a classical result about regularisation of variational inequalities involving a monotone operator, [4]. For $\lambda > 0$ the problem

(24) $\quad$ Find $p_\lambda \in \tilde{C}$, $\forall \eta \in \tilde{C} \quad \lambda \langle \eta - p_\lambda, p_\lambda \rangle + \langle \eta - p_\lambda, \mathcal{K}'(p_\lambda) \rangle \geq 0$

admits a unique solution $p_\lambda \in \tilde{C}$. Furthermore (23') admits a solution if and only if

(25) $\quad \exists L \ .\ \forall \lambda \quad |||p_\lambda||| \leq L$

and the solution of (23') is obtained as a limit of p_λ when $\lambda \to 0$. An estimate of $|||\tilde{p}_\lambda|||$ is obtained when $\eta = \bar{p}_\lambda$ in (24). Thanks to (20), (21) it comes $\lambda |||\tilde{p}_\lambda|||^2 + c|||\tilde{p}_\lambda|||^2 \leq C_1 |||\tilde{p}_\lambda|||$ which results in

(26) $\quad \exists B_1 > 0 \quad , \quad \forall \lambda > 0 \quad . \quad |||\tilde{p}_\lambda||| \leq B_1$

Proposition IV.4 (cf.[1]).

$H^1(\Omega)^2$ *is an ordered topological vector space sublattice of* $L^2(\Omega)^2$. $V \subset H^1(\Omega)^2$ *is also an o.t.v.s. and a lattice.* D V *is an o.t.v.s and a lattice for the order induced by the bijection D. B is an o.t.v.s. sublattice of* $L^2(\partial_4\Omega)$.*K,* Γ, C, *are lattices for the order of the respective spaces and contain the respective positive cones.*

Let $p \in H^1_{\#}((0,T),DV)$ (*i.e.* $p, \frac{\partial p}{\partial t} \in L^2_{\#}((0,T), DV)$). *The family* $\{p(t)\}_{t \in [0,T]}$ *has a greatest lower bound in* DV *denoted* $\inf\{p(t)\}$ *with the following properties :*

i) $\exists\ \mu > 0 \qquad ||\inf\{p(t)\}||_{DV} \leq \mu\ ||p||_{H^1_{\#}((0,T),\ DV)}$

ii) *If* $\forall t$, $p(t) \in C$ *then* $\inf\{p(t)\} \in C$

iii) *If* $p = \tilde{p} + \overline{p}$, *then* $\inf\{p(t)\} = \inf\{\tilde{p}(t)\} + \overline{p}$.

This proposition derives from a more general situation where DV may be replaced by an ordered and a lattice reflexive Banach space with some properties (see [3]).

Proposition IV.4 allow us to prove

Theorem IV.3 (Existence).

If the data e^o, s^o, *have derivatives* $\frac{\partial e^o}{\partial t}$, $\frac{\partial s^o}{\partial t}$ *in* $L^2_{\#}$, *the problem 3 has a solution* (e^*, p^*), *such that* $\lim_{\lambda \to 0} |||\tilde{p}^* - \tilde{p}_\lambda||| = 0$

Outline of the proof.

Proposition IV.1 allows to write (24) in the equivalent form

$$\text{a.e } t \in [0,T] \quad , \forall \eta \in C \quad , \quad (\eta - p_\lambda(t)).\ \sigma_\lambda(t) \geq 0$$

where $\sigma_\lambda = \mathcal{H}'(p_\lambda) + \lambda p_\lambda$. It results

$$\text{a.e } t \in [0,T] \quad , \nabla_h\ p_\lambda(t) \ .\ \nabla_h\ \sigma_\lambda(t) \leq 0$$

where $\nabla_h p_\lambda(t) = \dfrac{p_\lambda(t+h) - p_\lambda(t)}{h}$, $\nabla_h\ \sigma_\lambda(t) = \dfrac{\sigma_\lambda(t+h) - \sigma_\lambda(t)}{h}$

Then one obtains the following estimates taking into account the hypothesis on e^o, s^o, (15) and (18)

(27) $\exists\ C_1, C_2 > 0$, $\forall \lambda > 0$, $\forall h \in]0, h_o]$, $|||\nabla_h p_\lambda||| = |||\nabla_h\ \tilde{p}_\lambda||| \leq C_1 + C_2|||\tilde{p}_\lambda|||$

When $h \to 0$ and using (26) it comes

$\frac{\partial p_\lambda}{\partial t} \in L^2_{\#}$ and then

(28) $\qquad \exists\ B_2 > 0$ such that $\forall \lambda > 0\ |||\frac{\partial p_\lambda}{\partial t}||| \leq B_2$

So with (26), $p_\lambda \in H^1_{\#}((0,T),DV)$ and $\|\tilde{p}_\lambda\|_{H^1_{\#}} \leq (B_1^2 + B_2^2)^{\frac{1}{2}}$

It results from proposition IV.4.

(29) $\exists B_3 > 0 \quad , \quad \forall \lambda > 0 \quad , \quad \| \inf \{ \tilde{p}_\lambda(t) \} \| \leq B_3$

which implies

(30) $\exists L > 0 \quad , \quad \forall \lambda > 0 \quad ||| \tilde{p}_\lambda - \inf \{ \tilde{p}_\lambda(t) \} ||| \leq L \ ;$

furthermore $\hat{\eta}_\lambda = p_\lambda - \inf\{p_\lambda(t)\} = \tilde{p}_\lambda - \inf\{\tilde{p}_\lambda(t)\}$, and $|||\hat{\eta}_\lambda||| \leq L$. Obviously $\hat{\eta}_\lambda \in \tilde{C}$.

Then

$$\lambda < \hat{\eta}_\lambda - p_\lambda , p_\lambda > \geq < \inf\{p_\lambda(t)\} \ , \ \mathcal{K}'(p_\lambda) > \ = \ T \inf\{p_\lambda(t)\} [\bar{\sigma}^0 + Va^*] \geq 0$$

where the last inequality holds since $\inf \{p_\lambda(t)\} \in C$ and $\bar{\sigma}^0 + Va^* \in -C^0$. It results $|||p_\lambda||| \leq |||\hat{\eta}_\lambda||| \leq L$. There exists a sequence $(\lambda_n) \to 0$ such that $p_{\lambda_n} \to p^*$ weakly in $L^2_{\#}$. Then using the linearization lemma of Minty, the monotonicity and the continuity of $\mathcal{K}_1$, one proves that p^* is a solution of (23)'.

Let us put $\eta = p^*$ into (24), $\eta = p_\lambda$ into (23) and let us add the respective side, then

$$\lambda < p^* - p_\lambda , \ p_\lambda > \geq < p^* - p_\lambda \ , \ \mathcal{K}_1(p^* - p_\lambda) > \geq c |||\tilde{p}^* - \tilde{p}_\lambda|||^2 \geq 0$$

$$\lambda |||p^* - p_\lambda||| \ |||p_\lambda||| \ \geq \ c \ |||\tilde{p}^* - \tilde{p}_\lambda|||^2$$

Since $|||p_\lambda||| \leq L$, this last inequality implies that $\lim_{\lambda \to 0} |||\tilde{p}^* - \tilde{p}_\lambda||| = 0$.

Some easy remarks allow to prove the following

Theorem IV.4 (unicity).

a) Let (a_1, p_1, σ_1) be a solution of problem 1; then (a_2, p_2, σ_2) is another solution if and only if

$a_1 = a_2 \quad , \quad \tilde{p}_1 = \tilde{p}_2 \quad , \quad \sigma_1 = \sigma_2$

$p_2 \in \tilde{C}$

$(\bar{p}_1 - \bar{p}_2) \cdot \bar{\sigma}_1 = 0$

b) If the hypothesis of the existence theorem IV.3 are satisfied we can choose p_1 such that

$p_1(t) = \tilde{p}_1(t) - \inf\{\tilde{p}_1(t)\}$. *Then the previous conditions become* $a_1 = a_2$, $\tilde{p}_1 = \tilde{p}_2$, $\sigma_1 = \sigma_2$, $\bar{p}_2 - \bar{p}_1 \in C$, $(\bar{p}_2 - \bar{p}_1) . \bar{\sigma}_1 = 0$.

This means that every solution p_2 *must be written* $p_2 = \tilde{p}_1 - \inf\{\tilde{p}_1(t)\} + h$ *where* $h \in C$ *and* $h . \bar{\sigma}_1 = 0$.

BIBLIOGRAPHY.

[1] R.Bouc, G.Geymonat, M.Jean, B.Nayroles.
Problèmes hilbertiens unilatéraux pour les corps viscoélastiques fissurés.
En préparation.

[2] G.Duvaut, J.L.Lions.
Les inéquations en mécanique et en physique.
Dunod, Paris, 1972.

[3] M.Jean.
Un cadre abstrait pour l'espace vectoriel topologique ordonné $W^{1,p}(\Omega)$ et quelques uns de ses bornés.
Travaux du séminaire d'analyse convexe, Université des Sciences et Techniques du Languedoc, Montpellier 1975.

[4] G.Stampacchia.
Variational inequalities.
Proc.NATO, Adv. Study Inst., Venezia 1968, Oderisi Ed., 1969, 101-191.

ON THE NORM-DEPENDENCE OF THE CONCEPT OF STABILITY

H. Brauchli
Institute of Mechanics
Swiss Federal Institute
of Technology, Zurich
Switzerland

In order to define stability one needs to measure the deviation from the desired state of the system. This is usually done by introducing a metric, or, if the underlying space is linear, a norm. If the system is of finite order, all norms one might choose are equivalent and lead to the same definition of stability. In the case of a continuous body, however, the state space is infinite dimensional and inequivalent norms, leading to different stability concepts, do exist. Furthermore, as has been pointed out by Movchan [5], one may use a two-norm definition of stability, measuring initial and subsequent deviations of the system in two distinct norms.

Usually, a fixed norm is used to investigate stability of a given system. In mechanics, a natural choice seems to be the energy norm [3,4], although it may yield strange effects [2,6]. From a practical point of view, the energy norm seems satisfactory for elastic bodies under dead loading. Yet, for nonconservative systems and instationary loads, the question, how choosing a different norm would affect stability imposes itself.

It is natural to expect a simple behaviour for the norm-dependence of stability. This is actually the case if the two-norm definition is used and if one of the two norms is held fixed while the other one is varied. Replacing the second norm by a weaker one may result in changing an unstable system into a stable one but cannot destroy stability. The situation is similar, if the second norm is fixed, but the initial norm is to vary. Loosely speaking one may say that stability is improved by weakening the second norm and by strengthening the first one. Things may be more complex, however, if a single norm is used. Then a loss of stability may occur either by weakening or by strengthening the norm.

1. Definitions and Simple Conclusions

Let B denote a Banach space with elements x, c and let R^+ be the nonnegative reals. Let $\Phi : B \times R^+ \rightarrow B$ be a continuous mapping satisfying $\Phi(c,0)=c$. For a fixed c, $x(t)= \Phi(c,t)$ is a motion with initial point c. Finally, let $\|\cdot\|_\alpha$ and $\|\cdot\|_\beta$ denote two norms in B, the first stronger than (or equivalent to) the second. According to [1] and [5], the two-norm version of Lyapunov-stability will be:

Definition 1: The motion $x_0(t)= \Phi(c_0,t)$ is stable with respect to the norms $\|\cdot\|_\alpha$ and $\|\cdot\|_\beta$, if for every $\varepsilon > 0$ there is a $\delta > 0$, such that $\|c-c_0\|_\alpha < \delta$ implies $\|x(t)-x_0(t)\|_\beta < \varepsilon$ for all $t \geq 0$.

Definition 2: The motion $x_0(t)$ is asymptotically stable, if it is stable and if $\|c-c_0\|_\alpha < \delta$ implies $\lim_{t\to\infty} \|x(t)-x_0(t)\|_\beta =0$ for some $\delta > 0$.

As in [1], a motion is called weakly stable, if it is stable but not asymptotically stable. If the two norms coincide, the single norm definition of stability is obtained.

It is an immediate consequence of the definition, that a motion stable with respect to $\|\cdot\|_\alpha$ and $\|\cdot\|_\beta$ will also be stable with respect to $\|\cdot\|_\alpha$ and $\|\cdot\|_\gamma$, if $\|\cdot\|_\gamma$ is weaker than $\|\cdot\|_\beta$. The following example will show, that an unstable motion can be made stable by weakening the sdcond norm. The situation is similar, if the second norm is fixed and the first varied.

2. Example

For k=0,1,2,... consider the functions

$$y_k(t) = (t^k/k!) \exp(-t)$$

satisfying the relations

$$\sum y_k = 1 , \quad \sum k\, y_k = t , \quad \sum (1+k)^{-1} y_k = g(t) = (1-e^{-t})/t .$$

The summation is over all k. The function g(t) is regular in R^+ and monotonically decreasing from 1 to 0.

Let B be the Banach space of sequences x_k (k=0,1,2,...) with convergent norm

$$\|x\|_1 = \sum (1+k)\ x_k$$

and define two other norms,

$$\|x\|_0 = \sum x_k \ , \qquad \|x\|_{-1} = \sum (1+k)^{-1}\ x_k \ .$$

The functions y_k satisfy the differential equations

$$x_k' + x_k = x_{k-1} \ , \qquad k=0,1,\ldots \tag{1}$$

where $x_k=0$ for negative k for convenience. The general solution to (1) is

$$x(t) = \sum c_m\ Y^{(m)}(t) \tag{2}$$

where $Y^{(m)}$ denotes the sequence $\{y_{k-m}\}$. x(t) satisfies the initial condition x(0)=c.

A simple calculation shows that the following inequalities hold:

$$\|x\|_1 \leq \|c\|_1 + \|c\|_0\ t \ , \tag{3}$$

$$\|x\|_0 \leq \|c\|_0 \ , \tag{4}$$

$$\|x\|_{-1} \leq \|c\|_0\ g(t)\ . \tag{5}$$

For $c_k \geqslant 0$, (3) and (4) are equalities. Hence x(t) is unstable for the norms 1,1, it is weakly stable for the norms 1,0 and it is asymptotically stable for the norms 1,-1.

3. Integrals of Motion

In systems of finite order, the existence of a nondegenerate and time-independent integral of motion excludes the possibility of asymptotically stable equilibria. The above example shows, that this is no longer true in infinite spaces. System (1) admits the integral

$$H(x) = \sum x_k = \sum c_k \ . \tag{6}$$

But x=0 is asymptotically stable in the appropriate norms.

Generally, one can prove the following:

Theorem: Let x(t)=c be an equilibrium of a system with a linear and time-independent integral of motion h(x). Let x be asymptotically stable with respect to the norms $\|\cdot\|_\alpha$ and $\|\cdot\|_\beta$. Then there exist norms $\|\cdot\|_{\bar\alpha}$ and $\|\cdot\|_{\bar\beta}$, which make x weakly stable.

Define

$$\|x\|_{\bar\alpha} = \|x\|_\alpha + |h(x)| \quad ,$$

$$\|x\|_{\bar\beta} = \|x\|_\beta + |h(x)| \quad .$$

It can be shown that these are actually norms and that x(t)=c is then weakly stable. In case of a nonlinear, nondegenerate integral independent of time a similar theorem could be proved using metrics instead of norms.

References

[1] W. Hahn, Theorie und Anwendung der direkten Methode von Ljapunov. Springer-Verlag, Berlin, Göttingen, Heidelberg (1959).

[2] G. Hamel, Theoretische Mechanik. Springer-Verlag, Berlin, Göttingen, Heidelberg, 269 (1949).

[3] W.T. Koiter, The concept of stability of equilibrium for continuous bodies. Proc.Kon.Ned.Ak.Wet. B66, 173-177 (1963).

[4] W.T. Koiter, The energy criterion of stability for continuous elastic bodies. Proc.Kon.Ned.Ak.Wet. B68, 178-202 (1965).

[5] A.A. Movchan, Stability of processes with respect to two metrics. Prikl.Mat.Meh. 24, 988-1001 (1960) (Appl.Math.Mech. 24, 1506-1524).

[6] R.T. Shield and A.E. Green, On certain methods in the stability theory of continuous systems. Arch.Rat.Mech.Anal. 12, 354-360 (1963).

THE HODOGRAPH METHOD IN FLUID-DYNAMICS IN THE LIGHT OF VARIATIONAL INEQUALITIES

Haïm BREZIS (Paris) and Guido STAMPACCHIA (Pisa)

Introduction

This paper discusses the problem of a flow past a given profile with prescribed velocity at infinity. In 1973 we announced (see {4}) a result concerning the study in the hodograph plane of some two-dimensional subsonic flows past a given convex profile, by using variational inequalities. The aim of the present work is to give a proof of this result, as well as further properties of the solution, confining ourselves to the case of incompressible fluids, for simplicity. We plan to return to the general case in a forthcoming paper.

For the case of incompressible fluids, the existence and uniqueness of a flow past a given profile with prescribed velocity at infinity is well known. A complete bibliography can be found for example in the book of L.Bers {2}.

For the case of compressible fluids, the hodograph method has the notable advantage of "linearizing" the equations, but it leads - even in the incompressible case - to a free boundary value problem which may be considered as difficult as the original one.

Since no direct mathematical method was available, the difficulty was usually avoided by considering indirect problems, namely by determining a flow with a given hodograph.

L.Bers {2} expresses the feeling by saying : "In general the boundary conditions become extremely complicated by going over the hodograph plane, and this is even more pronounced if one uses the Legendre transform".

Our main purpose here is to show that this difficulty can be overcome in some simple cases by using variational inequalities instead of equations.

We would like to thank L.Nirenberg who has made crucial observations and has given us useful suggestions while we were writing this paper.

The plan is the following :

- § 1 - Statement of the problem in the physical plane and some classical results.
- § 2 - The hodograph transform.
- § 3 - The stream function in the hodograph plane.
- § 4 - The variational inequality in the hodograph plane.
- § 5 - Some hints for numerical computations.

Appendix : The smoothness of the stream function near the edges of P.

- § 1 - Statement of the problem in the physical plane and some classical results -

Let P be a closed convex profile in $\mathbb{R}^2$ which is symmetric with respect to the x-axis. We can always assume that the origin is an interior point of P . We denote by G the open set $\mathbb{R}^2 \setminus P$ and by $\vec{n}$ the outward normal to P . The study of an irrotational symmetric steady flow for an incompressible fluid with uniform velocity q_∞

at infinity in the direction of the x-axis leads to the following problem :

Find the velocity $\vec{q} = (q_1,q_2)$ defined on $\bar{G}$ and satisfying

(1.1) $\vec{q}$ is C^1 on G and continuous on $\bar{G}$

(1.2) $\operatorname{div} \vec{q} = \frac{\partial q_1}{\partial x} + \frac{\partial q_2}{\partial y} = 0$ in G

(1.3) $\operatorname{curl} \vec{q} = \frac{\partial q_2}{\partial x} - \frac{\partial q_1}{\partial y} = 0$ in G

(1.4) $\vec{q}.\vec{n} = 0$ on ∂G

(1.5) $q_1(x,y) \to q_\infty$ and $q_2(x,y) \to 0$ as $(x,y) \to \infty$

(1.6) $q_1(x,y) = q_1(x,-y)$ and $q_2(x,y) = -q_2(x,-y)$ (symmetry condition).

Assuming now that $\vec{q}$ satisfies (1.1) - (1.6) we introduce the complex function $V(z)$ by

(1.7) $V(z) = q_1(x,y)-i\, q_2(x,y)$ where $z = x+iy$.

By (1.2) and (1.3) the Cauchy-Riemann conditions are satisfied and thus V is holomorphic on G . Also, by (1.6) we have :

(1.8) $V(\bar{z}) = \overline{V(z)}$ in G .

Using (1.5) we see that $V(z) \to q_\infty$ as $|z| \to \infty$ and therefore V has an expansion at infinity :

$$V(z) = q_\infty + \sum_{n=1}^{\infty} \frac{a_n}{z^n} .$$

From (1.8) we obtain that $a_n \in \mathbb{R}$ for every $n \geq 1$.

On the other hand, by choosing a positive orientation on ∂G we have :

$$\int_{\partial G} V(z)dz = 2i\pi\, a_1 .$$

But

$$\int_{\partial G} V(z)dz = \int_{\partial G} (q_1-iq_2)(dx+idy) =$$

$$= \int_{\partial G} q_1dx + q_2dy + i\int_{\partial G} q_1dy - q_2dx .$$

Since by (1.4) $q_1dy - q_2dx = 0$, we infer that $\int_{\partial G} V(z)dz$ is real. Hence

(1.9) $$a_1 = 0 ,$$

which is the statement that symmetric flows have no circulation around P .

Consequently there is a (singlevalued) holomorphic function Φ in G such that $\Phi'(z) = V(z)$, and near infinity Φ has the expansion :

$$\Phi(z) = q_\infty z - \sum_{n=1}^{\infty} \frac{a_{n+1}}{nz^n}$$

Let us denote $\phi = \operatorname{Re} \Phi$ and $\psi = \operatorname{Im} \Phi$; ϕ and ψ are, respectively, the velocity potential and the stream function. The functions ϕ and ψ are harmonic on G and of class C^1 in $\bar{G}$ by (1.1). In addition, it follows from (1.8) that

(1.10) $$\Phi(\bar{z}) = \overline{\Phi(z)} .$$

Thus

(1.11) $$\psi(x,-y) = -\psi(x,y) \quad \text{on} \quad G$$

and in particular

(1.12) $$\psi(x,0) = 0 .$$

Also we have

(1.13) $$\phi_x = \psi_y = q_1 \quad \text{and} \quad \phi_y = -\psi_x = q_2 .$$

Moreover (1.4) express that the tangential derivative of ψ along ∂P vanishes so that

(1.14) $$\psi = 0 \quad \text{on} \quad \partial P .$$

For the sake of completeness we prove now that the problem (1.1) - (1.6) admits a unique solution.

Uniqueness follows immediately from the maximum principle applied to the difference $(\psi-\hat{\psi})$ of two stream functions. Indeed $(\psi-\hat{\psi})$ is harmonic in G , vanishes on ∂G and goes to zero at infinity.

We introduce the notation.

$$\partial P_+ = \partial P \cap \{z; \operatorname{Im} z \geq 0\} ,$$

and we assume now that ∂P_+ is of class $C^{1,\alpha}$ up to and including its end points denoted by A and B.

The existence follows easily from the Riemann Mapping Theorem. Indeed let f be the conformal mapping from G onto $\Delta=\{\zeta\in\mathbb{C} ; |\zeta|>1\}$ such that $f(\infty) = \infty$ and $f'(\infty) > 0$ i.e. f is one-to-one, onto, holomorphic on G , $f' \neq 0$, $|f(z)| \to \infty$ as $|z| \to \infty$ and $f'(z)$ converges to a positive limit as $|z| \to \infty$ (actually G is simply connected in S^2) .

Also observe that, because G is symmetric and the conformal mapping is unique, we have :

(1.15) $$f(\bar{z}) = \overline{f(z)}$$

In addition f is continuous on $\bar{G}$ and f is one to one from $\bar{G}$ onto $\bar{\Delta}$ since ∂G is a simple closed curve (see {7} p.367).

If we define now

(1.16) $$\Phi(z) = \frac{q_\infty}{f'(\infty)} \left[f(z) + \frac{1}{f(z)} \right]$$

then $V = \Phi'$ satisfies all the required properties. Indeed V is holomorphic and $V(\bar{z}) = \overline{V(z)}$ by (1.15), thus (1.2), (1.3) and (1.6) follow. Next we have :

(1.17) $$V(z) = \frac{q_\infty}{f'(\infty)}\ f'(z)\left[1 - \frac{1}{f(z)^2}\right]$$

and so $V(z) \to q_\infty$ as $|z| \to \infty$.

From (1.16) we deduce that :

(1.18) $$\psi = \text{Im}\ \Phi = \frac{q_\infty}{f'(\infty)}\ ,\ \text{Im}\ f(z)\left[1 - \frac{1}{|f(z)|^2}\right]\ ;$$

hence ψ is continuous on $\bar{G}$ and $\psi = 0$ on ∂G .

It remains to be shown that $\vec{q}$ is continuous on $\bar{G}$. We know from the Appendix that $\psi \in C^{1,\beta}(\bar{G})$ for some $0 < \beta < 1$ and thus $\vec{q} = (\psi_y, -\psi_x)$ lies in $C^{0,\beta}(\bar{G})$.

Finally observe that :

(1.19) $\vec{q}$ vanishes at A and B .

Indeed we have $q_2(x,0) = 0$, but also $\lim\limits_{\substack{z\to A\\ z\in\partial P}} \vec{q}(z).\vec{n}(z) = 0$

The following properties of ψ will be relevant later.

<u>Proposition 1.1.-</u> *We have* $\psi(z) > 0$ *for* $z \in G_+ = \{z\in G\ ;\ \text{Im}\ z>0\}$.

<u>Proof</u> Using (1.18) we see that $\psi(z)$ has the same signum as $\text{Im}\ f(z)$. However since $f'(\infty) > 0$, G_+ is mapped onto $\Delta_+ = \{z\in\Delta ; \text{Im}\ z>0\}$ and so $\text{Im}\ f(z) > 0$ for $z \in G_+$.

<u>Proposition 1.2.-</u> *We have* $\text{Re}\ V = \psi_y > 0$ *in* G , *and on* ∂G , *except at* A *and* B (*where* $V = 0$) .

<u>Proof</u> The function ψ is harmonic on G_+ and the minimum is achieved on ∂G_+ (where $\psi = 0$). It follows from Hopf maximum principle that $\frac{\partial\psi}{\partial n} > 0$ in ∂P_+ except at A and B (where the normal is not defined) ; since $\psi = 0$ on ∂P , we conclude that $\psi_y > 0$ on ∂P_+ except at A and B . We have also $\psi_y(x,0) > 0$ for $(x,0) \in G$. Finally $\Delta\psi_y = 0$ in G_+ and $\psi_y \geq 0$ on ∂G_+ imply again by the maximum principle that $\psi_y > 0$ on G_+ .

<u>- § 2 - The hodograph transform -</u>

We assume now that P is strictly convex.

<u>Theorem 2.1</u>

a) *The function* $q_1(x,0)$ *is decreasing from* q_∞ *to* 0 *as* x *goes from* $-\infty$ *to* A, *and increasing from* 0 *to* q_∞ *as* x *goes from* B *to* $+\infty$.

b) *The set* $V(\partial P_+)$ *is a simple closed curve ; we denote by* $\sum_+$ *the bounded open component of* $\mathbb{R}^2 \setminus V(\partial P_+)$.

Then $(0,q_\infty] \subset \sum_+$ *and the hodograph mapping* $z \longmapsto V(z)$ *is one to one from* G_+ *onto* $\sum_+ \setminus (0,q_\infty]$.

<u>Proof</u>

a) By Proposition 1.2 we know that $V \neq 0$ on P_+ , except at A and B . Since P is strictly convex and $\vec{q}$ is tangent to P , it is clear that $V(P_+)$ is a simple curve ; also $V(\partial P_+)$ is closed since $V(A) = V(B) = 0$.

The function V is holomorphic on G and $V(\partial G)$ consists of the two simple closed curves $V(\partial P_+)$ and $V(\partial P_-)$ (which are symmetric with respect to the x-axis). Therefore, every point in $\mathbb{C}$, not lying on $V(\partial G)$ has an index with respect to $V(\partial G)$ which assumes only the values 0,1 and 2. Hence the equation $V(z) = a$, for a given $a \notin V(\partial G)$ has *at most* two solutions. Actually we should remark that the above conclusion holds for bounded domains ; but we can pass to this case by an inversion ($z \mapsto \frac{1}{z}$) and take into account that $V(z) \rightarrow q_\infty$ as $|z| \to \infty$ (so that $V(\frac{1}{z})$ is holomorphic on the image $\hat{G}$ of G under the inversion). The conclusion a) follows directly.

Now we prove b). Since V is holomorphic, V is open on G ; but 0 is an interior point of $\hat{G}$, and thus $V(\infty) = q_\infty$ is an interior point of $V(G)$. Hence $q_\infty \notin V(\partial P_+)$ and $(0,q_\infty] \subset \sum_+$. Next, consider the domain G_+ ; its boundary is mapped by V onto $V(\partial P_+) \cup [0,q_\infty)$. Therefore every point in $\sum_+$ not lying on $(0,q_\infty)$ has index 1 with respect to $V(\partial G_+)$. Hence every point in $\sum_+ \setminus (0,q_\infty]$ is the image of exactly one point of G_+.

Finally $V(G_+) \subset \sum_+ \setminus (0,q_\infty|$; indeed :

a) $V(G_+) \subset \sum_+$ because V is open,

b) $V(G_+) \cap (0,q_\infty) = \emptyset$, otherwise some point on $(0,q_\infty)$ would have three preimages in G,

c) $V(G_+) \subset \sum_+ \setminus (0,q_\infty]$ since $V(G_+) \subset \sum_+ \setminus (0,q_\infty)$ and $V(G_+)$ is open.

- § 3 - The stream function in the hodograph plane -

We start with some notations. Let $\theta_A \in (0,\frac{\pi}{2}]$ (resp. $\theta_B \in [-\frac{\pi}{2},0)$) be the angle determined by the x-axis and the tangent to ∂P_+ at A (resp.B).

Suppose ∂P_+ is strictly convex and of class C^1 ; then for each $\theta \in (\theta_B,\theta_A)$ there is a unique point P on ∂P_+ where the tangent to ∂P_+ at P makes an angle θ with the x-axis ; we denote the coordinates of P by $[X(\theta),Y(\theta)]$. We assume that

(3.1) $X(\theta)$ and $Y(\theta)$ are $C^{1,\alpha}$ functions of $\theta \in [\theta_B,\theta_A]$.

This is equivalent to the assumption that ∂P_+ is of class $C^{2,\alpha}$ and that the radius of curvature of ∂P_+ is bounded.

We denote by $R(\theta)$ the algebraic radius of curvature of ∂P_+ at $[X(\theta),Y(\theta)]$ i.e.

(3.2) $$R(\theta) = - \sqrt{X'(\theta)^2+Y'(\theta)^2}$$

For $z \in \bar{G}_+$, $z \neq A$ and $z \neq B$, let

(3.3) $$W(z) = -i\,\overline{\mathrm{Log}}\,V(z) = -\mathrm{Arg}\,V(z) - i\log|V(z)|$$

where $-\theta_A < \mathrm{Arg}\,V(z) < -\theta_B$ (by Theorem 2.1). W is well defined (since $V \neq 0$) and W is antiholomorphic in G.

For a point $z = X(\theta) + i\,Y(\theta)$ on ∂P_+ we have

$$V(z) = |V(z)|e^{-i\theta} \qquad (\text{since } V = q_1 - iq_2 \text{ and } \vec{q}.\vec{n} = 0)$$

and thus $W(z) = \theta - i \log|V(z)|$.

We denote by Γ the curve $W(\partial P_+)$, so that Γ can be represented as

$$(3.4) \qquad \Gamma = \{\theta + i\ell(\theta) \;\; ; \;\; \theta_B < \theta < \theta_A\}$$

where $\ell(\theta) = -\log|V(X(\theta)+iY(\theta))|$.

Observe that :

$$(3.5) \qquad \ell(\theta) \text{ is of class } C^{1,\alpha} \text{ on } (\theta_B,\theta_A)$$

and

$$(3.6) \qquad \lim_{\substack{\theta\to\theta_B \\ \text{and} \\ \theta\to\theta_A}} \ell(\theta) = +\infty \; .$$

Indeed ψ is $C^{2,\alpha}$ on $\bar{G}$, except at A and B , since ∂P_+ is $C^{2,\alpha}$. Thus $V = \psi_y + i\psi_x$ is $C^{1,\alpha}$ on $\bar{G}$ except at A and B and $\ell(\theta)$ is $C^{1,\alpha}$ on (θ_B,θ_A) by assumption (3.1).

Property (3.6) follows from the fact that V is continuous on $\bar{G}$ and $V(A) = V(B) = 0$.

Theorem 2.1 implies that the mapping $z \longrightarrow W(z)$ is one to one from G_+ onto the open domain $\mathcal{O}$ defined by

$$\mathcal{O} = \{\theta+i\sigma \; ; \; \sigma>\ell(\theta) \text{ for } \theta\varepsilon(\theta_B,\theta_A) \text{ and } \theta \neq 0\} \cup \{i\sigma \text{ for } \sigma\varepsilon(\ell(0),\sigma_\infty)\}$$

where $\sigma_\infty = -\log q_\infty$.

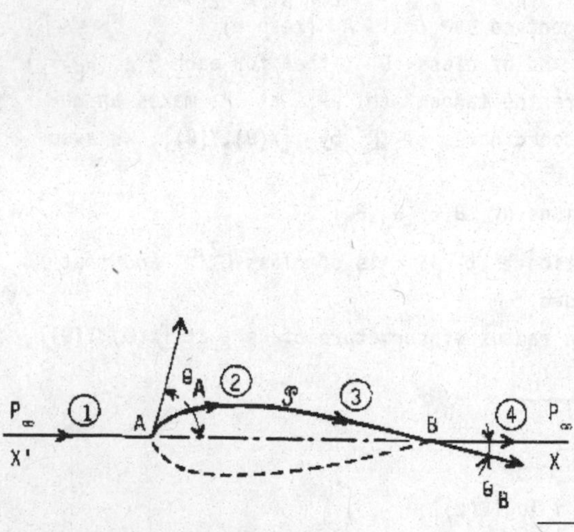

The (x,y) physical plane

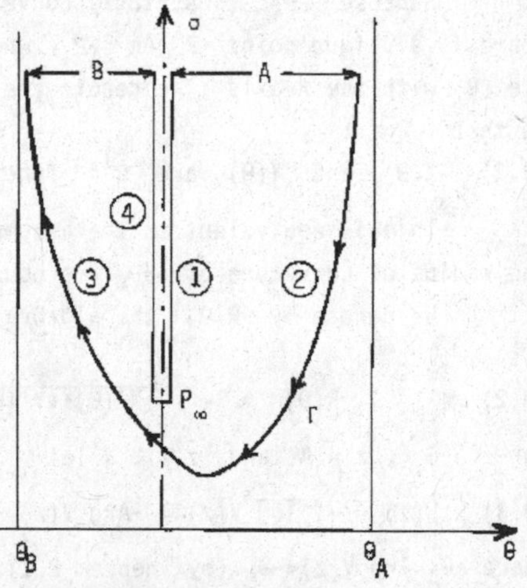

The (θ,σ) hodograph plane

Thus for $\theta + i\sigma \in O$ we can define :

(3.7) $$\tilde{\psi}(\theta,\sigma) = \psi(z)$$

where z is the unique point in G_+ such that $W(z) = \theta + i\sigma$.

Clearly $\tilde{\psi}(\theta,\sigma)$ is harmonic on O (since $\psi(z)$ is harmonic on G_+ and W is antiholomorphic). We extend $\tilde{\psi}$ by setting it to be 0 on $\bar{O}\setminus O$. It is easy to check that $\tilde{\psi}$ *is continuous on* $\bar{O}$ *except at the point* $[0,\sigma_\infty]$. Also $\tilde{\psi}(\theta,\sigma)$ is $C^{1,\alpha}$ up to Γ (since Γ is $C^{1,\alpha}$, $\Delta\tilde{\psi} = 0$ on O and $\tilde{\psi} = 0$ on Γ).

Near every point $[0,\sigma]$, $\sigma > \sigma_\infty$, the restrictions of $\tilde{\psi}(\theta,\sigma)$ to each side $\{\theta \geq 0\}$ and $\{\theta \leq 0\}$ are smooth.

Near the point $[0,\sigma_\infty]$, $\tilde{\psi}(\theta,\sigma)$ has a singularity ; the presence of the singularity "explains" why $\tilde{\psi}$ is not identically zero even though $\Delta\tilde{\psi} = 0$ on O , $\tilde{\psi} = 0$ on ∂O and $\tilde{\psi} \to 0$ at infinity.

Let us describe explicitly the singularity in case $P = \{z \in \mathbb{C} \; ; \; |z| \leq 1\}$ and $q_\infty = 1$. Then the conformal mapping reduces to $f(z) = z$ and thus :

$$\Phi(z) = z + \frac{1}{z}$$

$$\psi(x,y) = y(1 - \frac{1}{x^2+y^2})$$

$$V(z) = 1 - \frac{1}{z^2}$$

$$W(z) = -i\,\overline{\mathrm{Log}(1 - \frac{1}{z^2})} = \theta + i\sigma \quad .$$

Hence $1 - \frac{1}{z^2} = e^{-\sigma-i\theta}$ and in particular for $\theta = 0$ and $\sigma < 0$ we have $z = \frac{1}{\sqrt{e^{-\sigma}-1}}$ i.e. $x = 0$ and $y = \frac{1}{\sqrt{e^{-\sigma}-1}}$.

Therefore $\tilde{\psi}(0,\sigma) = \frac{1}{\sqrt{e^{-\sigma}-1}}(2-e^{-\sigma})$ and we see that $\tilde{\psi}(0,\sigma) \to +\infty$ as $\sigma \to \sigma_\infty = 0$.

The following result, which shows that $\tilde{\psi}(\theta,\sigma)$ decays very fast as $\sigma \to +\infty$, will be used later.

Proposition 3.1.- *There exists a constant* K *such that*

(3.8) $\tilde{\psi}(\theta,\sigma) \leq Ke^{-2\sigma}$ *for every* $[\theta,\sigma] \in O$ *with* $\sigma > \sigma_\infty + 1$.

Proof

Consider the open domain O' determined by

$O' = \{[\theta,\sigma]$ with $\sigma > \mathrm{Max}\{\ell(\theta),\sigma_\infty+1\}$ and $0 < \theta < \theta_A\}$.

We have :

$$\Delta(\tilde{\psi}(\theta,\sigma) - Ke^{-2\sigma}\sin 2\theta) = 0 \text{ on } O' .$$

On the other hand $\tilde{\psi}(\theta,\sigma) - Ke^{-2\sigma}\sin 2\theta \to 0$ as $\sigma \to +\infty$ (since the preimage z in G_+ of $\theta + i\sigma$ by W tends to A as $\sigma \to +\infty$) .

Next we have $\tilde{\psi}(\theta,\sigma) - Ke^{-2\sigma}\sin 2\theta \leq 0$ for $\sigma = \ell(\theta)$ and $\sigma > \sigma_\infty + 1$ (since

$\tilde{\psi}(\theta,\ell(\theta)) = 0$ and $\sin 2\theta \geq 0$).

Finally it is possible to choose K in such a way that :

(3.9) $\tilde{\psi}(\theta,\sigma) - Ke^{-2\sigma} \sin 2\theta \leq 0$ for $\sigma = \sigma_\infty + 1$ and $\ell(\theta) \leq \sigma_\infty + 1$.

Indeed we determine first a $\delta > 0$ so that

(3.10) $\ell(\theta) < \sigma_\infty + 1$ for $0 < \theta < \delta$

(3.11) $\ell(\theta) > \sigma_\infty + 1$ for $\theta_A - \delta < \theta < \theta_A$

(this is possible since $\ell(0) < \sigma_\infty$ and $\ell(\theta) \to +\infty$ as $\theta \to \theta_A$) .

By the smoothness of $\tilde{\psi}(\theta,\sigma_\infty+1)$ on $[0,\delta]$ and since $\tilde{\psi}(0,\sigma_\infty+1) = 0$, then is a K_1 such that :

(3.12) $\tilde{\psi}(\theta,\sigma_\infty+1) \leq K_1 e^{-2(\sigma_\infty+1)} \sin 2\theta$ for $\theta \in [0,\delta]$.

Next we choose a K_2 such that :

(3.13) $\tilde{\psi}(\theta,\sigma_\infty+1) \leq K_2 e^{-2(\sigma_\infty+1)} \sin 2\theta$ for $\theta \in [\delta, \theta_A - \delta]$.

Therefore (3.9) holds with $K = \mathrm{Max}\{K_1,K_2\}$. We conclude by the maximum principle that :

$$\tilde{\psi}(\theta,\sigma) - Ke^{-2\sigma} \sin 2\theta \leq 0 \quad \text{on } \mathcal{O}' .$$

The same argument is valid for $\theta \leq 0$.

- § 4 - The variational inequality in the hodograph plane -

We assume here as in § 3 that $X(\theta)$ and $Y(\theta)$ are $C^{1,\alpha}$.

For $[\theta,\sigma] \in \mathcal{O}$ we define :

(4.1) $$U(\theta,\sigma) = e^{-\sigma} \int_{\ell(\theta)}^{\sigma} e^{s} \tilde{\psi}(\theta,s)ds .$$

Since $\sigma > \ell(\theta)$ and $\tilde{\psi} > 0$ on $\mathcal{O}$ (by Proposition 1.1) we have :

(4.2) $U(\theta,\sigma) > 0$ on $\mathcal{O}$.

Next, observe that U is of class $C^{1,\alpha}$ in $\mathcal{O}$ and even up to Γ (since $\ell(\theta)$ is $C^{1,\alpha}$ and $\tilde{\psi}$ is $C^{1,\alpha}$ up to Γ).

We have :

(4.3) $U_\sigma = \tilde{\psi} - U$ on $\mathcal{O}$

(4.4) $U_\theta = e^{-\sigma} \int_{\ell(\theta)}^{\sigma} e^{s} \tilde{\psi}_\theta(\theta,s)ds$ on $\mathcal{O}$ (since $\tilde{\psi}(\theta,\ell(\theta))=0$)

(4.5) $U = U_\sigma = U_\theta = 0$ on Γ .

The following result provides a very useful expression for U.

<u>Theorem 4.1 -</u> *Let* $z = x + iy \in G_+$ *and let* $W(z) = \theta + i\sigma \in O$. *Then :*

$$(4.6)\quad U(\theta,\sigma) = \tilde{\psi}(\theta,\sigma) + e^{-\sigma}(x \sin\theta - y \cos\theta) - e^{-\sigma}(X(\theta)\sin\theta - Y(\theta)\cos\theta)$$

i.e.

$$(4.7)\quad U(\theta,\sigma) = \mathrm{Im}(\Phi(z) - z\Phi'(z)) - e^{-\alpha}(X(\theta)\sin\theta - Y(\theta)\cos\theta) .$$

<u>Proof</u> For a fixed $\theta \in (\theta_B,\theta_A)$ we denote by $a(s) + ib(s)$ the unique point in $\bar{G}_+$ such that :

$$(4.8)\quad W(a(s)+ib(s)) = \theta - is \quad \text{for} \quad s \geq \ell(\theta)$$

(and $\ell(0) \leq s < \sigma_\infty$ when $\theta = 0$) .

In other words we have :

$$(4.9)\quad e^{-s-i\theta} = V(a(s)+ib(s)) = \psi_y(a(s),b(s)) + i\psi_x(a(s),b(s)) .$$

Thus :

$$(4.10)\quad \psi_y(a(s),b(s)) = e^{-s} \cos\theta \quad , \quad \psi_x(a(s),b(s)) = -e^{-s}\sin\theta .$$

From the definition of $\tilde{\psi}$ we have :

$$(4.11)\quad U(\theta,\sigma) = e^{-\sigma} \int_{\ell(\theta)}^{\sigma} e^s \, \psi(a(s),b(s))ds =$$

$$= e^{-\sigma} \int_{\ell(\theta)}^{\sigma} \psi(a(s),b(s))d(e^s) =$$

$$= \psi(a(\sigma),b(\sigma)) - e^{-\sigma} \int_{\ell(\theta)}^{\sigma} |\psi_x(a(s),b(s))a'(s) + \psi_y(a(s),b(s)b'(s)| e^s ds$$

$$= \tilde{\psi}(\theta,\sigma) + e^{-\sigma} \int_{\ell(\theta)}^{\sigma} |\sin\theta \, a'(s) - \cos\theta \, b'(s)| ds \qquad \text{(by (4.10))} .$$

Therefore :

$$(4.12)\quad U(\theta,\sigma) = \tilde{\psi}(\theta,\sigma) + e^{-\sigma}\sin\theta[a(\sigma) - a(\ell(\theta))] - e^{-\sigma}\cos\theta[b(\sigma) - b(\ell(\theta))]$$

and (4.6) follows.

Finally observe that :

$$\Phi(z) - z\Phi'(z) = (\phi + i\psi) - (x+iy)(\psi_y + i\psi_x)$$

and hence

$$\mathrm{Im}[\Phi(z) - z\Phi'(z)] = \psi - x\psi_x - y\psi_y .$$

We deduce from (4.10) that (4.7) holds.

<u>Theorem 4.2 -</u> *The function* $U(\theta,\sigma)$ *is of class* $C^{2,\alpha}$ *on* O *up to* Γ *and satisfies :*

$$(4.13)\quad \frac{\partial^2 U}{\partial\theta^2} + \frac{\partial^2 U}{\partial\sigma^2} = \Delta U = -e^{-\sigma}R(\theta) \quad \text{on } O .$$

Proof The mapping $\theta + i\sigma \longmapsto z$ is antiholomorphic on $\mathcal{O}$ and the mapping $z \longmapsto \Phi(z) - i\Phi'(z)$ is holomorphic on G_+. It follows from (4.7) that $U(\theta,\sigma)$ has on $\mathcal{O}$ the same regularity as

$$\chi(\theta) = X(\theta)\sin\theta - Y(\theta)\cos\theta$$

and that

$$\Delta U = -\Delta(e^{-\sigma}\chi(\theta)) \quad .$$

However $\chi(\theta)$ is $C^{2,\alpha}$; indeed

$$\chi'(\theta) = X'(\theta)\sin\theta - Y'(\theta)\cos\theta + X(\theta)\cos\theta + Y(\theta)\sin\theta$$

$$= X(\theta)\cos\theta + Y(\theta)\sin\theta \quad (\text{since } \frac{Y'(\theta)}{X'(\theta)} = \text{tg}\theta) \quad .$$

Thus $\chi'(\theta)$ is $C^{1,\alpha}$ and

$$\chi''(\theta) = X'(\theta)\cos\theta + Y'(\theta)\sin\theta - \chi(\theta) = R(\theta) - \chi(\theta) \quad .$$

Therefore

$$\Delta U = -\Delta(e^{-\sigma}\chi(\theta)) = -e^{-\sigma}\chi''(\theta) - e^{-\sigma}\chi(\theta) = -e^{-\sigma}R(\theta) \quad .$$

Finally observe that from (4.3) U_σ is $C^{1,\alpha}$ on $\mathcal{O}$ up to Γ. Also, by (4.13), $U_{\theta\theta}$ is $C^{0,\alpha}$ on $\mathcal{O}$ up to Γ. Consequently U is $C^{2,\alpha}$ on $\mathcal{O}$ up to Γ.

Theorem 4.3 - *The function* $U(\theta,\sigma)$ *is continuous on* $\bar{\mathcal{O}}$ *provided we define*

(4.14) $$U(0,\sigma) = He^{-\sigma} \quad \text{for} \quad \sigma \geq \sigma_\infty$$

where $H = Y(0)$ *represents the height of* ∂P_+.

Proof As $[\theta,\sigma] \to [0,\sigma_\infty]$, the corresponding $z \in G_+$ satisfies $|z| \to \infty$. Now, recall the expansions

$$\Phi(z) = q_\infty z - \sum_{n=1}^{\infty} \frac{a_{n+1}}{nz^n}$$

$$z\Phi'(z) = q_\infty z + \sum_{n=1}^{\infty} \frac{a_{n+1}}{z^n} \quad .$$

Therefore $|\Phi(z) - z\Phi'(z)| \to 0$ as $|z| \to \infty$, and from (4.7) we conclude that

$$U(\theta,\sigma) \to Y(0)e^{-\sigma_\infty} = He^{-\sigma_\infty} \quad \text{as} \quad (\theta,\sigma) \to (0,\sigma_\infty) \quad .$$

Consider now a point $[0,\sigma_1]$ with $\sigma_1 > \sigma_\infty$; it has two preimages $[x_1,0]$ and $[x_2,0]$ in G.

When $[\theta,\sigma]$ lies in a neighborhood of $[0,\sigma_1]$, the corresponding $[x,y]$ lies either close to $[x_1,0]$ or close to $[x_2,0]$. In any case we see from (4.6) that $U(\theta,\sigma) \to Y(0)e^{-\sigma_1}$ as $[\theta,\sigma] \to [0,\sigma_1]$.

Lemma 4.4 - *As* $\sigma \uparrow +\infty$, $e^{\sigma}U(\theta,\sigma) \uparrow w(\theta)$ *where*

$$W(\theta) = \begin{cases} (x_A - X(\theta))\sin\theta + Y(\theta)\cos\theta & \text{for} \quad \theta > 0 \\ (x_B - X(\theta))\sin\theta + Y(\theta)\cos\theta & \text{for} \quad \theta < 0 \end{cases}$$

(x_A *and* x_B *denote the* x*-coordinates of* A *and* B) .

In particular $U(\theta,\sigma) \leq He^{-\sigma}$ *on* O .

Proof It follows from Proposition 3.1 that $e^{\sigma}\tilde{\psi}(\theta,\sigma) \to 0$ as $\sigma \to +\infty$. On the other hand it is clear that for z in some neighborhood of A (resp.B) in G_+ then $\theta = \text{Re}W(z) \geq 0$) (resp. $\theta \leq 0$).

Therefore for a fixed $\theta > 0$, the preimage z of $[\theta,\sigma]$ converges to A as $\sigma \to +\infty$. We conclude from (4.6) that $U(\theta,\sigma) \to (x_A - X(\theta))\sin\theta + Y(\theta)\cos\theta$ as $\sigma \to +\infty$. Observe that by (4.1), $e^{\sigma}U(\theta,\sigma)$ is increasing in σ . A similar argument holds for $\theta < 0$.

Finally, for all θ , we have

$$w(\theta) \leq Y(\theta)\cos\theta \leq H\cos\theta \leq H .$$

Remark - The function $w(\theta)$ has a very simple geometrical interpretation. For $\theta > 0$ (resp. $\theta < 0$) it is easy to check that $w(\theta)$ represents the distance between A (resp. B) and the tangent to ∂P_+ at $[X(\theta), Y(\theta)]$. Also it is clear that for $\theta>0$ and $\theta < 0$, $w''(\theta) + w(\theta) = -R(\theta)$ and $w(\theta_A) = w(\theta_B) = w'(\theta_A) = w'(\theta_B) = 0$.

Pick any $m < \min_{\theta_B<\theta<\theta_A} \ell(\theta)$ and let $\Omega = (\theta_B,\theta_A) \times (m,+\infty)$.

We extend U to Ω by choosing $U(\theta,\sigma) = 0$ on $\Omega\setminus\bar{O}$.

Lemma 4.5 - *The function* U *lies in* $H_0^1(\Omega)$.

Proof We already know (Theorem 4.2) that U is of class $C^{2,\alpha}$ on O up to Γ. Also the restrictions of U on each side $\{\theta < 0\}$ and $\{\theta > 0\}$ are $C^{2,\alpha}$ near every point $[0,\sigma_1]$ with $\sigma_1 > \sigma_\infty$. Thus it is only necessary to check that $U \in H^1(\Omega)$ near $\sigma = +\infty$ and near the singular point $[0,\sigma_\infty]$.

Let $\xi_0(\sigma)$ be a smooth function in R such that $0 \leq \xi_0 \leq 1$ and $\xi_0(\sigma) = 1$ for $\sigma \leq 0$, $\xi_0(\sigma) = 0$ for $\sigma \geq 1$; let $\xi_n(\sigma) = \xi_0(\sigma-n)$.

Let $\eta_0(\theta,\sigma)$ be a smooth function on $\mathbb{R}^2$ such that $0 \leq \eta_0 \leq 1$ and $\eta_0(\theta,\sigma)=1$ for $|\theta| + |\sigma-\sigma_\infty| > 1$, $\eta_0(\theta,\sigma) = 0$ for $|\theta| + |\sigma-\sigma_\infty| < \frac{1}{2}$; let $\eta_n(\theta,\sigma) = \eta_0(n\theta,n(\sigma-\sigma_\infty))$.

Finally define $\zeta_n(\theta,\sigma) = \xi_n(\sigma)\eta_n(\theta,\sigma)$.

We have, after integration by parts

$$\int_O |\text{grad}U|^2\zeta_n d\theta d\sigma = -\int_O \Delta U.U\zeta_n d\theta d\sigma - \int_O \text{grad}U.\text{grad}\zeta_n U d\theta d\sigma \tag{4.15}$$

On the other hand we get :

(4.16) $$\int_{O} \text{grad}U.\text{grad}\zeta_n U d\theta d\sigma = \frac{1}{2}\int_{O} \text{grad}U^2.\text{grad}\zeta_n d\theta d\sigma = -\frac{1}{2}\int_{O} U^2\Delta\zeta_n d\theta d\sigma \quad .$$

Combining (4.13), (4.15) and (4.16) we obtain :

(4.17) $$\int_{O} |\text{grad}U|^2\zeta_n d\theta d\sigma = \int_{O} R(\theta)e^{-\sigma}U\zeta_n d\theta d\sigma + \frac{1}{2}\int_{O} U^2\Delta\zeta_n d\theta d\sigma$$

$$\leq \frac{1}{2}H^2\int_{O} e^{-2\sigma}|\Delta\zeta_n| d\theta d\sigma \leq \frac{1}{2}H^2 e^{-2m}\int_{O} |\Delta\zeta_n| d\theta d\sigma$$

since $R(\theta) \leq 0$ and $0 \leq U(\theta,\sigma) \leq He^{-\sigma}$.

But $\int_{O} |\Delta\zeta_n| d\theta d\sigma$ remains bounded as $n\to+\infty$; indeed we have for n large enough

$$\Delta\zeta_n = n^2\Delta\eta_0(n\theta,n(\sigma-\sigma_\infty)) + \xi_0''(\sigma-n)$$

and therefore

$$\int_{O} |\Delta\zeta_n| d\theta d\sigma \leq n^2 \int_{\substack{O \\ \frac{1}{2n}<|\theta|+|\sigma-\sigma_\infty|<\frac{1}{n}}} |\Delta\zeta_0(n\theta,n(\sigma-\sigma_\infty))| d\theta d\sigma + \int_{\substack{O \\ n<\sigma<n+1}} |\xi_0''(\sigma-n)| d\theta d\sigma$$

$$\leq \pi \,||\Delta\zeta_0||_{L^\infty} + \pi||\xi_0''||_{L^\infty} \quad .$$

We are now ready to prove the main result.

Theorem 4.6 - *Let*

$\mathbb{K} = \{V(\theta,\sigma)\in H_0^1(\Omega)\ ;\ V\geq 0 \text{ in } \Omega \text{ and } V(0,\sigma)=He^{-\sigma} \text{ for } \sigma>\sigma_\infty\}$.

Then U *defined by* (4.1) *satisfies*

(4.18) $$\begin{cases} U \in \mathbb{K} \\ \int_\Omega\left[\frac{\partial U}{\partial\theta}\frac{\partial}{\partial\theta}(V-U) + \frac{\partial U}{\partial\sigma}\frac{\partial}{\partial\sigma}(V-U)\right]d\theta d\sigma \geq \int_\Omega R(\theta)e^{-\sigma}(V-U)d\theta d\sigma \quad , \ \forall\ V\in\mathbb{K} \end{cases}$$

i.e.

$$\underset{\mathbb{K}}{\text{Min}} \int_\Omega\left[|\frac{\partial V}{\partial\theta}|^2 + |\frac{\partial V}{\partial\sigma}|^2 - 2R(\theta)e^{-\sigma}V\right]d\theta d\sigma \ \textit{is achieved at}\ U\in\mathbb{K} \ .$$

Proof By Theorem 4.3 and Lemma 4.5 we already know that $U\in\mathbb{K}$. It is sufficient to show that (4.18) holds for all $V \in \mathbb{K}\cap L^\infty(\Omega)$. The general case then follows, for example, by truncation.

We have :

$$\int_\Omega \text{gradU.grad}(V-U)\zeta_n d\theta d\sigma = \int_{\mathcal{O}} \text{gradU.grad}(V-U).\zeta_n d\theta d\sigma =$$

$$= -\int_{\mathcal{O}} \Delta U(V-U)\zeta_n d\theta d\sigma - \int_{\mathcal{O}} \text{gradU.grad}\zeta_n (V-U) d\theta d\sigma \quad .$$

But from (4.13) we get :

$$-\int_{\mathcal{O}} \Delta U(V-U)\zeta_n d\theta d\sigma = \int_{\mathcal{O}} R(\theta)e^{-\sigma}(V-U)\zeta_n d\theta d\sigma \geq \int_\Omega R(\theta)e^{-\sigma}(V-U)\zeta_n d\theta d\sigma$$

since $R(\theta) < 0$, $U = 0$ on $\Omega\setminus\bar{\mathcal{O}}$ and $V \geq 0$ on Ω .

Therefore it remains only to show that :

(4.9) $$\int_{\mathcal{O}} \text{gradU.grad}\zeta_n (V-U) d\theta d\sigma \to 0 \quad \text{as} \quad n \to +\infty \quad .$$

However for n large enough we have $\text{grad}\zeta_n = \text{grad}\xi_n + \text{grad}\eta_n$.
Consider first :

$$\left|\int_{\mathcal{O}} \text{gradU.grad}\xi_n (V-U) d\theta d\sigma\right| = \left|\int_{n\leq\sigma\leq n+1} \text{gradU.grad}\xi_n (V-U) d\theta d\sigma\right|$$

$$\leq ||V-U||_{L^\infty} \left[\int_{n\leq\sigma\leq n+1} |\text{gradU}|^2 d\theta d\sigma\right]^{1/2} \sqrt{\pi}\,||\xi_0'||_{L^\infty} \to 0$$

as $n \to +\infty$ since $\text{gradU} \in L^2(\Omega)$.

Next,

$$\left|\int_{\mathcal{O}} \text{gradU.grad}\eta_n (V-U) d\theta d\sigma\right| = \left|\int_{|\theta|+|\sigma-\sigma_\infty|<\frac{1}{n}} \text{gradU.grad}\eta_n (V-U) d\theta d\sigma\right|$$

$$\leq ||V-U||_{L^\infty} \left[\int_{|\theta|+|\sigma-\sigma_\infty|<\frac{1}{n}} |\text{gradU}|^2 d\theta d\sigma\right]^{1/2} n\,||\text{grad}\eta_0||_{L^\infty} \frac{\sqrt{\pi}}{n} \to 0$$

as $n \to +\infty$ since $\text{gradU} \in L^2(\Omega)$.

This concludes the proof of (4.19) and Theorem 4.6 follows.

Conclusion - Consider now the variational inequality (4.18). Problem (4.18) is well formulated once Ω , σ_∞ , H and $R(\theta)$ are known. They depend only on properties of P (through θ_A, θ_B, H and $R(\theta)$), of q_∞ (since $\sigma_\infty = -\log q_\infty$) and of m (which enters in the definition of $\Omega = (\theta_B,\theta_A) \times (m,+\infty)$). Therefore, once we know P , q_∞ and any upper bound for $\underset{\partial P_+}{\text{Max}}|\vec{q}|$ we are able to state directly the variational inequality

(4.18) and get a unique solution U by standard results (see e.g.{8}).[1] Finally the curve which separates the two sets $\{U > 0\}$ and $\{U = 0\}$ provides the free boundary Γ and describes ipso facto the distribution of velocities along ∂P.

Remarks

1) A direct computation (see {6} p.46) shows that, along Γ, $\tilde{\psi}$ satisfies :

$$\tilde{\psi}(\theta,\ell(\theta)) = 0$$

$$\frac{\partial\tilde{\psi}}{\partial\sigma} = -\frac{R(\theta)e^{-\sigma}}{1+\left(\frac{d\ell}{d\theta}\right)^2} \quad , \qquad \frac{\partial\tilde{\psi}}{\partial\theta} = \frac{R(\theta)e^{-\sigma}\,\frac{d\ell}{d\theta}}{1+\left(\frac{d\ell}{d\theta}\right)^2}$$

In this form we get a free boundary value problem which is similar to the problem considered by Baiocchi {1}. In order to adapt his method to our case we were led to introduce U defined by (4.1).

2) A result similar to Theorem 4.6 can be obtained for some flows with cavitations (see {3}).

3) It turns out that U is very closely related to the Legendre transform of the stream function. More precisely the Legendre transform $\Psi(\theta,\sigma)$ is defined (see {5} p.249) by :

$$\Psi(\theta,\sigma) = e^{-\sigma}(y\cos\theta - x\sin\theta) - \tilde{\psi}(\theta,\sigma) \quad .$$

Therefore we deduce from (4.6) that :

$$U(\theta,\sigma) = -\Psi(\theta,\sigma) - e^{-\sigma}(X(\theta)\sin\theta - Y(\theta)\cos\theta) \quad .$$

Consequently Theorem 4.6 could be reformulated by saying that Ψ is the solution of some variational inequality (associated with the constraint $e^{-\sigma}(X(\theta)\sin\theta -$ $- Y(\theta)\cos\theta)$.

- § 5 - Some hints for numerical computations -

In order to compute a flow by the method described in Theorem 4.6 we have to get some upper bound for $\underset{\partial P_+}{\mathrm{Max}}|\vec{q}|$. Our next result provides such a bound.

Theorem 5.1 - *Suppose* $r = \underset{\theta_B \le \theta \le \theta_A}{\mathrm{Min}} |R(\theta)| > 0$; *then we have*

(5.1) $$\underset{\partial P_+}{\mathrm{Max}}|\vec{q}| \le q_\infty\, e^{-\sigma_0}$$

where $\sigma_0 < 0$ *is the unique solution of* $(1+\sigma_0)e^{-\sigma_0} = 1 - \frac{H}{r}$.

(1) Observe that coerciveness follows from Poincare's inequality since (θ_B,θ_A) is a bounded interval.

<u>Proof</u> Since m is any number satisfying $m < \operatorname{Min} \ell(\theta)$, we can always assume that $m < \sigma_\infty + \sigma_0$.

Define the function $h(\sigma)$ to be

$$h(\sigma) = \begin{cases} re^{-\sigma} + re^{-\sigma_\infty - \sigma_0}(\sigma - \sigma_\infty) + He^{-\sigma_\infty} - re^{-\sigma_\infty} & \text{for } \sigma \geq \sigma_\infty + \sigma_0 \\ 0 & \text{for } \sigma < \sigma_\infty + \sigma_0 \end{cases}$$

It follows from the definition of σ_0 that

$$h(\sigma_\infty) = He^{-\sigma_\infty}$$

$$h(\sigma_\infty + \sigma_0) = 0 \quad , \quad h'(\sigma_\infty + \sigma_0) = 0 \quad .$$

Also we have :

$$h''(\sigma) = r\, e^{-\sigma} \qquad \text{for} \quad \sigma > \sigma_\infty + \sigma_0$$

$$h(\sigma) \geq 0 \qquad \text{for} \quad \sigma > \sigma_\infty + \sigma_0$$

$$h(\sigma) \geq He^{-\sigma} \qquad \text{for} \quad \sigma \geq \sigma_\infty \quad ,$$

so that in particular :

$$h(\sigma) \geq U(\theta,\sigma) \qquad \text{for} \quad \sigma \geq \sigma_\infty \quad .$$

Clearly $V = \operatorname{Min}\{U,h\} \in \mathbb{K}$ and thus

$$a(U,V-U) \geq \int_\Omega R(\theta)e^{-\sigma}(V-U)\,d\theta d\sigma$$

where $\quad a(U_1,U_2) = \int_\Omega \operatorname{grad} U_1 . \operatorname{grad} U_2 \; d\theta d\sigma \quad .$

But $V-U = -(U-h)^+$ and thus

$$-a(U,(U-h)^+) \geq - \int_{\sigma<\sigma_\infty} R(\theta)e^{-\sigma}(U-h)^+ d\theta d\sigma \quad .$$

Therefore

$$a(U-h,(U-h)^+) \leq -a(h,(U-h)^+) + \int_{\sigma<\sigma_\infty} R(\theta)e^{-\sigma}(U-h)^+ d\theta d\sigma \quad .$$

On the other hand

$$a(h,(U-h)^+) = \int_{\sigma_\infty+\sigma_0<\sigma<\sigma_\infty} \frac{dh}{d\sigma}\,\frac{\partial}{\partial\sigma}\,(U-h)^+ d\theta d\sigma$$

$$= - \int_{\sigma_\infty+\sigma_0<\sigma<\sigma_\infty} \frac{d^2h}{d\sigma^2}\,(U-h)^+ d\theta d\sigma = - \int_{\sigma_\infty+\sigma_0<\sigma<\sigma_\infty} re^{-\sigma}(U-h)^+ d\theta d\sigma \quad .$$

Finally we obtain :

$$a(U-h),(U-h)^+) \leq \int_{\sigma_\infty+\sigma_0<\sigma<\sigma_\infty} (R(\theta)+r)e^{-\sigma}(U-h)^+ d\theta d\sigma + \int_{\sigma<\sigma_\infty+\sigma_0} R(\theta)e^{-\sigma}(U-h)^+ d\theta d\sigma$$

and hence $a(U-h,(U-h)^+) \leq 0$. Consequently $U \leq h$ and in particular $U = 0$ for $\sigma < \sigma_\infty + \sigma_0$.

We conclude that $\min_{\theta_B<\theta<\theta_A} \ell(\theta) \geq \sigma_\infty + \sigma_0$ and Theorem 5.1 follows.

Remark When P is a disk of radius r , we have $(1+\sigma_0)e^{-\sigma_0} = 0$ and so $\sigma_0 = -1$. We deduce from Theorem 5.1 that $\max_{\partial P_+}|\vec{q}| \leq eq_\infty$. Actually, a direct computation (see § 3) in this case leads to $\max_{\partial P_+}|\vec{q}| = 2q_\infty$.

For the purpose of numerical computations it is necessary to replace Ω by a bounded domain

$$\Omega_T = (\theta_B,\theta_A) \times (m,T) .$$

Therefore the question arises : what boundary condition should be imposed on the upper part of the boundary $(\theta_B,\theta_A) \times \{T\}$? By imposing a Dirichlet condition and a Neumann condition we get two different solutions.

Our last result shows that U lies in between the two.

Let $$a_T(U,V) = \int_{\Omega_T} \text{grad}U.\text{grad}V \, d\theta d\sigma ,$$

$$\mathbb{K}_T^1 = \left\{ \begin{array}{l} V \in H^1(\Omega_T) \; ; \; V \geq 0 \text{ on } \Omega_T \; , \; V(0,\sigma) = He^{-\sigma} \text{ for } \sigma_\infty<\sigma<T \\ V = 0 \text{ on } \{\theta_A\}\times(m,T)\cup\{\theta_B\}\times(m,T)\cup(\theta_B,\theta_A)\times\{m\} \end{array} \right\}$$

$$K_T^2 = \{V \in K_T^1 \; ; \; V(\theta,T) = w(\theta)e^{-T} \text{ for } \theta_B<\theta<\theta_A\}$$

Let $U^1 \in \mathbb{K}_T^1$ be the solution of

$$(5.2) \qquad a_T(U^1,V-U^1) \geq \int_{\Omega_T} R(\theta)e^{-\sigma}(V-U^1)d\theta d\sigma - \int_{\theta_B}^{\theta_A} He^{-T}(V-U^1)_{|\sigma=T} \, d\theta$$

for all $V \in \mathbb{K}_T^1$

and let $U^2 \in K_T^2$ be the solution of

$$(5.3) \qquad a_T(U^2,V-U^2) \geq \int_{\Omega_T} R(\theta)e^{-\sigma}(V-U^2)d\theta d\sigma \quad \text{for all } V \in \mathbb{K}_T^2 .$$

Theorem 5.2 - *On* Ω_T *we have* $U^1 \leq U \leq U^2$.

Proof Clearly $V = \min\{U^1,U\} \in \mathbb{K}_T^1$, and since $V - U^1 = -(U^1-U)^+$ we get from (5.2):

$$(5.4) \qquad a_T(U^1,(U^1-U)^+) \leq \int_{\Omega_T} R(\theta)e^{-\sigma}(U^1-U)^+ d\theta d\sigma - \int_{\theta_B}^{\theta_A} He^{-T}(U^1-U)^+_{|\sigma=T} \, d\theta .$$

On the other hand, it is easy to show, as in the proof of Theorem 4.6, that for all $V \in H^1(\Omega_T)$

(5.5) $$\int_{\Omega_T} \mathrm{grad}U.\mathrm{grad}V \, d\theta d\sigma = \int_{\theta_B}^{\theta_A} \frac{\partial U}{\partial \sigma} V_{|\sigma=T} d\theta + \int_{\Omega_T \cap \mathcal{O}} R(\theta)e^{-\sigma} V d\theta d\sigma \quad .$$

In particular :

$$a_T(U,(U^1-U)^+) = \int_{\Omega_T \cap \mathcal{O}} R(\theta)e^{-\sigma}(U^1-U)^+ d\theta d\sigma + \int_{\theta}^{\theta_A} \frac{\partial U}{\partial \sigma}(U^1-U)^+_{|\sigma=T} d\theta \quad .$$

However since the function $\sigma \mapsto e^{\sigma}U(\theta,\sigma)$ is nondecreasing we have $U + \frac{\partial U}{\partial \sigma} \geq 0$.

Therefore :

(5.6) $$a_T(U,(U^1-U)^+) \geq \int_{\Omega_T} R(\theta)e^{-\sigma}(U^1-U)^+ d\theta d\sigma - \int_{\theta_B}^{\theta_A} U(U^1-U)^+_{|\sigma=T} d\theta \quad .$$

Combining (5.4) and (5.6) we get :

$$a_T(U^1-U,(U^1-U)^+) \leq \int_{\theta_B}^{\theta_A} (U-He^{-T})(U^1-U)^+_{|\sigma=T} d\theta \leq 0 \quad .$$

Consequently $U^1 \leq U$.

Similarly $V = \mathrm{Max}\{U^2,U\} \in \mathbb{K}_T^2$ and since $V-U^2 = (U-U^2)^+$ we get from (5.3) :

(5.7) $$a_T(U^2,(U-U^2)^+) \geq \int_{\Omega_T} R(\theta)e^{-\sigma}(U-U^2)^+ d\theta d\sigma \quad .$$

On the other hand we have by (5.5) :

$$a_T(U,(U-U^2)^+) = \int_{\Omega_T \cap \mathcal{O}} R(\theta)e^{-\sigma}(U-U^2)^+ d\theta d\sigma$$

$$\geq \int_{\Omega_T} R(\theta)e^{-\sigma}(U-U)^2)^+ d\theta d\sigma \quad .$$

Hence

$$a_T(U-U^2),(U-U^2)^+) \leq 0 \qquad \text{and consequently} \quad U \leq U^2 \quad .$$

<u>Remark</u> One can also consider the variational inequality

$\tilde{U}^1 \in \mathbb{K}_T^1$

$$a_T(\tilde{U}^1,V-U^1) + \int_{\theta_B}^{\theta_A} \tilde{U}^1(V-\tilde{U}^1)_{|\sigma=T} d\theta \geq \int_{\Omega_T} R(\theta)e^{-\sigma}(V-\tilde{U}^1) d\theta d\sigma$$

for all $V \in \mathbb{K}_T^1$

and show, as in the proof of Theorem 5.2, that $\tilde{U}^1 \leq U$.

- Appendix - The smoothness of ψ near the edges of P -

We assume here that ∂P_+ is of class $C^{1,\alpha}$.

Lemma *Suppose* ψ *is a continuous function on* $\bar{G}_+$ *such that* $\Delta\psi = 0$ *on* $\bar{G}_+$,

$\psi = 0$ *on* ∂P_+ *and* $\psi = 0$ *on the x-axis.*

Then $\psi \in C^{1,\beta}(\bar{G}_+)$ *for some* $0 < \beta < 1$.

Proof We shall use the following well known result (Korn-Lichtenstein-Schauder). Suppose ψ is harmonic on a domain O ; let $P \in \partial O$ and let ω be a neighborhood of P such that $\partial O \cap \omega$ is of class $C^{1,\alpha}$. Suppose ψ is continuous on $\bar{O} \cap \omega$ and $\psi = 0$ on $\partial O \cap \omega$. Then ψ lies in $C^{1,\alpha}(\bar{O} \cap \omega)$.

In order to prove the lemma, all we have to show is that ψ lies in $C^{1,\beta}(\bar{G}_+)$ near the edges A and B . We choose for example B and shift the origin to B.

Consider the mapping $z \mapsto F(z) = z^k$ where $k = \frac{\pi}{\pi - |\theta_B|}$ and $z^k = |z|^k e^{ik\theta}$ with $z = e^{i\theta}$, $0 \leq \theta \leq \pi$; since $0 < |\theta_B| \leq \frac{\pi}{2}$ we have $1 < k \leq 2$.

Let $C = F(\partial P_+)$: since F is holomorphic, the function $\eta(\zeta) = \psi(F^{-1}(\zeta))$ is harmonic in the open domain above $C \cup [0,+\infty)$, continuous on the closure and vanishes on the boundary.

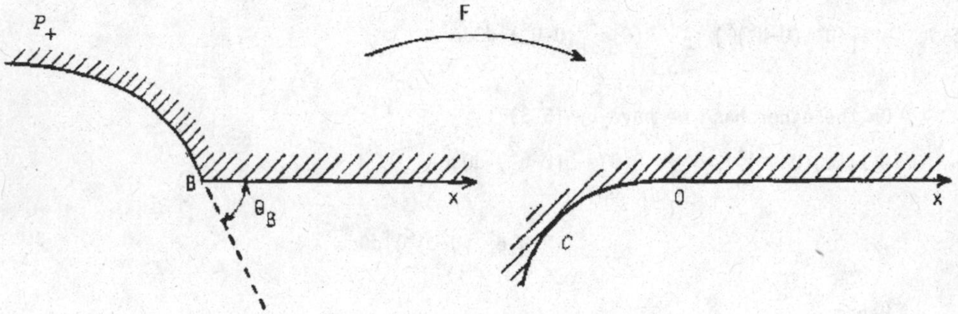

If we can show that $C \cup [0,+\infty)$ is $C^{1,\gamma}$ for some $0 < \gamma < 1$, we will be able to conclude that η is $C^{1,\gamma}$ (by the above mentioned result) and therefore $\psi(z) = \eta(F(z))$ will lie in some class $C^{1,\beta}$ since F is of class $C^{1,k-1}$) .

Near B , ∂P_+ admits a representation of the form $e^{i(\pi - |\theta_B|)}(t+ih(t))$ where $h(t)$ is a real valued function of class $C^{1,\alpha}$ on some interval $[0,\delta]$, and $h(0) = h'(0) = 0$.

Thus C admits the representation $-(t+ih(t))^k$ i.e. $x(t) = -\mathrm{Re}(t+ih(t))^k$, $y(t) = -\mathrm{Im}(t+ih(t))^k$.

Hence :

$$\frac{dy}{dx}(t) = \frac{\operatorname{Im}(1+i\,\frac{h(t)}{t})^{k-1}\,(1+ih'(t))}{\operatorname{Re}(1+i\,\frac{h(t)}{t})^{k-1}(1+ih'(t))} = h'(t)+(k-1)\frac{h(t)}{t} + O(t^{3\alpha})$$

so that $\left|\frac{dy}{dx}(t)\right| = O(t^{\alpha})$.

Since on the other hand we have $x(t) = -t^{K}(1+O(t^{2\alpha}))$ we conclude that $\left|\frac{dy}{dx}(x)\right| = O(|x|^{\alpha/k})$ and consequently $CU[0,+\infty)$ is of class $C^{1,\gamma}$ with $\gamma = \frac{\alpha}{k}$.

- References -

{1} C.BAIOCCHI, *Su un problema di frontiera libera connesso a questioni di idraulica.* Annali di Mat. Pura ed Applic. 92 (1972), p.107-127.

{2} L.BERS, *Mathematical aspects of subsonic and transonic gas dynamics,* Chapman and Hall, London 1958.

{3} H.BREZIS-G.DUVAUT, *Ecoulements avec sillages autour d'un profil symétrique sans incidence*, C.R.Acad.Sci.Paris, 276 (1973), p.875-878.

{4} H.BREZIS-G.STAMPACCHIA, *Une nouvelle méthode pour l'étude d'écoulements stationnaires,* C.R.Acad.Sci.Paris, 276 (1973), p.129-132.

{5} R.COURANT-K.O.FRIEDRICHS, *Supersonic flow and shock waves,* Interscience , New-York (1948).

{6} C.FERRARI-F.TRICOMI, *Aerodinamica transonica,* Cremonese Roma (1962) (English translation : *Transonic aerodynamics,* Acad. Press, New-York (1968)).

{7} E.HILLE, *Analytic function theory,* Boston, Ginn (1962).

{8} J.L.LIONS-G.STAMPACCHIA, *Variational inequalities,* Comm. Pure Appl. Math. 20 (1967) p.493-519.

A NEW FORMULATION OF DIPHASIC INCOMPRESSIBLE FLOWS IN POROUS MEDIA

G. CHAVENT
IRIA - LABORIA
DOMAINE DE VOLUCEAU - LE CHESNAY 78150

I - INTRODUCTION :

Polyphasic flows in porous media have not received much attention from mathematicians upon now. This may have come from the difficulty of establishing satisfying models, and because already the simplified model using the hypothesis of relative permeabilities and capillarity pressure is rather intricate (see for instance ref [6]). A mathematical study of the equation of interface beetwen the two fluids is made in ref [11].

The lack of rigorous mathematical formulation has proved to be an obstacle to the application to polyphasic flows of modern approximation and control methods based on fonctiónnal analysis (cf. [8] [9]). We give hereunder a new formulation of the equations of diphasic incompressible flows, which admits a variationnal version, and we give an existence theorem. This formulation happens to be identical in form to that of the miscible case, which we recall first.

II - THE PHYSICAL PROBLEM :

Let $\Omega \subset R^n$ be a porous body with boundary Γ and pressure independant porosity ϕ. We consider two incompressible liquids 1 and 2, miscible or immiscible, and denote by S either the concentration (miscible liquids) or the saturation (immiscible liquids) in the first component.

We consider now the displacement of liquid 1 by liquid 2. More precisely, we suppose

i) at initial time $t = 0$, the body Ω is saturated by the liquid 1, i.e. $S = 1$

ii) at every point s of the part Γ_1 of boundary Γ, we inject liquid 2 with a rate $q(s,t)$

iii) on the part Γ_2 of boundary Γ, the body is in contact with a fluid at a known pressure P_e (s t) and concentration (or saturation) S_e

iV) the part Γ_3 of the boundary Γ is closed.

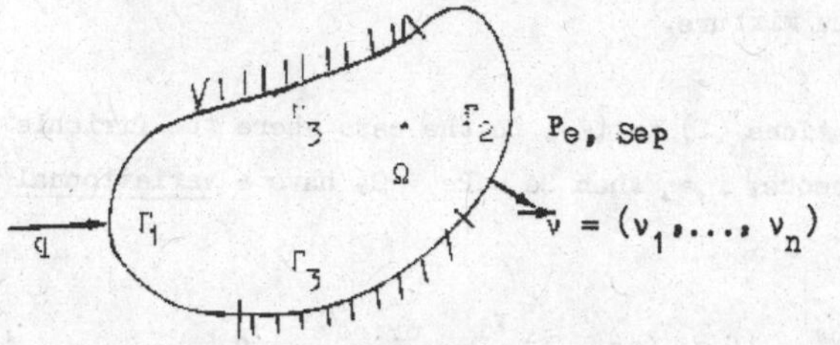

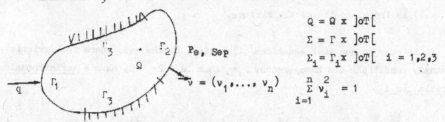

$$Q = \Omega \times]oT[$$
$$\Sigma = \Gamma \times]oT[$$
$$\Sigma_i = \Gamma_i \times]oT[\quad i = 1,2,3$$
$$\sum_{i=1}^{n} \nu_i^2 = 1$$

Remark 1 : For simplicity, we have neglected the gravity effect but it can be taken in account, see [1].

Remark 2 : We suppose the there are no sinks or wells in Ω. If not, they have to be modelized as a part of the boundary Γ_1.

III - THE MISCIBLE CASE:

1) Physical

For slow displacements, the well known equations of the concentration S and the pressure P in the liquid phase are (cf [2] [3] [6])

In $Q = \Omega \times]o,T[$:

$$(1) \quad 2\emptyset \frac{\partial S}{\partial t} - \sum_{i=1}^{n} \frac{\partial}{\partial x_i}\left(2D_i \frac{\partial S}{\partial x_i}\right) - \sum_{i=1}^{n} \frac{Ki}{\mu(S)} \frac{\partial P}{\partial x_i} \frac{\partial}{\partial x_i}(2S - 1) = 0$$

$$(2) \quad - \sum_{i=1}^{n} \frac{\partial}{\partial x_i}\left(\frac{Ki}{\mu(S)} \frac{\partial P}{\partial x_i}\right) = 0$$

Initial conditions :

$$(3) \quad S(x, 0) = 1 \text{ on } \Omega$$

Boundary conditions

$$(4) \quad S = 0, \quad \sum_{i=1}^{n} \frac{Ki}{\mu(S)} \frac{\partial P}{\partial x_i} \nu_i = q \quad \text{on } \Sigma_1$$

$$(5) \quad S = Se, \qquad P = Pe \qquad \text{on } \Sigma_2$$

$$(6) \quad \sum_{i=1}^{n} D_i \frac{\partial S}{\partial x_i} \nu_i = 0, \quad \sum_{i=1}^{n} \frac{Ki}{\mu(S)} \frac{\partial P}{\partial x_i} \nu_i = 0 \quad \text{on } \Sigma_3$$

where :

ϕ is the porosity of the porous media (constant)

D_i is the dispersion coefficient in the direction of the $i^{\underline{th}}$ axis

Ki is the permability of porous media " "

$\mu(S)$ is the viscosity ot the mixture.

It is easy to see that equations (1) to (6), in the case where the Dirichlet boundary conditions are homogeneous, i.e. when Se = Pe = 0, have a <u>variationnal formulation</u> :

$$(7) \quad 2\phi \int_\Omega \frac{\partial S}{\partial t} v + \int_\Omega \sum_{i=1}^{n} 2D_i \frac{\partial S}{\partial x_i}\frac{\partial v}{\partial x_i} + \int_\Omega \sum_{i=1}^{n} (2S-1)\frac{K_i}{\mu(S)}\frac{\partial P}{\partial x_i}\frac{\partial v}{\partial x_i} = 0 \qquad \forall v,\ v=0 \text{ on } \Gamma_1 \cup \Gamma_2$$

$$(8) \quad \int_\Omega \frac{K_i}{\mu(S)}\frac{\partial P}{\partial x_i}\frac{\partial w}{\partial x_i} = \int_{\Gamma_1} q\, w \qquad \forall w,\ w=0 \text{ on } \Gamma_2$$

$$(9) \quad S=0 \text{ on } \Sigma_1 \cup \Sigma_2 \quad P=0 \text{ on } \Sigma_2$$

$$(10) \quad S=1 \text{ on } \Omega \text{ at } t=0$$

2) <u>Existence theorem</u> :

Let :

$$(11)\begin{cases} V = \{v \in H^1(\Omega),\ v=0 \text{ on } \Gamma_1 \cup \Gamma_2\} \\ W = \{w \in H^1(\Omega),\ w=0 \text{ on } \Gamma_2\} \\ H = L^2(\Omega) \end{cases}$$

By identifying Hand to its dual H' and V and W to parts of H, we have

$$(12) \quad V \subset H \subset V', \quad W \subset H \subset W'$$

Consider the <u>non linear variationnal system</u> :

$$(13) \quad 2\phi\left(\frac{\partial S}{\partial t}, v\right) + \sum_{i=1}^{n}\int_\Omega A_i(x)a(S)\frac{\partial S}{\partial x_i}\frac{\partial v}{\partial x_i} + \sum_{i=1}^{n}\int_\Omega b(S)\,B_i(x)d(S)\frac{\partial P}{\partial x_i}\frac{\partial v}{\partial x_i} = 0 \qquad \forall v \in V, \text{ a.e. on }]oT[$$

$$(14) \quad \sum_{i=1}^{n}\int_\Omega B_i(x)d(S)\frac{\partial P}{\partial x_i}\frac{\partial w}{\partial x_i} = \int_{\Gamma_1} q\, w \qquad \forall w \in W, \text{ a.e. on }]oT[$$

$$(15) \quad S(0) = So$$

$$(16) \quad S(t) \in V \text{ and } P(t) \in W \text{ a.e. on }]oT[$$

which contains, as a special case, the system (7)-(10) if we choose

$$(17)\quad \begin{cases} A_i(x) = D_i & a(s) = 2 \\ B_i(x) = K_i(x) & b(S) = 2S-1 \\ d(S) = \dfrac{1}{\mu(S)} & S_o = 1 \end{cases}$$

Then we have the

<u>Theorem 1</u> Under hypotheses (11), (12) and :

(18) $\emptyset \in R$, $S_o \in L^2(\Omega)$, $q \in L^2(\Sigma_1)$, Ω regular and bounded

(19) $\begin{cases} A_i, B_i \in L^\infty(\Omega) \\ a, b, d \text{ are } \underline{\text{continuous}} \text{ and } \underline{\text{bounded}} \text{ applications of } R \text{ in } R \end{cases}$

(20) $\alpha > 0$ such that $A_i(x) \geq \alpha$, $B_i(x) \geq \alpha$ a.e. on Ω

$a(\xi) \geq \alpha$, $d(\xi) \geq \alpha$ $\forall \xi \in R$

the system (13)-(16) has at least a solution (S, P) such that

(21) $S \in L^2(oT\ ;\ V) \cap L^\infty(oT\ ;\ H)$, $\dfrac{dS}{dt} \in L^2(oT\ ;\ V')$

(22) $P \in L^2(oT\ ;\ W)$

If moreover there exists two values S min and S max such that

(23) $S \min \leq S\, o(x) \leq S \max$ a.e. on Ω

then the solution S of (13) satisfies :

(24) $S \min \leq S(x,t) \leq \max$ a.e. on Q

We shall omit for brevity the proof of this theorem. A complete proof (for the case including the gravity effects) will be found in [1]. It follows the proof given in [4] for a simpler case, the last part of the theorem is the classical maximum principle.

3) <u>Unicity theorem</u>

We have not been able to proove unicity under the same hypothesis that existence ; we know only the

<u>Theorem 2</u> : Under hypothesis of theo.1 and either

(25) d is a constant independant of S

or :

(26) Ω is one - dimensionnal

then the system (13) to (16) has a unique solution.

Proof : Hypothesis (25) makes pressure P independant of saturation, so that the system (13) (14) reduces to the equation (13).

Hypothesis (26) allows to integrate one time equation (14), so that the pressure P can be eliminated between (13) and (14), and once again the system reduces to only one equation.

In both cases, the proof given in [4] applies with minor adaptations.

IV - THE IMMISCIBLE CASE

Though the physical reality is very different from the last case (2 liquid phases instead of one), we shall demonstrate that, for incompressible fluids, the introduction of the classical hypothesis of relative permeability and capillarity pressure makes it possible to put the immiscible equations into the same form than the miscible equations.

1) The classical formulation - The two liquid phases are

- the non-wetting phase (phase 1 - for instance oil)
- the wetting phase (phase 2 _ for instance water)

and the main unknowns of the system are

$$(27)\quad \begin{cases} \text{- the pressure } P_i \text{ in the phase i, } i = 1,2 \\ \text{- the reduced saturation } S \text{ in phase 1} \end{cases}$$

Using the hypothesis of relative permeabilities and capillarity pressure, the equations are :

In $Q = \Omega \times]oT[$:

$$(28)\quad \frac{\partial}{\partial t}[\phi S] - \sum_{i=1}^{n} \frac{\partial}{\partial x_i}\left(\frac{K_i(x)\, k_1(S)}{\mu_1} \frac{\partial P_1}{\partial x_i}\right) = 0$$

$$(29)\quad \frac{\partial}{\partial t}[\phi (1-S)] - \sum_{i=1}^{n} \frac{\partial}{\partial x_i}\left(\frac{K_i(x)\, k_2(S)}{\mu_2} \frac{\partial P_2}{\partial x_i}\right) = 0$$

$$(30)\quad P_1 - P_2 = P_c\,(S)$$

Initial condition :

$$(31)\quad S(x,o) = 1$$

Boundary conditions

i) on Σ_1 :

(32) $$S = 0 \; , \quad \sum_{i=1}^{n}\left(\frac{K_i\, k_1}{\mu_1}\;\frac{\partial P_1}{\partial x_i} + \frac{K_i\, k_2}{\mu_2}\,\frac{\partial P_2}{\partial x_i}\right)\nu_i = q\,(s,t)$$

ii) on Σ_2 :

(33) $$S = 0 \; , \quad P_1 = P_2 = P_e$$

iii) on Σ_3 :

(34) $$\sum_{i=1}^{n} \frac{K_i\, k_1}{\mu_1}\,\frac{\partial P_1}{\partial x_i}\,\nu_i = 0 \; , \; \sum_{i=1}^{n} \frac{K_i\, k_2}{\mu_2}\,\frac{\partial P_2}{\partial x_i}\,\nu_i = 0$$

where :

- $\emptyset$ and K_i are the porosity and the permeability of porous media (as in eq.(1))
- k_j (S) j = 1,2 is the relative permeability of liquid j in presence liquid 1 at the saturation S (k_j vanishes when the phase j is at minimum saturation)
- μ_j j = 1,2 is the viscosity of the phase j
- P_c(S) is the capillarity pressure (vanishes at S = 0)

We first make somme comment about those equations :

. Eq.(28) (resp.(29)) disappears and pressure P_1 (resp.P_2) has no sense where $S \equiv 0$ (resp. $S \equiv 1$) , so that at a given place of Ω , the number of equations is not a priori known. This makes difficult a direct mathematical study of eq.(28)(29). Moreover, from a numerical point of view, there is no reason for taking P_1 or P_2 as main unknown - as both do not make sense at some places. The choice of P_2 which is often done - cf [10] do not keep the symmetry of eq.(28)(29).

. The boundary condition (33) on Σ_2 is physically satisfying after the breaktrough of the wetting phase, wich has to accumulate along Γ_2 up to its maximum saturation before it is allowed to get out of Ω (extremity effect). Before the breaktrough however, the condition S = 0 is met only for particular experimental conditions, the more common condition beeing S = 1 . Condition (33) was choosen for simplicity reasons, and because the influence on the solution of this simplification remains localized in the neighbourhood of Σ_2.

2) A new unknown : the "global pressure" P

Let us simplify the notations by setting :

$$(35)\quad \begin{cases} m(S) = \dfrac{k_1(S)}{\mu_1} + \dfrac{k_2(S)}{\mu_2} = \text{"global mobility"} \\ m_j(S) = \dfrac{k_j(S)}{\mu_j m(S)} = \text{reduced mobility of phase } j. \end{cases}$$

We give the

Définition : At every point $(x, t) \in Q$ with saturation S and pressure P_1 and P_2 in phases 1 and 2, we define the "global pressure" P by

$$(36)\quad P = \frac{1}{2}(P_1 + P_2) + \frac{1}{2}(\gamma(S) - \gamma(o))$$

where $\gamma(S)$ is the function :

$$(37)\quad \gamma(S) = \int_{\bar{S}}^{S} (m_1(S) - m_2(S)) \frac{dP_c}{dS}(S)\, d\, S$$

where $\bar{S}$ is some arbitrary point of $[0,1]$

We give now some physical interpretation of P ; by derivating (36) and (30) with respect to x_i we get

$$\frac{\partial P}{\partial x_i} = \frac{1}{2}\left(\frac{\partial P_1}{\partial x_i} + \frac{\partial P_2}{dx_i}\right) + \frac{1}{2}(m_1(S) - m_2(S))\left(\frac{\partial P_1}{\partial x_i} - \frac{\partial P_2}{\partial x_i}\right)$$

As $m_1 + m_2 = 1$ we get :

$$\frac{\partial P}{\partial x_i} = m_1 \frac{\partial P_1}{\partial x_i} + m_2 \frac{\partial P_2}{\partial x_i}$$

which may be rewritten

$$(38)\quad K_i m(S) \frac{\partial P}{\partial x_i} = K_i \frac{k_1(S)}{\mu_1} \frac{\partial P_1}{\partial x_i} + \frac{k_2(S)}{\mu_2} \frac{\partial P_2}{\partial x_i}$$

From (38) we see that the global pressure is pressure which would give, for a fluid with mobility $m(S)$ equal to the sum of the mobilities of the fluids 1 and 2, and for each direction x_i, a flow equal to the sum of the flows of fluids 1 and 2.

3) A new formulation - physical and variationnal - of the immiscible equations

We can now rewrite the equations (28) to (34) using this "global pressure " P :

(39) in $Q = \Omega \times]0T[$: Making the sum of (28) and (29) we get :

$$-\sum_{i=1}^{n} \frac{\partial}{\partial x_i}\left(K_i\, m(S) \frac{\partial P}{\partial x_i}\right) = 0$$

Making the difference of (28) and (29) we get :

$$2\phi \frac{\partial S}{\partial t} - \sum_{i=1}^{n} \frac{\partial}{\partial x_i}\left(K_i\, mm_1 \frac{\partial P_1}{\partial x_i} - K_i\, mm_2 \frac{\partial P_2}{\partial x_i}\right) = 0$$

Using the relation

$$m_1 \frac{\partial P_1}{\partial x_i} - m_2 \frac{\partial P_2}{\partial x_i} = 2\, m_1\, m_2 \left(\frac{\partial P_1}{\partial x_i} - \frac{\partial P_2}{\partial x_i}\right) + (m_1 - m_2)\left(m_1 \frac{\partial P_1}{\partial x_i} + m_2 \frac{\partial P_2}{\partial x_i}\right)$$

and (30), (38) we get :

$$(40) \quad 2\,\phi \frac{\partial S}{\partial t} - \sum_{i=1}^{n} \frac{\partial}{\partial x_i}\left[K_i m\, 2 m_1\, m_2 \frac{dP_c}{dS} \frac{\partial S}{\partial x_i}\right] - \sum_{i=1}^{n} \frac{\partial}{\partial x_i}\left[(m_1 - m_2)\, K_i m \frac{\partial P}{\partial x_i}\right] = 0$$

. Initial condition : as (31)

. Boundary conditions

i) on Σ_1 : (32) becomes, using (38) :

$$(41) \quad S = 0 \qquad \sum_{i=1}^{n} K_i m(S) \frac{\partial P}{\partial x_i} \nu_i = q\,(s,t)$$

ii) on Σ_2 : (33) becomes, thanks to the choice of the constant γ (o) in the definition (36) of P :

$$(42) \quad S = 0 \qquad P = P_e$$

iii) on Σ_3 : Making the sum and difference of the two equations (34) we get :

$$(43) \quad \sum_{i=1}^{n} K_i m\, 2 m_1\, m_2 \frac{dP_c}{dS} \frac{\partial S}{\partial x_i} \nu_i = 0 \;, \quad \sum_{i=1}^{n} K_i\, m(S) \frac{\partial P}{\partial x_i} \nu_i = 0$$

The equations (28)(29) are replaced by :

- a family of non-degenerate elliptic equations (39) (typically, m(S) does not vanish on [0,1]) giving the global pressure P, with coefficients depending on saturation.

- a parabolic equation (40) giving the saturations S, containing a degenerated diffusion term (typically $m_1\, m_2 \frac{dP_c}{dS}$ vanishes for S = 0 and S = 1) and a convection term introducing a coupling with above family of elliptic equations.

We can see that the equations (39)-(43) are exactly of the same form than the equations (1)-(5) of the miscible case.

As a consequence, the variationnal system (13)-(16) contains the immiscible equations (39)-(43) as a special case, obtained by setting

$$(44) \quad \begin{cases} A_i\,(x) = K_i\,(x) & a(S) = 2\, m(S)\, m_1(S)\, m_2(S) \dfrac{dP_c}{dS}(S) \\ B_i\,(x) = K_i\,(x) & b(S) = m_1(S) - m_2(S) \\ d\,(S) = m(S) & \end{cases}$$

An analogy between miscible and immiscible displacements, but without introduction of the global pressure, may be found in [5]. For the one -dimensionnal case ($n=1$), the pressure P can be eliminated between eq.(39) and (40), giving back a well known equation for one-dimensionnal diphasic flows (eq. E 43112 of ref [6]).

4) Existence theorem of a solution to the immiscible equations

The theorem 1 is not sufficient to show this existence, because the function a (S) defined by (44) does not meet hypothesis (20) of theo 1. We must give a theorem which allows the equation (40) to degenerate.

We first recall some results in the

Lemma : Let $F : \mathbb{R} \to \mathbb{R}$ be a Hölder-function of order $\theta \in]0,1[$, such that $F(0) = 0$. Then, for every $v \in W^{s,p}(\Omega)$, $0 < s <$ and $1 < p < +\infty$ we have :

$$(45)\quad \begin{cases} F(v) \in W^{\theta s,\frac{p}{\theta}}(\Omega) \\ \text{and} \quad \|F(v)\|_{W^{\theta s,\frac{p}{\theta}}(\Omega)} \leq \|v\|^{\theta}_{W^{s,p}(\Omega)} \|F\|_{Höl\,\theta} \end{cases}$$

If moroever Ω is regular and bounded :

(46) the injection of $W^{s,p}(\Omega)$ in $L^p(\Omega)$ is compact

Proof : One uses the caracterisation of the space $W^{s,p}(\Omega)$ (see for instance [7] t.I p.59) :

$$(47)\quad v \in W^{s,p}(\Omega) \Leftrightarrow \begin{cases} \int_\Omega |v(x)|^p dx < +\infty \quad \text{i.e.} \quad v \in L^p(\Omega) \\ \iint_{\Omega\times\Omega} \frac{|v(x)-v(y)|^p}{|x-y|^{n+sp}} dx dy < +\infty \end{cases}$$

Moroever, the quantity

$$(48)\quad \left(\|v\|^p_{L^p(\Omega)} + \iint_{\Omega\times\Omega} \frac{|v(x)-v(y)|^p}{|x-y|^{n+sp}} dx dy \right)^{\frac{1}{p}}$$

is a norm equivalent to the usual norm $\|v\|_{W^{s,p}(\Omega)}$

Then (45) is a direct consequence of (47)(48) and the definitions of Hölder functions. The result (46) can be deduced from Kondrachoff's theorem (cf [12]) using interpolation (cf [13]).

Theorem 3 : With hypothesis and notation (11)(12)(18)(19) and :

$$(49)\quad \begin{cases} \exists \alpha > o \text{ such that } & A_i(x) \geqslant \alpha \quad B_i(x) \geqslant \alpha \quad \text{a.e. on } \Omega \\ & a(\xi) \geqslant o\ ,\ d(\xi) \geqslant \alpha \quad \forall \xi \in R \end{cases}$$

$$(50)\quad \exists\ \theta \in]o,1[\text{ such that } \sup_{o \leqslant \zeta < \xi \leqslant 1} \frac{\xi-\zeta}{\left(\int_\zeta^\xi \sqrt{a(\tau)}d\tau\right)^\theta} < +\infty$$

$$(51)\quad o \leqslant S_o(x) \leqslant 1 \quad \text{a.e. on } \Omega$$

The variationnal systems (13)-(15) has a solution (S,P) such that

$$(52)\quad S \in L^\infty(Q) \qquad 0 \leqslant S(x,t) \leqslant 1 \text{ a.e. on } Q$$

$$(53)\quad \beta(S) \in L^2(oT\ ;V),\quad \frac{dS}{dt} \in L^2(oT\ ;V')$$

$$(54)\quad P \in L^2(oT\ ;W)$$

where β is the primitive of $\sqrt{a}$:

$$(55)\quad \beta(\xi) = \int_o^\xi \sqrt{a(\tau)}\,d\tau$$

<u>Proof</u> : We approach the degenerated equation by a sequence of non-degenerated equations by setting :

$$(56)\quad a_k(\xi) = a(\xi) + \frac{1}{k} \quad \forall \xi \in R \quad \forall k \in \mathbb{N}$$

The approached equations (13-k)-(16-k) have a solution (theo-1) (S_k, P_k) satisfying

$$(55)\quad 0 \leqslant S_k(x,t) \leqslant 1 \qquad \text{a.e. on } Q$$

$$S_k \in L^2(oT : v)$$

$$(56)\quad P_k \in L^2(oT : W) \quad \text{and } ||P_k||_{L^2(oT:W)} \leqslant \text{cst independant of } k$$

Making $v = S_k$ in (13-k) and $w = \gamma(S_k)$ in (14-k), where γ is a primitive of b, we get :

$$(57)\quad \phi\frac{d}{dt}|S_k(t)|^2_{L^2(\Omega)} + \frac{1}{k}\sum_{i=1}^n \left|\frac{\partial S_k}{\partial x_i}\right|^2_{L^2(\Omega)} + \sum_{i=1}^n \left|\sqrt{a(S_k)}\,\frac{\partial S_k}{dx_i}\right|^2_{L^2(\Omega)} + \sum_{i=1}^n \int_\Omega b(S_k)B_i\, d(S_k)\frac{\partial P_k}{\partial x_i}\,\frac{\partial S_k}{\partial x_i} = 0$$

$$(58)\quad \sum_{i=1}^n \int_\Omega B_i\, d(S_k)\,\frac{\partial P_k}{\partial x_i}\,\frac{\partial}{\partial x_i}[\gamma(S_k)] = 0$$

Using (58), the last term of (57) vanishes and we get the majoration, using notation (55) :

(59) $$||\beta(S_k)||_{L^2(oT\ ;\ V)} \leq \text{cst independant of } k.$$

From (56) and (59) we deduce, using equation (13-k) :

(60) $$\left\|\frac{dS_k}{dt}\right\|_{L^2(oT\ ;\ V')} \leq \text{cst independant of } k.$$

We deduce now from (59) a majoration on S_k. We have :

(61) $Q_k = \beta(S_k)$ is bounded in $L^2(oT\ ; V) \subset L^2(oT\ ; H^1(\Omega) \subset L^2(oT, W^{s,2}(\Omega))$

$\forall\ s \in\ ]o1[$

Then $S_k = \beta^{-1}(Q_k)$, and hypothesis (50) exactly says that β^{-1} is Hölder of order θ. Then using the lemma we get :

$$||S_k(t)||^{\frac{1}{\theta}}_{W^{\theta s,\frac{2}{\theta}}(\Omega)} \leq \text{cst}\ ||Q_k(t)||_{W^{s,2}(\Omega)}\ ||\beta^{-1}||^{\frac{1}{\theta}}_{\text{Hol}\ \theta}$$

which by squaring and integrating with respect to time gives:

$$||S_k||^{\frac{2}{\theta}}_{L^{\frac{2}{\theta}}(oT\ :\ W^{\theta s\frac{2}{\theta}}(\Omega)} \leq \text{cst}\ ||Q_k||^2_{L^2(oT\ ;\ W^{s,2}(\Omega)}\ ||\beta^{-1}||^{\frac{2}{\theta}}_{\text{Hol}\ \theta}$$

which, using (59) gives :

(62) $$||S_k||_{L^{\frac{2}{\theta}}(oT\ ;\ W^{\theta s,\frac{2}{\theta}}(\Omega)} \leq \text{cst independant of } k.$$

Using the second part of the lemma, we get

(63) $W^{\theta s,\frac{2}{\theta}}(\Omega) \subset L^{\frac{2}{\theta}}(\Omega)$ with compact injection

As $W^{\theta s,\frac{2}{\theta}}(\Omega) \subset L^{\frac{2}{\theta}}(\Omega) \subset L^2(\Omega) \subset V'$, it follows from (63) that, the injection of

(64) $$W(oT) = \left\{ v \in L^{\frac{2}{\theta}}(oT, W^{\theta s,\frac{2}{\theta}}(\Omega))\ ;\ \frac{dv}{dt} \in L^2(oT\ ;\ V') \right\}$$

in $L^{\frac{2}{\theta}}(oT\ ;\ L^{\frac{2}{\theta}}(\Omega))$ is compact

So we can choose a subsequence S_k', P_k' of S_k, P_k such that

(65) $P_k' \rightarrow P$ in $L^2(oT\ ; W)$ weakly (from(56))

(66) $S_k' \rightarrow S$ in $L^2(Q)$ strong and almost every where (from (62) and (64))

(67) $\frac{dS_k'}{dt} \rightarrow \frac{dS}{dt}$ in $L^2(oT, V')$ weakly (from(60))

(68) $a(S_k') \dfrac{\partial S_k'}{\partial x_i} \to \chi_i$ in $L^2(Q)$ weakly (from (59) and (19))

(69) $b(S_k')\, d(S_k') \dfrac{\partial P_k'}{\partial x_i} \to \varphi_i$ " " (from (56) and (19))

(70) $d(S_k') \dfrac{\partial P_k'}{\partial x_i} \to \varphi_i$ " " "

. As $a(S_k') \dfrac{\partial S_k}{\partial x_i} = \dfrac{\partial}{\partial x_i} \rho(S_k')$ (with $\rho' = a$) and as $\rho(S_k') \to \rho(S)$ in $L^2(Q)$ strongly ((55),(66) and Lebesque theorem), one sees that :

$$\chi_i = a(S) \frac{\partial S}{\partial x_i}$$

. Similarly

$$b(S_k')d(S_k') \to b(S)d(S) \text{ in } L^2(Q) \text{ strongly}$$

$$\frac{\partial P_k'}{\partial x_i} \to \frac{\partial P}{\partial x_i} \quad \text{in } L^2(Q) \text{ weakly}$$

so the product converges weakly in $L^1(Q)$ to $b(S)\, d(S) \dfrac{\partial P}{\partial x_i}$ and :

$$\varphi_i = b(S)\, d(S) \frac{\partial P}{\partial x_i}$$

. for the same reason

$$\Psi_i = d(S) \frac{\partial P}{\partial x_i}$$

So we can pass to the limit as $k' \to \infty$ in (13-k') and (14-k') and we see that (S,P) are solution of equations (13)-(16), and theorem 3 is proved.

Remark 4 : Condition (50) on the function a (S) is met for instance when

$$(71) \quad \begin{cases} \cdot\ a(\xi) > 0 \quad \forall\, \xi \in \,]0 1[\\ \cdot\ a(\xi) \text{ is equivalent to } \xi^{p_0} \text{ as } \xi \to 0,\ p_0 > 0 \\ \ a(\xi) \text{ ———————— } (1-\xi)^{p_1} \text{ as } \xi \to 1,\ p_1 > 0 \end{cases}$$

which takes in account a large class of degenerescence of the function a (S), including the physical case corresponding to (44).

Remark 5 : Unicity can be proved as for the miscible case.

VI - CONCLUSION :

We have given a rigorous formulation and an existence theorem for n-dimensional diphasic incompressible flows with relative permeabilitie and capillarity pressure.

This may be of pratical use for the approximation of multidimensionnal diphasic flows by finite elements methods as for the use of control theory. We are now investigating those two points, which have pratical applications in the oil industry (as secondary recuperation of oil in a field by water injection, automatic determination of relative permabilities from production datas).

An other point wich would deserve attention is the formulation of the boundary-condition on Γ_2 juste before the break through and its rigorous mathematical formulation.

Aknowledgment

I am very indebted to L.Tartar, who inspirated the proof of theo.3

REFERENCES :

1 - G.CHAVENT Etude Mathématique des Ecoulements Diphasiques Rapport Laboria, to appear.

2 - PEACEMAN D.W. and H.H. Rachford Jr, Numerical calculation of multidimensional miscible displacement Soc. Petrol. Eng. J, 2, 327, 1962.

3 - CHATWAL S.S., COX R.L, GREEN D.W., GHANDI B., Experimental and mathematical modeling of liquid-liquid miscible displacement in porous media Water resources research, Vol 9, n°5 1973.

4 - CHAVENT G, LEMONNIER P, Identification de la non-linéarité d'une équation parabolique quasilinéaire Applied Math. and Opt, Vol.1 n°2, 1974.

5 - LANTZ R.B., Rigorous calculation of miscible deplacement using immiscible reservoir simulators Soc. Petroleum Engineers, paper SPE 2594, 1969.

6 - C.MARLE Cours de production, Tome IV : les écoulements polyphasiques Editions Technip 1965.

7 - J.L.LIONS - E.MAGENES - Problèmes aux limites non homogènes et applications, Volume 1 - Dunod 1968.

8 - J.L.VERMUELEN Numerical simulation of edge water drive with well effect by Galerkin's method. Soc.Petroleum engineers, paper SPE 4634, 1973.

9 - G.CHAVENT, P.LEMONNIER - Estimation des permeabilités relatives et de la pression capillaire dans un écoulement diphasique, in "Lectures Notes in Economics and mathematical systems" Vol.107 - Springer.

10 - SONIER F, BESSET Ph, OMBRET R - "A numerical model of multiphase flow around a well". Soc et. Eng. J. Déc.1973.

11 - CANNON J.R, FASANO A, On the movement of two immiscible liquids in poroux media. to appear in SIAM.J. on Funct.Analysis.

12 - V.I.KONDRACHOFF "Sur certaines propriétés des fonctions de l'espace L^p" Doklady Akad. Nauk., 48(1945) 563-566.

13 - J.L.LIONS et J.PEETRE "Sur une classe d'espaces d'interpolation" Inst. Hautes Etudes, n°19, Paris (1964).

CONVERGENCE OF SOLUTIONS IN PROBLEMS OF ELASTIC PLASTIC TORSION OF CYLINDRICAL BARS

Michel Chipot

Université de Nancy I
UER Sciences Mathématiques
54 OOO - Nancy-France

1. NOTATIONS.

Let Ω be a bounded open set in $\mathbb{R}^n$ ($n \geq 2$, *note that the case n=2 is this one of elastic plastic torsion*) with smooth boundary Γ and $\mathcal{D}(\Omega)$ the space of infinitely differentiable real functions with compact support in Ω. If $L^2(\Omega)$ is the space of (classes of) square-integrable functions in Ω normed by

$$\|v\|_2 = \left(\int_\Omega |v(x)|^2 \, dx\right)^{1/2}$$

we note

$$H^1(\Omega) = \{v \in L^2(\Omega) \mid \frac{\partial v}{\partial x_i} \in L^2(\Omega) \quad \forall\, i = 1 - n\}$$

the derivation being taken in sense of distributions. Then $H^1(\Omega)$ with the norm

$$\|v\|_1 = \|v\|_2 + \sum_i \left\|\frac{\partial v}{\partial x_i}\right\|_2$$

is a Hilbert space and we denote by $H^1_o(\Omega)$ the closure of $\mathcal{D}(\Omega)$ in $H^1(\Omega)$ équiped with $\| \;\|_1$. We recall that in $H^1_o(\Omega)$, $\| \;\|_1$ is equivalent to the norm defined by

$$\|v\|_o = \left(\int_\Omega |\text{grad } v(x)|^2 \, dx\right)^{1/2}$$

and in which follows we shall only use this second norm.

2. A RESULT OF H. BREZIS AND M. SIBONY.

Let J be the functional defined in $H_o^1(\Omega)$ by

$$J(v) = \int_\Omega (|\text{grad } v(x)|^2 - 2\alpha\, v(x))\, dx$$

with α a constant > 0 and

$$K = \{v \in H_o^1(\Omega) \mid |\text{grad } v(x)| \leqslant 1 \quad \text{a.e in } \Omega\}$$

$$= \{v \in C\,(\bar\Omega) \mid v = 0 \text{ ou } \Gamma,\ |v(x) - v(y)| \leqslant |x - y|\ \forall\, x,y \in \Omega\}$$

where $C(\bar\Omega)$ is the space of continuous functions on $\bar\Omega$ closure in $\mathbb{R}^n$ of Ω. It is clear that K is a closed convex set included in $H_o^1(\Omega)$, then (see for instance [4]), there is a unique element $\theta \in K$ such that

$$J(\theta) \leqslant J(v) \qquad \forall\, v \in K$$

If we put now

$$K' = \{v \in H_o^1(\Omega) \mid \quad |v(x)| \leqslant \delta(x) \quad \text{a.e in } \Omega\}$$

with

$$\delta(x) = d(x,\Gamma) \qquad \forall\, x \in \Omega$$

we obtain another closed convex set of $H_o^1(\Omega)$ and there is a unique element $\theta' \in K'$ such that

$$J(\theta') \leqslant J(v) \qquad \forall\, v \in K'$$

H. Brezis and M. Sibony have proved in [1] that

$$\theta = \theta'$$

we deal now with a generalization of this result.

3. A GENERALIZATION.

For all $i \in \mathbb{N}$ let

$\Omega^i_1, \Omega^i_2, \Omega^i_p$ be smooth open sets in $\mathbb{R}^n$ such that

$$\bar{\Omega}^i_j \subset \Omega \qquad \forall\, i \in \mathbb{N} \qquad \forall\, j = 1 - p$$

$$\bar{\Omega}^i_j \cap \bar{\Omega}^i_k = \emptyset \qquad \forall\, j \neq k$$

(the role of i will be clear in part 4).

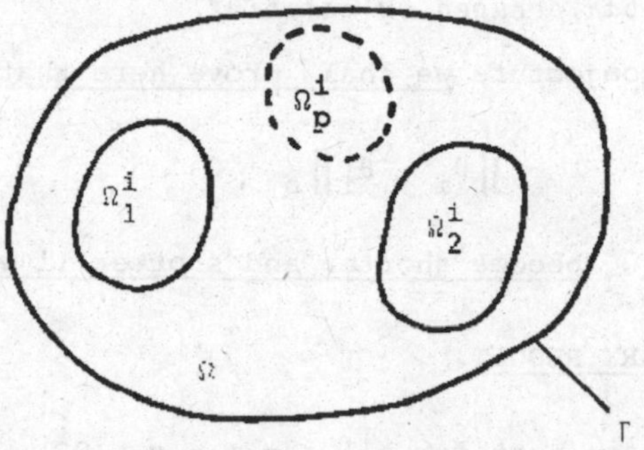

. δ_i the function defined in Ω by (see [5])

$$\delta_i(x) = \text{Min}\,[\,d(x,\Gamma),\ \underset{j}{\text{Min}}\ (d^i_j + d(x,\Omega^i_j))]$$

where d^i_j is the " shortest way to go from Ω^i_j to Γ without take in mind the passage in holes " that is to say :

$$d^i_j = \underset{i_1 - i_k}{\text{Min}}\ [\,d(\Omega^i_j, \Omega^i_{i_1}) + d(\Omega^i_{i_1}, \Omega^i_{i_2}) + \dots + d(\Omega^i_{i_k}, \Gamma)]$$

. $V_i = \{v \in H^1_o(\Omega) \mid v(x) = \text{cste} = C_j \text{ a.e in } \Omega^i_j\ \forall\, j = 1 - p\}$

(one can prove easely that V_i is a closed subspace of $H^1_o(\Omega)$).

. $K_i = K \cap V_i$

$$K'_i = \{v \in V_i \mid \quad |v(x)| \leq \delta_i(x) \text{ a.e in } \Omega\}$$

Then it is clear that K_i and K'_i are closed convex sets in $H^1_o(\Omega)$ and there is a unique element θ_i (Resp θ'_i) in K_i (Resp K'_i) such that

$$J(\theta_i) \leq J(v) \qquad \forall\, v \in K_i$$

$$(\text{ Resp} \quad J(\theta'_i) \leq J(v) \qquad \forall\, v \in K'_i\ )$$

<u>The problem of egality of θ_i and θ'_i is open</u>.

When $n = 2$, a such egality would be very usefull to describe the propagation of plasticity inside $\Omega - \bigcup_j \bar{\Omega}^i_j$ (see [5]), this open being the part of the section of the bar occuped by material.

According to this conjecture <u>we shall prove here that</u>

$$\|\theta_i - \theta'_i\|_o \to 0$$

<u>when the " holes " Ω^i_j become shorter and shorter (i varying)</u>.

4. <u>A PRELIMINARY RESULT</u>.

We suppose here that for all $j = 1 - p$, $(\Omega^i_j)_{i \in \mathbb{N}}$ is a monotone sequence of smooth open sets with the properties of part 3 and with intersection reduced to a point - ie.

$$\Omega^{i+1}_j \subset \Omega^i_j \qquad \forall\, j = 1 - p \qquad \forall\, i \in \mathbb{N}$$

$$\bigcap_i \bar{\Omega}^i_j = \omega_j \in \Omega \qquad \forall\, j = 1 - p$$

$(K_i)_{i \in N}$ is then a monotone sequence of closed convex sets of $H^1_o(\Omega)$ and if we put

$$K_\infty = \bigcup_i K_i$$

We have

<u>Proposition 1 : K_∞ is dense in K for the topology of $H_o^1(\Omega)$</u>

<u>Proof</u> : We give it with p = 1, the general result being in [2]

<u>Step 1</u> : Let k be in K with $k(\omega_1) \neq 0$ and

$$(1) \qquad \| k \|_{1,\infty} = \sup_{\substack{x,y \in \Omega \\ x \neq y}} \frac{|k(x) - k(y)|}{|x - y|} = \beta < 1$$

We can suppose that $k(\omega_1) > 0$ (if not $-k$ would be in the closure $\overline{K_\infty}$ of K_∞ and k also since K_∞ and then $\overline{K_\infty}$ are balanced).
We put then

$$M_n = \sup_{x \in B_0(\omega_1,\, 1/n)} k(x)$$

for sufficient large n (ie - such that the open ball $B_0(\omega_1, 1/n)$ with center in ω_1 and radius 1/n would be in Ω) and we define two functions Δ and Δ_n by

$$\Delta(x) = \begin{cases} k(\omega_1) - |x - \omega_1| & \text{for } |x - \omega_1| \leq k(\omega_1) \\ 0 & \text{for all other } x \end{cases}$$

$$\Delta_n(x) = \begin{cases} M_n & \text{for } |x - \omega_1| \leq 1/n \\ M_n + 1/n - |x - \omega_1| & \text{for } 1/n \leq |x - \omega_1| \leq M_n + 1/n \\ 0 & \text{for all other } x \end{cases}$$

We still note Δ and Δ_n the restrictions of this functions to Ω

From (1) we obtain

$$k(\omega_1) \leq \beta\, d(\omega_1, \Gamma)$$

Hence, for large n

$$M_n + 1/n < d(\omega_1, \Gamma)$$

and therefore $\Delta_n \in H_o^1(\Omega)$. Moreover when n goes to infinity it is easy to see that

$$\| \Delta_n - \Delta \|_o \to 0$$

Let $$E^+ = \{x \in \Omega \mid k(x) > 0\}$$

$$E^- = \{x \in \Omega \mid k(x) < 0\}$$

then the closed ball $B(\omega_1, k(\omega_1))$ is in E^+. (If it was not in E^+ we could choose $x \in B(\omega_1, k(\omega_1))$ with $k(x) \leq 0$. Hence $k(\omega_1) - k(x) \geq |\omega_1 - x|$ wich contradicts (1)). E^+ being open (k continuous, since $k \in K$) it is possible to find a neighbourhood of $B(\omega_1, k(\omega_1))$ in E^+ and for large n the support of Δ_n is in E^+. We put then

$$k_n = \text{Max}\,[\Delta_n, k]$$

The function k_n is constant on $B_0(\omega_1, 1/n)$ wich contains Ω_1^i for large i. Moreover for large n we have

$$\text{Supp}\, k_n \subset \bar{E}^+ \qquad \text{and} \qquad k_n \to \text{Max}\,[k,0] = k^+$$

(Since $\Delta_n \to \Delta$, we have $k_n \to \text{Max}\,[k,\Delta]$; but on $B(\omega_1, k(\omega_1))$, $k \geq \Delta$ - if not, (1) fails - thus $k_n \to k^+$).

The support of k_n being in $\overline{E^+}$ for large n, if we consider the function $k_n - k^-$ where $k^- = \text{Max}\,[-k,0]$ we have (see for instance [3])

$$\text{grad}\,(k_n - k^-)\,(x) = \begin{cases} \text{grad}\, k_n(x) & \text{on } \bar{E}^+ \\ \text{grad}\, k(x) & \text{on } E^- \end{cases}$$

that is to say, k_n being clearly in K,

$$|\text{grad}\,(k_n - k^-)\,(x)| \leq 1 \quad \text{a.e in } \Omega$$

Hence $k_n - k^- \in K_\infty$ and converge to k in $H_0^1(\Omega)$ wich achieves the proof in that case.

<u>Step 2</u> Let now $k \in K$ with $k(\omega_1) = 0$ and $\|k\|_{1,\infty} = \beta < 1$ then we claim that k is in $\overline{K_\infty}$. For this, let $d \in \mathcal{D}(\Omega)$ with $d(\omega_1) \neq 0$ and $\|d\|_{1,\infty} \leq (1 - \beta) / 2$. By first part of the proof we have k + d and k - d in $\overline{K_\infty}$ and $\overline{K_\infty}$ being convex

$$k = \frac{k + d}{2} + \frac{k - d}{2} \in \overline{K_\infty}$$

Hence the result.

Step 3 If now k is in K without stronger assumption on $\| \|_{1,\infty}$ then $\forall \beta < 1$, $\beta k \in K$ and we are in the case 1 or 2, thus $\beta k \in \overline{K_\infty}$ $\forall \beta < 1$ - But $\beta k \to k$ when $\beta \to 1$ wich concludes the proof.

Remark : Let F be a closed subset in Ω, then it is easy to prove by the same way that it is possible to approach the functions of K wich are constant on F by functions of K constant in a neighbourhood of F.

5. THE RESULT OF CONVERGENCE.

The open sets Ω_j^i being like in part 4 we claim :

Proposition 2 : When i goes to infinity θ_i and θ_i' have the same limit in $H_0^1(\Omega)$ wich is the function θ defined in part 2.

Proof : We have $\delta_i \leq \delta$ hence $K_i' \subset K'$ and

$$J(\theta) = J(\theta') \leq J(\theta_i') \qquad \forall i \in \mathbb{N}$$

note also that we have $K_i \subset K_i'$ (see for instance [5]) and then

$$J(\theta_i') \leq J(\theta_i) \qquad \forall i \in \mathbb{N}$$

hence

$$J(\theta) \leq J(\theta_i') \leq J(\theta_i) \qquad \forall i \in \mathbb{N} \tag{2}$$

The sequence K_i being monotone we have for all i

$$J(\theta) \leq J(\theta_{i+1}) \leq J(\theta_i) \leq J(v) \quad \forall v \in K_i$$

wich proves that the sequence $J(\theta_i)$ converges to ℓ such that

$$J(\theta) \leq \ell \leq J(v) \qquad \forall v \in K_\infty$$

But K_∞ being dense in K we have $\ell = J(\theta)$, hence θ_i and θ_i' (from (2)) being minimising sequences converge both to θ - Q E D.

BIBLIOGRAPHY

[1] H. Brezis et M. Sibony - Equivalence de deux inéquations variationnelles et applications - A.R.M.A. - vol. 41 - 1971 - p. 254-265.

[2] M. Chipot - Thèse de 3ème Cycle.

[3] J. Deny - J.L. Lions - Les espaces du type de Beppo-Levi - Ann Inst. Fourier - vol. 5 - 1953-54 - p. 305-370.

[4] I. Ekeland - R. Temam - Analyse convexe et problèmes variationnels - Dunod - Gauthier-Villars 1974.

[5] H. Lanchon - Torsion élastoplastique d'un arbre cylindrique de section simplement ou multiplement connexe - Journal de Méca. vol. 13 - 1974 - p. 267-320.

Work made in the L.E.M.T.A.
Laboratoire d'Energétique et de
Mécanique Théorique et Appliquée
2, rue de la Citadelle
B.P. 850
54 011 - Nancy-Cédex (France)

ON AN EVOLUTION PROBLEM IN LINEAR ACOUSTICS
OF VISCOUS FLUIDS

Jean COIRIER
Laboratoire de Mécanique
40 Avenue du Recteur Pineau
86022 Poitiers. France.

Abstract. - This paper is concerned with the study of simplified equations governing the laminar non-stationary motion of a viscous compressible gas in a cylindrical duct of finite length. We have chosen spaces and operators leading to the formulation of a well posed Cauchy - Hadamard problem. Galerkin's method has been used to prove the existence of the solution and to construct a sequence of approximations.

The theory has been applied to a circular cross-section duct open at one end and subjected at the other end to a sinusoidal pressure fluctuation.

§1 - Introduction

The problem of the propagation of sound in a viscous and heat conducing gas contained in a rigid cylindrical duct is a classical one. In 1868, Kirchhoff deduced from the equations of linear acoustics a complicated equation linking the propagation constant with the driving frequency.

After that, many authors published approximate solutions obtained either in an explicit form from simplified equations, or through a numerical process.

The two following assumptions are often made.

- The data at both ends of the duct are sinusoidal functions with respect to time.
- The transient part of the solution is neglected.

The object of this paper is to propose a method allowing to take into account the initial conditions as well as more general data at the ends of the duct. This work was done from simplified equations.

In paragraph 2, basic assumptions, equations, initial and boundary conditions are formulated. We then chose spaces and operators leading to a Cauchy-Hadamard problem.

Paragraph 3 is concerned with the existence and the unicity of the solution. To prove the existence, we use Galerkin's method. Since some sesquilinear forms associated with operators have no classical properties, the usual processes had to be adapted. One can then put forward a sequence of approximations which converge towards the solution. That solution depends continuously on the data.

The above results have been applied (§4) to the motion of a gas contained in a circular cross-section duct open at one end, the other end being subjected to a sinusoidal pressure fluctuation. Some of Galerkin's approximations have been compared with the approximate solution obtained for the first time by Kirchhoff for "wide" ducts.

§2 - Statement of the problem

2.1 Basic assumptions

Consider a cylindrical duct the inside of which is defined, in dimensionless form, by the open subset $\Omega \times \Sigma$ of $\mathbb{R}^3$, Ω being the interval $]0,\ell[$ of $\mathbb{R}$ and Σ a bounded domain of $\mathbb{R}^2$, with area S and regular boundary Σ^*.

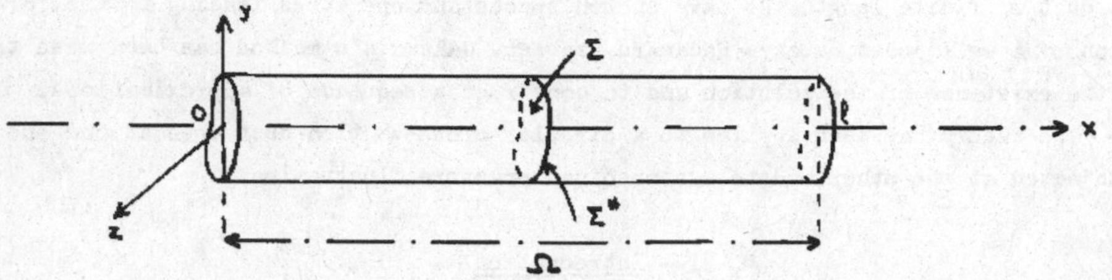

Our aim is to study laminar non-stationary motions of a viscous and heat conducing compressible gas contained in this duct. The main basic assumptions can be stated as follows.

- The motion can be linearized about a configuration of rest which is also the initial configuration.
- The transverse dimensions of the duct are small compared with its length. The longitudinal variations of functions characterizing the motion are small compared with their transverse variations.
- There are no shock waves.
- Gravitational forces can be neglected.
- The pressure is known at both ends of the duct. On the lateral surface, the velocity vanishes (the no-slip condition of the fluid at the wall) and the temperature is constant (the thermal conductivity of the fluid is assumed to be small compared with that of the wall).

2.2 Equations

Let $\mathcal{P}$ be the Prandtl number of the gas, γ its isentropic exponent, α and β its coefficients of thermal expansion at constant pressure and of pressure increase at constant volume respectively.

The study of the mentionned above motions leads to the following formal problem P_1.

Problem P1. - Find a function p on $]0,+\infty[\times \Omega$ into the set $\mathbb{C}$ of complex numbers and two functions u and θ on $]0,+\infty[\times \Omega \times \Sigma$ into $\mathbb{C}$ verifying, for any (t, x, σ) in $]0,+\infty[\times \Omega \times \Sigma$, the following equations

$$(1) \qquad S\frac{\partial p}{\partial t}(t,x) - \int_\Sigma \frac{\partial \theta}{\partial t}(t,x,\sigma)d\sigma + \int_\Sigma \frac{\partial u}{\partial x}(t,x,\sigma) = f^1(t,x),$$

$$(2) \qquad \frac{\partial u}{\partial t}(t,x,\sigma) - (\Delta_\sigma u)(t,x,\sigma) + \frac{\alpha}{\beta}\frac{\partial p}{\partial x}(t,x) = f^2(t,x,\sigma),$$

$$(3) \qquad C\frac{\partial \theta}{\partial t}(t,x,\sigma) - c((\Delta_\sigma \theta)(t,x,\sigma)) - \frac{\partial p}{\partial t}(t,x) = f^3(t,x,\sigma),$$

with $\qquad C = \frac{\gamma}{\gamma-1}$ and $c = \frac{C}{\mathcal{P}}$.

These functions should also satisfy the following conditions

$$\forall\, t \in (0,+\infty(, \quad p(t,0) = 0\,, \quad p(t,\ell) = 0,$$

$$\forall\, (t,x,\sigma) \in (0,+\infty(\times \Omega \times \Sigma^*, \quad u(t,x,\sigma) = 0, \quad \theta(t,x,\sigma) = 0,$$

$$\forall\, (x,\sigma) \in \Omega \times \Sigma,\ p(0,x) = 0, \quad u(0,x,\sigma) = 0, \quad \theta(0,x,\sigma) = 0,$$

where Δ_σ stands for the laplacian operator with respect to transverse variables, the functions f_i $(i = 1,2,3)$ being given.

The functions p, u and θ are respectively linked with the acoustic pressure, the longitudinal component of the velocity and the acoustic temperature.

Relations (2) and (3) are respectively derived from Navier-Stokes and energy equations after linear and "boundary-layer" type approximations, a Reynolds number having been assumed to be large compared with 1. Relation (1) is a consequence of continuity and state equations; the transverse components of the velocity were eliminated thanks to an integration over a cross-section of the duct (cf réf (1)).

Without limiting the generality of the subsequent results, it can be assumed that the gas is thermodynamically perfect ($\frac{\alpha}{\beta} = 1$).

2.3 Spaces and operators

a) Spaces H et V

Let E be a Hilbert space and U an open set of $\mathbb{R}^n$. We denote by $L^2(U;E)$ the space of (classes of) functions f from U into E, which are strongly measurable and such that

$$\|f\|_{L^2(U;E)} = \left(\int_U \|f(\omega)\|_E^2 \, d\omega\right)^{1/2}$$

exists. Provided with the scalar product

$$(f|g)_{L^2(U;E)} = \int_U (f(\omega)|g(\omega))_E \, d\omega,$$

$L^2(U;E)$ is a Hilbert space.

We set

$$L^2(U;\mathbb{C}) = L^2(U)$$

and we identify, for any open set U' of $\mathbb{R}^n$, the spaces $L^2(U\,;\,L^2(U'))$ and $L^2(U \times U')$.

Let us denote by

- $H^1(U)$ the space of the elements f of $L^2(U)$ the first distributional derivatives $\partial_i f$ $(i = 1,2,..n)$ of which, are in $L^2(U)$; we provide $H^1(U)$ with the scalar product

$$(f|g)_{H^1(U)} = (f|g)_{L^2(U)} + \sum_{i=1}^{n} (\partial_i f|\partial_i g)_{L^2(U)} ,$$

- $H^1_o(U)$ the closure of (U) in $H^1(U)$((U) is the space of complex infinitely functions with compact support in U); we provide $H^1_o(U)$ with the scalar product

$$(f|g)_{H_o^1(U)} = \sum_{i=1}^{n} (\partial_i f|\partial_i g)_{L^2(U)} .$$

Let H and V be the hilbertian direct sums

$$H = L^2(\Omega) \oplus L^2(\Omega \times \Sigma) \oplus L^2(\Omega \times \Sigma) , \quad V = H_o^{\,1}(\Omega) \oplus V(\Omega;\Sigma) \oplus V(\Omega;\Sigma)$$

with $V(\Omega;\Sigma) = L^2(\Omega;\, H_o^{\,1}(\Sigma))$.

It is verified that V is dense in H and that the injection of V into H is continuous.

b) Mappings A and B

We denote by A the linear mapping of H into H :

$$X = (p,u,\theta) \longrightarrow AX = (Sp - \int_\Sigma \theta(\sigma)d\sigma,\ u,\ C\theta - p) ,$$

and we denote by $B = B_1 + B_2$ the linear mapping of V into V' (antidual of V) defined for any $X = (p,u,\theta)$ of V by

$$B_1X = (0,\ -\Delta_\sigma u,\ -c\,\Delta_\sigma \theta),\quad B_2X = (\int_\Sigma \frac{\partial u}{\partial x}(\sigma)d\sigma,\ \frac{\partial p}{\partial x},\ 0).$$

The sesquilinear forms a, b_1, b_2 are classically defined by

$$\forall\ (X,Y) \in H \times H ,\quad a(X,Y) = (AX|Y)_H ,$$

$$\forall\ (X,Y) \in V \times V ,\quad b_i(X,Y) = \langle B_i X,Y \rangle \qquad \text{for } i = 1,2,$$

where $\langle , \rangle$ denote the duality between V and V'.

The following proposition can now be proved.

Proposition 1. - (i) The sesquilinear form a on H is hermitian, bounded and, if the constant C is strictly greater than 1, H-elliptic.

(ii) The sesquilinear form b_1 (resp. b_2) on V is hermitian, bounded, positive and degenerate.

Remark - Since $C = \frac{\gamma}{\gamma - 1}$ and $\gamma > 1$, the condition $C > 1$ always holds.

Proof. - Let $X_1 = (p_1,u_1,\theta_1)$ and $X_2 = (p_2,u_2,\theta_2)$ be two elements of H. Then

$$a(X_1,X_2) = (S\,p_1 - \int_\Sigma \theta_1(\sigma)d\sigma|p_2)_{L^2(\Omega)} + (u_1|u_2)_{L^2(\Omega\times\Sigma)} + (C\,\theta_1 - p_1|\theta_2)_{L^2(\Omega\times\Sigma)}$$

Now

$$(S\,p_1|p_2)_{L^2(\Omega)} = (p_1|p_2)_{L^2(\Omega\times\Sigma)},\quad (\int_\Sigma \theta_1(\sigma)d\sigma|p_2)_{L^2(\Omega)} = (\theta_1|p_2)_{L^2(\Omega\times\Sigma)}$$

The form a can thus be expressed by means of scalar products of elements of $L^2(\Omega\times\Sigma)$ in the following manner :

$$(4)\qquad a(X_1,X_2) = (p_1|p_2)_{L^2(\Omega\times\Sigma)} + (u_1|u_2)_{L^2(\Omega\times\Sigma)} + C(\theta_1|\theta_2)_{L^2(\Omega\times\Sigma)} - (\theta_1|p_2)_{L^2(\Omega\times\Sigma)} - (p_1|\theta_2)_{L^2(\Omega\times\Sigma)}$$

It is easy to see that a is hermitian and bounded on V.

Let us show that a is H-elliptic if C is strictly greater than 1.

From relation (4) it follows that

$$\forall X \in H,\ a(X,X) = \|p\|^2_{L^2(\Omega\times\Sigma)} + \|u\|^2_{L^2(\Omega\times\Sigma)} + C\|\theta\|^2_{L^2(\Omega\times\Sigma)} - 2\mathrm{R\acute{e}}(p|\theta)_{L^2(\Omega\times\Sigma)}.$$

Consequently, from Schwarz's inequality,

$$a(X,X) \geq \|p\|^2_{L^2(\Omega\times\Sigma)} + \|u\|^2_{L^2(\Omega\times\Sigma)} + C\|\theta\|^2_{L^2(\Omega\times\Sigma)} - 2\|p\|_{L^2(\Omega\times\Sigma)}\|\theta\|_{L^2(\Omega\times\Sigma)},$$

hence, since C is strictly greater than 1,

$$a(X,X) \geq \delta(\|p\|^2_{L^2(\Omega\times\Sigma)} + \|\theta\|^2_{L^2(\Omega\times\Sigma)}) + \|u\|^2_{L^2(\Omega\times\Sigma)}$$

for any positif number α verifying

$$\delta \geq \frac{(1+C) - \sqrt{(1-C)^2+4}}{2}$$

Since δ is less than 1, it holds

$$\forall X \in H,\quad a(X,X) \geq \delta\,\|X\|^2_H\ ,$$

Similarly, for any couple (X_1,X_2) in $V\times V$, we have

$$b_1\,(X_1,X_2) = (u_1|u_2)_{V(\Omega;\Sigma)} + c\,(\theta_1|\theta_2)_{V(\Omega;\Sigma)},$$

$$b_2\,(X_1,X_2) = -\,(u_1|\frac{\partial p_2}{\partial x})_{L^2(\Omega\times\Sigma)} + (\frac{\partial p_1}{\partial x}|u_2)_{L^2(\Omega\times\Sigma)}$$

The proof of properties (i) and (ii) is then easy to develop.

2.4 Formulation of the problem P_2

Let I be the interval $]0,T[$ of R (T finite) and let F be an element of $L^2(I;H)$

such that F' belongs to $L^2(I;H)$ (F' is the derivative of F in the space of the distributions defined on I and the range of which is H).

Problem P_2. - Find a (class of) function(s) X with the following properties

(4) $$X \in L^2(I;V) , \quad X' \in L^2(I;H) ,$$

(5) $$AX'(t) + BX(t) = F(t) \text{ for almost any } t \text{ of } I,$$

(6) $$X(0) = 0 .$$

Remark - Denote $\bar{I} = [0,T]$ and let $C(\bar{I};H)$ be the space of the continuous functions from $\bar{I}$ into H. We provide $C(\bar{I};H)$ with the norm

$$\|f\|_{C(\bar{I};H)} = \sup_{t\in I} \|f(t)\|_H$$

From (4), X belongs to $C(\bar{I};H)$, so that the condition (6) has meaning.

§3 - Existence and uniqueness of the solution. Galerkin's method.

We intend to establish the following proposition.

Proposition 2. - The problem P_2 has one and only one solution.

We prove the uniqueness classically, by means of the positivity of the form b on V as well as the hermitian symmetry and H - ellipticity of the form a on H.

The existence of a solution is proved by using Galerkin's method. As the form b_1 is not V-elliptic, the usual demonstrations must be adapted. We proceed in the following manner.

3.1 Galerkin's method

Let N^* be the set of the strictly positive integers. We shall denote by (f_n) every sequence of elements f_n in which n covers N^*.

Let $(e^1_{i_1})$ be a basis of $H_0^1(\Omega)$ orthonormal in $L^2(\Omega)$, $(e^2_{i_2})$ and $(e^3_{i_3})$ bases of $V(\Omega;\Sigma)$ orthonormal in $L^2(\Omega\times\Sigma)$. Such bases exist for $H_0^1(\Omega)$ and $V(\Omega;\Sigma)$ are separables.

The set $(\psi^\alpha_{i_\alpha})_{\substack{\alpha=1,2,3\\ i_\alpha \in N^*}}$ of the elements of V defined by

$$\psi^1_{i_1} = (e^1_{i_1}, 0, 0) , \quad \psi^2_{i_2} = (0, e^2_{i_2}, 0) , \quad \psi^3_{i_3} = (0, 0, e^3_{i_3})$$

is then a basis of V orthonormal in H that we can list as a sequence (ψ_n).

Let V_n be the linear subspace of V spanned by the n vectors ψ_i, for i = 1,2,.n. Let n_α (α = 1, 2, 3) be respectively the numbers of vectors $\psi^\alpha_{i_\alpha}$ belonging to the basis of V_n. We provide V_n with the topology induced by V.

The Galerkin's approximations $X_n = (p_n, u_n, \theta_n)$ are elements of V_n verifying

(7) $$\forall t \in \bar{I}, \forall Y \in V_n, \; a(X'_n(t),Y) + b(X_n(t),Y) = (F(t)|Y)_H ,$$

(8) $$X_n(0) = 0 \ .$$

These relations lead to the resolution of differential systems with constant coefficients and homogeneous initial conditions.

It is easy to prove the following result.

Proposition 3. - For any n of N^*, there exists one and only one element X_n of $C(\bar{I};V_n)$ verifying relations (7) and (8). Moreover X'_n and X''_n belong to $C(\bar{I};V_n)$ and $L^2(I;V_n)$ respectively.

3.2 "A priori" estimates

Proposition 4.(i) The sequences (X_n) and (X'_n) are bounded in $L^2(I;H)$ and $C(\bar{I};H)$.

(ii) The sequences (u_n), (u'_n), (θ_n) and (θ'_n) are bounded in $L^2(I;\ V(\Omega;\ \Sigma))$.

In order to establish the estimates relative to the sequences (X'_n), (u'_n) and (θ'_n), we have been led, as the form b is not hermitian, to suppose that F' belongs to $L^2(I;H)$ (cf paragraph 2.4). This condition enables us to derive each relation (7) in the distributional sense on I and to develop similar reasonings for the functions X_n and the functions X'_n. This process allows in particular to eliminate the anti-hermitian part of the form b.

As the form b_1 is not V-elliptic, we could not obtain an upper bound of the sequence (X_n) in $L^2(I;V)$, without introducing added assymptions concerning the basis (ψ_n) of V.

Let $(\phi_{i'_2})$ be an orthonormal basis of $L^2(\Omega)$ and $(h_{i''_2})$ a basis of $H^1_o(\Sigma)$ orthonormal in $L^2(\Sigma)$.

The set of the functions written as $\phi_{i'_2} \otimes h_{i''_2}$, when the couple $(i'_2,\ i''_2)$ covers $N^* \times N^*$, is a basis of $V(\Omega;\Sigma)$ which is orthonormal in $L^2(\Omega\times\Sigma)$. Let $\mathfrak{B}$ a bijection mapping $N^* \times N^*$ into N^*.

We make the two following assumptions.

$\mathcal{H}_1$ - We choose for elements of the basis $(e^2_{i_2})$ the functions

$$e^2_{i_2} = \phi_{i'_2} \otimes h_{i''_2}$$

with $i_2 = \mathfrak{B}(i'_2,i''_2)$.

$\mathcal{H}_2$ - There exists a subsequence (V_{n_m}) of the sequence (V_n) such that, for any m of N^*, for any t of $\bar{I}$, the element $R_{n_m}(t) = (0, \frac{\partial p_{n_m}}{\partial x}(t) \otimes h_{i''},0)$ of H be in V_{n_m}, $h_{i''}$ being an element of the sequence (h_i) such that $s_{i''} = \int_\Sigma h_{i''}(\sigma)d\sigma$ is non zero.

Proposition 5. - The subsequence (p_{n_m}) of the sequence (p_n) is bounded in $L^2(I;H^1_o(\Omega))$.

<u>Proof</u>. - Consider relation (7) verified by each function X_n :

(7) $$\forall t \in \bar{I},\ \forall Y \in V_n,\quad a(X'_n(t),Y) + b(X_n(t),Y) = (F(t)|Y)_H .$$

Let V_{n_k} be an element of the sequence (V_{n_m}) verifying the assumption $(\mathcal{R}_2)$. Let us choose $Y = R_{n_k}(t)$ in (7); then

$$b(X_{n_k}(t), R_{n_k}(t)) = (F(t)|R_{n_k}(t))_H - a(X'_{n_k}(t), R_{n_k}(t)).$$

Let us write the expression of $b(X_{n_k}(t), R_{n_k}(t))$ by means of the scalar product defined on $L^2(\Omega\times\Sigma)$

$$b(X_{n_k}(t), R_{n_k}(t)) = \left(\frac{\partial p_{n_k}}{\partial x}(t) \,\middle|\, \frac{\partial p_{n_k}}{\partial x}(t) \otimes h_{i''}\right)_{L^2(\Omega\times\Sigma)} + \left(u_{n_k}(t) \,\middle|\, \frac{\partial p_{n_k}}{\partial x}(t) \otimes h_{i''}\right)_{V(\Omega;\Sigma)} .$$

By using Schwartz's inequality we then get

$$|s_{i''}| \left\|\frac{\partial p_{n_k}}{\partial x}(t)\right\|^2_{L^2(\Omega)} \leq \|R_{n_k}(t)\|_H \, \|F(t)\|_H + M\|X'_{n_k}(t)\|_H$$
$$+ \|h_{i''}\|_{H^1_o(\Sigma)} \, \|u_{n_k}(t)\|_{V(\Omega;\Sigma)} \left\|\frac{\partial p_{n_k}}{\partial x}(t)\right\|_{L^2(\Omega)} ,$$

M being a positive real number such that, for every couple (X,Y) of H × H it holds

$$|a(X,Y)| < M \|X\|_H \|Y\|_H$$

Now

$$\|R_{n_k}(t)\|_H = \left\|\frac{\partial p_{n_k}}{\partial x}(t)\right\|_{L^2(\Omega)}$$

for $\|h_{i''}\|_{L^2(\Sigma)} = 1$.

Thus

$$\|p_{n_k}(t)\|^2_{H^1_o(\Omega)} \leq \frac{1}{|s_{i''}|} \|p_{n_k}(t)\|_{H^1_o(\Omega)} \; \|F(t)\|_H + M\|X'_{n_k}(t)\|_H + \|h_{i''}\|_{H^1_o(\Sigma)} \|u_{n_k}(t)\|_{V(\Omega;\Sigma)} ,$$

hence

$$\|p_{n_k}(t)\|^2_{H^1_o(\Omega)} \leq \frac{3}{2|s_{i''}|^2} \|F(t)\|^2_H + M\|X'_{n_k}(t)\|^2_H + \|h_{i''}\|^2_{H^1_o(\Sigma)} \|u_{n_k}(t)\|^2_{V(\Omega;\Sigma)} .$$

Let us integrate the above relation on I; it holds

$$\|p_{n_k}\|^2_{L^2(I;H^1_o(\Omega))} \leq \frac{3}{2|s_{i''}|^2} \|F\|^2_{L^2(I;H)} + M\|X'_{n_k}\|^2_{L^2(I;H)} + \|h_{i''}\|^2_{H^1_o(\Sigma)} \|u_{n_k}\|^2_{L^2(I;V(\Omega;\Sigma))} .$$

The proposition results from the fact that the sequences (u_n) and (X'_n) are bounded in $L^2(I;V(\Omega;\Sigma)$ and $L^2(I;H)$ respectively (cf proposition 4).

There remains to indicate how to construct a sequence (V_{n_m}) verifying the assumption $(\mathcal{H}_2)$.

The function $p_n(t)$ can be written as

$$p_n(t) = \sum_{i_1=1}^{n_1} p_{n_{i_1}}(t)\, e^1_{i_1}.$$

Hence we infer

$$\frac{\partial p_n}{\partial x}(t) = \sum_{i_1=1}^{n_1} p_{n_{i_1}}(t) \frac{\partial e^1_{i_1}}{\partial x}.$$

It is then natural to introduce the following assumption

$\mathcal{H}_3$ - <u>The set</u> $(\frac{\partial e^1_{i_1}}{\partial x})$ <u>is a basis of</u> $L^2(\Omega)$, <u>and for any k of</u> N^*, <u>the elements</u> $(\phi_i)_{i=1,2,..k}$ <u>and</u> $(\frac{\partial e^1_i}{\partial x})_{i=1,2,..k}$ <u>of</u> $L^2(\Omega)$ <u>span the same linear subspace of</u> $L^2(\Omega)$.

It is thus sufficient, to get a sequence (V_{n_m}) verifying the assumption $(\mathcal{H}_2)$, to extract in a strictly increasing order from the sequence (V_n) elements (V_{n_m}) for which the set of the n vectors $(\psi^2_{i_2})_{i=1,2,..n_2}$ includes the n_1 vectors $(0, \phi_{i'_2} \otimes h_{i''}, 0)_{i'_2=1,2,..n_1}$.

<u>Remark</u> - When the set $(\frac{\partial e^1_{i_1}}{\partial x})$ is orthogonal in $L^2(\Omega)$, we can take

$$\phi_{i'_2} = \left\| \frac{\partial e^1_{i'_2}}{\partial x} \right\|^{-1}_{L^2(\Omega)} \frac{\partial e^1_{i'_2}}{\partial x}, \quad i'_2 \in N^*.$$

This situation is found when we choose as a basis $(e^1_{i_1})$ of $H^1_o(\Omega)$ the sequence of the eigenfunctions

$$x \to \sqrt{\frac{2}{\ell}} \sin \frac{i_1 \pi x}{\ell} \tag{9}$$

of the mapping $\frac{\partial^2}{\partial x^2}$ of $H^1_o(\Omega)$ into $H^{-1}(\Omega)$.

This remark reveals that the assumption $(\mathcal{H}_3)$ is little restricted.

As the sequence (p_{n_m}) is bounded in $L^2(I;H^1_o(\Omega))$, the following proposition holds.

<u>Proposition 6</u>. - <u>The sequence</u> (X_{n_m}) <u>is bounded in</u> $L^2(I;V)$,

3.3 <u>Convergence of the approximations to the solution of the problem.</u>

<u>Proposition 7</u>. - <u>There exists an element X of</u> $L^2(I;V)$ <u>such that X' belongs to</u> $L^2(I;H)$ <u>and there exists a subsequence</u> (Y_k) <u>of the sequence</u> (X_{n_m}) <u>with the following properties.</u>

(i) (Y_k) converges weakly to X in $L^2(I;V)$,

(ii) (Y_k) converges weakly to X in $L^2(I;H)$,

(iii) (Y'_k) converges weakly to X' in $L^2(I;H)$,

(iv) $(Y_k(0))$ converges weakly to $X(0)$ in H.

The proof of this proposition is done, thanks to the estimates proved in the above paragraph, by using the weak sequential compactness of the unit ball of a Hilbert space.

Then we show that the element X of $L^2(I;V)$ thus put forward is solution of problem P_2, which completes the proof of proposition 2.

We can then prove that

- the sequence (X_{n_m}) itself converges weakly to X in $L^2(I;V)$,
- for any t of $\bar{I}$, the sequence $(X_{n_m}(t))$ converges to $X(t)$ for the normed topology of H,
- the mapping $F \to X$ of $L^2(I;H)$ into $L^2(I;H)$ is continuous.

§4 - Application to some motions in a circular cross-section duct

Consider a circular cross-section duct. Let R_o, r, ν be respectively the internal radius, the dimensionless coordinate in the radial direction and the kinematic viscosity of the gas.

The duct us open at one end and subjected at the other end to a sinusoidal pressure fluctuation the pulsation of which is ω.

The dimensionless pulsation being taken as equal to 1, we seek a function written as

$$X(t,x,r) = \tilde{X}(\sigma)\, e^{i(t-Kx)},$$

(K belonging to C) verifying equations (1) to (3) and the only boundary conditions. It is then possible to determine, for large values of the dimensionless number $R = R_o \sqrt{\frac{\omega}{\nu}}$, approximate expressions for the wave-number $K_1 = \text{Ré}\, K$ and for the attenuation coefficient $K_2 = \text{Im}\, K$ which confirm those found by Kirchhoff (cf réf (2) p 319-326). These expressions lead to approximations of the steady-state part of the solution which are compared with those derived from Galerkin's method.

We take as a basis $(e^1_{i_1})$ the sequence of functions (9) and as a basis (h_i) of $H^1_o(\Sigma)$ the eigenfunctions of the mapping Δ_σ of $H^1_o(\Sigma)$ into $H^{-1}(\Sigma)$.

We can extract from the sequence of the linear subspaces V_n of V, a subsequence $(V_{\ell(2\ell+1)})_{\ell \in N^*}$ verifying assumption $(\mathcal{H}_2)$ and leading to the resolution of differential systems splitting up into ℓ systems of $(2\ell+1)$ equations each.

For a diatomic gas such as $\gamma = 1,4$, $\mathcal{P} = 0,84$, the steady-state parts of the first approximations have been expressed in an explicit form. For $4 \leq R \leq 100$ and

$0{,}7 \leqslant \ell \leqslant 6$ we get a good agreement with the Kirchhoff approximate solution. To obtain a better agreement over a wider range of values of ℓ, it would be necessary to introduce into the elements of the basis of V_n a fairly high number of functions related to transverse variables.

Conclusion. - We can thus resolve, in a theoretical and constructive manner, a wide class of acoustics problems by taking into account viscosity and heat conduction effects. We intend now to apply the same method to an example enabling us to compare the transient and steady-state parts of the solution. It would be also interesting to prove the convergence of Galerkin's approximations in the case of more general boundary and initial conditions.

References.

1. J-L. PEUBE. Propagation des ondes dans une conduite cylindrique de section quelconque (Colloque Euromech, 23, Zurich, avril 1971).

2. Lord RAYLEIGH. The Theory of Sound, tome II. Dover Publications. 1896.

3. J-L. LIONS et E. MAGENES. Problèmes aux limites non homogènes et applications, tome I. Dunod, Paris. 1968.

4. C. ZWIKKER et C. KOSTEN. Sound Absorbing Materials. Amsterdam : Elsevier. 1949.

5. H. TIJDEMAN. On the propagation of sound waves in cylindrical tubes. Journal of Sound and Vibration 39 (1) p 1-33. 1975.

<u>ON THE MECHANICS OF MATERIALS WITH FADING MEMORY</u>

Bernard D. Coleman
Department of Mathematics, Mellon Institute of Science
Carnegie-Mellon University, Pittsburgh, Pennsylvania

Often, when considering the dynamical behavior of materials with memory, one encounters functional-differential equations which can be cast in the form

$$\frac{dx(\tau)}{d\tau} = f(x^\tau, \tau); \tag{1}$$

here $x(\tau)$ is, for each τ, in a Banach space V; x^τ, called the "history of x up to time τ", is the V-valued function on $\mathbb{R}^+ = [0,\infty[$ defined by

$$x^\tau(s) = x(\tau - s), \qquad s \in \mathbb{R}^+; \tag{2}$$

and f is a preassigned function mapping $\mathfrak{S} \times [0,T]$ into V, with $T > 0$ and $\mathfrak{S}$ a subset of a Banach function space $\mathfrak{U}$ formed from V-valued functions on $\mathbb{R}^+$.

Let us here assume that the space $\mathfrak{U}$ of histories is, for some $p \geq 1$, an $\mathcal{L}_p$-space with a norm of the form

$$\|\Phi\|^p = \int_0^\infty |\Phi(s)|^p d\mu(s), \tag{3}$$

where $|\cdot|$ is the norm on V, and μ is a positive non-trivial measure defined on all bounded Borel subsets of $\mathbb{R}^+$; $\mathfrak{U}$ is formed from the set $\mathcal{U}$ of those strongly μ-measureable functions Φ: $\mathbb{R}^+ \to V$ for which $\|\Phi\|^p$ exists; the elements u of $\mathfrak{U}$ are the equivalence classes obtained by calling equivalent those functions Φ_α, Φ_β in $\mathcal{U}$ with $\|\Phi_\alpha - \Phi_\beta\| = 0$. It is supposed that an open connected subset S of V has been specified, and $\mathfrak{S}$ is taken to be the subset of $\mathfrak{U}$ formed from all S-valued functions in $\mathcal{U}$.

Given $\mathfrak{h}$ in $\mathfrak{S}$ and t in $(0,T]$, we say that a function x: $[0,t] \to S$ is a <u>solution up to</u> t <u>of</u> (1) <u>with initial history</u> $\mathfrak{h}$ if x is differentiable on $[0,t]$ and can be extended to $]-\infty,t]$ in such a way that: (a) x^0, "the history of x up to time zero", belongs to the class $\mathfrak{h}$ in $\mathfrak{S}\subset\mathfrak{U}$, and (b) the equation (1) holds for each τ in $[0,t]$.

As in my earlier papers with Victor J. Mizel [1] - [4], I shall here assume that μ in the formula (3) has the following two properties:

(I) If Φ is in $\mathcal{U}$ and σ in $\mathbb{R}^+$, then the function $\Phi^{(\sigma)}$, defined by

$$\Phi^{(\sigma)}(s) = \begin{cases} \Phi(0), & s\in[0,\sigma], \\ \Phi(s-\sigma), & s\in]\sigma,\infty[, \end{cases}$$

is also in $\mathcal{U}$; furthermore, if Φ_α and Φ_β in $\mathcal{U}$ are such that $\|\Phi_\alpha - \Phi_\beta\| = 0$, then $\|\Phi_\alpha^{(\sigma)} - \Phi_\beta^{(\sigma)}\| = 0$ for all $\sigma \geq 0$.

(II) If Φ is in $\mathcal{U}$ and σ in $\mathbb{R}^+$, then the function $\Phi_{(\sigma)}$, defined by

$$\Phi_{(\sigma)}(s) = \Phi(s+\sigma), \qquad s\in\mathbb{R}^+,$$

is in $\mathcal{U}$.

In reference [1] it is shown that it follows from (I) and (II) that μ has an atom at $s=0$ and is absolutely continuous on $\mathbb{R}^{++} =]0,\infty[$ with respect to Lebesgue measure; hence (I) and (II) imply that the norm defined in (3) is equivalent to a norm of form

$$\|\Phi\|^p = |\Phi(0)|^p + \int_0^\infty |\Phi(s)|^p k(s)ds, \tag{4}$$

where k is a non-negative Lebesgue measurable function on $\mathbb{R}^{++}$ equal to the Radon-Nikodym derivative (with respect to Lebesgue measure) of the restriction of μ to $\mathbb{R}^{++}$. Properties (I) and (II) of μ imply, further, that if μ is not identically zero on all subsets of $\mathbb{R}^{++}$, then k is positive almost everywhere (with respect to Lebesgue measure) on $\mathbb{R}^{++}$. A positive function k which when put into equation (4) insures that (I) and (II) hold, i.e. a function $k: \mathbb{R}^{++} \to \mathbb{R}^{++}$ which is the Lebesgue-Radon-Nikodym derivative of the restriction to $\mathbb{R}^{++}$ of a Borel measure μ with properties (I) and (II), is called an <u>influence function</u>. A locally summable function $k: \mathbb{R}^{++} \to \mathbb{R}^{++}$ is an influence function if and only if

$$\operatorname*{ess.\,sup}_{s\in\mathbb{R}^{++}} \frac{k(s+\sigma)}{k(s)} < \infty \quad \text{and} \quad \operatorname*{ess.\,sup}_{s\in\mathbb{R}^{++}} \frac{k(s)}{k(s+\sigma)} < \infty \quad \text{for every } \sigma \geq 0;$$

(here "ess. sup" means "essential supremum with respect to Lebesgue measure").

For each influence function k there are positive numbers a, b, c, and d such that

$$ae^{-bs} < k(s) < ce^{ds}.$$

These and other properties of influence functions are derived in [1] and discussed in [5]. The following functions, with ν and β arbitrary constants, are examples of influence functions:

$$k(s) = s^{\nu}, \qquad k(s) = (1+s)^{\nu} e^{\beta s}, \qquad k(s) = \frac{1+\alpha \sin^2 s}{1+s^2}, \qquad \alpha > -1.$$

For early proposals that norms equivalent to those of the form (4) should be employed to obtain non-linear generalizations of the linear theory viscoelasticity, see references [6] - [9]. In the usual treatments of fading memory spaces and non-linear viscoelasticity, it is assumed that $\mathfrak{U}$ contains non-zero constant functions and that the norm $\|\cdot\|$ has a certain "relaxation property"; such assumptions, which, of course, place restrictions on the behavior of k(s) as $s \to \infty$, have not been stated here, because they are not needed for the validity of the theorems I am about to state.

David R. Owen and I have recently developed a theory of initial value problems for equations of the form (1), assuming that f is continuous on its domain $\mathfrak{S} \times [0,T]$ and that f is locally Lipschitz continuous in its first argument (uniformly in the second argument) in the sense that for each $\mathfrak{h}$ in $\mathfrak{S}$ there are positive numbers $\delta = \delta(\mathfrak{h})$ and $L = L(\mathfrak{h})$ such that for each τ in [0,T] and each pair of elements $\mathfrak{h}'$, $\mathfrak{h}''$ of $\mathfrak{S}$ with $\|\mathfrak{h}' - \mathfrak{h}\| \leq \delta$ and $\|\mathfrak{h}'' - \mathfrak{h}\| \leq \delta$, there holds

$$|f(\mathfrak{h}'',\tau) - f(\mathfrak{h}',\tau)| \leq L\|\mathfrak{h}'' - \mathfrak{h}'\|.$$

Our principal results are the following [5].

Theorem 1 (local existence). For each $\mathfrak{h}$ in $\mathfrak{S}$, there is a t in]0,T] such that (1) has a solution up to t with initial history $\mathfrak{h}$.

Theorem 2 (conditional global existence). Let $\mathfrak{h}$ be in $\mathfrak{S}$ and t in]0,T], and suppose there is a solution up to t of (1) with initial history $\mathfrak{h}$. There then exists a positive number $\delta = \delta(\mathfrak{h}, t)$ such that, for each $\mathfrak{h}'$ in $\mathfrak{S}$ with $\|\mathfrak{h}' - \mathfrak{h}\| < \delta$, there is a solution up to t of (1) with initial history $\mathfrak{h}'$.

Theorem 3 (continuous dependence on the initial history). If $\mathfrak{h}$ is in $\mathfrak{S}$, t is in $]0,T]$, and x is a solution up to t of (1) with initial history $\mathfrak{h}$, then there are positive numbers δ and M such that for each pair $(\mathfrak{h}',t')$ in $\mathfrak{S}\times]0,t]$ with $\|\mathfrak{h}'-\mathfrak{h}\| < \delta$, there holds

$$\sup_{\tau\in[0,t']} |x'(\tau) - x(\tau)| \leq M\|\mathfrak{h}' - \mathfrak{h}\|,$$

whenever x' is a solution up to t' with initial history $\mathfrak{h}'$.

As a corollary to Theorem 3, we have

Theorem 4 (uniqueness). For each pair $(\mathfrak{h},t)$ in $\mathfrak{S}\times]0,T]$, there is at most one solution up to t of (1) with initial history $\mathfrak{h}$.

Theorems partly analogous to those listed above have been obtained by Delfour and Mitter [10], employing assumptions about f in equation (1) distinct from, but yet somewhat resembling, those used here. The class of function spaces considered here differs from that employed in [10]. I believe that the present class is appropriate to problems in which one seeks non-linear generalizations of the classical linear theories of viscoelasticity and dielectric dispersion.

Examples of dynamical systems obeying hypotheses closely related to those made here may be found in references [4], [5], [11,], [12], and [13]. The theory of fading memory, i.e. the class of function spaces $\mathfrak{U}$, introduced in [2] and [3] and employed in [4], [11], [12], and [13], is more general than that used here, but the $\mathcal{L}_p$-spaces defined by the present formula (3) form an important subclass of those covered by the general theory.

References

[1] Coleman, B. D., & V. J. Mizel, Norms and semi-groups in the theory of fading memory, Arch. Rational Mech. Anal. 23, 87-123 (1966).

[2] Coleman, B. D., & V. J. Mizel, A general theory of dissipation in materials with memory, ibid. 27, 255-274 (1967).

[3] Coleman, B. D., & V. J. Mizel, On the general theory of fading memory, ibid. 29, 18-31 (1968).

[4] Coleman, B. D., & V. J. Mizel, On the stability of solutions of functional-differential equations, ibid. 30, 173-196 (1968).

[5] Coleman, B. D., & D. R. Owen, On the initial value problem for a class of functional-differential equations, ibid. 55, 275-299 (1974).

[6] Coleman, B. D., & W. Noll, An approximation theorem for functionals, with applications in continuum mechanics, ibid. 6, 355-370 (1960).

[7] Coleman, B. D., & W. Noll, Foundations of linear viscoelasticity, Rev. Mod. Phys. 33, 239-249 (1961); errata: ibid. 36, 1103.

[8] Coleman, B. D., Thermodynamics of materials with memory, Arch. Rational Mech. Anal. 17, 1-46 (1964).

[9] Coleman, B. D., On thermodynamics, strain impulses, and viscoelasticity, ibid. 17, 230-254 (1964).

[10] Delfour, M. C., & S. K. Mitter, Hereditary differential systems with constant delays, I. General Case, J. Diff. Eqs. 12, 213-235 (1972).

[11] Coleman, B. D., & E. H. Dill, On the stability of certain motions of incompressible materials with memory, Arch. Rational Mech. Anal. 30, 197-224 (1968).

[12] Coleman, B. D., Thermodynamics of discrete mechanical systems with memory, Adv. Chem. Phys. 24, 95-154 (1973).

[13] Coleman, B. D., & D. R. Owen, A mathematical foundation for thermodynamics, Arch. Rational Mech. Anal. 54, 1-104 (1974).

CONTRACTION SEMIGROUPS AND TREND TO EQUILIBRIUM IN CONTINUUM MECHANICS

C. M. Dafermos
Lefschetz Center for Dynamical Systems
Brown University
Providence, R. I. 02912/USA

1. Introduction

The laws of continuum mechanics and thermodynamics often induce damping mechanisms in the nature of viscosity, diffusion, thermal dissipation, etc. Usually these mechanisms manifest themselves in the equations of evolution with the existence of a functional with physical interpretation (e.g., energy, entropy, etc.) which is decreasing along motions and may thus serve as a Liapunov functional.

The natural question is whether dissipation drives the system to equilibrium as time tends to infinity. When the damping mechanism is so strong that the rate of decay of the Liapunov functional is negative definite, one can usually establish by elementary means exponential decay to equilibrium. Quite often, however, the damping mechanism is weak so that the rate of decay of the Liapunov functional is only negative semidefinite. In these cases the problem of asymptotic behavior is deeper and the analysis required for its study rather subtle. Even when there is trend to equilibrium, it is no longer necessarily exponentially fast.

Among the methods proposed for investigating asymptotic behavior, the topological dynamics approach seems the most powerful and elegant. Mechanical or thermodynamical processes are identified with trajectories of a dynamical system on an appropriate function space and the study of asymptotic behavior is reduced to the study of the structure of ω-limit sets of these trajectories. The method, introduced by LaSalle [1] for autonomous systems of ordinary differential equations, has been extended in several directions and has found a wide variety of applications (e.g. [2,3,4]).

Here we will illustrate this approach by means of several

examples of which some are new while others have been studied earlier by various techniques. We have selected our examples so that the resulting dynamical systems are contraction semigroups in Hilbert space because, as shown by Dafermos and Slemrod [5], in this setting the method is particularly simple. (The Liapunov functional induces the "natural" norm.)

Our examples include: vibrations of an elastic membrane with viscous boundary support which constitutes the damping mechanism; motion of a mixture of two linear elastic materials (damping mechanism induced by diffusion); linear thermoelasticity (damping mechanism induced by thermal dissipation); linear viscoelasticity of the Boltzmann type (damping mechanism induced by viscosity).

In the above examples we establish trend to equilibrium generically but we demonstrate that, for some nongeneric combinations of material constants and reference configuration, motions behave asymptotically as undamped oscillations (almost periodic functions).

2. Asymptotic Behavior of Nonlinear Contraction Semigroups

In a Hilbert space H (state space) we consider the initial value problem

$$\begin{aligned} \dot{\chi}(t) &\in A\chi(t) \\ \chi(0) &= \psi \end{aligned} \tag{2.1}$$

where A is a (generally nonlinear, multivalued) maximal dissipative operator, i.e. $D(A)$ is dense in H, $R(I-A) = H$, and

$$\langle \phi^*-\psi^*, \phi-\psi \rangle \leq 0$$

for any $\phi,\psi \in D(A)$, $\phi^* \in A\phi$, $\psi^* \in A\psi$. Under these conditions, $-A$ generates a contraction semigroup T on H and the (weak) solution of (2.1) is $\chi(t) = T(t)\psi$ (for a systematic exposition see [6]). We also assume that $A^{-1}0 = \{0\}$ so that 0 is the only fixed point of T.

In the terminology of classical topological dynamics, the set $\gamma(\psi) = \bigcup_{t\geq 0} T(t)\psi$ is the orbit through ψ and $\omega(\psi) = \bigcap_{t\geq 0} \overline{\gamma(T(t)\psi)}$ is the (possibly empty) ω-limit set of ψ. The structure of ω-limit sets was investigated in [5] and from that paper we borrow the following results:

Definition 2.1. A set $X \subset H$ is called

(i) minimal under T if for every $\psi \in X$, $\overline{\gamma(\psi)} = X$.

(ii) strongly invariant under T if for each $t \in R^+$, $T(t)$ is a homeomorphism of X onto X so that the restriction of T on X can be extended as a group on X.

(iii) equi-almost periodic under T if it is strongly invariant under T and for each $\varepsilon > 0$ there is $\ell_\varepsilon > 0$ such that every interval of R of length ℓ_ε contains a point t with the property

$||T(t)\psi - \psi|| < \varepsilon$ for all $\psi \in X$.

Theorem 2.1. If for some $\psi \in H$, $\gamma(\psi)$ is precompact in H, then

(i) $\omega(\psi)$ is nonempty, connected, compact, minimal, equi-almost periodic under T and $T(t)\psi \to \omega(\psi)$, as $t \to \infty$;

(ii) $\omega(\psi)$ lies on a sphere centered at 0 and $0 \in \overline{co}\ \omega(\psi)$;

(iii) there is a linear group $\hat{T}$ of isometries, defined on the closed subspace of H spanned by $\overline{co}\ \omega(\psi)$, which coincides with T on $\overline{co}\ \omega(\psi)$.

If, in addition, $\psi \in D(A)$, then

(iv) $\overline{co}\ \omega(\psi) \subset D(A)$ and $A^0\omega(\psi)$ is compact, minimal, and equi-almost periodic under T. (As usual, A^0 denotes the minimal section of A);

(v) the infinitesimal generator $\hat{A}$ of $\hat{T}$ is a linear operator which coincides with A^0 on $\overline{co}\ \omega(\psi)$.

Theorem 2.2. If $(I-A)^{-1}$ is a compact operator, then for every $\psi \in H$, $\gamma(\psi)$ is precompact in H.

We assume now that $\psi \in D(A)$ and $\gamma(\psi)$ is precompact in H. Let $\hat{\psi} \in \omega(\psi)$. On account of Theorem 2.1,

$$T(t)\hat{\psi} = \hat{T}(t)\hat{\psi} \sim \sum_k e^{i\nu_k t}\phi_k$$

where for each k, $i\nu_k$ is a purely imaginary eigenvalue of $\hat{A}$ and ϕ_k is an associated eigenvector. Thus, if ν stands for the typical ν_k and ϕ for the corresponding ϕ_k,

$$\hat{A}\phi = i\nu\phi. \tag{2.2}$$

Furthermore, if $\chi \in \overline{co}\ \omega(\psi)$, Theorem 2.1 gives

$$\langle A\chi,\chi\rangle = \frac{1}{2}\frac{d}{dt}\ ||\hat{T}(t)\chi||^2\Big|_{t=0} = 0. \tag{2.3}$$

Equation (2.3) is used to delimit $D(\hat{A})$. Equation (2.2) will usually admit only the trivial solution in which case $\omega(\psi) = \{0\}$ and $T(t)\psi \to 0$ as $t \to \infty$. This holds even if $\psi \notin D(A)$ because, since T is a contraction semigroup, the set of $\psi \in H$ with the property $T(t)\psi \to 0$, as $t \to \infty$, is closed in H. It is remarkable that although T is in general nonlinear its asymptotic behavior is characterized by the linear operator $\hat{A}$.

Remark 2.1. As shown in [5], identical results hold for the nonautonomous system

$$\dot{\chi}(t) \in A\chi(t) + f(t)$$

where $f(t) \in L^1(R^+;H)$.

Remark 2.2. If A is a linear maximal dissipative operator on H, then A is also maximal dissipative on each one of the spaces H_k, $k = 1,2,\ldots$, defined as the set $D(A^k)$ equipped with the inner product $\langle\phi,\psi\rangle_k = \langle (I-A)^k\phi, (I-A)^k\psi\rangle$. Thus, in this case one may establish trend to equilibrium in norms stronger than the norm of H.

3. Elastic Membrane with Viscous Support

As a first application consider the equation

$$\ddot{u} = \Delta u - u \tag{3.1}$$

in an open bounded subset Ω of R^m with smooth boundary $\partial\Omega$. The damping mechanism is induced by the boundary condition

$$\frac{\partial u}{\partial n} = -k(\dot{u}) \tag{3.2}$$

where k is a continuously differentiable, strictly increasing function with bounded derivative and $k(0) = 0$.

In order to apply the method of Section 2, we rewrite (3.1) as a first order system

$$\dot{u} = v$$
$$\dot{v} = \Delta u - u.$$

The phase space H is $H^1(\Omega) \times H^0(\Omega)$ with inner product

$$\langle (u,v),(\hat{u},\hat{v})\rangle = \int_\Omega (u\hat{u}+\nabla u\cdot\nabla\hat{u}+v\hat{v})dx.$$

The domain $D(A)$ of the infinitesimal generator A is the set of $(u,v) \in H^2(\Omega) \times H^1(\Omega)$ with $\frac{\partial u}{\partial n} = -k(v)$ on $\partial\Omega$ in $H^{1/2}(\partial\Omega)$. For (u,v), $(\hat{u},\hat{v}) \in D(A)$ a simple computation gives

$$\langle A(u,v) - A(\hat{u},\hat{v}),(u-\hat{u},v-\hat{v})\rangle = -\oint_{\partial\Omega} [k(v) - k(\hat{v})](v-\hat{v})dS \leq 0 \tag{3.3}$$

so that A is dissipative. It is also easily verified that $A^{-1}0 = \{0\}$, $I - A$ is surjective, and $(I-A)^{-1}$ is compact. Therefore, the results of Section 2 apply. On account of (2.3) and (3.3), (2.2) with $\phi = (u,v)$, here takes the form

$$\begin{aligned} v &= i\nu u \\ \Delta u - u &= i\nu v \end{aligned} \quad \text{in } \Omega$$
$$v = 0,\ \frac{\partial u}{\partial n} = 0 \quad \text{on } \partial\Omega$$

Eliminating v,

$$\begin{aligned} \Delta u + (\nu^2-1)u &= 0 \quad \text{in } \Omega \\ u = 0,\ \frac{\partial u}{\partial n} &= 0 \quad \text{on } \partial\Omega. \end{aligned} \tag{3.4}$$

By uniqueness of solution of the Cauchy problem, zero is the only solution of (3.4) and this implies that all solutions of (3.1), (3.2) decay to zero as $t \to \infty$.

For $m = 2$, (3.1), (3.2) may be interpreted as the equations of motion of an elastic membrane with interior elastic support and boundary viscous support. Although this model is of no particular interest in mechanics it has been presented here because it provides an instructive demonstration of a weak damping mechanism which is nevertheless able to drain out all the energy and thus drive the membrane to equilibrium.

4. Mixtures of Elastic Materials

In this section we discuss the asymptotic behavior of solutions of the equations of motion of a mixture of two linear, homogeneous, isotropic elastic materials. We refer to [7] for the development of the model and to [8] for a discussion of asymptotic stability by a less efficient method.

The reference configuration of the mixture is an open bounded subset Ω of R^n ($n = 1,2$ or 3). The two components have densities ρ and ρ^*, respectively. The state of the mixture is characterized by four n-vector fields (u,u^*,v,v^*) where u,u^* are the displacements and v,v^* the momenta of the two components.

The equations of evolution read

$$\dot{u} = \rho^{-1}v$$

$$\dot{u}^* = \rho^{*-1}v^*$$

$$\dot{v} = \lambda\nabla\cdot\nabla u + \mu\nabla\nabla\cdot u + \kappa\nabla\cdot\nabla u^* + \sigma\nabla\nabla\cdot u^* - \alpha(\rho^{-1}v - \rho^{*-1}v^*)$$

$$\dot{v}^* = \lambda^*\nabla\cdot\nabla u^* + \mu^*\nabla\nabla\cdot u^* + \kappa\nabla\cdot\nabla u + \sigma\nabla\nabla\cdot u - \alpha(\rho^{*-1}v^* - \rho^{-1}v)$$

where $\alpha > 0$ and $\lambda,\mu,\lambda^*,\mu^*,\kappa,\sigma$ are material constants satisfying the strong ellipticity condition

$$\lambda > 0,\ \lambda + \mu > 0,\ \lambda\lambda^* - \kappa^2 > 0$$

$$(\lambda+\mu)(\lambda^*+\mu^*) - (\kappa+\sigma)^2 > 0.$$

We prescribe boundary conditions

$$u = 0,\quad u^* = 0 \quad \text{on} \quad \partial\Omega.$$

The state space H is $[H_0^1(\Omega)]^n \times [H_0^1(\Omega)]^n \times [H^0(\Omega)]^n \times [H^0(\Omega)]^n$ with inner product

$$\langle(u,u^*,v,v^*),(\hat{u},\hat{u}^*,\hat{v},\hat{v}^*)\rangle = \int_\Omega \{\rho^{-1}v\cdot\hat{v} + \rho^{*-1}v^*\cdot\hat{v}^*$$

$$+ \lambda\nabla u : \nabla\hat{u} + \lambda^*\nabla u^* : \nabla\hat{u}^* + \mu(\nabla\cdot u)(\nabla\cdot\hat{u}) + \mu^*(\nabla\cdot u^*)(\nabla\cdot\hat{u}^*)$$

$$+ \kappa\nabla u^* : \nabla\hat{u} + \kappa\nabla u : \nabla\hat{u}^* + \sigma(\nabla\cdot u^*)(\nabla\cdot\hat{u}) + \sigma(\nabla\cdot u)(\nabla\cdot\hat{u}^*)\}dx.$$

The domain of the infinitesimal generator is $D(A) = [H^2(\Omega) \cap H_0^1(\Omega)]^n \times [H^2(\Omega) \cap H_0^1(\Omega)]^n \times [H_0^1(\Omega)]^n \times [H_0^1(\Omega)]^n$. For $(u,u^*,v,v^*) \in D(A)$, a simple computation gives

$$\langle A(u,u^*,v,v^*),(u,u^*,v,v^*)\rangle = -\alpha\int_\Omega |\rho^{-1}v-\rho^{*-1}v^*|^2 dx \le 0 \tag{4.1}$$

which shows that A is dissipative. Employing the basic theory of elliptic systems, it is easy to show that $A^{-1}0 = \{0\}$, $I - A$ is surjective, and $(I-A)^{-1}$ is compact. Thus the results of Section 2 apply. If we set $\phi = (u,u^*,v,v^*)$ and use (4.1), we deduce from (2.2), (2.3) that $u^* = u$, $v = i\nu\rho u$, $v^* = i\nu\rho^* u$ and u satisfies

$$\begin{aligned}(\lambda+\kappa)\nabla\cdot\nabla u + (\mu+\sigma)\nabla\nabla\cdot u + \rho\nu^2 u = 0\\ (\lambda^*+\kappa)\nabla\cdot\nabla u + (\mu^*+\sigma)\nabla\nabla\cdot u + \rho^*\nu^2 u = 0.\end{aligned} \tag{4.2}$$

The overdetermined system (4.2) admits in general only the trivial solution in which case the diffusive force drives the mixture to equilibrium. On the other hand, it is easy to construct combinations of reference configuration and material constants for which (4.2) admits nontrivial solutions. In these nongeneric situations, only part of the energy is dissipated by diffusion. As time tends to infinity, the motions of the two components of the mixture approach the same almost periodic function and asymptotically diffusion wanes. We shall not proceed to details here because the example of Section 5 has the same flavor and is more instructive.

5. Linear Thermoelasticity

The asymptotic behavior of solutions of the equations of linear thermoelasticity has been discussed in [9,3]. For simplicity we consider here the homogeneous isotropic case. However, the same approach works also in the general situation.

The reference configuration of the body is a bounded open subset Ω of R^n $(n = 1,2$ or $3)$ with smooth boundary $\partial\Omega$. The state at time t is characterized by the displacement n-vector field $u(x,t)$, the momentum n-vector field $v(x,t)$ and the scalar temperature field $\theta(x,t)$. The evolution equations are

$$\begin{aligned}\dot u &= \rho^{-1}v\\ \dot v &= \mu\nabla\cdot\nabla u + (\lambda+\mu)\nabla\nabla\cdot u + m\nabla\theta\\ \dot\theta &= c^{-1}k\Delta\theta + c^{-1}\rho^{-1}\Theta m\nabla\cdot v.\end{aligned} \tag{5.1}$$

Equations $(5.1)_2$ and $(5.1)_3$ express conservation of linear momentum and energy. Here $\rho > 0$ is the density of the body, $c > 0$ the specific heat per unit volume, $\Theta > 0$ the reference temperature, $k > 0$ the heat conductivity, m the stress-temperature coefficient, and λ,μ the

Lame moduli which satisfy the strong ellipticity condition $\mu > 0$, $\lambda + 2\mu > 0$.

For definiteness we prescribe boundary conditions

$$u = 0, \quad \theta = 0 \quad \text{on } \partial\Omega.$$

The state space H is $[H_0^1(\Omega)]^n \times [H^0(\Omega)]^n \times H^0(\Omega)$ with inner product

$$<(u,v,\theta),(\hat{u},\hat{v},\hat{\theta})> = \int_\Omega \{\rho^{-1}v\cdot\hat{v} + \mu\nabla u : \nabla\hat{u} + (\lambda+\mu)(\nabla\cdot u)(\nabla\cdot\hat{u}) + c\Theta^{-1}\theta\hat{\theta}\}dx.$$

The domain of the infinitesimal generator A is $D(A) = [H^2(\Omega) \cap H_0^1(\Omega)]^n \times [H_0^1(\Omega)]^n \times [H^2(\Omega) \cap H_0^1(\Omega)]$. For $(u,v,\theta) \in D(A)$, a simple computation gives

$$<A(u,v,\theta),(u,v,\theta)> = -k\Theta^{-1}\int_\Omega |\nabla\theta|^2 dx \leq 0 \tag{5.2}$$

so that A is dissipative. Employing the basic theory of elliptic systems it is easy to verify that $A^{-1}0 = \{0\}$, $I - A$ is surjective, and $(I-A)^{-1}$ is compact. Thus, the results of Section 2 apply. For $\phi = (u,v,\theta)$, and on account of (5.2), (2.2), (2.3) yield $\theta = 0$, $v = i\nu\rho u$, and u satisfies

$$\left.\begin{array}{l} \mu\nabla\cdot\nabla u + \nu^2\rho u = 0 \\ \nabla\cdot u = 0 \end{array}\right\} \text{ in } \Omega \tag{5.3}$$

$$u = 0 \quad \text{on } \partial\Omega. \tag{5.4}$$

Generically, (5.3), (5.4) admit only the zero solution. To see this note that if the scalar eigenvalue problem

$$\mu\Delta\Psi + \nu^2\rho\Psi = 0 \quad \text{in } \Omega \tag{5.5}$$

$$\Psi = 0 \quad \text{on } \partial\Omega \tag{5.6}$$

has only distinct eigenvalues, then any solution of $(5.3)_1$, (5.4) must be of the form $u = \Psi w$ where w is a constant n-vector and Ψ satisfies (5.5), (5.6). Furthermore, by $(5.3)_2$, $w\cdot\nabla\Psi = 0$. But by virtue of (5.6), this implies that w is tangent to $\partial\Omega$ at every point which is of course impossible unless $w = 0$.

For $n = 2$, any solution of (5.3), (5.4) is induced by a scalar function Φ: $u_1 = \partial\Phi/\partial x^2$, $u_2 = -\partial\Phi/\partial x^1$, where

$$\mu\Delta\Phi + \nu^2\rho\Phi = 0 \quad \text{in } \Omega \tag{5.7}$$

$$\nabla\Phi = 0 \quad \text{on } \partial\Omega. \tag{5.8}$$

By the uniqueness of solution of the Cauchy problem and the analyticity of solutions of (5.7), it follows that any open subset of $\partial\Omega$ determines the possible solutions of (5.7), (5.8). It is instructive to pursue this further. Assuming that $\partial\Omega$ is analytic at a point $\hat{x}$, we construct with the help of the Cauchy-Kowalewski theorem an analytic transformation $x = x(y)$ such that locally $\partial\Omega$ is mapped into the line $y^1 = a =$ constant and

$$\frac{\partial g^{11}}{\partial y^2} = 0, \quad \frac{\partial}{\partial y^2}\;\frac{1}{\sqrt{g}}\left[\frac{\partial(\sqrt{g}\,g^{11})}{\partial y^1} + \frac{\partial(\sqrt{g}\,g^{21})}{\partial y^2}\right] = 0 \qquad (5.9)$$

where g^{ij} are the components of the metric tensor induced by the curvilinear coordinate system (y^1, y^2) and $g = \det[g^{ij}]^{-1}$. In the y-system (5.7) becomes

$$\frac{\mu}{\sqrt{g}} \sum_{i,j} \frac{\partial}{\partial y^i}\left(\sqrt{g}\; g^{ij}\,\frac{\partial\Phi}{\partial y^j}\right) + \nu^2\rho\Phi = 0. \qquad (5.10)$$

Taking the derivative of (5.10) with respect to y^2 and using (5.9) and (5.8) we deduce that $\Psi \overset{\text{def}}{=} \partial\Phi/\partial y^2$ satisfies a second order elliptic equation together with initial conditions $\Psi(a,y^2) = 0$, $\partial\Psi(a,y^2)/\partial y^1 = 0$. Thus Ψ vanishes identically and $\Phi = \Phi(y^1)$ satisfies the ordinary differential equation

$$\mu g^{11}\Phi'' + \frac{\mu}{\sqrt{g}}\left[\frac{\partial(\sqrt{g}\,g^{11})}{\partial y^1} + \frac{\partial(\sqrt{g}\,g^{21})}{\partial y^2}\right]\Phi' + \nu^2\rho\Phi = 0 \qquad (5.11)$$

with initial condition $\Phi'(a) = 0$. A simple computation shows that, in the y-coordinate system, the contravariant components of the solution u of (5.3) induced by Φ are $u^1 = 0$, $u^2 = -g^{-1/2}\Phi'$. Thus u is tangential to the family of curves $y^1 =$ constant.

It is now possible to identify domains Ω for which (5.7), (5.8) admit only the trivial solution. For example, if in a neighborhood of $\hat{x}$, $\partial\Omega$ is a straight line, the y-system is Cartesian in which case (5.11) becomes $\mu\Phi'' + \nu^2\rho\Phi = 0$ and its solution $\Phi = b\cos[\nu\rho^{1/2}\mu^{-1/2}(y^1-a)]$ does not satisfy boundary conditions (5.8) for any bounded subset of R^2, unless $b = 0$.

On the other hand, it is remarkable that there are domains in which (5.7), (5.8), and hence also (5.3), (5.4), have nontrivial solutions. For example, if Ω is the circle of radius a, we may select as our y-system a polar coordinate system in which case (5.11) reduces to Bessel's equation

$$\mu\Phi'' + \frac{\mu}{r}\,\Phi' + \nu^2\rho\Phi = 0.$$

Thus, we have a sequence $\{\Phi_k\}$ of solutions, $\Phi_k = J_0(\nu_k\rho^{1/2}\mu^{-1/2}r)$, where J_m denotes the Bessel function of order m and ν_k is the kth positive root of the equation $J_1(\nu\,\rho^{1/2}\mu^{-1/2}a) = -J_0'(\nu\,\rho^{1/2}\mu^{-1/2}a) = 0$. The functions Φ_k induce solutions u_k of (5.3), (5.4) with physical components $u_k = (0, J_1(\nu_k\rho^{1/2}\mu^{-1/2}r))$.

From the above discussion it becomes clear that generically

thermal dissipation drains out the energy and drives thermodynamic processes to equilibrium. In exceptional situations, however, thermodynamic processes behave asymptotically as isothermal, isochoric, undamped oscillations.

6. Linear Viscoelasticity of the Boltzmann Type

The asymptotic behavior of solutions of the equations of linear viscoelasticity has been studied by a variety of methods (e.g. [10, 4, 11]). For simplicity we consider here a one-dimensional homogeneous body with density $\rho > 0$ and constitutive equation

$$\sigma(x,t) = cu_x(x,t) - \int_{-\infty}^{t} g(t-\tau)u_x(x,\tau)d\tau$$

where g is a continuously differentiable function satisfying

$$g(\xi) \geq 0,\ g'(\xi) \leq 0,\ a \overset{\text{def}}{=} c - \int_0^\infty g(\xi)d\xi > 0. \qquad (6.1)$$

The mechanistic interpretation of these assumptions is familiar.

The reference configuration is the interval $\Omega = (b,d)$ and we prescribe boundary conditions

$$u(b,t) = u(d,t) = 0, \quad t \in R.$$

The state of the body at time t is characterized by the displacement $u(x,t)$, the momentum $v(x,t)$, and the history of displacement $w(x,\xi,t) \overset{\text{def}}{=} u(x,t-\xi)$, $\xi \in R^+$. The equation of motion is $\rho\ddot{u} = \sigma_x$ so that the evolution equations can be written in the form

$$\begin{aligned} \dot{u} &= \rho^{-1}v \\ \dot{v} &= cu_{xx} - \int_0^\infty g(\xi)w_{xx}d\xi \qquad (6.2) \\ \dot{w} &= -\frac{\partial w}{\partial \xi}\ . \end{aligned}$$

The state space H is $H_0^1(\Omega) \times H^0(\Omega) \times L_g^2(R^+;\ H_0^1(\Omega))$ where L_g^2 denotes g-weighted L^2 space, with inner product

$$<(u,v,w),(\hat{u},\hat{v},\hat{w})> = \int_\Omega \left\{au_x\hat{u}_x + \rho^{-1}v\hat{v} + \int_0^\infty g(\xi)[u_x-w_x][\hat{u}_x-\hat{w}_x]d\xi\right\}dx.$$

The domain $D(A)$ of the infinitesimal generator A is the set of $(u,v,w) \in H$ with $v \in H_0^1(\Omega)$, $\partial w/\partial\xi \in L_g^2(R^+;\ H_0^1(\Omega))$, $w(\cdot,0) = u(\cdot)$, and $cu_{xx} - \int_0^\infty g(\xi)w_{xx}d\xi \in H^0(\Omega)$. Using $(6.1)_3$, it is easily seen that $A^{-1}0 = \{0\}$. A simple computation shows that for $(u,v,w) \in D(A)$,

$$<A(u,v,w),(u,v,w)> = \frac{1}{2}\int_0^\infty\int_\Omega g'(\xi)[u_x-w_x]^2dxd\xi \leq 0 \qquad (6.3)$$

so that A is dissipative.

In order to determine the range of $I - A$, we consider the system

$$u - \rho^{-1}v = \hat{u}$$

$$v - cu_{xx} + \int_0^\infty g(\xi)w_{xx}d\xi = \hat{v} \qquad (6.4)$$

$$w + \frac{\partial w}{\partial \xi} = \hat{w}, \quad w(\cdot,0) = u(\cdot)$$

where $(\hat{u},\hat{v},\hat{w}) \in H$. Integrating $(6.4)_3$ we obtain

$$w(\cdot,\xi) = u(\cdot)e^{-\xi} + \int_0^\xi e^{\zeta-\xi}\hat{w}(\cdot,\zeta)d\zeta. \qquad (6.5)$$

Furthermore, if $u \in H_0^1(\Omega)$, then $w(\cdot,\xi) \in L_g^2(R^+;\, H_0^1(\Omega))$ because one easily obtains from $(6.4)_3$ the integral

$$\int_0^\infty\int_\Omega g(\xi)w_x^2dxd\xi = \frac{1}{2}g(0)\int_\Omega u_x^2dx + \frac{1}{2}\int_0^\infty\int_\Omega g'(\xi)w_x^2dxd\xi + \int_0^\infty\int_\Omega g(\xi)w_x\hat{w}_xdxd\xi$$

which, in turn, yields, by virtue of (6.1), the estimate

$$\int_0^\infty\int_\Omega g(\xi)w_x^2dxd\xi \le g(0)\int_\Omega u_x^2dx + \int_0^\infty\int_\Omega g(\xi)\hat{w}_x^2dxd\xi.$$

Going back to $(6.4)_3$ we then deduce $\partial w/\partial\xi \in L_g^2(R^+;\, H_0^1(\Omega))$. Substituting into $(6.4)_2$ v from $(6.4)_1$ and w from (6.5), we obtain

$$\rho u - [c - \int_0^\infty e^{-\xi}g(\xi)d\xi]u_{xx} = \hat{v} + \rho\hat{u} - \int_0^\infty e^{-\xi}g(\xi)\int_0^\xi e^{\zeta}\hat{w}_{xx}(\cdot,\zeta)d\zeta d\xi. \qquad (6.6)$$

It can be shown that the right-hand side of (6.6) is in $H^{-1}(\Omega)$. Furthermore, the bracket on the left-hand side of (6.6) is positive by virtue of (6.1). Therefore, there is a solution $u \in H_0^1(\Omega)$ of (6.6). Finally, v is determined from $(6.4)_1$ and is in $H_0^1(\Omega)$. Thus, $I - A$ is surjective and A is maximal dissipative.

Unfortunately, $(I-A)^{-1}$ is not compact and thus we cannot employ Theorem 2.2. Nevertheless, precompactness of orbits can be established by the following procedure: We denote by $\mathscr{C}$ the set of $(\hat{u},\hat{v},\hat{w}) \in D(A)$ with $\hat{w} \in L^\infty(R^+;\, H^2(\Omega))$. For a fixed $(\hat{u},\hat{v},\hat{w}) \in \mathscr{C}$, let $(u,v,w)(t) \overset{def}{=} T(t)(\hat{u},\hat{v},\hat{w})$. Then, for any $t \in R^+$, $(u,v,w)(t) \in D(A)$ and $||A(u,v,w)(t)|| \le ||A(\hat{u},\hat{v},\hat{w})||$. In particular, $v(\cdot,t) \in C(R^+;\, H_0^1(\Omega))$, $\partial w(\cdot,\cdot,t)/\partial\xi \in C(R^+;\, L_g^2(R^+;\, H_0^1(\Omega)))$, and

$$cu_{xx}(\cdot,t) - \int_0^t g(t-\tau)u_{xx}(\cdot,\tau)d\tau \in C(R^+;\, H^0(\Omega)).$$

Inverting the Volterra integral operator (e.g. by the standard Picard iteration scheme) and using (6.1) we obtain the estimate

$$\sup_{R^+}||u_{xx}(\cdot,t)||_{H^0(\Omega)} \le \frac{1}{a}\sup_{R^+}||cu_{xx}(\cdot,t) - \int_0^t g(t-\tau)u_{xx}(\cdot,\tau)d\tau||_{H^0(\Omega)}$$

which implies $u(\cdot,t) \in C(R^+;\, H^2(\Omega))$. From the above information we infer that $\{(u,v,w)(t)|t \in R^+\}$ is precompact in H. We have thus proved that the orbit through any point of $\mathscr{C}$ is precompact in H.

Now since $T(t)$ is a contraction semigroup, it can be shown easily [12, Prop. 4.3] that the set of points which generate precompact orbits is closed in H. But $\mathscr{L}$ is dense in H so that the orbit through any point of H is precompact and the results of Section 2 apply.

For $\phi = (u,v,w)$ (2.2) takes the form

$$\rho^{-1}v = i\nu u$$
$$cu_{xx} - \int_0^\infty g(\xi)w_{xx}d\xi = i\nu v \tag{6.7}$$
$$-\frac{\partial w}{\partial \xi} = i\nu w, \quad w(\cdot,0) = u(\cdot).$$

Moreover, by virtue of (6.3), (2.3) yields

$$g'(\xi)(u_{xx}-w_{xx}) = 0 \tag{6.8}$$

in $H^{-1}(\Omega)$ for almost all ξ. Integrating (6.8) over R^+, integrating by parts, and using $(6.7)_3$, we deduce

$$\int_0^\infty g(\xi)w_{xx}d\xi = 0 \tag{6.9}$$

in $H^{-1}(\Omega)$. On account of $(6.7)_1$ and (6.9), $(6.7)_2$ becomes $cu_{xx} + \rho\nu^2 u = 0$ so that u is an eigensolution of a second order elliptic equation. From $(6.7)_3$, $w(\cdot,\xi) = \exp(-i\nu\xi)u(\cdot)$. Substituting w into (6.8) we deduce $u = 0$ which in turn implies $v = 0$, $w = 0$. Thus (6.7), (6.8) admit only the trivial solution and this establishes trend to equilibrium in one-dimensional viscoelasticity of the Boltzmann type.

The above procedure works equally well in the n-dimensional situation where the constitutive equation is

$$\sigma_{ij}(x,t) = \sum_{k,\ell}\left\{C_{ijk\ell}(x)\frac{\partial u_k}{\partial x^\ell} - \int_{-\infty}^t G_{ijk\ell}(x,t-\tau)\frac{\partial u_k}{\partial x^\ell}d\tau\right\}.$$

The necessary assumptions are $C_{ijk\ell} = C_{k\ell ij}$, $G_{ijk\ell} = G_{k\ell ij}$ as well as the analogs of (6.1); they all admit natural mechanistic interpretations. One establishes trend to equilibrium generically.

References

[1] LASALLE, J.P., The extent of asymptotic stability. Proc. Nat. Acad. Sci. USA, 46(1960), 363-365.

[2] HALE, J.K., Dynamical systems and stability. J. Math. Anal. Appl., 26(1969), 39-59.

[3] SLEMROD, M. and INFANTE, E.F., An invariance principle for dynamical systems on Banach space. Instability of Continuous Systems (H. Leipholz, Ed.), pp. 215-221. Springer-Verlag, Berlin 1971.

[4] DAFERMOS, C.M., Asymptotic stability in viscoelasticity. Arch. Rat. Mech. Analysis, 37(1970), 297-308.

[5] DAFERMOS, C.M. and SLEMROD, M., Asymptotic behavior of nonlinear contraction semigroups. J. Functional Analysis, 13(1973), 97-106.

[6] BREZIS, H., Opérateurs Maximaux Monotones et Semi-groupes de Contraction dans les Espaces de Hilbert. North-Holland Publishing Co., Amsterdam 1973.

[7] STEEL, T.R., Applications of a theory of interacting continua, Quart. J. Mech. Appl. Math. 20(1967), 57-72.

[8] DAFERMOS, C.M., Wave equations with weak damping. SIAM J. Appl. Math. 18(1970), 759-767.

[9] DAFERMOS, C.M., On the existence and the asymptotic stability of solutions to the equations of linear thermoelasticity. Arch. Rat. Mech. Analysis 29(1968), 241-271.

[10] DAFERMOS, C.M., An abstract Volterra equation with applications to linear viscoelasticity. J. Diff. Eqs. 7(1970), 554-569.

[11] MCCAMY, R.C. and WONG, J.S.W., Stability theorems for some functional differential equations. Trans. A.M.S. 164(1972), 1-37.

[12] DAFERMOS, C.M., Semiflows associated with compact and uniform processes. Math. Systems Theory 8(1974), 142-149.

Acknowledgment

This research was supported in part by the National Science Foundation under grant GP-28931, the Office of Naval Research under contract ONR N-1467-AD-101000907 and the U.S. Army Research Office under contract DAHCO4-75-G-0077.

THE BUCKLING OF A THIN ELASTIC PLATE SUBJECTED TO UNILATERAL CONDITIONS

Claude DO
Université de Nantes (Ecole Nationale Supérieure de Mécanique),
et Laboratoire de Mécanique Théorique associé au CNRS, Université Paris VI - France -

1 - INTRODUCTION - The phenomena of buckling concern numerous structures composed of beams, plates, shells, ... These phenomena have been studied for a long time under imposed conditions which remain bilateral. In mathematical terms, these phenomena are, in general, described by eigenvalue problems for partial differential equations. A general treatment of these questions is found in [1].

Yet, situations exist in which the structures undergo unilateral conditions. This is the case, for example, of a tank placed upon a rigid ground : the plate, which schematize the bottom of the tank, can undergo stresses which produce buckling, necesseraly unilateral; a solution to this problem is proposed in [2] and [3].

The present paper concerns the buckling of a thin elastic plate subjected to unilateral conditions. These conditions can occur in the domain occupied by the plate as well as on the edge. We have used von Karman's non-linear description of plates : the need to take into consideration a non-linear theory to describe the buckling is well know in the classical bilateral situation.

From a mathematical point of view, the problem is stated in terms of eigenvalues for a variational inequality in an Hilbert space Z ; we must find a real λ and ζ in Z such that

$$((I-\lambda L)\zeta + C\zeta, z-\zeta) + \psi(z) - \psi(\zeta) \geqslant 0 \quad \forall z \in Z ;$$

I is the identity on Z, L is a linear operator, C is a non-linear operator, ψ is a convex function; $\zeta = 0$ is a solution for all λ, and the problem consists in finding the non-zero solutions for the suitable values of λ. Consequently, it is an existence theorem of non-trivial solutions which describes the buckling phenomenon. The bilateral case ($\psi = 0$) has been studied by numerous authors, notably M.S. Berger [6], M.S. Berger and P. Fife [19]; the case L = 0 has been studied by G. Duvaut and J.L. Lions [4], M. Potier-Ferry [7], for the von Karman's equations; by G. Duvaut and J.L. Lions [6] for the Love-Kirchhoff's equations (C = 0).

The question of the bifurcation from the trivial solution is very interesting : the notion of critical loading is directly connected with this problem (see [1]). Results have been obtained in this way for some types of unilateral conditions; they will be published separately (see [2], [3], for a special case).

The eigenvalue problems for variational inequalities has recently taken up. The case when ψ is the indicator of a closed convex cone of Z is treated in [2] and [3]. A more systematic study is undertaken in [9], ..., [12], working up monotone operators; but this framework does not suit our problem concerning plates : C is not monotone (except when C = 0, which corresponds to the linear description of Love-Kirchhoff; but it is know that the linear theory cannot describe buckling phenomena). In the following study, it is supposed that C possesses a property of compactness (actually satisfied in the framework of von Karman's theory of plates); but the analysis which is proposed necessitates nevertheless that the domain of ψ be a cone. The methods consists in adapting the study of [6] when ψ is differentiable; if this is not the case, we proceed by regularization.

The plan of the paper is as follows :

2 - The physical problem
3 - Variational formulation
4 - The abstract problem : formulation and results
5 - Proofs

2 - THE PHYSICAL PROBLEM - In its natural state, the plate fills an open bounded region in the $x = (x_\alpha)$ plane ([1]), with sufficiently smooth boundary Γ; let be n the unit outward normal to Γ, τ the tangent vector deducted from n by a $+\pi/2$ rotation. The characteristic parameters of the plate are : its thickness h; its flexural rigidity constant D; and the elasticity constants $a_{\alpha\beta\gamma\delta}$, with the following properties :

$$(2.1)\quad \left| \begin{array}{l} a_{\alpha\beta\gamma\delta} = a_{\beta\alpha\gamma\delta} = a_{\gamma\delta\alpha\beta} , \\ a_{\alpha\beta\gamma\delta} A_{\alpha\beta} A_{\gamma\delta} \geq a_0 A_{\alpha\beta} A_{\alpha\beta} , \ \forall A \text{ , symetric tensor ;} \quad a_0 > 0 \quad (^2). \end{array} \right.$$

The deformed state is characterized by the horizontal and vertical displacements, respectively $u = (u_\alpha)$ and ζ. p is the normal loading surfacic density. The equilibrium is described by the following von Karman's equations, where U is the non-linear deformations tensor of the von Karman's theory; σ is the plane stress tensor (see [13]):

$$(2.2)\quad D\Delta^2\zeta - h(\sigma_{\alpha\beta}\zeta_{,\beta})_{,\alpha} = p, \quad (^2)$$

$$(2.3)\quad \sigma_{\alpha\beta,\beta} = 0,$$

$$(2.4)\quad \sigma_{\alpha\beta} = a_{\alpha\beta\gamma\delta} U_{\gamma\delta},$$

$$(2.5)\quad U_{\gamma\delta} = \varepsilon_{\gamma\delta}(u) + \frac{1}{2}\zeta_{,\gamma}\zeta_{,\delta} \ ; \qquad \varepsilon_{\gamma\delta} = \frac{1}{2}(u_{\gamma,\delta} + u_{\delta,\gamma})$$

Green's formula for plates, proved in [5], is

$$(2.6)\quad a(\zeta,z) + \int_\Omega h\,\sigma_{\alpha\beta}\zeta_{,\alpha}z_{,\beta}\,dx = \int_\Omega pz\,dx + \int_\Gamma h\,\sigma_{\alpha\beta}\zeta_{,\alpha}n_\beta z\,d\Gamma + \int_\Gamma F(\zeta)z\,d\Gamma - \int_\Gamma M(\zeta)\frac{\partial z}{\partial n}\,d\Gamma \ ;$$

where z is a sufficiently regular function and where we put

$$(2.7)\quad a(\zeta,z) = D\int_\Omega \{\zeta_{,11}z_{,11} + \zeta_{,22}z_{,22} + \nu(\zeta_{,11}z_{,22} + \zeta_{,22}z_{,11}) + 2(1-\nu)\zeta_{,12}z_{,12}\}\,dx,$$

$$(2.8)\quad M(\zeta) = -D\{\Delta\zeta + (1-\nu)(2\zeta_{,12}n_1n_2 - \zeta_{,11}n_2^2 - \zeta_{,22}n_1^2)\},$$

$$(2.9)\quad F(\zeta) = -D\{\frac{\partial\Delta\zeta}{\partial n} + (1-\nu)\frac{\partial}{\partial\tau}[(\zeta_{,22} - \zeta_{,11})n_1n_2 + \zeta_{,12}(n_1^2 - n_2^2)]\}.$$

In the same way, we obtain from (2.3), for each sufficiently smooth $v = (v_\alpha)$:

$$(2.10)\quad \int_\Omega \sigma_{\alpha\beta}\varepsilon_{\alpha\beta}(v)\,dx = \int_\Gamma \sigma_{\alpha\beta}v_\alpha n_\beta\,d\Gamma.$$

Thus, various boundary value problems associated with the von Karman's plate theory are exhibited by relations (2.6),(2.10).

The buckling phenomenon we are studing in this paper are produced by edge loading in the plane of the plate; thus, we impose boundary conditions of the form

$$(2.11)\quad \sigma_{\alpha\beta}n_\beta = \lambda t_\alpha \quad \text{on } \Gamma,$$

([1]) Greek indices take values 1 and 2

([2]) We use the summation convention on the repeated indices and the notation $f_{,\alpha} = \dfrac{\partial f}{\partial x_\alpha}$

where $t = (t_\alpha)$ are forces acting on Γ, and λ is a real parameter characterizing the magnitude of the boundary loading. On the contrary, works of G. Duvaut & J.L. Lions [4], M. Potier-Ferry [8] exclude the buckling phenomenon and consider only $u = 0$ or $\sigma_{\alpha\beta} n_\beta = 0$.

Remark 2.1 - The buckling phenomenon can be produced by stress sources (heating, for example) from inside the plate; then, we must replace (2.3) by $\sigma_{\alpha\beta,\beta} = \lambda \Sigma_\alpha$. The present work can be adapted to this situation.

We still have to write the boundary conditions for the vertical deflection ζ, and the imposed unilateral conditions. We distinguish two situations, according to the way the unilateral conditions are written in Ω or on the boundary.

PROBLEM 2.2 - Conditions on the deflection in Ω. The plate is clamped :

(2.12) $\zeta = \frac{\partial \zeta}{\partial n} = 0$ on Γ.

Elsewhere, normal loading possesses a superpotential j, which is proper convex and lower semi-continuous (l.s.c.) function from $\mathbb{R}$ into $(-\infty, +\infty]$ (see [14]) :

(2.13) $-p \in \partial j(\zeta)$,

where $\partial j(\xi) = \{y \in R \,/\, j(\eta) \geqslant j(\xi) + y(\eta - \xi)\ \forall \eta \in R\}$; obviously, j can be function of x in Ω. In order to get an eigenvalue problem, we suppose $\zeta = 0$ solution for all read λ, i.e.

(2.14) $0 \in \partial j(0)$.

If we have to take into account given normal loading, we write $j(\xi) = -\xi f + J(\xi)$, where f is a numerical function on Ω which represents normal loading, and where J is a convex l.s.c. function; (2.14)becomes $f \in \partial J(0)$.

Example 1 - We take $J = 0$; then $f = 0$ by (2.14). It is the classical bilateral case as studied in [6].

Example 2 - J is the indicator of $\mathbb{R}_+$: $J(\xi) = +\infty$ if $\xi < 0$, $J(\xi) = 0$ if $\xi \geqslant 0$. (2.13) gives $\zeta \geqslant 0$, $f_1 = p - f \geqslant 0$, $\zeta f_1 = 0$. These conditions express that the *plate is layed on a rigid support* ($\zeta \geqslant 0$); f_1 is the reaction of the support on the plate. (2.14)imposes $f \leqslant 0$ in Ω, i.e. the external strengths tend to apply the plate on its support.

Example 3 - The plate is layed on an *elastic support:* we take $J(\xi) = -k\xi^2$ if $\xi \leqslant 0$, $J(\xi) = 0$ if $\xi \geqslant 0$ ($k > 0$). Then $f = 0$ in Ω by (2.14).

PROBLEM 2.3 - Unilateral conditions at the edge. We suppose that the vertical strengths density F (defined in (2.9))and the density of moments M on the tangent τ (defined in (2.8)) possess a superpotentiel, respectively k and l, l.s.c., proper convex functions on $\mathbb{R}_+$ into $(-\infty, +\infty]$:

(2.15) $-F(\zeta) \in \partial k(\zeta)$ on Γ,

(2.16) $M(\zeta) \in \partial l\left(\frac{\partial \zeta}{\partial n}\right)$ on Γ.

The surfacic density p is independent of ζ. The following hypothesis ensure that $\zeta = 0$ is solution for all real λ :

(2.17) $p = 0$ in Ω,

(2.18) $0 \in \partial k(0)$, $0 \in \partial l(0)$.

Obviously, k and l can be functions of x in Γ. In the sequel, we suppose that the plate is simply supported on a part $\Gamma_1 \subset \Gamma$, that is (2.19) or (2.20).

(2.19) $x \in \Gamma_1 \implies k(\xi) = 0$ if $\xi = 0$ and $k(\xi) = +\infty$ if $\xi \neq 0$;

(2.20) $\zeta = 0$ on Γ_1.

Example 1 - *Displacement bounded on a side* on $\Gamma_2 \subset \Gamma$: $\zeta \geqslant 0$ on Γ_2; then for $x \in \Gamma_2$, $k(\xi) = +\infty$ if $\xi < 0$, $k(\xi) = 0$ if $\xi \geqslant 0$.

Example 2 - *Displacement with friction* on $\Gamma_2 \subset \Gamma$. We take for $x \in \Gamma_2$, $k(\xi) = -\alpha\xi$ if $\xi \leq 0$, $k(\xi) = \alpha\xi$ if $\xi \geq 0$ $(\alpha > 0)$, and we obtain on Γ_2:

$$|F(\zeta)| \leq \alpha \; ; \; |F(\zeta)| < \alpha \Longrightarrow \zeta = 0; F(\zeta) = -\alpha \Longrightarrow \zeta \geq 0; F(\zeta) = \alpha \Longrightarrow \zeta \leq 0.$$

Example 3 - *Plastic hinge* (or rotation with friction on Γ). The plate is simply supported ($\zeta = 0$ on Γ) and we take $l(\xi) = -\alpha\xi$ if $\xi \leq 0$, $l(\xi) = \alpha\xi$ if $\xi \geq 0$ $(\alpha > 0)$. From (2.16), we obtain :

$$|M(\zeta)| \leq \alpha \; ; \; |M(\zeta)| < \alpha \Longrightarrow \frac{\partial\zeta}{\partial n} = 0 \; ; M(\zeta) = -\alpha \Longrightarrow \frac{\partial\zeta}{\partial n} \leq 0 \; ; M(\zeta) = \alpha \Longrightarrow \frac{\partial\zeta}{\partial n} \geq 0.$$

Results 2.4 - The precise formulation of results obtained for the problems 2.2 and 2.3 constitutes our theorem 4.3. Provisionally, we state the following. Hypothesis are :

(2.21) $\exists q > 0$ such that $j(t\xi) \leq t^q j(\xi)$ $\forall \xi \in \mathbb{R}$, $t \geq 1$ (and the same for k and l).

In the case of problem 2.3, we suppose that Γ_1 is non rectilinear and has a positive measure. Then, for each of the two problems, we have the following conclusions :

1°) $\zeta = 0$ is unique solution for $\lambda'^* \leq \lambda \leq \lambda^*$, where $\lambda'^* < 0 < \lambda^*$;

2°) there exists a family of non-trivial solutions $(\lambda_r, u_r, \zeta_r, r > 0)$, $\zeta_r \neq 0$, indexed by positive reals. (This conclusion is an existence theorem which describes the buckling phenomenon).

3 - VARIATIONAL FORMULATION - The previous problems will be studied in functional spaces described in [4] : $V = (H^1(\Omega))^2$ is the set of kinematically admissible horizontal displacements ; the vertical deflection ζ will be found in a closed subspace Z of $H^1(\Omega)$.

We begin by eliminating u in relations (2.2), ... , (2.5), (2.11); first, one resolve the plane problem of elasticity : find $u^0 \in V$ such that :

(3.1) $\quad \sigma^0_{\alpha\beta,\beta} = 0$, $\sigma^0_{\alpha\beta} = a_{\alpha\beta\gamma\delta}\, \varepsilon_{\gamma\delta}(u^0)$, in Ω ; $\sigma^0_{\alpha\beta} n_\beta = t_\alpha$ on Γ.

This problem has a solution, unique except rigid (plane) displacement, if the total force and the total moment resulting are zero ; but $\sigma^0_{\alpha\beta}$ is unique because $\varepsilon_{\gamma\delta}$ is zero on rigid plane displacements.

Remark 3.1 - The previous result supposes that $t_\alpha \in H^{-1}(\Omega)$, $\alpha = 1,2$. Then, we have $\sigma^0_{\alpha\beta} \in L^2(\Omega)$, $\alpha, \beta = 1,2$. But it is possible to assume more regularity for $\sigma^0_{\alpha\beta}$ if the data t_α are more regular (by regularity theorems for elliptic problems; see [16], [17], for example).

Let z fixed in $H^2(\Omega)$. Using [5], there exists $v = v(z) \in Z$ such that :

(3.2) $\quad S_{\alpha\beta,\beta} = 0$, $S_{\alpha\beta} = a_{\alpha\beta\gamma\delta}(\varepsilon_{\gamma\delta}(v) + \frac{1}{2} z_{,\gamma} z_{,\delta})$, in Ω ; $S_{\alpha\beta} n_\beta = 0$ on Γ,

as before, $S_{\alpha\beta} = S_{\alpha\beta}(v)$ is unique. In addition, $S_{\alpha\beta}(z) \in L^2(\Omega)$ because $H^2(\Omega) \subset W^{1,4}(\Omega)$; but this inclusion is compact and we deduce (3.3); elsewhere, we get (3.4).

(3.3) The mapping $z \to S_{\alpha\beta}(z)$ is compact from $H^2(\Omega)$ into $L^2(\Omega)$ $\quad (\alpha, \beta = 1,2)$.

(3.4) $S_{\alpha\beta}(\tau z) = \tau^2 S_{\alpha\beta}(z)$ $\quad \forall \tau \in \mathbb{R}$, $z \in H^2(\Omega)$; $\alpha, \beta = 1,2$.

Let $u \in V$, $\zeta \in H^2(\Omega)$ a solution of (2.2), ... , (2.6), (2.12) ; the previous study of systems (3.1), (3.2) shows that :

(3.5) $\quad u = \lambda u^0 + v(\zeta)$, $\sigma_{\alpha\beta} = \lambda\sigma^0_{\alpha\beta} + S_{\alpha\beta}(\zeta)$.

Now, let Z be a closed subspace of $H^2(\Omega)$; we choose in Z a scalar product (,) such that the associated norm is equivalent to the initial norm. Assume that $\sigma^0_{\alpha\beta}$ is bounded in Ω (remark 3.1). Then, for each $z \in Z$, we have $|\int_\Omega \sigma^0_{\alpha\beta} z_{,\alpha} \bar{z}_{,\beta} dx| \leq c \, \|z\|_{W^{1,4}(\Omega)} \|\bar{z}\|$, and there exists a bounded linear operator L on Z with properties (3.6), (3.7); in order to simplify, we shall assume (3.8 [1]).

---

[1] Operators L and C have be studied by M,S. Berger [6], in a related way.

(3.6) $\int_\Omega h\,\sigma^0_{\alpha\beta}\, z_{,\alpha}\, \bar z_{,\beta}\, dx = (L z, \bar z) \quad \forall z, \bar z \in Z,$

(3.7) L is self-adjoint and compact,

(3.8) $(L z, z) \geq 0$; $(L z, z) \neq 0$ if $z \neq 0$.

In the same way, there exists a nonlinear operator C with the three properties

(3.9) $(Cz, \bar z) = \int_\Omega h\, S_{\alpha\beta}(z)\, z_{,\alpha}\, \bar z_{,\beta}\, dx \quad \forall z, \bar z \in Z,$

(3.10) C is compact (by (3.3)),

(3.11) C is cubic : $C(\tau z) = \tau^3 C z$; $\| Cz \| \leq c \, \| z \|^3$ (by (3.4)).

Lemma 3.2 - The functional $z \to (Cz,z)$ is differentiable on Z ; its differential is the linear functional $h \to 4\,(Cz,h)$.

Proof - Let f,g fixed in $H^2(\Omega)$; as in (3.2), there exists $v = v(f,g) \in V$ such that

$S_{\alpha\beta} = a_{\alpha\beta\gamma\delta}(\varepsilon_{\gamma\delta}(v) + \frac{1}{2} f_{,\gamma}\, g_{,\delta})$, $S_{\alpha\beta,\beta} = 0$ in Ω, $S_{\alpha\beta} n_\beta = 0$ on Γ.

From symmetry properties (2.1) and reciprocity theorem (see [18]) we obtain

$\int_\Omega S_{\alpha\beta}(f,g)\, \bar f_{,\alpha}\, \bar g_{,\beta}\, dx = \int_\Omega S_{\alpha\beta}(\bar f, \bar g)\, f_{,\alpha}\, g_{,\beta}\, dx, \; \forall f, \bar f, g, \bar g \in H^2(\Omega).$

The proof can now be completed without any difficulty.

VARIATIONAL FORMULATION FOR THE PROBLEM 2.2 - Let $Z = H^2_0(\Omega)$; on this space, the continuous bilinear form $a(z_1, z_2)$ is coercive. We choose on Z the scalar product defined by this form and we write :

(3.12) $a(z_1, z_2) = (z_1, z_2)$

For $z \in Z$, we define

(3.13) $\varphi(z) = \int_\Omega j(z)\,dx$, if $j(z) \in L^1(\Omega)$, $\varphi(z) = +\infty$ otherwise:

(2.14) give $\varphi(z) \geq \varphi(\zeta) - \int_\Omega p\,(z - \zeta)\,dx$, $\forall z \in Z$. φ is a l.s.c. proper convex function (see [15]). Besides (2.14) implies (3.14), and we can always assume (3.15):

(3.14) $0 \in \partial\varphi(0)$,

(3.15) $\varphi(0) = 0$.

Green's formula (2.6) leads to the variational formulation of problem 2.2. We summarize :

Problem 3.3 - *Let L defined in (3.6), with properties (3.7), (3.8); C defined in (3.9), with properties (3.10), (3.11); φ defined in (3.13), with properties (3.14), (3.15). Find $\lambda \in \mathbb{R}$, $\zeta \in Z$ such that :*

(3.16) $((I - \lambda L)\zeta + C\zeta, z - \zeta) + \varphi(z) - \varphi(\zeta) \geq 0 \quad \forall z \in Z$ (1)

VARIATIONAL FORMULATION FOR THE PROBLEM 2.3 - We assume

(3.17) $\zeta = 0$ on $\Gamma_1 \subset \Gamma$,

where Γ_1 is nonrectilinear and has a positive measure. Let

(3.18) $Z = \{ z \in H^2(\Omega) / z = 0 \text{ on } \Gamma_1 \}$;

(3.17) ensure that $a(z_1, z_2)$ is coercive on Z (see [5], ch.4, rem. 4.5).

---

(1) I is the identity mapping on Z

We choose this bilinear form as scalar product on Z and we put

$a(z_1, z_2) = (z_1, z_2)$.

For $z \in Z$, we define :

$$(3.19)\ \psi(z) = \begin{cases} \int_\Gamma k(z)d\Gamma + \int_\Gamma l(\frac{\partial z}{\partial n})d\Gamma, \text{if } k(z) \text{ and } l(\frac{\partial z}{\partial n}) \in L^1(\Gamma), \\ +\infty \text{ otherwise.} \end{cases}$$

Function ψ is convex ; besides, it is l.s.c. on Z : if $\lim z_k = z$ in Z, then $\lim z_k = z$ and $\lim \frac{\partial z_k}{\partial n} = \frac{\partial z}{\partial n}$ in $L^2(\Gamma)$ and almost everywhere on Γ, at least for some subsequence; one concludes by using the Fatou's lemma. In addition, (2.18) implies (3.20) and we can always assume (3.21) :

(3.20) $0 \in \partial \psi(0)$,

(3.21) $\psi(0) = 0$.

By (2.15) and (2.16), we obtain

$$\psi(z) \geqslant \psi(\zeta) - \int_\Gamma (z-\zeta) F(\zeta) d\Gamma + \int_\Gamma \frac{\partial}{\partial n}(z-\zeta) M(\zeta) d\Gamma, \quad \forall z \in Z.$$

Green's formula (2.7) leads to the variational formulation of problem 2.3 ; we summarize :

Problem 3.4 - *Let L, defined in (3.6), with properties (3.7), (3.8); C, defined in (3.9), with properties (3.10), (3.11); ψ, defined in (3.19), with properties (3.17), (3.20), (3.21).*

Find $\lambda \in \mathbb{R}$, $\zeta \in Z$ such that :

(3.22) $((I - \lambda L)\zeta + C\zeta, z - \zeta) + \psi(z) - \psi(\zeta) \geqslant 0 \quad \forall z \in Z$.

4 - THE ABSTRACT PROBLEM : FORMULATION AND RESULTS - In the present section, we describe an abstract problem (problem 4.1), containing the concrete cases of § 3 ; but other applications can be given (see [2]). The results are stated precisely in theorem 4.3.

Z is a real Hilbert space in which scalar product and norm are denotes by (,) and $\| \ \|$ respectively. The data are L, C, ψ, with the following hypothesis.

(4.1) L is a self-adjoint compact operator on V;

(4.2) $(Lz, z) \geqslant 0 \quad \forall z \in Z$; $(Lz,z) > 0$ if $z \neq 0$.

(4.3) C is a bounded continuous non-linear operator on V;

(4.4) $\exists$ a functional Φ on V, Gateaux-differentiable, such that $\Phi(0)=0$ and $\Phi'(z) = Cz$;

one says C is a « variational operator » ; observe that

(4.5) $\Phi(z) = \int_0^1 (C(tz), z)dt \quad \forall z \in Z \quad (\Phi(z) = \frac{1}{4}(Cz,z)$ in section 3);

(4.6) $C(0) = 0$;

(4.7) $(Cz,z) \geqslant 0 \quad \forall z \in Z$;

since $\Phi(0) = 0$, (4.7) implies the following

(4.8) $\Phi \geqslant 0$;

(4.9) $(Cz,z) = 0$ imply $z = 0$ [1];

(4.10) C is compact ;

---

[1] This condition is satisfied when C and Z are defined as in section 3 (see [20]).

(4.11) $\exists p > 0$ such that $\|Cz\| < c_1 \|z\|^p \quad \forall z \in Z.$

(4.12) ψ is a l.s.c. convex function on Z ;

(4.13) $\psi(0) = 0$;

(4.14) $\psi \geq 0$;

by these two relations, we obtain

(4.15) $0 \in \partial\psi(0)$;

(4.16) $\mathrm{dom}\,\psi = \{z \in Z / \psi(z) < +\infty\} \neq \{0\}$;

(4.17) $\exists q > 0$ such that $\psi(tz) \leq t^q \psi(z) \quad \forall z \in Z$ and $t \geq 1$ ([1]);

(4.18) $\forall B$, bounded in Z, ψ is bounded on $B \cap \mathrm{dom}\,\psi$ ([2]).

Problem 4.1 - *Find $\lambda \in \mathbb{R}$, $\zeta \in Z$ such that*

(4.19) $((I - \lambda L)\zeta + C\zeta, z - \zeta) + \psi(z) - \psi(\zeta) \geq 0 \quad \forall z \in Z$

Remark 4.2 - (4.19) is equivalent to $-((I - \lambda L)\zeta + C\zeta) \in \partial\psi(\zeta)$.

Observe that (4.15) express that $\zeta = 0$ is solution $\forall \lambda \in \mathbb{R}$. Thus (4.19) is an eigenvalue problem for a variational inequation.

Theorem 4.3 - *In the scheme which is described by (4.1), ... , (4.18), we get :*

1°) $\zeta = 0$ is the unique solution if $\lambda \not< \lambda^*$, where λ^* is the smallest characteristic value of L.

2°) There exists a family of nontrivial solutions $(\lambda_r, \zeta_r, r > 0)$, $\zeta_r \neq 0$, indexed by all positive reals. More precisely, for each $r > 0$, there exist a solution (λ_r, ζ_r) such that

$(L\zeta_r, \zeta_r) = \sup\limits_{\partial A_r} (Lz, z)$, with $\partial A_r = \{z \in Z / \frac{1}{2}\|z\|^2 + \Phi(z) + \psi(z) = r\}$

In addition, $\lim\limits_{r \to 0} \zeta_r = 0$, $\lim\limits_{r \to \infty} \zeta_r = +\infty$ *, in Z.*

Remark 4.4 - The previous theorem precisely states result 2.4. More particularly, λ'^* is the largest negative characteristic value of L ; here, $\lambda'^* = -\infty$ because of the simplifier hypothesis (3.7) or (4.2).

5 - PROOFS - For the uniqueness, we choose $z = \zeta/2 \neq 0$ in (4.19). By (4.7), (4.9), (4.13), (4.14), we obtain $\lambda(L\zeta, \zeta) > \|\zeta\|^2$; then $\lambda > 0$ and $\lambda^{-1} < \|\zeta\|^{-2}(L\zeta, \zeta) < \lambda^{*-1}$.

The existence result in theorem 4.3 is proved in three stages. First, we study the problem when ψ is differentiable : in this case, we adapt M.S. Berger's work [7]. In the second stage ψ is any function : we use a regularity process. Third, we show that these solutions can be indexed by the positive reals.

To begin with, let us introduce some notations. We introduce the functional F, then sets A_r, ∂A_r for each $r > 0$:

(5.1) $F(z) = \frac{1}{2}\|z\|^2 + \Phi(z) + \psi(z)$,

(5.2) $A_r = \{z \in Z / F(z) \leq r\}$, $\partial A_r = \{z \in Z / F(z) = r\}$

Lemma 5.1 - *With the hypothesis of the section 4, and for each fixed $r > 0$,*

(i) *A_r is bounded ;*

(ii) *0 is not in ∂A_r ;*

(iii) *$\forall z \in \mathrm{dom}\,\psi$, $z \neq 0$, $\exists t > 0$ such that $tz \in \partial A_r$ (and $t \geq 1$ if $z \in A_r$).*

([1]) This condition imposes that dom ψ is a cone with 0 as vertex.

([2]) This condition results from (2.21), when ψ is defined as in section 3.

First stage - ψ is differentiable on Z. Thus $\partial\psi$ is univalued on Z and (remark 4.2) we must find (λ,ζ) such that $(I-\lambda L)\zeta + C\zeta + \partial\psi(\zeta) = 0$. Introduce the

Problem 5.2 - *Let* $r > 0$ *fixed. Find* $\zeta_r \in \partial A_r$ *such that* $(L\zeta_r, \zeta_r) = \sup_{\partial A_r} (Lz,z)$.

Since F is differentiable, we obtain the

Lemma 5.3 - *If* ζ_r *is solution of problem 5.2, then* ζ_r *is non-trivial solution of the problem 4.1, with a suitable* $\lambda = \lambda_r$.

Then, when ψ is differentiable, theorem 4.3 is a consequence of the :

Lemma 5.4 - *For each* $r > 0$, *problem 5.2 possesses a solution* ζ_r. *In addition,* $\lim_{r\to 0} \zeta_r = 0$ *and* $\lim_{r\to\infty} \|\zeta_r\| = \infty$.

Proofs - 1°) (Lz,z) is bounded on the bounded set ∂A_r. Let $(z_n, n \geqslant 0)$ a maximizing sequence which we can choose to be weakly convergent with weak limit z. Then $\lim (L z_n, z_n) = (Lz,z)$ by (4.1); $\lim \Phi(z_n) = \Phi(z)$ by (4.5) and (4.10). The semicontinuity properties of ψ and of the norm for the weak topology implies $F(z) \leqslant r$, i.e. $z \in A_r$. Since by construction $z \neq 0$, there exists $t \geqslant 1$ such that $t z \in \partial A_r$. But $(L(tz),tz) = t^2 (Lz,z)$ and the extremum impose $t = 1$ and $z \in \partial A_r$. Then $z = \zeta_r$ is a solution.

2°) $\|\zeta_r\|^2 \leqslant 2r$, therefore $\lim_{r\to 0} \zeta_r = 0$. Furthermore, F is a sum of bounded functions; then $\lim_{r\to\infty} \|\zeta_r\| = \infty$.

Second stage - ψ is no longer supposed to be differentiable. $A = \partial\psi$ is a maximal monotone multivalued operator. We recall some properties of such operators (see [15] for more details). For each $\varepsilon > 0$, $J_\varepsilon = (I+\varepsilon A)^{-1}$, the resolvent of A, is a contraction on all Z. The Yosida-approximation of A is defined by (5.3) and there exists a convex function ψ_ε, which is Frechet-differentiable on Z, and satisfies (5.4); it is characterized by (5.5), with the property (5.6).

(5.3) $\quad A_\varepsilon = \varepsilon^{-1}(I - J_\varepsilon)$,

(5.4) $\quad \partial\psi_\varepsilon = A_\varepsilon$,

(5.5) $\quad \psi_\varepsilon(z) = \inf_{v\in Z} \{\frac{1}{2\varepsilon}\|v-z\|^2 + \psi(v)\} = \frac{1}{2\varepsilon}\|z - J_\varepsilon z\|^2 + \psi(J_\varepsilon z)$,

(5.6) $\quad \psi_\varepsilon(z) \nearrow \psi(z)$ increasingly as ε decreases to 0.

It is now easy to verify that the hypothesis (4.13), (4.14), (4.17) for ψ imply the same properties for ψ_ε. Therefore, there exist (by the first stage for each $r > 0$, $\zeta^\varepsilon \in \partial A_r^\varepsilon$, $\lambda^\varepsilon \in \mathbb{R}$ (obviously, F_ε and ∂A_r^ε are defined as in (5.1), (5.2) by replacing ψ by ψ_ε) such that

(5.7) $\quad ((I - \lambda^\varepsilon L)\zeta^\varepsilon + C\zeta^\varepsilon, z - \zeta^\varepsilon) + \psi_\varepsilon(z) - \psi_\varepsilon(\zeta^\varepsilon) \geqslant 0 \quad \forall z \in Z.$

A priori estimates - Since $\zeta^\varepsilon \in \partial A_r^\varepsilon$, we have the estimations

(5.8) $\quad \|\zeta^\varepsilon\| \leqslant c_1 ;\ \psi_\varepsilon(\zeta^\varepsilon) \leqslant c_2 ;\ \frac{1}{2\varepsilon}\|\zeta^\varepsilon - J_\varepsilon\zeta^\varepsilon\| \leqslant c_3$.

We put (5.9) (where the inequality comes from (4.2) and from lemma 5.1)

(5.9 $\quad c = \sup_{A_r} (Lz, z) > 0$.

Since $A_r \subset A_r^\varepsilon$ by (5.6), we have

(5.10) $\quad (L\zeta^\varepsilon, \zeta^\varepsilon) \geqslant c > 0$.

Choosing $z = 2\zeta_\varepsilon$ in (5.7), we obtain

$$\lambda^\varepsilon (L\zeta^\varepsilon, \zeta^\varepsilon) \leqslant \|\zeta^\varepsilon\|^2 + (C\zeta^\varepsilon, \zeta^\varepsilon) + \psi_\varepsilon(2\zeta^\varepsilon) - \psi_\varepsilon(\zeta^\varepsilon).$$

Since $\lambda^\varepsilon > 0$, and by (4.11), (4.17) (1), (5.8), (5.10), we have

(5.11) λ^ε is bounded in $\mathbb{R}$.

Limit process - The estimates (5.8) and (5.11) imply that there exists a subsequence, which we still call ε with limit 0, such that

$$\lim \zeta^\varepsilon = \zeta \text{ in the weak topology of } Z \text{ ; } \lim \lambda^\varepsilon = \lambda \text{ in } \mathbb{R}.$$

In addition, $\lim J_\varepsilon \zeta^\varepsilon = \zeta$ weakly in Z, by (5.8). Elsewhere, $\psi(J_\varepsilon \zeta^\varepsilon) \leqslant \psi_\varepsilon(\zeta^\varepsilon)$ by (5.5) and $\psi_\varepsilon(z) \leqslant \psi(z)$ by (5.6), therefore

$$((I - \lambda^\varepsilon L)\zeta^\varepsilon + C\zeta^\varepsilon, z - \zeta^\varepsilon) + \psi(z) - \psi(J_\varepsilon \zeta^\varepsilon) \geqslant 0 \quad \forall z \in Z.$$

Compactness hypothesis (4.1), (4.10), and semi continuity of ψ and of the norm, imply

$$\|\zeta\|^2 + \psi(\zeta) \leqslant \liminf \{\|\zeta^\varepsilon\|^2 + \psi(J_\varepsilon \zeta^\varepsilon)\}$$
$$\leqslant (\zeta, z) + \lambda(L\zeta, z - \zeta) + (C\zeta, z - \zeta) + \psi(z) \quad \forall z \in Z.$$

Therefore $\zeta = 0$ is solution, and $\zeta \neq 0$ by (5.10).

Third stage - We get the following for the solution ζ displayed in the second stage :

Lemma 5.5 - $\zeta \in \partial A_r$, $(L\zeta, \zeta) = \sup_{\partial A_r} (Lz, z)$.

Proof - First, $\zeta \in \partial A_r$: $\frac{1}{2}\|\zeta\|^2 + \psi(\zeta) \leqslant \liminf \{\frac{1}{2}\|\zeta^\varepsilon\|^2 + \psi(J_\varepsilon \zeta^\varepsilon)\} \leqslant r - \Phi(\zeta)$; therefore $\zeta \in \partial A_r$. Elsewhere, by (5.9) and (5.10), $(L\zeta, \zeta) = \sup_{A_r} (Lz, z)$. If $\zeta \notin \partial A_r$, there exists $t > 1$ such that $t\zeta \in \partial A_r$ (lemma 5.1); but $(L(t\zeta), t\zeta) > (L\zeta, \zeta)$; such an inequality contradicts the previous property of a maximum. Finally, $\zeta_r \in \partial A_r$ implies $\lim_{r \to 0} \zeta_r = 0$; elsewhere F is bounded on dom $\psi \cap B$, for all bounded B in Z, by (4.18), therefore $\lim F(\zeta_r) = +\infty$ and also $\lim \|\zeta_r\| = +\infty$ when $r \to \infty$.

REFERENCES

[1] B. BUDIANSKY, Theory of buckling and post-buckling behavior of elastic structures, Advance in applied Mechanics, 14, Chia Shun Yih ed., Academic Press (1974).

[2] Cl. DO, Problèmes de valeurs propres d'inéquations variationnelles; application aux plaques minces, International Congress of Mathematicians, Vancouver (1974).

[3] Cl. DO, Problèmes de valeurs propres pour une inéquation variationnelle sur un cône et application au flambement unilatéral d'une plaque mince, C.R. Acad. Sc., A, 280, p. 45-48 (1975).

[4] G. DUVAUT and J.L. LIONS, Problèmes unilatéraux dans la théorie de la flexion forte des plaques, J. Méca., 13, N° 1, p. 51-74 (1974).

[5] G. DUVAUT and J.L. LIONS, Les inéquations en Mécanique et en Physique, Dunod, Paris (1972).

[6] M.S. BERGER, On von Karman's Equations and the buckling of thin elastic plate, I, the clamped plate, Comm. Pure Appl. Math, XX, p. 687-719 (1967).

[7] M.S. BERGER, A bifurcation theory for nonlinear elliptic partial differential equations and related systems, Bifurcation theory and nonlinear eigenvalue problems, J.B. Keller & S. Antman ed., Benjamin (1969).

(1) It is essentially now that we use the fact that dom Ψ is a cone.

[8] M. POTIER-FERRY, Problèmes unilatéraux en théorie des plaques non-linéaires, Thèse de 3ème cycle, Univ. de Paris VI (1974).

[9] J.P. DIAS, Vairational inequalities and eigenvalue problems for nonlinear maximal monotone operators in a Hilbert space, à paraître dans Amer. J. Maths.

[10] J.P. DIAS, Un théorème de Sturm-Liouville pour une classe d'opérateurs non-linéaires maximaux monotones, J. Math. Anal. Appl. 47, p. 400-405 (1974).

[11] H. BEIRAO-DA-VEIGA and J.P. DIAS, Sur la surjectivité de certains opérateurs non-linéaires liés aux inéquations variationnelles, à paraître dans Bolletino della Unione Matematica Italiana.

[12] J.P. DIAS and J. HERNANDEZ, A Sturm-Liouville theorem for some odd multivalued maps, à paraître dans Proceedings Amer. Math. Soc.

[13] L. LANDAU and E. LIFCHITZ, Théorie de l'élasticité, Physique théorique , t. VII, Editions Mir,Moscou (1967).

[14] J.J. MOREAU, La notion de sur-potentiel et les liaisons unilatérales en élastostatique, C.R. Acad. Sc.,A.,267 p.954-957 (1968).

[15] H. BREZIS, Monotonicity methods in Hilbert space and some applications to nonlinear differential equations, Contribution to nonlinear functional analysis, E.H. Zarantonello ed., Acad. Press, p. 101-156 (1971).

[16] J.L. LIONS and E. MAGENES, Problèmes aux limites non homogènes et applications. Dunod, Paris (1968).

[17] J. NECAS, Les méthodes directes dans la théorie des équations elliptiques, Acad. Tchécoslovaque des Sciences, Prague (1967).

[18] P. GERMAIN, Mécanique des milieux continus, Masson, Paris (1962).

[19] M.S. BERGER and P. FIFE, von Karman's equations and the buckling of a thin elastic plate, II, plate with general edge conditions, Comm. Pure Appl. Math., Vol. XXI, p. 227-241 (1968).

[20] M. POTIER-FERRY, Charges limites et théorème d'existence en théorie des plaques élastiques, C.R. Acad. Sc., A, 280, p.1317-1320, and p. 1385-1387 (1975).

PROBLEMES DE CONTACT ENTRE CORPS SOLIDES DEFORMABLES

Georges DUVAUT

I.M.T.A. - Université Pierre et Marie Curie
et Lab.Ass. C.N.R.S. N°229
4, Place Jussieu
75230 PARIS CEDEX 05
FRANCE

1 - Introduction.

On se propose l'étude des déformations dans un demi-espace sur lequel s'appuie un solide rigide convexe. Dans les cas où le demi-espace est élastique le problème aux déplacements a été étudié par l'auteur (G.DUVAUT [1]) et par M.LOPPIN [1]. Il correspond au problème de HERTZ lorsque le solide de révolution est une sphère. Dans ce cas un calcul numérique de la solution est en cours de réalisation par M.LOPPIN[1], en vue de comparer les résultats avec la solution approchée donnée par HERTZ.

Nous généralisons ici les résultats aux situations où le demi-espace est non-élastique, plastique de type HENCKY ou viscoélastique.

Ce travail conduit sur le plan mathématique à des inéquations qui présentent des particularités assez intéressantes. Sur le plan mécanique, outre son intérêt direct, il constitue une étude préliminaire pour les cas où le solide rigide serait en roulement sur le demi-espace.

2 - Formulation générale du problème.

Désignons par Ω le demi-espace déformable défini par

(2.1) $$x_3 > 0 ,$$

dans le repère orthonormé $0\, x_1\, x_2\, x_3$, où $0\, x_3$ est vertical descendant. Le solide Ω_1 est convexe et on désigne par Γ_1 l'ensemble des points du plan $x_3 = 0$ qui sont sur la verticale d'un point de Ω_1, soit

(2.2) $$\Gamma_1 = \left\{ (x_1, x_2, 0) \;\middle|\; \exists\, x_3 \leq 0 \,,\ (x_1, x_2, x_3) \in \Omega_1 \right\}$$

Sous l'effet d'une force verticale $F > 0$, le solide rigide Ω_1 est appliqué contre Ω en le déformant. Désignons par $u = (u_1, u_2, u_3)$ le champ de déplacements infinitésimaux de Ω et par $U > 0$ le déplacement vertical de Ω_1. Si nous désignons par

(2.3) $$-x_3 = \varphi(x_1 , x_2)$$

l'équation, définie sur Γ_1, de la calotte inférieure de Ω_1 avant déformation et en supposant le contact établi, c'est à dire

(2.4) $$\exists\, (x_1, x_2) \quad \text{tel que} \quad \varphi\, (x_1, x_2) = 0 .$$

Nous supposons φ continue et bornée sur Γ_1. Nous allons formuler les conditions aux limites de ce problème dans le cadre habituel des déplacements infinitésimaux, c'est à dire que les conditions aux limites qui s'appliquent dans la réalité aux points des milieux déformés seront écrites aux points des milieux non déformés.

Nous désignons par (σ_{ij}) le tenseur des contraintes dans $\bar{\Omega}$ (fermeture de Ω). Nous avons alors

(2.6) $$\sigma_{i3} = 0 \quad \text{sur} \quad \Gamma - \Gamma_1$$

Sur Γ_1, du fait de la non-interpénétration, nous avons

(2.7) $$U \leq u_3 + \varphi \quad \text{sur} \quad \Gamma_1$$

ce qui permet de distinguer sur Γ_1 une zone de contact et une zone de non-contact. Sur la zone de non-contact on a

(2.9) $$U < u_3 + \varphi \quad , \sigma_{13} = \sigma_{23} = \sigma_{33} = 0 .$$

Sur la zone de contact, on a

(2.10) $$U = U_3 + \varphi \quad , \quad \sigma_{33} \leq 0 ,$$

et si on suppose le contact sans frottement - hypothèse habituelle dans ce type de problème - on a de plus,

(2.11) $$\sigma_{31} = \sigma_{32} = 0 \ .$$

L'équilibre du système entraine enfin

(2.12) $$\int_{\Gamma_1} \sigma_{33} \ dx_1 \ dx_2 = - F \ .$$

On doit ajouter à ces conditions les équations d'équilibre (*)

(2.13) $$\frac{\partial}{\partial x_j} \sigma_{ij} = 0 \qquad \text{dans } \Omega ,$$

les équations de comportement qui seront précisées ultérieurement et les conditions à l'infini, qui seront en fait imposées par le cadre fonctionnel dans lequel on cherche la solution, et qui expriment qu'en un certain sens les déplacements et les déformations tendent vers zéro à l'infini.

3 - Formulation variationnelle du problème aux déplacements.

3.1 Cadre fonctionnel.

Introduisons l'espace $\mathcal{D}(\bar{\Omega})$ des fonctions indéfiniment différentiables et à support compact dans $\bar{\Omega}$. Muni de la norme $\mathcal{E}(v)$

(3.1) $$\mathcal{E}(v) = \left\{ \int_{\Omega} \mathcal{E}_{ij}(v) \ \ \mathcal{E}_{ij}(v) \ \ dx \right\}^{1/2} \quad , \ v = (v_1, v_2, v_3)$$

(3.2) $$\mathcal{E}_{ij}(v) = \frac{1}{2} \left(\frac{\partial v_i}{\partial x_j} + \frac{\partial v_j}{\partial x_i} \right)$$

l'espace $\left[\mathcal{D}(\bar{\Omega})\right]^3$ est préhilbertien. Par complétion on obtient un espace de Hilbert W. On utilisera les deux lemmes suivants (**)

<u>Lemme 3.1</u>. L'espace W s'identifie algébriquement et topologiquement à

(3.3) $$\left\{ v \ \middle| \ v \in \left[L^6(\Omega)\right]^3 , \ \ \mathcal{E}_{ij}(v) \in L^2(\Omega) \right\} .$$

<u>Lemme 3.2</u>. L'application

(3.4) $$v \longrightarrow v|_{\Gamma}$$

définie sur $\left[\mathcal{D}(\Omega)\right]^3$ se prolonge en une application linéaire continue de W dans $\left[L^4(\Gamma)\right]^3$.

(*) On n'écrit pas les forces de pesanteur car on ne s'intéresse qu'aux surcontraintes créées par l'appui du corps rigide.

(**) Ces résultats sont des conséquences de J.L.LIONS [1].

3.2 Formulation variationnelle.

Introduisonsle convexe K par

(3.6) $$K = \left\{ (v , V) \;\middle|\; v \in W , V \in \mathbb{R} , V \leq v_3 + \varphi \text{ sur } \Gamma_1 \right\}.$$

Il est clair que K est convexe fermé dans $W \times \mathbb{R}$. Supposons que (u, U) soit solution classique du problème posé au N°2 ; alors

(3.7) $$(u , U) \in K .$$

Soit (σ_{ij}) le champ de contraintes correspondant. Multiplions (2.12) par $v_i - u_i$, où $(v, V) \in K$. Il vient, après intégration par parties

(3.8) $$\int_\Omega \sigma_{ij} \varepsilon_{ij} (v - u) \, dx = \int_\Gamma \sigma_{ij} n_j (v_i - u_i) \, dx_1 \, dx_2 = -\int_{\Gamma_1} \sigma_{33} (v_3 - u_3) \, dx_1 \, dx_2 .$$

Le membre de droite peut alors être transformé de la manière suivante,

(3.9) $$-\int_{\Gamma_1} \sigma_{33} (v_3 - u_3) \, dx_1 \, dx_2 = -\int_{\Gamma_1} \sigma_{33} (v_3 - V + U - u_3 + V - U) dx_1 dx_2 \geq -\int_{\Gamma_1} \sigma_{33} (V - U) \, dx_1 \, dx_2 = F (V - U).$$

en tenant compte de (2.7) - (2.9), (2. 11). Nous avons alors

Propriété 3.1.

Si (u , U) et (σ_{ij}) sont les déplacements et les contraintes, solution classique du problème posé au N°2, alors

(3.10) $$(u , U) \in K$$

(3.11) $$\int_\Omega \sigma_{ij} \varepsilon_{ij} (v - u) \, dx \geq F (V - U) , \quad (v, V) \in K.$$

4 - Cas élastique.

Si le demi-espace Ω est élastique la loi de comportement qui relie σ_{ij} et $\varepsilon_{ij}(u)$ est de la forme,

(4.1) $$\sigma_{ij} = a_{ijkh} \varepsilon_{kh}(u)$$

où les coefficients d'élasticité a_{ijkh} satisfont les propriétés habituelles de symétrie et ellipticité (G.DUVAUT et J.L.LIONS [1])

(4.2) $$a_{ijkh} = a_{khij} = a_{jikh}$$
$$\exists \alpha > 0 , \text{ tel que } a_{ijkh} \tau_{ij} \tau_{kh} \geq \alpha \tau_{ij} \tau_{ij}, \forall \tau_{ij} = \tau_{ji}.$$

En posant

(4.3) $$a(u, v) = \int_{\Omega} a_{ijkh}\, \varepsilon_{ij}(u)\, \varepsilon_{kh}(v)\, dx$$

la condition (3.11) s'écrit

(4.4) $$a(u, v - u) \geqslant F(V - U), \quad (v, V) \in K.$$

Propriété 4.1.

Si (u , U) est un champ de déplacements, solution classique du problème posé au N°2 dans le cas élastique, alors (u, U) minimise dans K la fonctionnelle

(4.5) $$I(v, V) = \frac{1}{2} a(v, v) - F.V.$$

On peut alors établir le

Théorème 4.1.

Il existe un unique $(u, U) \in K$ satisfaisant (4.4).

Démonstration.

i) Existence.

Il suffit de démontrer que

(4.6) $$I(v, V) \longrightarrow +\infty \quad \text{si} \quad (v, V) \longrightarrow +\infty \text{ dans } K.$$

Comme en tout point de Γ_1 on a

(4.7) $$V \leqslant v_3 + \varphi$$

on en déduit par intégration sur Γ_1,

(4.8) $$V(\text{mes}\,\Gamma_1) \leqslant \int_{\Gamma_1} v_3\, dx_1\, dx_2 + \int_{\Gamma_1} \varphi\, dx_1\, dx_2 .$$

puis par des transformations simples et l'application du lemme 3.2

(4.9) $$V \leqslant C_1 + C_2\, \mathcal{E}(v)$$

où C_1 et C_2 sont des constantes positives. Il en résulte que

(4.10) $$\mathcal{E}(v) + V \leqslant C_1 + (C_2 + 1)\, \mathcal{E}(v)$$

et donc que $\mathcal{E}(v) + V \to +\infty$ entraine que $\mathcal{E}(v) \to +\infty$. Par ailleurs

(4.11) $$I(v, V) \geqslant \frac{\alpha}{2} \left[\mathcal{E}(v)\right]^2 - F\,[C_1 + (C_2 + 1)\, \mathcal{E}(v)],$$

ce qui établit (4.6).

ii) Unicité.

Soient (u , U) et (u^*, U^*) deux solutions. Choisissons $(v, V) = (u^*, U^*)$ (resp (u, U)) dans l'inéquation (4.4) relative à (u ,U) resp (u^*, U^*) et ajoutons membre à membre. Il vient

(4.12) $$a(u - u^*, u - u^*) \quad 0$$

ce qui entraine $u = u^*$ et par suite $U = U^*$ car les minima sont égaux.

Remarque 4.1.

On peut, par un calcul formel, montrer qu'inversement l'élément $(u, U) \in K$ satisfaisant (4.4) est solution, dans un certain sens, du problème physique posé dans le cas élastique.

Une démonstration rigoureuse demande l'utilisation d'un théorème de trace pour les éléments de W, question que nous ne voulons pas aborder ici.

5 - Cas élastoplastique de type Hencky.

5.1. Loi de comportement.

Désignons par $\hat{K}$ un ensemble convexe du sous-espace de $\mathbb{R}^9$

(5.1) $$E = \left\{ \tau \mid \tau = (\tau_{ij}) \ ; \ \tau_{ij} \in \mathbb{R} \ ; \ i, j = 1, 2, 3 \ ; \ \tau_{ij} = \tau_{ji} \right\} ,$$

et par $\chi_{\hat{K}}$ sa fonction indicatrice. La loi de comportement élastoplastique du type Hencky s'écrit,

(5.2) $$\varepsilon_{ij}(u) = A_{ijkh}\, \sigma_{kh} + \partial \chi_{\hat{K}}(\sigma) ,$$

où $\partial \chi_{\hat{K}}(\sigma)$ désigne le sous-différentiel de la fonction convexe $\chi_{\hat{K}}$ (cf. J.J.MOREAU [1]), ou de manière équivalente

(5.3) $$\begin{aligned} &(\sigma_{ij}) \in \hat{K} \\ &\left(\varepsilon_{ij}(u) - A_{ijkh}\sigma_{kh}\right)\left(\tau_{ij} - \sigma_{ij}\right) \leqslant 0 , \ \forall (\tau_{ij}) \in \hat{K} . \end{aligned}$$

Les coefficients A_{ijkh} peuvent être considérés comme constituant une matrice inverse de la matrice a_{ijkh} introduite au N° 4 et possédant le même type de propriétés de symétrie et ellipticité.

Remarque 5.1.

Si $\hat{K} = E$, alors on retrouve le cas élastique.

5.2. Formulation variationnelle.

On reprend le problème posé au N°2 avec la loi de comportement (5.3) définie ci-dessus. Nous allons en donner une formulation variationnelle d'un type de celle obtenue en 3.2.

Introduisons l'ensemble convexe Σ par,

(5.4) $$\Sigma = \Big\{ (\tau_{ij}) \ \Big| \ \tau_{ij} \in L^2(\Omega) \ ; \ i, j = 1, 2, 3 \ ; \ \tau_{ij} = \tau_{ji} \ ;$$
$$(\tau_{ij}) \in \hat{K} \text{ pour tout } x \in \Omega \text{ p.p. } ;$$

$$\frac{\partial}{\partial x_j}\tau_{ij} = 0 \text{ dans } \Omega,$$

$$\tau_{13} = \tau_{23} = 0 \text{ sur } \Gamma,$$

$$\tau_{33} = 0 \quad \text{sur } \Gamma - \Gamma_1,$$

$$\tau_{33} \leqslant 0 \quad \text{sur } \Gamma_1,$$

$$\left.\int_{\Gamma_1} \tau_{33}\, dx_1\, dx_2 = -F. \right\}$$

ce qui a un sens : en effet les éléments $\tau_{i3}\big|_\Gamma$ appartiennent à $H^{-1/2}_{Loc}(\Gamma)$, donc sont des distributions sur Γ. Comme τ_{33} est à support borné (contenu dans Γ_1), c'est une distribution intégrable ce qui donne un sens à la dernière ligne. De plus Σ est fermé dans $[L^2(\Omega)]^9$.

Supposons alors que (σ_{ij}) soit champ de contraintes, solution classique régulière. On aura

(5.5) $$(\sigma_{ij}) \in \Sigma.$$

Appliquant alors (5.3) avec $(\tau_{ij}) \in \Sigma$ et intégrant sur Ω, on obtient,

(5.6) $$A(\sigma, \tau - \sigma) \geqslant \int_\Omega (\tau_{ij} - \sigma_{ij}) \frac{\partial u_i}{\partial x_j}\, dx, \quad \forall \tau \in \Sigma$$

où on a posé

(5.7) $$A(\sigma, \tau) = \int_\Omega A_{ijkh}\, \sigma_{kh}\, \tau_{ij}\, dx.$$

Intégrant par parties le deuxième membre de (5.6) on obtient,

(5.8) $$A(\sigma, \tau - \sigma) \geqslant -\int_{\Gamma_1} (\tau_{33} - \sigma_{33})\, u_3\, dx_1\, dx_2 = X$$

en désignant par X le 2ème membre, qu'on peut aussi écrire

(5.9) $$X = -\int_{\Gamma_1} (\tau_{33} - \sigma_{33})(u_3 + \varphi - U - \varphi)\, dx,\, dx_2 \geqslant \int_{\Gamma_1} \varphi\, (\tau_{33} - \sigma_{33})\, dx_1\, dx_2 .$$

d'où l'énoncé -

<u>Propriété 5.1.</u>

Si $\sigma = (\sigma_{ij})$ est solution classique assez régulière du problème posé au N°2 avec la loi de comportement élastoplastique (5.3), alors

(5.10) $$\sigma \in \Sigma$$

(5.11) $$A(\sigma,\tau-\sigma) \geqslant \int_{\Gamma_1} \varphi(\tau_{33} - \sigma_{33})\, dx_1\, dx_2, \quad \forall\, \tau \in \Sigma .$$

5.3. Existence et unicité d'une solution.

Théorème 5.1.

Il existe un unique σ satisfaisant (5.10) (5.11).

Démonstration.

Il suffit de montrer que la fonctionnelle strictement convexe

(5.12) $$J(\tau) = \frac{1}{2} A(\tau,\tau) - \int_{\Gamma_1} \varphi \;\tau_{33}\, dx_1\, dx_2$$

Tend vers $+\infty$ si la norme de τ dans $\left(L^2(\Omega)\right)^9$ tend vers $+\infty$ avec $\tau \in \Sigma$.

Or pour $\tau \in \Sigma$ on a

$$\int_{\Gamma_1} \varphi(-\tau_{33})\, dx_1\, dx_2 \leqslant C_1 \int_{\Gamma_1} (-\tau_{33})\, dx_1\, dx_2 = C_1\, F$$

où C_1 est un majorant de φ. Grâce à la coercivité de A (σ, τ) sur $\left[L^2(\Omega)\right]^9$ il en résulte la propriété souhaitée.

Remarque 5.1.

Lorsque le convexe $\hat{K}$ est quelconque, la recherche du champ de déplacements associé au champ de contraintes σ est un problème ouvert, ce qui est habituel avec cette loi de comportement.

6 - Cas viscoélastique.

6.1 - Loi de comportement.

Nous choisirons la loi de viscoélasticité particulière

(6.1) $$\sigma_{ij} = a_{ijkh}\, \varepsilon_{kh}(u) + 2\mu\, \varepsilon_{ij}^{D}\left(\frac{\partial u}{\partial t}\right)$$

où

$$\varepsilon_{ij}^{D} = \varepsilon_{ij} - \frac{1}{3}\, \varepsilon_{kk}\, \delta_{ij}.$$

Le tenseur des contraintes est constitué d'une partie élastique caractérisée par les coefficients d'élasticité a_{ijkh} et une partie visqueuse proportionnelle au déviateur du tenseur des vitesses de déformation ; le coefficient positif μ est une viscosité.

6.2. Formulation variationnelle.

La formulation variationnelle générale (3.10) (3.11) devient ici, si $\left(u(t)\ , U(t)\right)$ est une solution pour $t \in (0, T)$ où t est le temps,

(6.3) $$\left(u(t)\ , U(t)\right) \in K \ , \quad \forall\, t \in (0\ , T)\ , \ \text{p.p.}$$

(6.4) $\quad a\Big(u(t)\ ,\ v - u(t)\Big) + 2\mu\int_{\Omega} \mathcal{E}^{D}_{ij}\left(\frac{\partial u}{\partial t}\right) \mathcal{E}_{ij}(v - u)\,dx \geqslant F\big(V - U(t)\big)$

Pour compléter l'ensemble des conditions du problème, on doit imposer des conditions initiales qui sont

(6.5) $\quad u(0) = 0 \quad , \quad U(0) = 0\ .$

6.3. Existence et unicité d'une solution.

Théorème 6.1.

Il existe une unique solution $\Big(u(t)\ ,\ U(t)\Big)$ du système de conditions (6.3) (6.4) (6.5).

Démonstration de l'unicité.

Soient $(u\ ,\ U)$ et (u^{*}, U^{*}) deux solutions. On choisit $v = u^{*}$, $V = U^{*}$ (resp $v = u$, $V = U$) dans l'inéquation (6.4) relative à (u, U) $\big($resp $(u^{*}\ ,\ U^{*})\big)$ et on ajoute membre à membre les inégalités obtenues. Il vient immédiatement

(6.6) $\quad u(t) = u^{*}(t)\ ,\quad \forall\, t \in (0\ ,\ T)\ .$

Utilisant (6.6) dans chacune des inégalités écrites précédemment on a de suite

(6.7) $\quad U^{*} - U \geqslant 0 \quad \text{et} \quad U^{*} - U \leqslant 0$

ce qui établit l'unicité.

Démonstration de l'existence.

Nous procédons par régularisation-pénalisation, estimations a priori et double passage à la limite. Nous ne donnerons que les grandes lignes de la démonstration.

i) Régularisation-pénalisation.

Soient η et ε deux scalaires positifs destinés à tendre vers zéro et soit β un opérateur de pénalisation sur K. Nous cherchons une solution $(u_{\eta\varepsilon}\ ,\ U_{\eta\varepsilon})$ satisfaisant

(6.6) $$\eta \frac{dU_{\eta\varepsilon}}{dt}V + \eta\, U_{\eta\varepsilon}V + \eta\int_{\Omega} \varepsilon_{ij}\left(\frac{\partial u_{\eta\varepsilon}}{\partial t}\right)\varepsilon_{ij}(v)\,dx + \frac{1}{\varepsilon}\Big[\beta\big(u_{\eta\varepsilon}, U_{\eta\varepsilon}\big),(v,V)\Big] +$$
$$a(u_{\eta\varepsilon}, v) + 2\mu\int_{\Omega} \varepsilon^{D}_{ij}\left(\frac{\partial u_{\eta\varepsilon}}{\partial t}\right)\varepsilon_{ij}(v)\,dx = FV,\ \forall (v,V)\in W\times\mathbb{R}.$$

(6.7) $\quad u_{\eta\varepsilon}(0) = U_{\eta\varepsilon}(0) = 0\ .$

L'opérateur de pénalisation choisi ici est

(6.8) $\quad [\,\beta\,(u\ ,\ U)\ ,\ (v\ ,\ V)] = -\int_{\Gamma_1} (u - U + \varphi)^{-}\ (v - V)\ d\Gamma.$

Les équations (6.6) (6.7) constituent un système parabolique monotone coercif sur l'espace $W \times \mathbb{R}$ qui possède une solution unique $(u_{\eta\varepsilon}, U_{\eta\varepsilon})$ dans la classe

(6.9) $$(u_{\eta\varepsilon}, U_{\eta\varepsilon}) , \frac{d}{dt}(u_{\eta\varepsilon}, U_{\eta\varepsilon}) \in L^2(0, T ; W \times \mathbb{R}).$$

ii) Estimations a priori (I).

Choisissant $(v , V) = u_{\eta\varepsilon}(t) , U_{\eta\varepsilon}(t)$ dans (6.6) et intégrant sur $(0 , t)$, on en déduit, grâce au choix (6.8) de l'opérateur β, des estimations sur

$$\sqrt{\eta}\, U_{\eta\varepsilon} , \quad u_{\eta\varepsilon} \text{ et } \frac{1}{\varepsilon}\beta(u_{\eta\varepsilon}, U_{\eta\varepsilon})$$

indépendantes de η et ε.

iii) Premier passage à la limite.

Faisant tendre η vers zéro, on montre grâce aux estimations (I) que

(6.10) $$(u_{\eta\varepsilon}, U_{\eta\varepsilon}) \longrightarrow (u_\varepsilon, U_\varepsilon)$$

dans $L^2(0\ T ; W \times \mathbb{R})$, où $(u_\varepsilon, U_\varepsilon)$ satisfait

(6.11) $$a(u_\varepsilon, v) + \frac{1}{\varepsilon}[\beta(u_\varepsilon, U_\varepsilon), (v, V)] + + 2\mu\int_\Omega \varepsilon_{ij}^D\left(\frac{\partial u_\varepsilon}{\partial t}\right)\varepsilon_{ij}(v)\,dx = FV, \quad (v, V) \in W\times\mathbb{R}.$$

(6.12) $$u(0) = 0.$$

iv) Estimation a priori (II)

Faisant $t = 0$ dans (6.11) on obtient que

$$\varepsilon_{ij}^D\left(\frac{\partial u_\varepsilon}{\partial t}(0)\right) = 0.$$

Dérivons alors (6.11) par rapport à t et choisissons

$$v = \frac{\partial u_\varepsilon}{\partial t} , \quad V = \frac{\partial U_\varepsilon}{\partial t}.$$

Après intégration en t, on obtient une estimation sur $\varepsilon_{ij}^D\left(\frac{\partial u_\varepsilon}{\partial t}\right)$ indépendante de ε.

v) Deuxième passage à la limite.

Grâce aux estimations a priori (I) et (II), on montre que

$$u_\varepsilon \longrightarrow u$$

$$U_\varepsilon \longrightarrow U$$

$$\varepsilon_{ij}^D\left(\frac{\partial u_\varepsilon}{\partial t}\right) \longrightarrow \varepsilon_{ij}^D\left(\frac{\partial u}{\partial t}\right)$$

dans les espaces correspondant aux estimations.

Choisissant alors $v - u_\varepsilon$, $V - U_\varepsilon$ avec $(v , V) \in K$ comme fonction test dans (6.11), on passe à la limite pour montrer que (u , U) satisfait (6.3) (6.4), ce qui achève la démonstration.

BIBLIOGRAPHIE

G.DUVAUT

[1] Conférence au Colloque Franco-polonais de Rhéologie - Nice, (1974).

G.DUVAUT et J.L.LIONS.

[1] Les inéquations en Mécanique et en Physique - Dunod, 1972.

J.L.LIONS

[1] Problèmes aux limites dans les équations aux dérivées partielles - Montréal, (1962).

M.LOPPIN

[1] Thèse de 3ème Cycle (en cours). Université P.et M.Curie -

J.J.MOREAU

[1] Fonctionnelles convexes. Collège de France. 1966-1967.

ON THE EXISTENCE AND UNIQUENESS OF A WARPENING FUNCTION IN THE ELASTIC - PLASTIC TORSION OF A CYLINDRICAL BAR WITH MULTIPLY CONNECTED CROSS - SECTION

C. Gerhardt *

0. INTRODUCTION.

In recent years the elastic - plastic torsion of a cylindrical bar has been of considerable interest for many mathematicians and mechanists. In the case of a bar with simply connected cross - section important contributions are due to TING [9,10,11,12], BREZIS [2], and BREZIS & STAMPACCHIA [3]. They could show the regularity of the stress components and the existence of a unique displacement vector.

In contrast to the torsion of a bar with simply connected cross - section the elastic - plastic torsion problem is much more involved when the cross - section is multiply connected in view of the non - homogeneous boundary conditions in that case. Using an idea of COURANT [4] LANCHON [7] could determine the stress components in a weak sense as the gradient of the solution to a variational problem with constraints. The $C^{1,\alpha}$ - regularity of that solution has been proved by us in [5] using an abstract regularity theorem of BREZIS & STAMPACCHIA [3].

Up to now it has been an open problem to determine the displacement vector in the case of a multiply connected cross - section. It is the aim of this paper to solve this problem. The way we do that gives a new physical interpretation of the elastic - plastic torsion problem, namely, we shall treat this problem within the framework of the <u>nonlinear</u> <u>elasti-</u>

---

* During the preparation of this article the author was supported by the SFB 40 at the University of Bonn.

city theory of hardening materials, i.e. the elastic - plastic torsion problem may be looked at as a nonlinear problem without constraints instead of a linear problem with constraints.

1. STATEMENT OF THE PROBLEM AND NOTATIONS.

Let $x = (x^1, x^2, x^3)$ be the Euclidean space coordinates. We consider the torsion of a cylindrical bar of arbitrary cross - section. Let the lower end of the bar be clamped in the plane $x^3 = 0$, and suppose that a constant torque is applied to the other end. Let the x^3 - axis be parallel to the generators of the cylinder. The cross - section Ω is supposed to be a multiply connected domain with finitely many holes Ω_k, $k = 1,\dots,N$. We assume moreover that the respective boundaries $\Gamma_k = \partial\Omega_k$ satisfy $\Gamma_k \cap \Gamma_1 = \emptyset$ for $k \neq 1$.

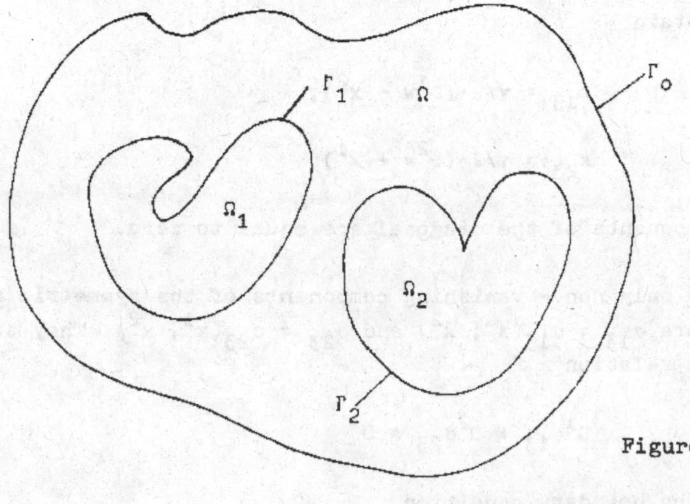

Figure 1

The boundary of Ω is the union of the disjoint family $\{\Gamma_o, \Gamma_1,\dots, \Gamma_N\}$. We assume that $\partial\Omega$ is a Lipschitz boundary which satisfies the following outward sphere condition : for any boundary point x_o there is a ball B of fixed radius R such that the intersection of $\bar{\Omega}$ and $\bar{B}$ consists of x_o alone.

For later use we define

$$(1.1)\qquad \Omega^* = \Omega \cup \bigcup_{k=1}^{N} \overline{\Omega}_k .$$

Following the hypotheses of St. VENANT we assume that the cross - sections of the bar rotate in their planes, but are warped in the direction of the x^3 - axis. Thus, the components of the displacement vector are

$$(1.2)\qquad v_1 = -\gamma x^2 x^3,\quad v_2 = \gamma x^1 x^3,\quad v_3 = \gamma w(x^1, x^2),$$

where γ is the torsion per unit length of the bar, and w is an unknown function, the so - called warpening function .

The components of the strain tensor $\varepsilon = (\varepsilon_{ij})$ are defined through the relation

$$(1.3)\qquad \varepsilon_{ij} = 1/2\cdot(D^i v_j + D^j v_i).$$

Hence we obtain

$$(1.4)\qquad \begin{aligned} \varepsilon_{13} &= \gamma/2\cdot(D^1 w - x^2), \\ \varepsilon_{23} &= \gamma/2\cdot(D^2 w + x^1), \end{aligned}$$

and all components of the diagonal are equal to zero.

The only non - vanishing components of the symmetric stress tensor $\sigma = (\sigma_{ij})$ are $\sigma_{13} = \sigma_{13}(x^1, x^2)$ and $\sigma_{23} = \sigma_{23}(x^1, x^2)$. They satisfy the equilibrium relation

$$(1.5)\qquad D^1\sigma_{13} + D^2\sigma_{23} = 0$$

in Ω, and the boundary condition

$$(1.6)\qquad \sigma_{13}\,\nu_1 + \sigma_{23}\,\nu_2 = 0$$

on $\partial\Omega$, where $\nu = (\nu_1, \nu_2)$ is the exterior normal vector of $\partial\Omega$.

We assume the yield criteria of v. MISES [8]

$$(1.7)\qquad T^2 = |\sigma_{13}|^2 + |\sigma_{23}|^2 \leq 1,$$

where the elastic range is defined by the strict inequality sign. T is called the tangential stress intensity.

The elastic - plastic torsion problem consists in finding a warpening function w and a stress tensor σ such that the relations (1.4) - (1.7) and

(1.8) $$\varepsilon_{ij} = \frac{\lambda}{2G}\sigma_{ij}$$

are satisfied, where G is a positive constant, the shear modulus, and $\lambda \in L^\infty(\Omega)$ has the property

(1.9) $$\lambda = \begin{cases} 1, & T < 1 \\ \geq 1, & T = 1. \end{cases}$$

λ is a Lagrange multiplier.

To determine σ LANCHON solved the variational problem

(1.10) $$\int_{\Omega^*} |Dv|^2\, dx - 4G\gamma \cdot \int_{\Omega^*} v\, dx \to \min \text{ in } K_1,$$

where K_1 is the convex set

(1.11) $$K_1 = \{v \in H_0^{1,2}(\Omega^*) : |Dv| \leq 1,\ v_{|\Omega_k} = \text{const},\ k = 1,\dots,N\}.$$

If u is the solution to this problem, then σ is determined through the definition

(1.12) $$\sigma_{13} = D^2 u, \quad \sigma_{23} = -D^1 u$$

in Ω; u is called the stress function.

According to the result of [5] u belongs to $H^{2,p}_{loc}(\Omega)$ for any p, $1 \leq p < \infty$; precisely we proved

(1.13) $$\Delta u \in L^\infty(\Omega).$$

The crucial step in solving the elastic - plastic torsion problem completely is to determine w and λ, such that the relations (1.4) and (1.8), or equivalently,

$$\lambda \cdot D^2 u = G\gamma \cdot (D^1 w - x^2)$$

(1.14)

$$-\lambda \cdot D^1 u = G\gamma \cdot (D^2 w + x^1)$$

are satisfied.

We attack this problem by treating it as the limit case of a sequence of nonlinear problems without constraints, namely, we shall approximate the elastic - plastic behaviour of the material by the behaviour of hardening materials.

Let Γ be the <u>shear strain intensity</u>

(1.15) $$\Gamma^2 = 1/4 \cdot (|\varepsilon_{13}|^2 + |\varepsilon_{23}|^2).$$

Then, for elastic - plastic materials, the dependence between T and Γ is given by

(1.16) $$T = \begin{cases} G\Gamma, & \text{if } T < 1 \\ 1, & \text{if } T = 1, \end{cases}$$

i.e. the dependence is not invertible.

For hardening materials Γ can be expressed as a function of T

(1.17) $$\Gamma = \frac{g(T^2)}{G} \cdot T,$$

where the real function $g = g(t)$ satisfies

(1.18) $$g \geq 1 \quad \text{and} \quad \frac{dg}{dt} \geq 0.$$

The relation between ε and σ is then of the form

(1.19) $$\varepsilon_{ij} = \frac{g(T^2)}{2G} \sigma_{ij}.$$

For hardening materials the stress function u is determined as the solution of the variational problem

(1.20) $$\int_{\Omega^*} \int_0^{|Dv|^2} g(t)\, dt dx - 4G\gamma \cdot \int_{\Omega^*} v\, dx \to \min \text{ in } K_2,$$

$$K_2 = \{v \in H_0^{1,2}(\Omega^*) : v_{|\Omega_k} = \text{const},\ k = 1,\dots,N\}.$$

The solution u then satisfies the nonlinear differential equation in Ω

$$(1.21) \qquad - D^i(g(|Du|^2)D^i u) - 2G\gamma = 0$$

and the free boundary conditions

$$(1.22) \qquad \int_{\Gamma_k} g(|Du|^2)\cdot Du\cdot\nu \, dH_1 = 2G\gamma\cdot|\Omega_k|$$

for k = 1,...,N, where $|\Omega_k|$ is the Lebesgue measure of Ω_k, and where for positive r H_r denotes the r - dimensional Hausdorff measure.

To determine the warpening function w in this case we have to integrate the following system of first order partial differential equations

$$(1.23) \qquad \begin{aligned} g(|Du|^2)\cdot D^2u &= G\gamma\cdot(D^1w - x^2) \\ -g(|Du|^2)\cdot D^1u &= G\gamma\cdot(D^2w + x^1). \end{aligned}$$

But necessary and sufficient conditions to solve this system are just the relations (1.21) and (1.22). Thus, for hardening materials the torsion problem is completely solvable.

Our plan is to approximate the non - invertible relation (1.16) by injective relations valid for hardening materials. This is indicated in Figure 2.

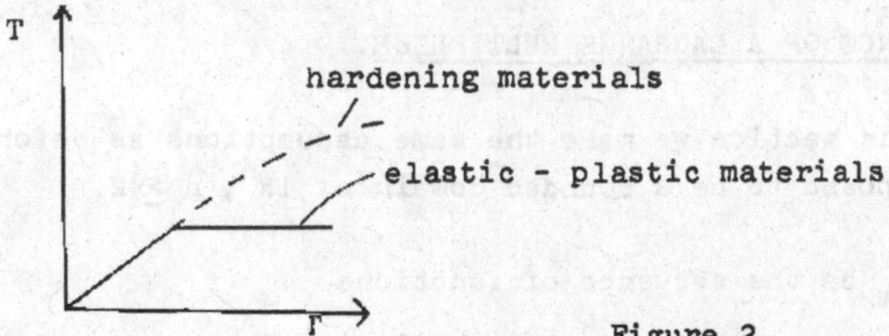

Figure 2

In doing so, we see that the corresponding functions g_ε converge towards the <u>monotone</u> graph

$$(1.24) \qquad g_0(t) = \begin{cases} 1, & t < 1 \\ [1,\infty), & t = 1. \end{cases}$$

Let u_ε be the corresponding solutions of the approximating problems. We are going to show in the next section, that we can approximate the graph g_0 such that the u_ε's are uniformly bounded in $H^{1,2}(\Omega) \cap H^{2,2}_{loc}(\Omega)$, and that they converge to the solution u of the variational problem (1.10). Moreover, the sequence $g_\varepsilon(|Du_\varepsilon|^2)$ is uniformly bounded and a subsequence converges weakly in $L^2(\Omega)$ to a function λ satisfying $\lambda(x) \in g_0(|Du|^2)$ for a.e. x.

Furthermore, if w_ε are the corresponding warpening functions, so that the relations

$$(1.25) \qquad \begin{aligned} g_\varepsilon(|Du_\varepsilon|^2)\cdot D^2 u_\varepsilon &= G\gamma\cdot(D^1 w_\varepsilon - x^2) \\ -g_\varepsilon(|Du_\varepsilon|^2)\cdot D^1 u_\varepsilon &= G\gamma\cdot(D^2 w_\varepsilon + x^1) \end{aligned}$$

are valid, then it follows from the preceding estimates that a subsequence of the w_ε's converges uniformly to a Lipschitz function w being the warpening function for the elastic - plastic torsion problem.

The stress function u also solves the differential equation

$$(1.26) \qquad - D^i(\lambda D^i u) - 2G\gamma = 0,$$

or writing it more suggestively,

$$(1.27) \qquad - D^i(g_0(|Du|^2)D^i u) - 2G\gamma \ni 0.$$

2. THE EXISTENCE OF A LAGRANGE MULTIPLIER.

In this section we make the same assumptions as before, except that Ω is supposed to be a bounded domain of IR^n, $n \geq 2$.

Let g_ε be the sequence of functions

$$(2.1) \qquad g_\varepsilon(t) = \begin{cases} 1, & t \leq 1 \\ e^{m/\varepsilon\cdot(t\ -\ 1)}, & t \geq 1, \end{cases}$$

where the positive constant m is to determined later. The g_ε's satisfy the condition (1.18) and approximate the monotone graph g_0.

Consider the variational problems

(2.2) $$\int_{\Omega^*} \int_0^{|Dv|^2} g_\varepsilon(t)\,dt\,dx - 4G\gamma\cdot\int_{\Omega^*} v\,dx \to \min \text{ in } K_2.$$

Let u_ε be the solutions of these problems. Then the sequence $\{u_\varepsilon\}$ is bounded in $H_0^{1,2}(\Omega^*)$, and hence a subsequence, not relabled, converges weakly to some element $u_0 \in K_2$. We assert

(2.3) $$u_0 = u,$$

where u is the solution of problem (1.10).

To prove this, we observe that

(2.4) $$u_\varepsilon \to u_0 \text{ in } L^2(\Omega^*).$$

Thus, for each k, k = 1,...,N,

(2.5) $$c_k^\varepsilon = u_\varepsilon|_{\Omega_k}$$

is bounded. Hence, we deduce

(2.6) $$|c_k^\varepsilon - c_l^\varepsilon| \le \text{const}\cdot\text{dist}(\Gamma_k, \Gamma_l)$$

for k, l = 0,...,N, where c_0^ε is equal to zero.

From the estimate (2.6) we immediately conclude (cf. Theorem 2.1 below for similar considerations) that $|Du_\varepsilon|_{\partial\Omega}$ is uniformly bounded, and hence $|Du_\varepsilon|_\Omega$.

We finally affirm

(2.7) $$|Du_0|_\Omega \le 1.$$

Suppose for a moment that this estimate would be valid. Then u_0 would belong to K_1, and would therefore be a solution of problem (1.10). The assertion (2.3) would then follow from the uniqueness of the solution.

To prove (2.7), choose $\rho > 1$ and $\alpha > 0$ arbitrarily. Then it follows from the minimum property of u_ε that

(2.8) $$|E_\varepsilon| = |\{|Du_\varepsilon| > \rho\}| < \alpha \quad \text{for a.e.}\,\varepsilon.$$

Hence, we conclude

$$(2.9)\qquad \int_\Omega \max\{|Du_\varepsilon| - \rho, 0\}\, dx = \int_E \{|Du_\varepsilon| - \rho\}\, dx \le \text{const}\cdot\alpha$$

for those values of ε, on account of the uniform boundedness of $|Du_\varepsilon|_\Omega$.

Since the function $t \to \max\{t - \rho, 0\}$ is convex, we then obtain

$$(2.10)\qquad \int_\Omega \max\{|Du_0| - \rho, 0\}\, dx \le \text{const}\cdot\alpha,$$

from which the result follows in view of the arbitrariness of α and ρ.

This result implies especially

$$(2.11)\qquad c_k^\varepsilon \to c_k \quad \text{for } k = 0,\dots,N,$$

where c_k are the boundary values of u on Γ_k.

We now make the following fundamental assumption, namely, we suppose

$$(2.12)\qquad |c_k - c_l| < \text{dist}(\Gamma_k, \Gamma_l) \quad \text{for } k \neq l.$$

This inequality is trivially satisfied if we replace the strict inequality sign by " $\le$ ".

The strict inequality means physically that there are no " plastic arcs " connecting two different components of the boundary of Ω. This condition is always satisfied if the cross - section is simply connected, or if the torsion angle is sufficiently small, since the plastic parts spread out from the boundary and vary continuously with γ.

Combining the relations (2.11) and (2.12) we immediately conclude that the inequality

$$(2.13)\qquad |c_k^\varepsilon - c_l^\varepsilon| \le \text{dist}(\Gamma_k, \Gamma_l)$$

is valid for a.e. ε.

Now we are able to prove the main theorem

THEOREM 2.1. - Let Ω satisfy an outward sphere condition of radius R. Let u resp. u_ε be the solutions to the variational problems (1.10) resp. (2. 2) and suppose inequality (2.12) to be valid. Then, we can demonstrate the following propositions

(i) $\limsup |Du_\varepsilon|_\Omega \leq 1$,
(ii) $g_\varepsilon(|Du_\varepsilon|^2) \leq$ const,
(iii) $||u_\varepsilon|| \leq$ const in $H^{2,2}_{loc}(\Omega)$,
(iv) $u_\varepsilon \to u$ in $H^{1,2}(\Omega)$,
(v) $g_\varepsilon(|Du_\varepsilon|^2) \rightharpoonup \lambda \in L^\infty(\Omega)$ in $L^2(\Omega)$,

where

(vi) $\lambda(x) \in g_0(|Du|^2)$ for a.e. x.

From these relations we finally conclude

$$- D^i(\lambda D^i u) - 2G\gamma = 0. \tag{2.14}$$

PROOF : To prove (i) it will be sufficient to estimate $\limsup |Du_\varepsilon|_{\partial\Omega}$.

For each boundary point $x_0 \in \partial\Omega$ we shall construct barrier functions δ^-_ε, δ^+_ε satisfying

$$- D^i(g_\varepsilon(|D\delta^+_\varepsilon|^2)D^i\delta^+_\varepsilon) - 2G\gamma \geq 0, \tag{2.15}$$

$$- D^i(g_\varepsilon(|D\delta^-_\varepsilon|^2)D^i\delta^-_\varepsilon) - 2G\gamma \leq 0, \tag{2.16}$$

$$\limsup \max\{|D\delta^-_\varepsilon|_\Omega, |D\delta^+_\varepsilon|_\Omega\} \leq 1, \tag{2.17}$$

$$\delta^-_\varepsilon(x) \leq u_\varepsilon(x) \leq \delta^+_\varepsilon(x) \quad \text{for all } x \in \partial\Omega, \tag{2.18}$$

and

$$\delta^-_\varepsilon(x_o) = \delta^+_\varepsilon(x_o) = u_\varepsilon(x_o). \tag{2.19}$$

From the maximum principle we then conclude

$$|Du_\varepsilon|_\Omega \leq |Du_\varepsilon|_{\partial\Omega} \leq \max\{|D\delta^-_\varepsilon|_\Omega, |D\delta^+_\varepsilon|_\Omega\}. \tag{2.20}$$

To construct the barriers let $x_0 \in \partial\Omega$ be arbitrary, and let B be a ball of radius R touching Ω at x_0 from the exterior. We suppose that B is centered in the origin. Let d be a constant such that

$$d \geq |x| - R \quad \text{for all } x \in \Omega. \tag{2.21}$$

We then define

$$\delta_\varepsilon^+(x) = \delta_\varepsilon(x) + u_\varepsilon(x_o), \tag{2.22}$$

where

$$\delta_\varepsilon(x) = \frac{e^{\varepsilon d}}{\varepsilon}\{1 - e^{-\varepsilon(|x| - R)}\}. \tag{2.23}$$

δ_ε^- is similar defined

$$\delta_\varepsilon^-(x) = -\delta_\varepsilon(x) + u_\varepsilon(x_o). \tag{2.24}$$

In the following we shall only consider the upper barrier δ_ε^+; the considerations for δ_ε^- are identical.

From the definition of δ_ε we immediately conclude

$$D^i\delta_\varepsilon(x) = e^{\varepsilon(d + R - |x|)}\cdot x^i|x|^{-1} \tag{2.25}$$

and

$$D^iD^j\delta_\varepsilon = e^{\varepsilon(d + R - |x|)}\cdot\{\frac{\delta^{ij}}{|x|} - \frac{x^ix^j}{|x|^3} - \varepsilon\cdot\frac{x^ix^j}{|x|^2}\} \tag{2.26}$$

Thus, we obtain

$$|D\delta_\varepsilon^+| = e^{\varepsilon(d + R - |x|)} \tag{2.27}$$

and

$$D^iD^j\delta_\varepsilon^+\cdot D^i\delta_\varepsilon^+\cdot D^j\delta_\varepsilon^+ = -\varepsilon\cdot e^{3\varepsilon(d + R - |x|)}. \tag{2.28}$$

We therefore deduce

$$\begin{aligned} - D^i(g_\varepsilon(|D\delta_\varepsilon^+|^2)D^i\delta_\varepsilon^+) - 2G\gamma &= g_\varepsilon\cdot\{-\Delta\delta_\varepsilon^+ + 2m\cdot e^{3\varepsilon(d + R - |x|)}\}- \\ &- 2G\gamma \geq 2m - n/R - 2G\gamma > 0 \end{aligned} \tag{2.29}$$

for small values of ε and sufficiently large m.

In view of these relations the conditions (2.15) - (2.19) are satisfied. Hence, the estimates (2.20) and (2.27) yield

$$|Du_\varepsilon|_\Omega \leq e^{\varepsilon d} \tag{2.30}$$

from which proposition (ii)

$$\lim\sup g_\varepsilon(|Du_\varepsilon|^2) \leq e^{2md} \tag{2.31}$$

is easily derived.

Proposition (iii) is an immediate consequence of the following lemma

LEMMA 2.1. - Let g satisfy the condition (1.18), and let $u \in C^3(\Omega) \cap C^{0,1}(\bar{\Omega})$ be a solution of the partial differential equation

$$(2.32)\qquad - D^i(g(|Du|^2)D^i u) - \mu = 0,$$

where $\mu \in L^\infty(\Omega)$ is given. Then, for any compact domain Ω', $\Omega' \subset\subset \Omega$, the estimate

$$(2.33)\qquad ||u||_{2,2,\Omega'} \leq \text{const}$$

is valid, where the constant depends on Ω', $|g(|Du|^2)|_\Omega$, $|Du|_\Omega$, and $|\mu|_\Omega$.

PROOF OF LEMMA 2.1 : Let $\xi \in H^{1,2}_{loc}(\Omega)$ and $\eta \in C^\infty_c(\Omega)$. Multiplying the equation (2.32) with $\xi\eta^2$ and integrating by part yield

$$(2.34)\qquad \int_\Omega gD^i u\{D^i\xi\eta^2 + 2\xi D^i\eta\eta\}\, dx = \int_\Omega \mu\xi\eta^2\, dx.$$

Setting $\xi = - D^r D^r u$ for some fixed number r, $1 \leq r \leq n$, we derive

$$(2.35)\qquad \int_\Omega g\{|DD^r u|^2\eta^2 + 2\eta D^i u(D^r\eta D^i D^r u - D^i\eta D^r D^r u)\}\, dx + \int_\Omega 2g'|D^i D^r u D^i u|^2\eta^2\, dx = - \int_\Omega \mu D^r D^r u\eta^2\, dx$$

The assertion is now obvious in view of (1.18).

To prove (iv) we observe that u_ε converges uniformly to u and that

$$(2.36)\qquad - D^i(g_\varepsilon(|Du|^2 D^i u) = - \Delta u.$$

Thus, we obtain

$$(2.37)\qquad \int_\Omega |D(u - u_\varepsilon)|^2\, dx \leq \int_\Omega \{g_\varepsilon(|Du|^2)D^i u - g_\varepsilon(|Du_\varepsilon|^2)D^i u_\varepsilon)\}D^i(u - u_\varepsilon)\, dx$$
$$\leq \int_\Omega |\Delta u - 2G\gamma|\cdot|u - u_\varepsilon|\, dx + \text{const}\cdot\int_{\partial\Omega}|u - u_\varepsilon|\, dH_{n-1}$$

from which the result follows.

The assertion (v) is an immediate consequence of (ii). The crucial step is to prove (vi) : Let $G = \{x \in \Omega : |Du(x)| < 1\}$. Then

$$g_\varepsilon(|Du_\varepsilon|^2) \to 1 \quad \text{for a.e. } x \in G \tag{2.38}$$

in view of (iv). The result now follows from Lebesgue's dominated convergence theorem.

The final relation (2.14) is derived from (iv) and (v).

3. THE EXISTENCE OF A WARPENING FUNCTION.

Let us return to the physical case n = 2. As we have seen in Section 1 there are functions w_ε such that

$$\begin{aligned} g_\varepsilon(|Du_\varepsilon|^2)D^2u_\varepsilon &= G\gamma(D^1w_\varepsilon - x^2) \\ - g_\varepsilon(|Du_\varepsilon|^2)D^1u_\varepsilon &= G\gamma(D^2w_\varepsilon + x^1). \end{aligned} \tag{3.1}$$

Thus, $|Dw_\varepsilon|$ is uniformly bounded. On the other hand we know that the w_ε's itselves are uniformly bounded, since they can be expressed as integrals of bounded functions. Therefore, a subsequence converges to some function w satisfying the system (1.14).

4. THE UNIQUENESS OF THE LAGRANGE MULTIPLIER.

We shall show that the condition (2.12) guaranteeing the existence of a Lagrange multiplier λ will also serve for proving the uniqueness of λ. The warpening function is then uniquely determined up to an additive constant.

The proof is a slight modification of BREZIS' proof [1] who derived the same result in the case of a simply connected domain.

We assume in the following that $\Omega \subset \mathbb{R}^n$, $n \geq 2$, and that $\partial\Omega$ is of class C^2, so that $u \in H^{2,2}(\Omega)$ (extend the result of Theorem 2.1 up to the boundary, or cf. [5]).

Suppose there were two different multipliers λ_1 and λ_2. Set $\mu = \lambda_1 - \lambda_2$. Let $E = \{|Du| < 1\}$ and $P = \{|Du| = 1\}$. Then μ is identically zero on E, and we have to show that this is also true on P.

We deduce from (2.14)

$$(4.1) \qquad - D^i(\mu D^i u) = 0.$$

Let k, $0 \leq k \leq N$, be given, and let U_k be neighbourhood of Γ_k. We are going to prove

$$(4.2) \qquad \mu_{|\Omega \cap U_k} \equiv 0$$

if U_k is sufficiently small.

Let h be an arbitrary smooth function with support in U_k, and set

$$(4.3) \qquad \zeta_k(x) = \int_{c_k}^{u(x)} h(x + (t - u)Du)\, dt.$$

Then $\zeta_k = 0$ on Γ_j for each j : for j = k this is trivial, and for $j \neq k$ we observe

$$(4.4) \qquad x + (t - u(x))Du(x) \notin \Gamma_k$$

if $x \in \Gamma_j$ and $t \in [c_k, c_j]$ in view of (2.12). Thus, $\zeta_k \in H_0^{1,2}(\Omega)$ and BREZIS' proof is applicable yielding

$$(4.5) \qquad \mu_{|\Omega \cap U} \equiv 0$$

in some neighbourhood U of $\partial\Omega$.

Finally, let h be a smooth function and η, $0 \leq \eta \leq 1$, be such that $\eta = 1$ on $\Omega - U$ and $\eta = o$ on $\partial\Omega$. Set

$$(4.6) \qquad \zeta(x) = \int_0^{u(x)} h(x + (t - u)Du)\, dt.$$

Then, multiplying (4.1) with $\zeta\eta$ we immediately conclude

$$(4.6) \qquad \int_\Omega \mu \cdot D^i u \cdot D^i \zeta \, dx = 0,$$

hence the result (cf. [1]).

REFERENCES

1. BREZIS, H.: Multiplicateur de Lagrange en torsion elasto plastique. Arch. Rat. Mech. Analysis 49, 32 - 40 (1972).
2. BREZIS, H. & M. SIBONY: Equivalence de deux inéquations variationelles et applications. Arch. Rat. Mech. Analysis 41, 254 - 265 (1971).
3. BREZIS, H. & G. STAMPACCHIA: Sur la regularité de la solution d'inéquations elliptiques. Bull. Soc. Math. France 96, 153 - 180 (1968).
4. COURANT, R.: Variational methods for the solution of problems of equilibrium and vibrations. Bull. Amer. Math. Soc. 49, 1 - 23 (1943).
5. GERHARDT, C.: Regularity of solutions of nonlinear variational inequalities with a gradient bound as constraint. Arch. Rat. Mech. Analysis, to appear 1975
6. KACHANOV, L.M.: Foundations of the theory of plasticity. Amsterdam: North - Holland 1971.
7. LANCHON, H.: Torsion élastoplastique d'un arbre cylindrique de section simplement ou multiplement connexe. Thèse Université de Paris VI (1972).
8. MISES, R. von: Three remarks on the theory of the ideal plastic body. Reissner Anniversary Volume, p. 415 - 419. Ann Arbor, Michigan: Edwards 1949.
9. TING, W.T.: Elastic - plastic torsion of a square bar. Trans. Amer. Math. Soc. 123, 369 - 401 (1966).
10. TING, W.T.: Elastic - plastic torsion problem II. Arch. Rat. Mech. Analysis 25, 342 - 366 (1967).
11. TING, W.T.: Elastic - plastic torsion III. Arch. Rat. Mech. Analysis 34, 228 - 244 (1969).
12. TING, W.T.: Elastic - plastic torsion of simply connected cylindrical bars. Indiana Univ. Math. J. 20, 1047 - 1076 (1971).

Claus Gerhardt

Math. Institut

der Universität

Wegelerstr. 10

D - 53 BONN

A METHOD FOR COMPUTING THE EIGENFREQUENCIES OF AN ACOUSTIC RESONATOR

J.P. GREGOIRE*, J.C. NEDELEC**, J. PLANCHARD*

*Electricité de France - Direction des Etudes et Recherches
1, Avenue du Général de Gaulle - 92140 CLAMART (FRANCE)
**Ecole Polytechnique (PALAISEAU)

I. - INTRODUCTION

Ω denotes an acoustic resonator containing a compressible fluid ; Γ is its wall. Let Γ_o be a (small) hole drilled in Γ, by which the fluid is excited. u is the small pressure fluctuation about an equilibrium state ; the eigenmodes with pulsation ω must satisfy

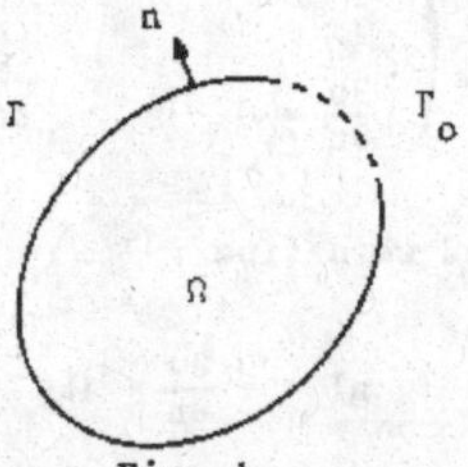

Fig. 1

(1.1) $$\Delta u + \frac{\omega^2}{c^2} u = 0 \text{ in } \Omega, \quad (c \text{ : sound velocity})$$

with the boundary conditions

(1.2) $$\begin{cases} u = 0 \quad \text{on } \Gamma_o, \\ \dfrac{\partial u}{\partial n} = 0 \quad \text{on } \Gamma_1 = \Gamma - \Gamma_o \ \left(\dfrac{\partial}{\partial n} \text{ : exterior normal derivative} \right). \end{cases}$$

It is known that there exists an enumerable infinite set of eigenpulsations $\omega_n \to \infty$ as $n \to \infty$. The aim of this paper is to describe a method to compute eigenfrequencies and modes.

In the case of acoustic pipe of length ℓ, let

$$u_i\ (x,t) = e^{i\omega\,(t+\frac{x}{c})},\ u_r\ (x,t) = e^{i\omega\,(t-\frac{x}{c}+\beta)},\ u = u_i + u_r$$

be respectively incident, reflected and resultant plane waves. Then we have :

$$\begin{cases} g = \left(\dfrac{\partial u}{\partial x} + \dfrac{i\omega}{c} u \right)_{x=\ell} = 2\, i\, \dfrac{\omega}{c}\, u_i\ (\ell), \\ g' = \left(- \dfrac{\partial u}{\partial x} + \dfrac{i\omega}{c} u \right)_{x=\ell} = -\, 2\, i\, \dfrac{\omega}{c}\, u_r\ (\ell). \end{cases}$$

Standard books on acoustics state that ω is an eigenfrequency of the pipe open at $x=\ell$ if :

$$u_i(\ell) + u_r(\ell) = 0 \quad \text{(then } g'=g\text{)}.$$

In the same way, ω is a resonance frequency of the pipe closed at $x=\ell$ if :

$$u_i(\ell) = u_r(\ell) \qquad \text{(then } g'= -g\text{)}.$$

The idea (due to P. CASEAU [4]) is to generalize this fact to the multidimensional case. Let k $(= \frac{\omega}{c})$ be fixed $\neq 0$, and u be the solution of :

$$(1.3)\quad \begin{cases} \Delta u + k^2 u = 0 & \text{in } \Omega, \\ \dfrac{\partial u}{\partial n} + iku = g & \text{given on } \Gamma_o, \\ \dfrac{\partial u}{\partial n} = 0 & \text{on } \Gamma_1. \end{cases}$$

and we define :

$$(1.4)\qquad g' = -\frac{\partial u}{\partial n} + iku \quad \text{on } \Gamma_o.$$

g and g' are analogous respectively to the incident and reflected waves. g' depends linearly on g ($g' = Bg$). It will be shown that B is a unitary operator defined for all $k = 0$.

Relation between eigenvalues of Δ and those of B

Let μ and g be eigenvalue and eigenvector of B. So, if u is a solution of (1.3) corresponding to g, we have :

$$(1.5)\qquad \frac{\partial u}{\partial n} + \sigma u = 0 \quad \text{on } \Gamma_o \quad \text{with}$$

$$(1.6)\qquad \sigma = ik\,\frac{\mu-1}{\mu+1} \quad \text{if } \mu \neq -1 \ (\sigma \text{ is real}).$$

If $\mu = +1$ (resp. $\mu = -1$), k^2 is an eigenvalue of Δ corresponding to Neumann conditions (resp. Dirichlet conditions) on Γ_o. In general case, k^2 is an eigenvalue for boundary conditions (1.4).

Conversely, if k^2 is eigenvalue of Δ for boundary conditions (1.4), the number

$$\mu = \frac{ik+\sigma}{ik-\sigma}$$

is an eigenvalue of B (corresponding to k).

The method consists in finding the values of k for which an eigenvalue of B is equal to -1 (if we are interested in Dirichlet conditions). In the following section we discuss the existence of the operator B and its spectrum.

II - EXISTENCE OF B AND STUDY OF ITS SPECTRUM

Notations : Ω is a bounded open set of R^n, with its boundary Γ sufficiently smooth.

L^2 (Ω) is the functional Hilbert space for the scalar product :

$$(u,v) = \int_{\Omega} u(x)\,\overline{v}(x)\,dx \qquad (\text{norm } |v|).$$

H^1 (Ω) is the Sobolev space with the scalar product :

$$((u,v)) = (u,v) + \sum_{i=1}^{n} \left(\frac{\partial u}{\partial x_i}, \frac{\partial v}{\partial x_i}\right) \quad (\text{norm } ||v||).$$

L^2 (Γ_o) is the Hilbert space for :

$$(u,v)_{\Gamma_o} = \int_{\Gamma_o} u(x)\,\overline{v(x)}\,d\Gamma_o \qquad (\text{norm } |v|_{\Gamma_o}).$$

Putting :

$$a(u,v) = \sum_{i=1}^{n} \left(\frac{\partial u}{\partial x_i}, \frac{\partial v}{\partial x_i}\right),$$

the problem (1.3) is equivalent to the variational equation (see [9]) :

$$(2.1) \qquad a(u,v) - k^2 (u,v) + ik\,(u,v)_{\Gamma_o} = (g,v)_{\Gamma_o},\ \forall v \in H^1(\Omega).$$

Lemma : Suppose that u is a solution of :

$$(2.2) \qquad \begin{cases} \Delta u + k^2 u = 0 & \text{in } \Omega, \quad k \neq 0, \\ \dfrac{\partial u}{\partial n} = 0 & \text{on } \Gamma. \end{cases}$$

If $u|_{\Gamma_o} = 0$ and meas $(\Gamma_o) > 0$, then $u \equiv 0$.

Proof : see [5] .

Proposition 1 : For all real $k \neq 0$ and $g \in L^2 (\Gamma_o)$, equation (2.1) has an unique solution.

Proof :

Uniqueness : We put $g = 0$, then

$$a (u,u) - k^2 |u|^2 + ik |u|^2_{\Gamma_o} = 0.$$

Thus u vanishes on Γ_o, and by the lemma, $u = 0$.

Existence : We "construct" an approximate solution u_m by the Galerkin's method with the complete basis of the Neumann eigenfunctions ω_j; u_m is a linear combination of the m first ω_j and satisfies :

$$\text{(2.3)} \qquad a (u_m,\omega_j) - k^2 (u_m,\omega_j) + ik (u_m,\omega_j)_{\Gamma_o} = (g,\omega_j)_{\Gamma_o} \quad j = 1,2... m.$$

By the lemma and the chosen basis, u_m exists. Since :

$$k |u_m|^2_{\Gamma_o} = \text{Im } (g,u_m)_{\Gamma_o},$$

the sequence $|u_m|_{\Gamma_o}$ is bounded independently on m. Let us suppose that :

$$|u_m|_{\Gamma_o} \to \infty \quad \text{as } m \to \infty,$$

and define :

$$\tilde{u}_m = u_m / |u_m| \quad (|\tilde{u}_m| = 1).$$

Hence $a (\tilde{u}_m,\tilde{u}_m)$ is bounded since $|\tilde{u}_m|_{\Gamma_o} \to 0$. According to the Rellich's compactness theorem, there exist $\tilde{u} \in H^1 (\Omega)$ and a subsequence (still denoted by $\tilde{u}_m$) such that

$$\tilde{u}_m \to \tilde{u} \quad \text{weakly in } H^1 (\Omega), \text{ and strongly in } L^2 (\Omega).$$

Thus, by (2.3), $\tilde{u}$ is a solution of (2.2). But $\tilde{u}$ vanishes on Γ_o, and the lemma implies that $\tilde{u} = 0$ on Ω, in contradiction with $|\tilde{u}_m| = 1$. Consequently, all the sequence $|u_m|$ is bounded (and also $\|u_m\|$). Applying the usual compactness argument and going to the limit in (2.3), we can show the existence of the solution u.

Moreover

$$u_m \to u \quad \text{strongly in } H^1(\Omega).$$ ▨

Proposition 1 and the relation :

$$g' = -g + 2\,ik\,u_{|\Gamma_o}$$

involve the existence of operator B. It is easy to see that B is an integral operator (use Green's function). It can be shown that B is infinitely differentiable with respect to the parameter k [5] .

Proposition 2 : B is an unitary operator (for $L^2(\Gamma_o)$ - norm).

Proof Observe that :

$$|g|^2_{\Gamma_o} = \int_{\Gamma_o} \left(\left|\frac{\partial u}{\partial n}\right|^2 + k^2\,|u|^2 \right) d\Gamma_o = |g'|^2_{\Gamma_o}$$ ▨

Since B is unitary, its spectrum belongs to the unit-circle and consists of eigenvalues and a continuous spectrum (see [10]).

Proposition 3 : B has an infinite set of enumerable eigenvalues μ whose -1 is the cluster point (eventually -1 belongs to the continous spectrum). The μ's are analytic functions of k.

Proof : Let μ belong to the spectrum ; there exists a sequence g_n with $|g_n|_{\Gamma_o} = 1$, such that :

$$|(B-\mu)\,g_n|_{\Gamma_o} = |-(1+\mu)\,g_n + ik\,u_n|_{\Gamma_o} \to 0 \quad \text{as } n \to \infty,$$

where u_n is the solution of (2.1) corresponding to $g = g_n$.

Suppose $\mu+1 \neq 0$. Obviously $|u_n|_{\Gamma_o}$ is bounded, and so $\|u_n\|$. Thus, by compactness theorem, after taking out a subsequence, there exist g and μ such that :

$$\begin{cases} g_n \to g & \text{weakly in } L^2(\Gamma_o), \\ u_n \to u & \text{weakly in } H^1(\Omega), \text{ strongly in } L^2(\Omega) \text{ and } L^2(\Gamma_o). \end{cases}$$

Then :

$$|u|_{\Gamma_o} = |1+\mu| \;/\; 2\;|k|$$

and :

$$g = \frac{2ik}{\mu+1}\, u|_{\Gamma_o}$$

is different from zero. Going to the limit in the variational equation, we obtain that u is an eigenfunction of Δ with the boundary conditions :

$$\frac{\partial u}{\partial n} + \quad ik\,\frac{\mu-1}{\mu+1}\, u = 0 \quad \text{on } \Gamma_o.$$

Consequently, g is eigenvector of B associated to μ. Now, if k^2 is not an Neumann eigenvalue of Δ, let us consider the compact self-adjoint operator H defined by :

$$Hg = u|_{\Gamma_o},$$

where u is a solution of :

$$\begin{cases} \Delta u + k^2 u = 0 & \text{in } \Omega, \\ \dfrac{\partial u}{\partial n} = g \quad \text{on } \Gamma_o, & \dfrac{\partial u}{\partial n} = 0 \quad \text{on } \Gamma_1. \end{cases}$$

H has an enumerably infinite sequence of eigenvalues $^{-1}/\sigma_m$ which converge towards zero as $m \to \infty$. Hence :

$$\frac{\partial u_m}{\partial n} + \sigma_m u_m = 0 \quad \text{on } \Gamma_o,$$

and the numbers :

$$\mu_m = \frac{ik-\sigma_m}{ik+\sigma_m}$$ are eigenvalues of B. $\mu_m \to -1$ because $|\sigma_m| \to \infty$.

When k^2 is a Neumann eigenvalue, we simply assume g to be orthogonal (in space $L^2(\Gamma_o)$) to the eigenspace associated to k^2. Since H depends analytically on k, the same holds for the μ's (see [7]).

Behaviour of the μ's as k varies

For all real number $\sigma \geq 0$, the eigenvalues $\lambda_n(\sigma)$ of

$$(2.4)\quad \begin{cases} \Delta u + \lambda_n(\sigma)\, u = 0 \quad \text{on } \Omega, \\ \dfrac{\partial u}{\partial n} = 0 \quad \text{on } \Gamma_1, \ \dfrac{\partial u}{\partial n} + \sigma u = 0 \text{ on } \Gamma_0 \end{cases}$$

are increasing continuous functions of σ and vary between $\lambda_n(0)$ (Neumann cond.) and $\lambda_n(\infty)$ (Dirichlet cond.).

For each $k = \sqrt{\lambda_n(\sigma)}$, the number

$$(2.5)\qquad \mu_n = \frac{ik+\sigma}{ik-\sigma} = -e^{2\, i \arg\, (\sigma + i \sqrt{\lambda_n(\sigma)})}$$

is an eigenvalue of B (with the same multiplicity as k^2). Relation (2.5) shows that the eigenvalues μ turn clockwise round the origin of the complex plane. On the other hand, for each $k > 0$, there is a finite number of values of $\sigma > 0$ for which k^2 is eigenvalue of (2.4). It follows from (2.5), that B has only a finite number of μ with $\text{Im}\, \mu < 0$ (the number of μ with $\text{Im}\, \mu > 0$, is infinite). The following corollary results from the fact that -1 is the cluster point of the μ's, and the spectrum of B is enumerable.

Corollary : (2.4) has an infinite number of eigenvalues for all σ (negative or positive). These eigenvalues are strictly increasing functions of σ.

Computation of B

It is very advantageous to split up Ω into elementary domains Ω_i of small size. The operators B_i are constructed for each Ω_i by finite difference or element techniques. B is obtained by assembling the differents Ω_i and using interfaces conditions. If Γ_0 is small compared with the whole boundary, the computing time of μ becomes negligible. Moreover, identical elements have the same matrices B_i (case of repetitive structure) : it follows that the assembling algorithm is very fast. On the other hand, this method is well adapted to find eigenfrequencies lying in given intervals. It is important to note that B is independent of the boundary conditions for which we search eigenfrequencies : this is of interest if the boundary conditions are not well known.

Numerical examples

a) Ω is a square of side length equal to π.

Γ_o is drilled at the middle of a side and meas $(\Gamma_o) = \pi/5$. The Neumann eigenvalues of Δ are :

$$\lambda_{m,n} = m^2 + n^2 \quad m,n = 0,1,2...$$

The variations of the μ's are illustrated in fig. 2 :

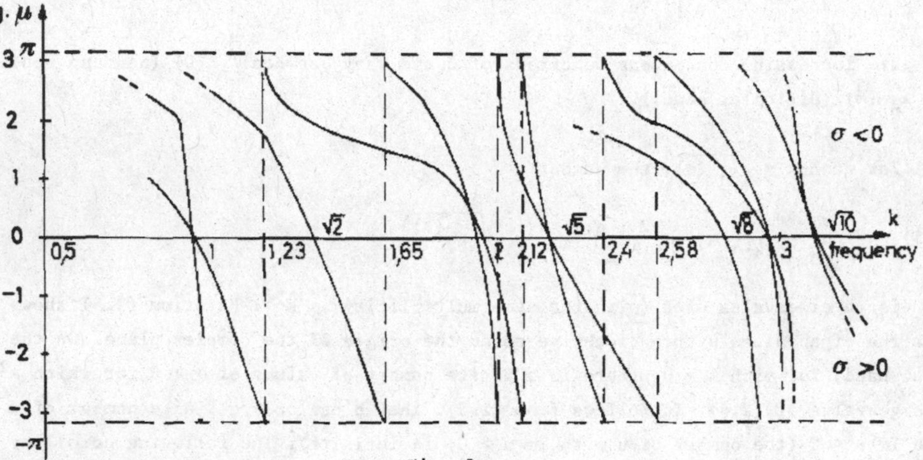

Fig. 2

b) Ω is a plane domain with repetitive structure.

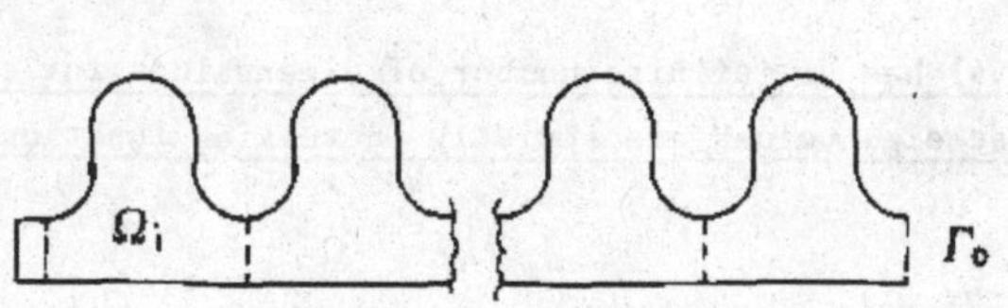

Fig. 3

For sound speed $c = 340$ m/s, we have computed the eigenfrequencies (Dirichlet conditions on Γ_o) between 100 and 2500 Hz (see [1]) :

$f_1 = 105{,}7$ Hz $\quad f_2 = 317$ Hz $\quad f_3 = 524$ Hz $\quad f_4 = 712$ Hz and so on...

III - EXTENSION OF THE METHOD

We can extend this process to other mechanical problems.

- Vibrations of an elastic body : we put, if u is the displacement vector :

(3.1) $$\sum_{i,j,k,l} \frac{\partial}{\partial x_i} (a_{ijkl}\, \varepsilon_{kl}(u))\, \vec{e}_j + k^2 \vec{u} = 0 \text{ in } \Omega,$$

with the boundary conditions :

$$\vec{t}\,(u) + ik\vec{u} = \vec{g} \quad \text{on } \Gamma \quad \text{where} \tag{3.2}$$

$$\begin{cases} \varepsilon_{kl}\,(u) = \frac{1}{2}\left(\frac{\partial u_k}{\partial x_l} + \frac{\partial u_l}{\partial x_k}\right) \\ \vec{e}_j \ : \ \text{unit-vector of } x_j\text{-axis} \\ \vec{t}\,(u) = \sum\limits_{ijkl} a_{ijkl}\,\varepsilon_{kl}\,(\vec{u})\cos(\vec{n},\vec{e}_i)\,\vec{e}_j \end{cases}$$

The operator B is defined by :

$$\vec{g}' = B\vec{g} = -\vec{t}\,(u) + ik\vec{u}$$

- <u>Oscillations of the free surface of a liquid contained in a rotating basin</u> : the eigenfrequencies k are given by [8] :

$$\begin{cases} \Delta u - \omega^2 u + k^2 u = 0 \quad \text{in } \Omega \quad (\omega \text{ angular velocity of the basin}) \\ \frac{\partial u}{\partial n} + \frac{i\omega}{k}\frac{\partial u}{\partial s} = 0 \quad \text{on } \Gamma \ \left(\frac{\partial}{\partial s} : \text{tangential derivative}\right). \end{cases} \tag{3.3}$$

This problem has infinitely eigenfrequencies k greater than ω (see [2]).

We put for each $k > \omega$ [6] :

$$\begin{cases} \Delta u + (k^2-\omega^2)\,u = 0 \quad \text{in } \Omega, \\ \frac{\partial u}{\partial n} + \frac{i\omega}{k}\frac{\partial u}{\partial s} + iku = g \text{ on } \Gamma \text{ and } \ g' = Bg = -\frac{\partial u}{\partial n} - \frac{i\omega}{k}\frac{\partial u}{\partial s} + iku. \end{cases} \tag{3.4}$$

- <u>Oscillations of a liquid contained in an elastic tank placed in a gravity field $\vec{g}$</u>

The equations of free oscillations are [3] :

$$\begin{cases} \Delta\omega = 0 \quad \text{in } \Omega_o \qquad (\omega : \text{velocity potential}), \\ \frac{\partial \omega}{\partial n} = \frac{k^2\omega}{g} \quad \text{on } \Gamma_o, \\ \frac{\partial \omega}{\partial n} = \vec{u}.\vec{n} \quad \text{on } \Sigma, \qquad (\Sigma = \partial\Omega_1 \cap \partial\Omega_o) \end{cases} \tag{3.5}$$

$$
(3.6)\quad
\begin{cases}
\sum_j \frac{\partial}{\partial x_j}\sigma_{ij}(\vec{u}) = k^2 u_i & \text{in } \Omega_1 \quad (\vec{u} : \text{displacement vector field}) \\
\sum_j \sigma_{ij}(\vec{u})\, n'_j = 0 & \text{on } \Gamma_1 \quad i = 1,2,\ldots n \\
\sum_j \sigma_{ij}(\vec{u})\, n'_j = p\, n'_i & \text{on } \Sigma \text{ , with} \\
\sigma_{ij}(\vec{u}) = \sum_{kl} a_{ijkl}\, \varepsilon_{kl}(\vec{u}) & \\
p = k^2\omega - (\vec{u}.\vec{n})\,(\vec{g}.\vec{n}), &
\end{cases}
$$

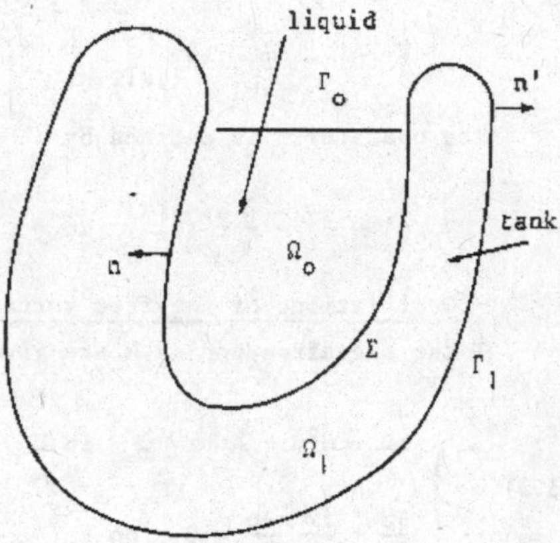

Fig. 4

Then, we put with (3.6)

$$
(3.7)\quad
\begin{cases}
\Delta\omega = 0 \quad \text{in } \Omega, & \\
\frac{\partial\omega}{\partial n} - \frac{k^2\omega}{g} + ik\omega = f \quad \text{on } \Gamma_o, & \\
\frac{\partial\omega}{\partial n} = \vec{u}.\vec{n} \quad \text{on } \Sigma. & \\
f' = Bf = -\frac{\partial\omega}{\partial n} + \frac{k^2\omega}{g} + ik\omega \quad \text{on } \Gamma_o \quad (\text{definition of B}). &
\end{cases}
$$

<u>REFERENCES</u>

[1] G. BAYLAC - J.P. GREGOIRE - C. PAILLY - J. PLANCHARD
"Noise induced stresses in expansion joints".
ASME Vibrations Conference - WASHINGTON - September 1975.

[2] C. BELLEVAUX - M. MAILLE
"Existence et continuité des solutions du problème des oscillations libres d'un bassin en rotation".
C.R.A.S. PARIS - t. 270 (1970) p. 1622-25.

[3] H. BERGER - J. BOUJOT - R. OMAYON
"Un problème spectral en mécanique des vibrations" : calcul des réservoirs élastiques partiellement remplis de liquide".
O.N.E.R.A. Report n° 1364 F (1974).

[4] P. CASEAU - Manuscript note - ELECTRICITE DE FRANCE (1970).

[5] J.P. GREGOIRE - J.C. NEDELEC - J. PLANCHARD
"Problèmes relatifs à l'équation de Helmholtz".
E.D.F. - Bulletin de la Direction des Etudes et Recherches.
Série C. Math. n° 2 (1974) p. 15-22.

[6] J.P. GREGOIRE - J.C. NEDELEC - J. PLANCHARD
"Une méthode de résolution de quelques problèmes aux valeurs propres de l'hydrodynamique".
To appear in "Bulletin de la Direction des Etudes et Recherches. EDF".

[7] KATO - "Perturbation Theory of linear Operator"
SPRINGER - VERLAG.

[8] LAMB
"Hydrodynamics" - CAMBRIDGE University Press.

[9] J.L. LIONS - E. MAGENES
"Problème aux limites non-homogènes" - DUNOD.

[10] RIECZ - NAGY
"Leçons d'Analyse Fonctionnelle" - GAUTHIERS-VILLARS.

SECONDARY BIFURCATION OF A STEADY SOLUTION INTO AN INVARIANT TORUS FOR EVOLUTION PROBLEMS OF NAVIER-STOKES' TYPE.

Gérard IOOSS

Institut de Mathématiques et Sciences Physiques

Parc Valrose .06034 NICE (FRANCE)

I. Introduction.

E. HOPF [1] and L. LANDAU [2] imagined a process of successive bifurcations, with an increasing complexity, to explain a wide class of turbulent flows. The first bifurcation occurs when a steady flow becomes unstable while a characteristic parameter λ of the problem crosses a first critical value λ_0 . In this case, it is known ([3],[4],[5],[6]) that in general it appears a periodic solution, stable if this one exists while the steady solution is unstable. In this paper, we study the next bifurcation, i.e. we look for what happens when the previous periodic solution becomes itself unstable. HOPF and LANDAU have given the idea that a quasi-periodic solution with two fundamental periods has to appear, and a formal development by D. D. JOSEPH [7] supports this belief.

Here, it is mathematically shown that, the bifurcated solution is in general on a two-dimensional torus, this one is stable if it exists while the periodic solution is unstable , moreover it is invariant by the dynamical system. This results are in good agreement with the formal results of [7] , but the quasi-periodicity of the solution stays as an open problem.

II. Statement of the problem.

1) Formulation in a special case.

Let us consider an evolution problem

(1) $$\frac{dV}{dt} = F(V,\lambda) \quad ,$$

where F is an unbounded non linear operator, in a certain Hilbert space H, where we look for $t \longmapsto V(t)$ continuous in the domain $\mathcal{D}$ of the operator F, and where λ is a real parameter. We do, in the following, certain assumptions on F, satisfied for a great number of problems governed by the Navier-Stokes equations. Effectively, if we consider the flow of a viscous incompressible fluid in a bounded regular domain $\Omega \subset \mathbb{R}^3$ or $\mathbb{R}^2$, we have the system

(2) $$\begin{cases} \left.\begin{array}{l} \dfrac{\partial V}{\partial t} + (V\cdot\nabla)V + \nabla p = \nu \Delta V + f \\ \nabla \cdot V = 0 \end{array}\right\} \text{in } \Omega \\ V|_{\partial\Omega} = a \quad , \quad \text{where } \displaystyle\int_{\partial\Omega} a.n\, ds = 0 \quad , \end{cases}$$

where ν is the reciprocal of the Reynolds number and if f and a are respectively a given steady external force and a steady boundary datum. Problems of thermal convection also obeys systems like (2) (see [5]). In infinite domains, where spatial periodicity may be assumed, Ω is a period cell. Problems of magnetohydrodynamic flows enter in our frame (see [8],[9]). Let us also note that in some problems the shear stress is prescribed on a non-deflecting surface : the condition $V|_{\partial\Omega} = a$ is replaced by $V.n|_{\partial\Omega} = 0$ and the tangential stress is given and is proportional to $[(\nabla V + {}^t\nabla V).n] \wedge n$ where n is the exterior normal on $\partial\Omega$ and ${}^t\nabla$ denotes the transposed tensor gradient.

Let us denote by λ a characteristic parameter of the considered problem, such as ν^{-1} for instance, or any other parameter occuring from a or f.

Now, we assume that there exists a T-periodic in t self-excited solution of (2) : $(t,x) \longmapsto [V_0(\lambda,t,x) , p_0(\lambda,t,x)]$, $t \in \mathbb{R}$, $x \in \Omega$, given by the bifurcation of a steady flow (see [5]) for $\lambda > \lambda_0$.

Let us note $V(t,x) = V_0(\lambda,t,x) + U(t,x)$

and $u(t) = U(t,\cdot)$,

and let us introduce in the case of the system (2), the following Hilbert spaces

$$H = \left\{ U \in [L^2(\Omega)]^3 ;\ \nabla\cdot U = 0 ,\ U\cdot n|_{\partial\Omega} = 0 \right\} ,$$

$$\mathcal{D} = \left\{ U \in [H^2(\Omega)]^3 ;\ \nabla\cdot U = 0 ,\ U|_{\partial\Omega} = 0 \right\} ,$$

$$K = \left\{ U \in [H^1(\Omega)]^3 ;\ \nabla\cdot U = 0 ,\ U\cdot n|_{\partial\Omega} = 0 \right\} ,$$

where $H^m(\Omega)$ is the classical Sobolev space. The definitions of the function spaces appropriate to spatially periodic functions do not give any trouble [5].

If the boundary conditions are not those of (2), we have to do the necessary change in $\mathcal{D}$.

The perturbation u satisfies an equation in H of the following form

$$\frac{du}{dt} = \mathcal{A}_\lambda(t)\, u + M_\lambda(t; u) , \tag{3}$$

where, in the case of system (2), we have $\forall U \in \mathcal{D}, \forall t \in \mathbb{R}$

$$\mathcal{A}_\lambda(t)\, u = \Pi \left[\nu \Delta U - (U\cdot\nabla) V_0(\lambda,t) - (V_0(\lambda,t)\cdot\nabla) U \right] \in H ,$$

and $$M_\lambda(t; u) = -\Pi \left[(U\cdot\nabla) U \right] \in K .$$

The operator Π is the orthogonal projection in $[L^2(\Omega)]^3$ onto H (see [10], [11]).

In the following, we express some assumptions for the system (1), which are in fact properties for the operators $\mathcal{A}_\lambda$ and M_λ, satisfied in the case of system (2).

2) Functional setting.

We have 3 Hilbert spaces with the continuous imbeddings

$$\mathcal{D} \hookrightarrow K \hookrightarrow H .$$

Then, we look for solutions u of (3) in $C^0(\mathbb{R}_+; \mathcal{D}) \cap C^1(\mathbb{R}_+; H)$, where $C^m(a,b; \mathfrak{h})$ is the Banach space of bounded continuous functions, with bounded continuous derivatives up to the order m , in $[a,b]$, taking values in the Hilbert space $\mathfrak{h}$. The norm is the usual one (supremum norm on all derivatives of order $\leq m$). We denote by $\mathcal{L}(\mathfrak{h}_1;\mathfrak{h}_2)$ [resp $\mathcal{L}(\mathfrak{h})$] the Banach spaze of the linear bounded operators from $\mathfrak{h}_1$ in $\mathfrak{h}_2$ (resp if $\mathfrak{h}_1 = \mathfrak{h}_2 = \mathfrak{h}$).

Now we assume the following properties :

P. 1 The linear operator $\mathcal{A}_\lambda(t)$ can be written

$$\mathcal{A}_\lambda(t) = A_\lambda + B_\lambda(t)$$

where

i) $[A_\lambda]_{\lambda \in D_0}$ is an holomorphic family of (A) type in H, of domain $\mathcal{D}$, D_0 being a domain of $\mathbb{C}$. A_λ is the infinitesimal generator of an holomorphic semi-group in H, and it has a compact resolvent in H, $\forall \lambda \in D_0$.

Moreover $\forall T_1 < +\infty, \exists c > 0$ and $\alpha \in [0,1[$ such that

$$\|e^{A_\lambda t}\|_{\mathcal{L}(K;\mathcal{D})} \leq c\, t^{-\alpha} \quad , \; t \in]0,T_1] .$$

ii) $(\lambda,t) \longmapsto B_\lambda(t)$ is analytic : $D_0 \times \mathbb{R} \longrightarrow \mathcal{L}(\mathcal{D};K)$ and $t \longmapsto B_\lambda(t)$ is T-periodic.

P. 2 The non linear operator $M_\lambda(t;\cdot)$ is such that :

i) $(\lambda,t,u) \longmapsto M_\lambda(t;u)$ is analytic : $\mathbb{C} \times \mathbb{R} \times \mathcal{D} \longrightarrow K$ for $\lambda \in D_0$, $u \in$ neighbourhood of 0 in $\mathcal{D}$.

ii) $t \longmapsto M_\lambda(t;u)$ is T-periodic.

iii) $\exists \gamma > 0$ such that $\|M_\lambda(t;u)\|_K \leq \gamma \|u\|^2_{\mathcal{D}}$, for u in a neighbourhood of 0 in $\mathcal{D}$.

P. 3 The operators $A_\lambda, B_\lambda, M_\lambda$ are real if λ is real. This implies that we can consider the spaces H, K, $\mathcal{D}$ as the complexified spaces of some real Hilbert spaces, in the aim to be able to speak on real vectors.

P. 4 The solution $u = 0$ of (3) corresponds to a T-periodic solution $t \longmapsto v_0(\lambda,t)$ of an autonomous system (1). Moreover $(\lambda,t) \longmapsto v_0(\lambda,t) - v_0(\lambda,0)$ is analytic : $D_0 \times \mathbb{R} \longrightarrow \mathcal{D}$.

Properties P. 1, P. 2, P. 3 are satisfied in the case of system (2). Indeed, for properties of A_λ, see [5] and note that $\alpha = 3/4 < 1$. Moreover, if we assume that the T-periodic solution v_0 is such that $\lambda \longmapsto v_0(\lambda,\cdot)$ (resp. $\lambda \longmapsto p_0(\lambda,\cdot)$) is analytic from a real interval I in $C^0(0,T;[H^2(\Omega)]^3)$

(resp. $C^0[0,T\,;\,H^1(\Omega)]$), then a theorem of regularity shows that we have in fact

$$(\lambda,t) \longmapsto (v_0(\lambda,t)\,,\,p_0(\lambda,t))$$

analytic $I \times]0,T[\longrightarrow [H^2(\Omega)]^3 \times H^1(\Omega)$. A part of this result is implicit in [12] , a complete and shorter proof is in [13] . Hence, we have the property ii) of P. 1 and P. 4 . For P. 2 we see that, in the example, $M_\lambda(t\,;u)$ is independant of (λ,t) and quadratic in u . The property is then clear if we remark that the imbedding theorems of Sobolev show that if $U \in [H^2(\Omega)]^3$, $\Omega \subset \mathbb{R}^3$, then $(U.\nabla)U \in [H^1(\Omega)]^3$.

3) Precise statement of the problem.

Let us assume that the solution $v_0(\lambda,\cdot)$ is stable for $\lambda < \lambda_1$ and that it becomes unstable for $\lambda > \lambda_1$. For the study of the stability we have to use here a generalisation of the Floquet's theory in infinite dimension for the linearized problem.

To this aim we consider the linearized problem

$$(4)\qquad \begin{cases} \dfrac{dv}{dt} = \mathcal{A}_\lambda(t)\, v \\ v(\tau) = v_0 \in \mathcal{D}\,, \end{cases}$$

where we look for $t \longmapsto v(t)$ continuous : $[\tau,+\infty[\longrightarrow \mathcal{D}$, with a continuous derivative : $[\tau,+\infty[\longrightarrow H$. It can be shown that (4) admits a unique solution

$$(5)\qquad v(t) = S_\lambda(t,\tau)\,v_0\,,$$

where the operator $S_\lambda(t,\tau)$ has the same regularity properties as $e^{A_\lambda(t-\tau)}$.

For example we have

$$S_\lambda(\tau,\tau) = 1\,,\; S_\lambda(t,\tau) \text{ compact in } \mathcal{D} \text{ for } t > \tau\,,$$

but the semi-group property is replaced by

$$S_\lambda(t,\tau) = S_\lambda(t,\eta)\cdot S_\lambda(\eta,\tau) \quad \text{for } t \geq \eta \geq \tau \qquad \text{(see [15])}$$

For the proofs we use $B_\lambda(t)$ as a perturbation term of A_λ and we solve, by a fixed point theorem, the equation

$$v(t) = e^{A_\lambda t} v_0 + \int_0^t e^{A_\lambda(t-\tau)} B_\lambda(\tau)\, v(\tau)\, d\tau ,$$

on an interval $[0,T_1]$, and then we translate the origin. Let us denote by $S_\lambda(t) \equiv S_\lambda(t,0)$. Now we have a fundamental property thanks to the T-periodicity of $\mathcal{A}_\lambda$:

(6) $$S_\lambda(t+T) = S_\lambda(t).S_\lambda(T) \quad \text{in } \mathcal{L}(\mathcal{D}) \text{ for } t \geqslant 0 .$$

This identity shows that to know the value of the spectral radius of $S_\lambda(T)$ with respect to 1 is essential for the study of the linear stability. As the spectrum of $S_\lambda(T)$ only contains isolated eigenvalues (except perhaps 0), we have to compare the moduli of the eigenvalues of $S_\lambda(T)$ with 1 . For a practical study of this spectrum we can use the following result : let us define $\Sigma = \{ \sigma \in \mathbb{C}$; the equation $\frac{dv}{dt} + \sigma v = \mathcal{A}_\lambda(t) v$ admits a non trivial, continuous in $\mathcal{D}$, T-periodic solution $v \}$.

If $\sigma \in \Sigma$, then $\zeta = e^{\sigma T}$ is an eigenvalue of $S_\lambda(T)$. Conversely if $\zeta \neq 0$ is an eigenvalue of $S_\lambda(T)$, then $\sigma = \frac{1}{T} \operatorname{Log} \zeta + 2i\frac{k\pi}{T} \in \Sigma, \forall k \in \mathbb{Z}$.

These results were pointed out first in [14] , all justifications being in [15] .

Now, we know, by assumption and by the identity (6), that for $\lambda < \lambda_1$ the eigenvalues of $S_\lambda(T)$ are of moduli $\leqslant 1$ and that some eigenvalues cross the unit circle while λ crosses λ_1 ,(we can also say that some $\sigma \in \Sigma$ cross the imaginary axis).

III. The Poincaré map.

1) Solution of the Cauchy problem.

The Cauchy problem (3) with $u(0) = u_0 \in \mathcal{D}$ admits a unique solution, continuous from $[0,T_1]$ in $\mathcal{D}$, where $T_1 > T$, if $\|u_0\|_{\mathcal{D}}$ is small enough (see [15]). Indeed to see this we have to write (3) into the equivalent form

$$(7) \qquad u(t) = S_\lambda(t)\,u_0 + \int_0^t S_\lambda(t,\tau)\,M_\lambda[\tau\,;u(\tau)]\,d\tau$$

and to apply the implicit function theorem in $C^0(0,T_1\,;\mathcal{D})$. We obtain [15]:

Theorem 1.

For a fixed $T_1 < \infty$, there is a $\mathcal{V}(0) = \{u_0\,;\|u_0\|_{\mathcal{D}} \leq \delta_0\}$ such that, for $u_0 \in \mathcal{V}(0)$, there exists a unique solution u of (3), with $u(0) = u_0$, in $C^0(0,T_1\,;\mathcal{D})$. Moreover $(\lambda, u_0) \longmapsto \mathcal{U}(\cdot,\lambda,u_0) = u$ is analytic : $I \times \mathcal{V}(0)$ in $C^0(]0,T_1]\,;\mathcal{D})$, and $\frac{\partial u}{\partial u_0}(t,\lambda,0) = S_\lambda(t)$. All terms of the Taylor series at the neighbourhood of $(\lambda_1, 0)$ can be explicited.

In fact, we need a better regularity in t for the solution u of (3). Then we can show [13] .

Theorem 2 of regularity.

If the properties P. 1, P. 2, P. 3, P. 4 are verified, then the map $(t,\lambda,u_0) \longmapsto \mathcal{U}(t,\lambda,u_0)$ is analytic : $\mathbb{R} \times \mathbb{C} \times \mathcal{D} \longrightarrow \mathcal{D}$ for $(t,\lambda,u_0) \in\,]0,T_1] \times D_0 \times \mathcal{V}(0)$.

Corollary.

The map $(\lambda, t) \longmapsto S_\lambda(t)$ is analytic : $D_0 \times \mathbb{R}_+ \longrightarrow \mathcal{L}(\mathcal{D})$. The proof of theorem 2 is based on a change of scaling in t , to make to appear an additional parameter which is shown to occur analytically in the solution of (3).

2) The eigenvalue 1 of $S_\lambda(T)$.

One of the main facts in our problem is that 1 is always an eigenvalue of $S_\lambda(T)$. This is due to the fact that $t \longmapsto v_0(\lambda, t+\delta)$ is a T-periodic solution of (1) (or (2)) $\forall \delta \in \mathbb{R}$. Indeed, by the property P. 4, the map $t \longmapsto \frac{\partial v_0}{\partial t}(\lambda,\cdot)$ is analytic : $\mathbb{R} \longrightarrow \mathcal{D}$ and we can derive, with respect to t , (1) or (2) where $V = v_0(\lambda,\cdot)$. We obtain

$$\frac{\partial}{\partial t}\left[\frac{\partial}{\partial t}\, v_0(\lambda,t)\right] = \mathcal{A}_\lambda(t)\left[\frac{\partial}{\partial t}\, v_0(\lambda,t)\right] ,$$

and then $$\frac{\partial}{\partial t}\, v_0(\lambda,0) = \frac{\partial}{\partial t}\, v_0(\lambda,T) = S_\lambda(T)\left[\frac{\partial}{\partial t}\, v_0(\lambda,0)\right] .$$

This shows that $\frac{\partial}{\partial t}\, v_0(\lambda,0)$ is an eigenvector of $S_\lambda(T)$ for the eigenvalue 1 .

To eliminate this parasitic eigenvalue we have to construct a map replacing the map $u_0 \longmapsto \mathcal{U}(T,\lambda,u_0)$ whose derivative at 0 is $S_\lambda(T)$ and has 1 as an eigenvalue. Then we do the assumption :

H.1. 1 is a simple eigenvalue of $S_\lambda(T)$ for λ near λ_1.

Hence, the projection* operator P_λ, which commutes with $S_\lambda(T)$, and corresponds with the eigenvalue 1, depends analytically on λ.

3) The Poincaré map.

Let us consider u_0 in a good neighbourhood of 0 in $\mathcal{D}$, and consider the equation

$$(8) \qquad P_\lambda\left[\mathcal{U}(\tau,\lambda,u_0) + v_0(\lambda,\tau) - v_0(\lambda,0)\right] = 0 ,$$

where we look for τ near T. We can write (8) $f(\tau,\lambda,u_0) = 0$, with

$$f(T,\lambda,0) = 0 , \quad \frac{\partial f}{\partial \tau}(T,\lambda,0) = \frac{\partial v_0}{\partial \tau}(\lambda,0) \neq 0 .$$

By the property P. 4 and theorem 2, we know that f is analytic, then, by the implicit function theorem there exists a unique solution τ of (8) near T and $(\lambda,u_0) \longmapsto \tau(\lambda,u_0)$ is analytic $\mathbb{C}\times\mathcal{D} \longrightarrow \mathbb{R}$ for $\lambda \in D_0$ and u_0 in a neighbourhood of 0 in $\mathcal{D}$. Moreover we have $\tau(\lambda,0) = T$. The Poincaré map is then defined by

$$(9) \qquad u_0 \longmapsto \Phi_\lambda(u_0) = \mathcal{U}\left[\tau(\lambda,u_0),\lambda,u_0\right] + v_0\left[\lambda,\tau(\lambda,u_0)\right] - v_0(\lambda,0).$$

Now, the map Φ_λ leave invariant a neighbourhood of 0 in $(1-P_\lambda)\mathcal{D}$ and we have

Lemma 1.

The map $(\lambda,u_0) \longmapsto \Phi_\lambda(u_0)$ is analytic : $D_0 \times \mathcal{V}(0) \longrightarrow \mathcal{D}$ and the derivative of Φ_λ at the origin is the restriction of $S_\lambda(T)$ in $(1-P_\lambda)\mathcal{D}$.

Proof. The analyticity follows from the property P. 4, the theorem 2 and the analyticity of τ. Now we have

$$D\,\Phi_\lambda(0) = \frac{\partial \mathcal{U}}{\partial \tau}(T,\lambda,0)\cdot\frac{\partial \tau}{\partial u_0}(\lambda,0) + \frac{\partial \mathcal{U}}{\partial u_0}(T,\lambda,0) + \frac{\partial v_0}{\partial \tau}(\lambda,T)\cdot\frac{\partial \tau}{\partial u_0}(\lambda,0) ,$$

---

* For the explicit expression of P_λ see [15].

and using the derivation of $f(\tau(\lambda,u_o),\lambda,u_o)=0$ with respect to u_o, we obtain $D\Phi_\lambda(0) = (1-P_\lambda)S_\lambda(T)$.

Remark : the geometric meaning of our Poincaré map is indicated on the figure 1.

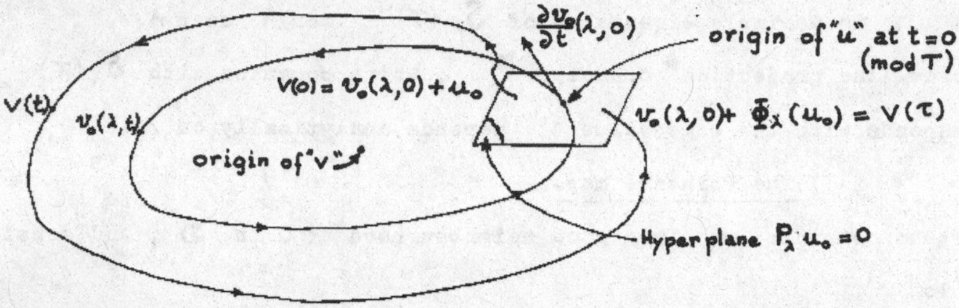

fig. 1.

Lemma 2.

If $\Phi_\lambda^p(u_o)$ belongs to a good neighbourhood of 0 for $p \leq n$, then

$$\Phi_\lambda^p(u_o) = \mathcal{U}\left(\sum_{k=0}^{p-1}\tau_k,\lambda,u_o\right) + v_o\left(\lambda,\sum_{k=0}^{p-1}\tau_k\right) - v_o(\lambda,0),$$

where $\tau_k = \tau\left[\lambda,\Phi_\lambda^k(u_o)\right]$, $k \geq 0$.

Proof. By the uniqueness of the solution τ of (8) such that τ is near pT, and the remark that $\sum_{k=0}^{p-1}\tau_k\Big|_{u_o=0} = pT$.

Moreover, the asymptotic behaviour of a solution $t \longmapsto u(t)$ of (3) can be studied using $\Phi_\lambda^n(u_o)$, $n \to \infty$, if $\Phi_\lambda^n(u_o)$ stays in a neighbourhood of 0 in $\mathcal{D}$, because $\sum_{k\in\mathbb{N}}\tau_k = +\infty$.

Lemma 3.

Let the spectral radius spr $\left[D\Phi_\lambda(0)\right] < 1$, then the cycle v_o is asymptotically stable.

Proof. For $\|u_o\|_{\mathcal{D}}$ small enough, it can be shown easily that $\Phi_\lambda^n(u_o)\xrightarrow[n\to\infty]{}0$.
In fact we have a more precise result in [5] :

$\exists\,\delta_o>0$; $\forall\, V(0)$ such that $\|V(0)-v_o(\lambda,\alpha_o)\|_{\mathcal{D}} \leq \delta_o$, then $\exists\,\alpha_\ell$ such that $\|V(t)-v_o(\lambda,t+\alpha_\ell)\|_{\mathcal{D}}\xrightarrow[t\to\infty]{}0$ exponentially.

IV. Bifurcation into a torus.

By assumption, when λ crosses λ_1, the cycle v_0 becomes unstable ; this leads to the following assumption :

H. 2. There exists two and only two conjugated simple eigenvalues of $D\Phi_{\lambda_1}(0)$ of moduli 1 , noted $\zeta_0, \overline{\zeta}_0$ such that $\zeta_0^n \neq 1$ for n = 1,2,3,4,5. When λ crosses λ_1, these two eigenvalues cross the unit circle.

1) Case when H. 2 is not satisfied.

Let us assume that there exists only one eigenvalue (1 or -1) on the unit circle, or two conjugated eigenvalues $(\zeta_0, \overline{\zeta}_0)$ such that $\zeta_0^n = 1$ for the opera- $D\Phi_{\lambda_1}(0)$. In this case we can look for a non trivial fixed point of $u_0 \longmapsto \Phi_\lambda^n(u_0)$ in the space $(1-P_\lambda)\mathcal{D}$. We obtain a classical steady bifurcation problem because the operator $1 - D\Phi_\lambda^n(0) = [1 - S_\lambda(nT)](1-P_\lambda)$ is not one to one for $\lambda = \lambda_1$. A very similar problem is solved in [16]. Here the solution corresponds to a new bifurcated periodic solution near λ_1 for (1), with a period near n T : the period is $\sum_{k=0}^{n-1} \tau_k = nT + o(1)$.

2) Existence of an invariant "circle" for Φ_λ .

Let H. 2 be verified, then there exists for λ near λ_1, two conjugated simple eigenvalues $\zeta_1(\lambda)$, $\overline{\zeta}_1(\lambda)$ of $D\Phi_\lambda(0)$, and the function ζ_1 is analytic : $\zeta_1(\lambda) = \zeta_0[1 + (\lambda-\lambda_1)\zeta_1 + O(\lambda-\lambda_1)^2]$. Now we assume

H. 3. $\mathcal{R}e\, \zeta_1 \neq 0$.

This assumption, with H. 2, gives $\mathcal{R}e\, \zeta_1 > 0$.

Now, the Lemma 1 and the assumptions H. 2 and H. 3 induct us in the frame of a theorem of RUELLE-TAKENS [17]. First we have to reduce the problem into a two-dimensional one by the "center manifold theorem", for which a detailed proof is in [18]. This locally invariant manifold by Φ_λ is attractive. It is obtained by a fixed point theorem in a good functional space. Once reduced into a two-dimensional one, the map Φ_λ can be converted into a normal form in $\mathbb{C}$:

$$(10) \quad z \longmapsto Z = (|\zeta(\lambda)| - f_1(\lambda)|z|^2)\, z\, e^{i(\mathrm{Arg}\,\zeta(\lambda) + f_2(\lambda)|z|^2)} + O(|z|^5) ,$$

where f_i are real continuous functions and $\zeta(\lambda) = \zeta_1(\lambda) + O(\lambda-\lambda_1)^2$. On the truncated normal form it is easily shown that there exists, following the sign of $f_1(\lambda_1)$, an invariant circle for $\lambda > \lambda_1$, or for $\lambda < \lambda_1$. In fact by the theorem in [17] and [18] we have the

Lemma 4.

$\exists$ a neighbourhood of λ_1, such that if $f_1(\lambda_1)>0$ there exists in a neighbourhood of 0 in $(1-P_\lambda)\mathcal{D}$, an invariant attracting "circle" Γ_λ for the map Φ_λ, (only) for $\lambda > \lambda_1$; if $f_1(\lambda_1) < 0$ the invariant "circle" is repelling and exists only for $\lambda < \lambda_1$. The diameter of Γ_λ is of order $|\lambda-\lambda_1|^{1/2}$.

Remark. The coefficient $f_1(\lambda_1)$ can be calculated explicitly (see [15] and [13]).

Now, we obtain easily the

Theorem 3.

Let us assume H.1, H. 2, H. 3 be realised and the coefficient $f_1(\lambda_1) \neq 0$. Then, there exists a neighbourhood of λ_1 such that, only for $\lambda>\lambda_1$ or for $\lambda < \lambda_1$, following $f_1(\lambda_1) > 0$ or <0, there is a "circle" Γ_λ in $(1-P_\lambda)\mathcal{D}$ such that the set

$$\mathcal{C} = \{ v(t) = v_0(\lambda,t) + u(t) \ ; \ t \in [0,\tau(\lambda,u_0)], u_0 \in \Gamma_\lambda,$$

$$t \longmapsto u(t) \text{ is the solution of (3), continuous in } \mathcal{D}\text{, with } u(0) = u_0 \}$$

is invariant by the dynamical system (1). When it exists for $\lambda > \lambda_1$, this bifurcated torus is attractive (stable) whereas in the case when it exists for $\lambda < \lambda_1$ it is repelling.

Remark. A solution of (1) such that $v(0) \in \mathcal{C}$, has a principal part which is quasi-periodic with the two fundamental periods T and $2\pi T/\mathrm{Arg}\,\zeta_0$ (see [15] and [13]).

BIBLIOGRAPHY.

[1] E. HOPF, Berichten der Math-Phys. Kl. Sächs. Akad. Wiss. Leipzig 94, 1-22 (1942).

[2] L. LANDAU, C. R. Acad. Sci. U.S.S.R. 44, 311-314 (1944).

[3] D. D. JOSEPH and D. H. SATTINGER, Arch. Rat. Mech. Anal. 45, 79-109 (1972).

[4] V. I. IUDOVICH, Prikl. Mat. Mek. 35, 638-655 (1971), and Prikl. Mat. Mek. 36, 450-459 (1972).

[5] G. IOOSS, Arch. Rat. Mech. Anal. 47, 301-329 (1972).

[6] J. MARSDEN, Bull. Am. Math. Soc. 79, 3, 537-541 (1973).

[7] D. D. JOSEPH, Lecture Notes in Math. N°322, 130-158. Berlin-Heidelberg-New-york : Springer 1973.

[8] O. A. LADYZHENSKAYA, Zap. Nauch. Sem. Lomi, 38, 46-93, Leningrad (1973).

[9] G. DURAND, Thèse de 3e cycle, Pub. Math. d'Orsay N°128 (1975).

[10] R. TEMAM, Lecture Notes N°9 , University of Maryland, Dep^{t} of Math. (1973).

[11] O. A. LADYZHENSKAYA, The Mathematical Theory of Viscous Incompressible flow. New-York, Gordon and Breach, 1963.

[12] H. FUJITA and T. KATO, Arch. Rat. Mech. Anal. 16, 269-315 (1964).

[13] G. IOOSS (to appear).

[14] V. I. IUDOVICH, Dokl. Akad. Nauk, 195, 292-295 (1970).

[15] G. IOOSS, Arch. Rat. Mech. Anal. (to appear in 1975).

[16] G. IOOSS, Arch. Mech. Stosowanej, 26, 795-804 (1974).

[17] D. RUELLE and F. TAKENS, Comm. Math. Phys. 20, 167-192 (1971).

[18] O. E. LANFORD III, Lecture Notes in Math. N°322, 159-192. Berlin- Heidelberg - New-York : Springer 1973.

A BASIC OPEN PROBLEM IN THE THEORY OF ELASTIC STABILITY

W.T. KOITER

1. Introduction and summary

Thermodynamics has provided a secure foundation for the energy criterion of stability of equilibrium for an elastic body under conservative (static) external loads. A positive definite potential energy functional in the L_2-space of admissible displacement fields in the vicinity of the equilibrium configuration, where this energy has the value zero, has been established as a sufficient condition for stability in the dynamical sense of Liapounoff [7, 10, 14]. In the conventional application of the stability criterion the requirement on the potential energy functional is weakened to the requirement of a weak proper relative minimum. From a practical point of view the results obtained from this relaxed criterion are entirely satisfactory. It is a largely open question, however, what restrictions (on the initial disturbances or otherwise) have to be imposed in order to achieve that the relaxed criterion does indeed ensure the stability of equilibrium in a strict mathematical sense. It is the purpose of the present paper to draw once again the attention of mathematicians to this basic open problem and to outline some avenues to be explored in an attempt to achieve a satisfactory solution.

2. The energy criterion of stability

The equilibrium configuration of a three-dimensional elastic body at the uniform temperature of its surroundings, which has to be investigated as to its stability, will be called the fundamental state I. A material point is identified by its radius vector $\underset{\sim}{x}$ with rectangular Cartesian components x_i (i = 1, 2, 3) in this fundamental state. We shall assume that the body is properly supported in its fundamental state such that rigid-body displacements from this state I are excluded by the supports.

Let $\underset{\sim}{u}(\underset{\sim}{x})$ denote an arbitrary admissible displacement field with components $u_i(\underset{\sim}{x})$ which carries the body from the equilibrium configuration I to some adjacent state II. The (additional) deformation due to the displacement field $\underset{\sim}{u}(\underset{\sim}{x})$ is described by the symmetric (additional) strain tensor

$$\gamma_{ij} = \frac{1}{2}\left[u_{i,j} + u_{j,i} + u_{h,i}u_{h,j}\right] = \theta_{ij} + \frac{1}{2}u_{h,i}u_{h,j}, \tag{2.1}$$

where we have adopted the summation convention for a repeated subscript and the notation that a comma preceding a subscript j denotes the partial derivative with respect to the coordinate x_j. The linearized (additional) strain tensor is denoted by θ_{ij}.

The elastic potential per unit volume of the undeformed body W is specified by the free energy density per unit volume of the undeformed body at the uniform and constant temperature T_I of the surroundings. It is a function of the strain tensor $W(\underset{\sim}{\gamma}(\underset{\sim}{x}))$, and we assume this function has continuous derivatives with respect to the strain components (2.1) up to any order required in the analysis. We also assume that this function is written symmetrically with respect to γ_{ij} and γ_{ji}. The increment of the elastic potential in the transition from the fundamental state I to the adjacent state II is given by

$$W_{II} - W_I = \left(\frac{\partial W}{\partial \gamma_{ij}}\right)_I \gamma_{ij} + \frac{1}{2}\left(\frac{\partial^2 W}{\partial \gamma_{ij}\partial \gamma_{k\ell}}\right)_I \gamma_{ij}\gamma_{k\ell} + \ldots \; . \tag{2.2}$$

This expansion terminates with the quadratic terms in the case of materials which obey the generalisation of Hooke's law to finite deformations in the form that the elastic potential is a homogeneous quadratic function of the (total) strain components.

Let ρ_o and ρ denote the mass densities in the undeformed state and the fundamental state I respectively. The (symmetric) stress tensor $\underset{\sim}{S}$ in state I is specified by its components

$$S_{ij} = \frac{\rho}{\rho_o}\left(\frac{\partial W}{\partial \gamma_{ij}}\right)_I . \tag{2.3}$$

The tensor of elastic moduli in state I for infinitesimal strain increments is defined by

$$E_{ijk\ell} = \frac{\rho}{\rho_o}\left(\frac{\partial^2 W}{\partial \gamma_{ij}\partial \gamma_{k\ell}}\right)_I . \tag{2.4}$$

It satisfies the symmetry relations

$$E_{ijk\ell} = E_{jik\ell} = E_{ij\ell k} = E_{k\ell ij} . \tag{2.5}$$

For the sake of simplicity we shall restrict our discussion to the case of so-called dead loading in which the potential energy of the external loads in an adjacent configuration II is a linear functional of the displacement field $\underset{\sim}{u}(\underset{\sim}{x})$. This linear energy functional of the external loads is cancelled by the linear term in the increment of the total elastic energy due to the same displacement field $\underset{\sim}{u}(\underset{\sim}{x})$, because state I is a configuration of equilibrium. The increment in total potential energy in the transition from the fundamental state I to the adjacent configuration II is thus specified by the energy functional

$$P[\underset{\sim}{u}(\underset{\sim}{x})] = \int \left(\frac{1}{2} S_{ij} u_{h,i} u_{h,j} + \frac{1}{2} E_{ijk\ell} \gamma_{ij} \gamma_{k\ell} + \ldots \right) dv, \tag{2.6}$$

where the integral is extended over the volume of the body in the fundamental state I. The second variation of the energy is obtained immediately from (2.5), if we replace the strain components in the second term by their linearized approximations θ_{ij}

$$P_2[\underset{\sim}{u}(\underset{\sim}{x})] = \int \left(\frac{1}{2} S_{ij} u_{h,i} u_{h,j} + \frac{1}{2} E_{ijk\ell} \theta_{ij} \theta_{k\ell} \right) dv. \tag{2.7}$$

Many writers prefer a slightly different form for this second variation [e.g. 4]

$$P_2[\underset{\sim}{u}(\underset{\sim}{x})] = \int \frac{1}{2} d_{ijk\ell} u_{i,j} u_{k,\ell} dv, \tag{2.8}$$

where the tensor

$$d_{ijk\ell} = \delta_{ik} S_{j\ell} + E_{ijk\ell} \tag{2.9}$$

exhibits only the symmetry property (2.5) with respect to the pair of subscripts i,j and k,ℓ.

Any definition of the stability of equilibrium in the dynamical sense of Liapounoff requires the specification of suitable norms to measure both the magnitude of an initial disturbance and the distance from the equilibrium configuration in the motion following upon the initial disturbance. For the admissible displacement field itself it is often convenient, and physically attractive, to employ the L_2-norm, and the basic theorem of elastic stability is usually formulated in terms of this special norm [7, 10, 14]. However, this special choice is by no means necessary. The stability theorem remains valid for any norm for the displacement field itself, say $||\underset{\sim}{u}||$, which satisfies the requirements of a normed linear function space.

The potential energy functional (2.6) is called positive definite with respect to the norm $||\underset{\sim}{u}||$, if it satisfies the inequality

$$P[\underset{\sim}{u}(\underset{\sim}{x})] \geqq d(c) \quad \text{for} \quad ||\underset{\sim}{u}|| = c, \tag{2.10}$$

where d(c) is a positive non-decreasing function of c in some interval $0 < c < c_1$. The basic stability theorem is now expressed by the statement that a positive definite potential energy functional ensures the stability of the equilibrium configuration [7, 10, 14]. The magnitude of the initial disturbance is here restricted by the magnitude of the initial value of the norm $||\underset{\sim}{u}||$ as well as by the magnitude of the total energy imparted to the system by the initial disturbance.

It is not so easy to establish the converse of the stability theorem. It would be expressed by the statement that an indefinite potential energy functional, i.e. a functional $P[\underset{\sim}{u}(\underset{\sim}{x})]$ with negative values for $||\underset{\sim}{u}|| = c$ no matter how small the positive number c is selected, implies instability of the equilibrium configuration. A theorem to this effect has apparently only been proved in the presence of a positive definite energy dissipation in any motion of the body, and on the additional assumption that the energy functional $P[\underset{\sim}{u}(\underset{\sim}{x})]$ has no negative stationary value for any displacement field with a norm $||\underset{\sim}{u}||$ less than some positive number c_2 [7, 10]. These restrictions do not seem too serious from a physical point of view. Henceforward we shall therefore also take for granted that an indefinite potential energy functional implies instability.

3. The relaxed form of the energy criterion

The energy criterion of stability is conventionally applied in the relaxed form that a weak proper minimum of the potential energy functional is a sufficient condition for stability of the equilibrium configuration. This relaxed version of the criterion is formulated in terms of displacement fields with a uniform bound on the tensor of displacement gradients: $|\nabla \underset{\sim}{u}| = (u_{i,j} u_{i,j})^{1/2} < g$, where g is a positive number. A necessary condition for stability is the existence of a positive number g with the property that

$$P[\underset{\sim}{u}(\underset{\sim}{x})] \geqq 0 \tag{3.1}$$

for all admissible displacement fields $\underset{\sim}{u}(\underset{\sim}{x})$ with $|\nabla \underset{\sim}{u}| < g$ everywhere. This condition is also taken to be sufficient, if the equality sign in (3.1) applies only for $\underset{\sim}{u}(\underset{\sim}{x}) \equiv 0$.

An immediate consequence of the inequality (3.1) is the criterion of the second variation. A necessary condition for stability is the inequality

$$P_2 [\underset{\sim}{u}(\underset{\sim}{x})] \geqq 0 \tag{3.2}$$

which must hold for all admissible displacement fields $\underset{\sim}{u}(\underset{\sim}{x})$. This condition is usually again taken to be also sufficient, if the equality sign in (3.2) applies only for $\underset{\sim}{u}(\underset{\sim}{x}) \equiv 0$.

Nearly all investigations of elastic stability, in particular the calculation of critical loads on structures which are liable to buckling, are based on the criterion (3.2) for the second variation. At a critical load, however, where the equality sign in (3.2) holds for one or more linearly independent displacement fields, the so-called buckling modes, we have to return to the more general requirement (3.1). The result of such an extended analysis allows predictions on the post-buckling behaviour of the structure and on its sensitivity to imperfections [5, 6].

In the light of our discussion in the previous section there is, of course, no question about the necessity of conditions (3.1) and (3.2) for stability in the dynamical sense of Liapounoff. On the other hand, there exists no mathematical justification for their interpretation as sufficient conditions. It is true that a sharpened version of (3.2) in the form [14]

$$P_2|\underset{\sim}{u}(\underset{\sim}{x})| \geqq C||\underset{\sim}{u}||^2,$$

where C is a positive constant, would indeed be a sufficient condition, if the potential energy functional (2.6) would be twice continuously differentiable in the Fréchet sense. We are not aware, however, of any successful attempt to establish the required property of differentiability in the Fréchet sense.

The lack of a sound mathematical justification of the relaxed energy criterion of stability was already recognized at an early date [e.g. 2]. We also refer to the extensive article by Knops and Wilkes [4] for a detailed discussion of the problem. Nevertheless engineers have continued to rely on this relaxed criterion, and with complete success. We are not aware of any structural failure due to a reliance on the relaxed energy criterion rather than on the mathematically fully justified criterion (2.10) of a positive definite energy functional. It seems to us that we are faced here with an open basic problem in the mathematical theory of elastic stability where it is unable to explain the facts of experience. In the next section we shall discuss some avenues which might be explored to resolve this problem.

4. Some avenues to be explored

In Hellinger's famous article [2] it was already observed that the problem of a justification of the criterion of the second variation does not arise in the case of Euler's one-dimensional theory of the elastica. The reason for the absence of our problem in that case was also pin-pointed by Hellinger, viz. the presence of a positive definite term in the second derivative of the deflection in the integrand of the energy functional. A similar situation was exploited later in the two-dimensional theory of flat plates and shallow shells [9]. Here again it has been shown that the criterion of a positive definite second variation ensures stability in the dynamical sense of Liapounoff. Analogous results have been obtained in the general three-dimensional theory on the assumption that the elastic potential does not depend only on the strain components themselves, but also on their gradients $\gamma_{ij,k}$, and these additional terms are assumed to be positive definite. This assumption results in additional terms in the integrand of the potential energy functional which are positive definite in the second derivatives $u_{i,jk}$ of the displacement field and they permit a rigorous justification of the criterion of the second variation [7]. Even if we are not entirely happy about the somewhat artificial nature of our assumption, as far as we are aware it is the only approach up to now which has achieved a definite result.

Recent attempts to justify the criterion of the second variation by Naghdi and Trapp [12], and by Como and Grimaldi [1] do not seem to us to be promising. The basic theorem 4.1 in [12] includes in assumption (c) effectively the standard assumption of the relaxed energy criterion that a weak proper minimum of the potential energy functional would ensure stability in the dynamical sense of Liapounoff. Como and Grimaldi claim in [1] that it would be possible to select the higher order terms in the strain components (represented by the dots in (2.6)) in such a way that the functional would be twice continuously differentiable in the Fréchet sense in a space with the energy norm of the linear theory of elasticity, equivalent to

$$||\underset{\sim}{u}|| = \left[\frac{1}{V}\int \theta_{ij}\theta_{ij}dv\right]^{1/2} . \qquad (4.1)$$

So far they have offered no valid proof of this claim, and Martini has shown that it cannot be correct [11]. Writing the functional (2.6) in the form

$$P[\underset{\sim}{u}(\underset{\sim}{x})] = \int F(\nabla\underset{\sim}{u})\, dv, \qquad (4.2)$$

he has proved that the function $F(\nabla\underset{\sim}{u})$ must be a quadratic function in order that the functional (4.2) be twice continuously differentiable in the space with norm (4.1).

Significant results have been achieved by Shield in the linear theory of elastic stability by suitable requirements on the smoothness of the initial disturbance [13]. Linear theory is equivalent to (2.7) or (2.8) as the complete potential energy functional, and it is by no means evident that similar smoothness requirements would have the same effective results in the nonlinear theory corresponding to the functional (2.6).

In fact, John's recent results on the development of singularities in nonlinear wave propagation [3], in particular for elastic waves, do not hold much promise for an effective solution of our problem only by imposing smoothness restrictions on the initial disturbance.

A more effective approach is, perhaps, to abandon the concept of a body which behaves purely elastically. The motion of an actual material always involves a dissipation of energy, even if the body behaves reversibly for infinitely slow deformations. We have already accepted this physical fact in section 2, where we have appealed to such energy dissipation in our recognition of the energy criterion as a necessary condition for stability. It is indeed quite conceivable that a proper allowance for dissipative effects, possibly combined with appropriate smoothness requirements on the initial disturbance, will ultimately establish the criterion of the second variation as a sufficient condition for stability in the dynamical sense of Liapounoff.

5. Concluding remarks

The present situation with respect to the energy criterion in its relaxed form in which a weak minimum of the potential energy functional is accepted as a sufficient condition for stability is hardly better than it was ten years ago. As a sobering moral to mechanical scientists and mathematicians alike we quote from an earlier paper [8] a paragraph which would not require any change if it were written now:

> We are now in a position to try and formulate a fair appraisal of the energy criterion of elastic stability in its conventional form that a weak minimum of the potential energy functional would represent a both necessary and sufficient condition for the stability of an equilibrium configuration. There is no question about the validity of this criterion as a necessary condition for stability, in the sense that an indefinite (or negative definite) energy in the class of displacement fields with sufficiently small displacement gradients at every point of the body implies instability, if the presence of some damping is admitted. The validity of the energy criterion in its conventional form as a sufficient condition for stability has not been proved or disproved for simple elastic materials. On the other hand, this criterion is capable of a rigorous proof for materials with a suitably modified stored energy function. Moreover, the breakdown of the argument in attempts to prove the criterion for simple materials occurs under circumstances which are open to suspicion from the physical point of view. Finally, in spite of the large number of applications of the conventional naive energy criterion, we are not aware of any incriminating evidence against this

criterion resulting from this extensive experience. It seems to us that engineering science is well advised to continue its reliance on the energy criterion in its traditional form. At the same time it may be hoped that theorists will continue their search for conditions under which a rigorous proof of this criterion may be given for simple materials.

References

1. M. COMO and A. GRIMALDI. Stability, buckling and post-buckling of elastic structures. Part 1, Definition and regularization of the potential energy. Report No. 1, Department of Structures, University of Calabria (June 1974).
2. E. HELLINGER. Die allgemeine Ansätze der Mechanik der Kontinua. Enc. math. Wiss. IV-4, 601-694, in particular 653-654, Leipzig (1914).
3. F. JOHN. Formation of singularities in one-dimensional nonlinear wave propagation. Comm. Pure & Appl. Math. 27, 377-405 (1974).
4. R.J. KNOPS and E.W. WILKES. Theory of elastic stability. Handb. Phys. VI a/3, 125-302 (1973).
5. W.T. KOITER. On the stability of elastic equilibrium. Thesis Delft, H.J. Paris, Amsterdam (1945). English translations issued as NASA TT F 10-833 (1967) and AFFDL Report 70-25 (1970).
6. W.T. KOITER. Elastic stability and post-buckling behaviour. Proc. Symp. Nonlinear Problems, Univ. Wisc. Press, 257-275 (1963).
7. W.T. KOITER. The energy criterion of stability for continuous elastic bodies. Proc. Kon. Ned. Ak. Wet. B68, 178-202 (1965).
8. W.T. KOITER. Purpose and achievements of research in elastic stability. Proc. 4-th Tech. Conf. Soc. Eng. Sci. Gordon & Breach, London (1966), pp. 197-218.
9. W.T. KOITER. A sufficient condition for the stability of shallow shells. Proc. Kon. Ned. Ak. Wet. B70, 367-375 (1967).
10. W.T. KOITER. Thermodynamics of elastic stability. Proc. 3. Can. Congr. Appl. Mech., Calgary, 29-37 (1971).
11. R. MARTINI. Private Communication (1975).To be published in Proc.Kon.Ned.Ak.Wet.
12. P.M. NAGHDI and J.A. TRAPP. On the general theory of stability for elastic bodies. Arch. Rat. Mech. Anal. 51, 165-191 (1973).
13. R.T. SHIELD. On the stability of linear continuous systems. ZAMP 16, 649-686 (1965).
14. C.C. WANG and C. TRUESDELL. Introduction to rational elasticity, in particular section VII-7. Noordhoff, Leiden (1973).

SOME APPLICATIONS AND METHODS OF NONLINEAR FUNCTIONAL ANALYSIS IN FINITE DISPLACEMENT PLATE THEORY

Franz Labisch
Lehrstuhl für Mechanik II
Ruhr-Universität Bochum / BRD

1. Introduction

In the well known KIRCHHOFF plate theory geometric and physical linearity is assumed. The bending and the stretching of the plate are independent of each other. Each problem can be reduced to a linear partial differential equation with appropriate boundary conditions. Well known powerful methods of linear functional analysis can be used. If the shape of the plate and the boundary conditions are not too complicated, there exists at least one solution. It can be approximated with any desired accuracy. Global and pointwise error bounds can be specified [1 - 5].

If the magnitude of the deflection is of the same order as the plate thickness, the stretching and the bending couple with each other, and can no longer be treated independently. The linear HOOKE's law remains valid, however a geometrically nonlinear problem is obtained. Solutions and error bounds of the linear mathematical problem exist, but they can not be used as an adequate approach. Based on more general assumptions, the non-linear problem can be reduced to the two coupled non-linear differential equations presented in 1910 by T. v. KÁRMÁN [6]. They contain the linear biharmonic operator and quadratic terms in the second derivatives. Due to the nonlinearity a straightforward solution to this problem proved to be extremely difficult. In order to surmount this a weak solution involving two complementary functionals is sought. Following the methods described by NOBEL [7], SEWELL [8] and RALL [9] and based on the proof of existence given by KNIGHTLY [10] and DUVAUT et LIONS [11] this paper attempts to provide a simple and constructive treatment of the mentioned problem. Using complementary error bounds a statement of the error of the deflection is obtained [12, 13]. Stability problems are to be excluded.

2. Basic Notions

Indices notation and the summation convention is used. The ranges of the subscripts are:

$i,j = 1,2,3$; $\alpha,\beta = 1,2$; $\ell,m = 1,2,\dots 8$;.

F_o denotes a region in the real Euclidean space E_2 with boundary C_o, closure $\bar{F}_o = F_o \cup C_o$

and points $(x_\alpha) \in \overline{F}_0$. For sake of simplicity let C_0 be decomposed into two, not necessarily disjoint, parts $C_0 = C_{ov} \cup C_{op}$. In the Lagrangian description $\overline{F}_0$ coincides with the undeformed middle plane of the considered thin elastic plate of constant thickness h. All quantities can be expressed in terms of $(x_\alpha) \in \overline{F}_0$ only. $(\ )_{,\alpha}$ stands for partial differentiation with respect to x_α. A Cartesian coordinate system x_i will be used and we denote by

u_i the displacement vector of the middle plane
$M_{\alpha\beta}$ Piola components of the stress moments
$N_{\alpha\beta}$ Piola components of the stress resultants
$N_{\alpha 3} = N_{\alpha\beta} u_{3,\beta}$ vertical stress components
$p(x_\alpha)$ the lateral load per unit area
n_α the unit outer normal to C_0
s_α the unit tangent vector to C_0
$N_{ni} = n_\alpha N_{\alpha i}$; $M_{\beta n} = n_\alpha M_{\beta\alpha}$; $Q_{n3} = M_{\alpha n,\alpha}$; $M_{ns} = s_\beta M_{\beta n}$; $M_{nn} = n_\alpha M_{\alpha n}$;
E YOUNG's modulus, ν POISSONS ratio,
$D = \frac{Eh}{12(1-\nu^2)}$ bending rigidity.

Prescribed external forces on C_{op} and prescribed displacement quantities on C_{ov} are indicated by a star. These quantities are complementary if C_{op} and C_{ov} are not disjoint. The notation T^* will also be used for the adjoint operator to the linear operator T.

Consider two real Hilbert spaces H_u, H_p with elements U, P and inner products (.) and <.> respectively, and a linear operator $T : H_p \to H_u$ with the adjoint operator $T^*; H_u \to H_p$
such that

$$(U,TP) = < T^*U,P > \tag{2.1}$$

for all $U \in H_u$, $P \in H_p$.
Elements of the Cartesian product space $H = H_u \times H_p$ will be denoted by $k = \binom{U}{P}$. H is a Hilbert space for the inner product {.} definded by

$$\{k_1,k_2\} = (U_1,U_2) + < P_1,P_2 > \tag{2.2}$$

3. Variational Functionals for the Finite Displacement Plate Theory

With these preliminaries we assume

$$U = (u_\alpha,\ u_3\ ;\ v_\alpha) \tag{3.1}$$

$$P = (N_{\alpha\beta},\ N_{\alpha 3}\ ;\ M_{\alpha\beta}), \tag{3.2}$$

U and P defined on F_0, C_{ov} and C_{op} as column vectors, see [8].

From KIRCHHOFF's hypothesis

$$v_\alpha = - u_{3,\alpha} . \tag{3.3}$$

With the inner products

$$(\underset{1}{U},\underset{2}{U}) = \int_{F_o} [\underset{1}{u}_i \underset{2}{u}_i + \underset{1}{v}_\alpha \underset{2}{v}_\alpha] \, d F_o + \int_{C_{ov}} [\underset{1}{u}_i \underset{2}{u}_i + \underset{1}{v}_\alpha \underset{2}{v}_\alpha] \, d C_{ov} + \int_{C_{op}} [\underset{1}{u}_i \underset{2}{u}_i + \underset{1}{v}_\alpha \underset{2}{v}_\alpha] \, d C_{op} ,$$

$$< \underset{1}{P},\underset{2}{P} > = \int_{F_o} [\underset{1}{N}_{\alpha i} \underset{2}{N}_{\alpha i} + \underset{1}{M}_{\alpha\beta} \underset{2}{M}_{\alpha\beta}] \, d F_o + \int_{C_{ov}} [\underset{1}{N}_{\alpha i} \underset{2}{N}_{\alpha i} + \underset{1}{M}_{\alpha\beta} \underset{2}{M}_{\alpha\beta}] \, d C_{ov} + \int_{C_{op}} [\underset{1}{N}_{\alpha i} \underset{2}{N}_{\alpha i} + \underset{1}{M}_{\alpha\beta} \underset{2}{M}_{\alpha\beta}] \, d C_{op}$$

and the operators

$$T^* = \begin{bmatrix} \frac{\partial}{\partial x_\alpha} \\ -n_\alpha \\ 0 \end{bmatrix} \quad ; \quad T = \begin{bmatrix} -\frac{\partial}{\partial x_\alpha} \\ 0 \\ n_\alpha \end{bmatrix} \begin{matrix} F_0 \\ \text{on } C_{ov} \\ C_{op} \end{matrix} , \tag{3.4}$$

the adjointness relation (2.1) follows immediately from GAUSS' theorem.
Introduce the functional

$$R(k) = -\int_{F_0} [w - p u_3] \, d F_o + \int_{F_o} [N_{\alpha i} u_{i,\alpha} + M_{\alpha\beta} v_{\beta,\alpha}] \, d F_0 + \int_{C_{op}} [N^*_{n\alpha} u_\alpha + Q^*_{n3} u_3 + M^*_n v_\beta] \, d C_{op}$$
$$- \int_{C_{ov}} [N_{ni} u^*_i - Q_{n3}(u_3 - u^*_3) + M_{n\alpha} v^*_\alpha] \, d C_{ov} \tag{3.5}$$

with the energy density

$$w = \frac{Eh}{2(1-\nu)} [(u_{1,1} + \frac{u^2_{3,1}}{2})^2 + (u_{2,2} + \frac{u^2_{3,2}}{2})^2 + 2 (u_{1,1} + \frac{u^2_{3,1}}{2})(u_{2,2} + \frac{u^2_{3,2}}{2}) \tag{3.6}$$
$$+ \frac{1-\nu}{2} (u_{1,2} + u_{2,1} + u_{3,1} u_{3,2})^2] + \frac{D}{2} [u^2_{3,11} + u^2_{3,22} + 2\nu u_{3,11} u_{3,22} + 2(1-\nu) u^2_{3,12}]$$

Let us consider the two functionals

$$I_u = < T^* U, P > - R(k) \tag{3.7}$$

$$I_c = (U, TP) \quad - R(k) \tag{3.8}$$

It may be shown that in some subspaces of H_u and H_p, I_u and I_c are complementary functionals. From (3.1 - 3.7) we see that (3.7) is the known displacement functional

$$I_u = \int_{F_0} [w - p u_3] \, d F_0 - \int_{C_{op}} [N^*_{n\alpha} u_\alpha + Q^*_{n3} u_3 - M^*_{n\alpha} u_{3,\alpha}] \, d C_{op}$$
$$- \int_{C_{ov}} [N_{n\alpha}(u_\alpha - u^*_\alpha) + (Q_{n3} + N_{n3})(u_3 - u^*_3) - M_{n\alpha}(u_{3,\alpha} - u^*_{3,\alpha})] \, d C_{ov} \tag{3.9}$$

To obtain numerical results we substitute trial functions strongly fulfilling the geometric boundary conditions on C_{ov}

$$u_i = u_i^* \; ; \; u_{3,n} = u_{3,n}^* \tag{3.10}$$

Then (3.9) is expressed only in terms of the displacement components and its partial derivatives up to the second order. Thus taking into account (2.1) and (3.1) - (3.8), see [14]

$$TP - \frac{\partial R(k)}{\partial U} = 0 \tag{3.11}$$

gives the linear equations of equilibrium

$$N_{\alpha\beta,\alpha} = 0 \tag{3.12}$$

$$M_{\alpha\beta,\alpha\beta} + N_{\alpha 3,\alpha} + p = 0 \tag{3.13}$$

and the static boundary conditions on C_{op}

$$N_{n\alpha} = N_{n\alpha}^* \quad ; \quad M_{nn} = M_{nn}^* \quad ; \quad N_{n3} + Q_{n3} + M_{ns,s} = Q_{n3}^* + M_{ns,s}^* \tag{3.14}$$

A substitution of the complementary energy density

$$W_c = N_{\alpha\beta} u_{\beta,\alpha} + M_{\alpha\beta} v_{\beta,\alpha} + N_{\alpha 3} u_{3,\alpha} - W \tag{3.15}$$

into (3.8), application of GAUSS' theorem and formulas (3.1 - 3.5) give the complementary functional proposed by STUMPF [14]

$$\begin{aligned} I_c = &- \int_{F_o} [N_{\alpha\beta,\alpha} u_\beta + (N_{\alpha 3,\alpha} + M_{\alpha\beta,\alpha\beta} + p) u_3] \, dF_o - \int_{F_o} W_c \, dF_o \\ &+ \int_{C_{op}} [(N_{n\beta} - N_{n\beta}^*) u_\beta + (N_{n3} + Q_{n3} - Q_{n3}^*) u_3 - (M_{n\alpha} - M_{n\alpha}^*) u_{3,\alpha}] \, dC_{op} \\ &+ \int_{C_{ov}} [N_{n\alpha} u_\alpha^* + (Q_{n3} + N_{n3}) u_3^* - M_{n\alpha} u_{3,\alpha}^*] \, dC_{ov} \, . \end{aligned} \tag{3.16}$$

Admissible trial functions for (3.16) shall fulfill strongly the equilibrium conditions (3.12 - 3.13) and the static boundary conditions (3.14).
I_c may be expressed in terms of P only and analogously to (3.11)

$$T^*U - \frac{\partial R(k)}{\partial P} = 0 \tag{3.17}$$

yields the relations between the Piola stress components and the derivatives of the displacement vector

$$N_{\alpha\beta} = \frac{\partial W}{\partial u_{\beta,\alpha}} \quad ; \quad N_{\alpha 3} = \frac{\partial W}{\partial u_{3,\alpha}} \quad ; \quad M_{\alpha\beta} = \frac{\partial W}{\partial v_{\beta,\alpha}} \quad , \tag{3.18}$$

and the geometric boundary conditions (3.10) on C_{ov}.

If Eqs. (3.18) can be inverted, the compatibility condition

$$N_{11,22} - 2N_{12,12} + N_{22,11} + Eh(u_{3,11}u_{3,22} - u^2_{3,12}) = 0 \qquad (3.19)$$

follows and the complementary energy density may be written

$$W_c = \frac{1}{2Eh}(N_{11} + N_{22} - 2\nu N_{11}N_{22} + 2(1+\nu)N_{12}) + \frac{1}{2}[N_{13}\varphi_1(P) + N_{23}\varphi_2(P)]$$
$$+ \frac{6}{Eh^3}[M^2_{11} + M^2_{22} - 2\nu M_{11}M_{22} + 2(1+\nu)M^2_{12}] \qquad (3.20)$$

with

$$\varphi_1(P) = \frac{N_{13}N_{22} - N_{23}N_{12}}{N_{11}N_{22} - N^2_{12}} \quad , \quad \varphi_2(P) = \frac{N_{23}N_{11} - N_{13}N_{12}}{N_{11}N_{22} - N^2_{12}} \quad . \qquad (3.21)$$

By the introduction of a stress function $F(x_\alpha)$

$$N_{11} = hF_{,22} \; ; \; N_{22} = hF_{,11} \; ; \; N_{12} = -hF_{,12} \qquad (3.22)$$

it may be easily seen, that (3.13) and (3.19) become the well known von KÁRMÁN Eqs.

$$F_{,1111} + 2F_{,1122} + F_{,2222} = E(u^2_{3,12} - u_{3,11}u_{3,22})$$
$$u_{3,1111} + 2u_{3,1122} + u_{3,2222} = \frac{h}{D}(\frac{p}{h} + F_{,22}u_{3,11} + F_{,11}u_{3,22} - 2F_{,12}u_{3,12}) \qquad (3.23)$$

If the nonlinear terms are disregarded, we obtain the two uncoupled linear equations for the linear plate bending and stretching.

4. Uniqueness of the Solution

Now we introduce for brevity

$$d_1 = u_{1,1} \; ; \; d_2 = u_{2,2} \; ; \; d_3 = u_{1,2} + u_{2,1} \; ; \; d_4 = u_{3,1} \; ; \; d_5 = u_{3,2}$$
$$d_6 = u_{3,11} \; ; \; d_7 = u_{3,22} \; ; \; d_8 = u_{3,12} \qquad (4.1)$$

and denote by $\hat{H}_u$ the set of geometrically admissible functions with a positive second order FRÉCHET derivative

$$U \in \hat{H}_u : \partial^2(I_u)(\eta_1, \eta_2) > 0 \qquad (4.2)$$

$\hat{H}_u$ os not empty. We obtain

$$\partial^2(I_u)(\eta_1, \eta_2) = \int_{F_o} \frac{\partial^2 W}{\partial d_\ell \partial d_m} \delta d_\ell \delta d_m \, dF_o \; ; \qquad (4.3)$$

where

$$\frac{\partial^2 W}{\partial d_\ell \partial d_m} = \begin{bmatrix} A, 0 \\ 0, B \end{bmatrix} . \tag{4.4}$$

A denotes the matrix

$$A = \frac{Eh}{1-\nu^2} \begin{bmatrix} 1 \;, & \nu \;, & 0 \;, & d_4 \;, & \nu d_5 \\ \nu \;, & 1 \;, & 0 \;, & \nu d_4 \;, & d_5 \\ 0 \;, & 0 \;, & \frac{1-\nu}{2} \;, & \frac{1-\nu}{2} d_5 \;, & \frac{1-\nu}{2} d_4 \\ d_4 \;, & \nu d_4 \;, & \frac{1-\nu}{2} d_5 \;, & a_1 \;, & a_3 \\ \nu d_5 \;, & d_5 \;, & \frac{1-\nu}{2} d_4 \;, & a_3 \;, & a_2 \end{bmatrix} \tag{4.5}$$

with the abreviations

$$a_1 = d_1 + \nu d_2 + \frac{3}{2} d_4^2 + \frac{1}{2} d_5^2 \; ; \; a_2 = \nu d_1 + d_2 + \frac{1}{2} d_4^2 + \frac{3}{2} d_5^2$$

$$a_3 = \frac{1-\nu}{2} d_3 + d_4 \cdot d_5 \; .$$

A simple calculation shows, that A is positive definite if and only if

$$N_{11} > 0 \; ; \; N_{11} N_{22} - N_{12}^2 > 0 \tag{4.6}$$

hold. All elements of the matrix

$$B = D \begin{bmatrix} 1, & \nu, & 0 \\ \nu, & 1, & 0 \\ 0, & 0, & 2(1-\nu) \end{bmatrix} \tag{4.7}$$

are constants. This matrix occurs also in the linear bending problem and is always positive definite. It follows, that (4.6) become sufficient conditions for the positive definiteness of (4.4).

A stationary point $U = U_o$ for the functional (3.9) is sought. Therefore the first derivative in the sense of Frêchet is set equal to zero

$$0 = \partial(I_u)\,(\eta) = \int_{F_o} [\frac{\partial W}{\partial d_\ell} \delta d_\ell - p \delta u_3] \, d F_o - \int_{C_{op}} [N^*_{n1} \delta u_1 + N^*_{n2} \delta u_2 + Q_{n3} \delta u_3 - M^*_{n1} \delta d_4 - M^*_{n2} \delta d_5] \, d C_{op} \tag{4.8}$$

Let $I_o = I_o(U_o)$ denote the value of I_u for the stationary point U_o, and $\tilde{I}$ the value in some neighborhood of the stationary point. A formal TAYLOR expansion up to a remainder term with a second order derivative in the sense of FRÉCHET implies

$$\tilde{I}_u - I_o = \frac{1}{2} \partial^2 (I_u)(\eta_1, \eta_2) \tag{4.9}$$

where the value of the second derivative is taken in some neighborhood of the stationary point. Strict convexity of I_u and uniqueness of U_o, at least for the subdomain in which (4.6) hold, follow.

Passing again to (3.16) we observe that, at least in this mentioned subdomain, all conditions to invert (3.18) are fulfilled. In the class of static admissible functions substitution into (3.16) leads to a functional in terms of Piola stress components only, for which a stationary P_o is sought. Using the notation

$$S_\ell \equiv (N_{11}, N_{22}, N_{12}, N_{13}, N_{23} \; ; M_{11}, M_{22}, M_{12})$$

the first derivative is set equal to zero:

$$0 = \partial(I_c)(\Theta) \equiv - \int_{F_o} \frac{\partial W_c}{\partial S_\ell} \delta S_\ell \, d F_o + \int_{C_{ov}} [\delta N_{n\alpha} u^*_\alpha + (\delta Q_{n3} + \delta N_{n3}) u^*_3 - \delta M_{n\alpha} u^*_{3,\alpha}] \, d C_{ov} \tag{4.10}$$

Analogously to (4.3) we obtain for the second derivative

$$\partial^2(I_c)(\Theta_1, \Theta_2) = - \int_{F_o} \frac{\partial^2 W_c}{\partial S_\ell \partial S_m} \delta S_\ell \delta S_m \, d F_o \quad , \tag{4.11}$$

and denote by $\hat{H}_p \subset H_p$ the set of static admissible functions with positive second derivative

$$P \in \hat{H}_p : \int_{F_o} \frac{\partial^2 W_c}{\partial S_\ell \partial S_m} \delta S_\ell \delta S_m \, d F_o > 0 \tag{4.12}$$

It is easy to see that the inequalities (4.6) become sufficient conditions for the strict concavity of (3.16). In the corresponding subspace of H_p uniqueness of the solution P_o follows.

Equation (3.11) arises from a minimum principle for the functional (3.9), (3.17) from a maximum principle for the functional (3.16). Hence Eqs. (3.11), (3.17) arise from complementary principles.
In some surrounding ball of the stationary point k_o H, see [9]

$$I_o = \max_P \min_U f(k) = \min_U \max_P f(k) \tag{4.13}$$

where

$$f(k) = < T^*U, P > - R(k) = (U, TP) - R(k) = \frac{1}{2} < T^*U, P > + \frac{1}{2} (U, TP) - R(k) \quad , \tag{4.14}$$

so that k_o is a minimax point.
By Eqs. (3.18) and the inverse to them a mapping of some subspace of H into itself

$$U = \Phi(P) \quad ; \quad P = \Psi(U) \tag{4.14}$$

is defined. In [10, 11] the existence of a solution k_0 was shown. The uniqueness follows from the strict concavity and strict convexity. Based on this theory it was easy to obtain numerical results for upper bounds $\tilde{I}$ and lower bounds $\tilde{\tilde{I}}$ [12].

5. Other Complementary Functionals

The functional (3.9) is expressed in terms of the displacement components and their partial derivatives only, the functional (3.16) in terms of the Piola stress components only. In both cases a RAYLEIGH-RITZ method leads to an approach for the solution and to upper and lower bounds for I_0.
Now we introduce the functional

$$I_\gamma(k) \equiv \gamma I_u + (1-\gamma) I_c \qquad (5.1)$$

where γ is a real constant.
Here U and P are to vary independently and a GALERKIN method has to be applied. Regarding (2.1) we obtain the same stationary point k_0 and the same value I_0. As an immediate consequence for

$\gamma \leq 0$ a maximum principle
$\gamma \geq 1$ a minimum principle
$0 < \gamma < 1$ a stationary principle only

follows. In particular for $\gamma = 0$, I_c and for $\gamma = 1$, I_u is obtained.

6. Estimates for the Deflection Error

The matrix B (4.7) may be split up

$$B = D\begin{bmatrix} 1-\nu^2, & 0, & 0 \\ 0, & 0, & 0 \\ 0, & 0, & 0 \end{bmatrix} + D\begin{bmatrix} \nu^2, & \nu, & 0 \\ \nu & 1, & 0 \\ 0\,, & 0, & 2(1-\nu) \end{bmatrix} \qquad (6.1)$$

where the second matrix is positive semidefinite.
Employing the difference $\tilde{I} - \tilde{\tilde{I}}$ called the global error bound and the formal TAYLOR expansion (4.9) we conclude:

$$\tilde{I} - \tilde{\tilde{I}} \geq \tilde{I} - I_0 = \frac{1}{2}\int_{F_0} \frac{\partial^2 W}{\partial d_\ell \partial d_m}\,\delta d_\ell \delta d_m \, dF_0 \geq \frac{1-\nu^2}{2} D \int_{F_0} (\delta u_{3,11})^2 \, dF_0 \, . \qquad (6.2)$$

Analogous formulas for $\delta u_{3,22}$ and $\delta u_{3,12}$ follow. These yield the L^2-norm estimates for the second partial derivatives of the error of the deflection Δu_3:

$$\int_{F_0} (\Delta u_{3,11})^2 \, dF_0 \leq \frac{2}{D(1-\nu^2)}(\tilde{I}-\tilde{\tilde{I}}); \int_{F_0} (\Delta u_{3,22})^2 \, dF_0 \leq \frac{2}{D(1-\nu^2)}(\tilde{I}-\tilde{\tilde{I}}); \int_{F_0} (\Delta u_{3,12})^2 dF_0 \leq \frac{1}{D(1-\nu)}(\tilde{I}-\tilde{\tilde{I}}). \qquad (6.3)$$

Now we assume, that the plate is supported in such a way, that the exact value of the deflection u_3 is known at least in one point and in one point the exact value of $u_{3,1}$ or $u_{3,2}$ is known. Taking into account that $C^2(\Omega)$ form a dense set in $W_{2,2}(\Omega)$, [15] and using twice POINCARE's inequality we obtain L^2-norm estimates for the first derivatives

$$\int_{F_0} (\Delta u_{3,1})^2 \, dF_0 \le C_1(F_0) \; (\tilde{I}-\tilde{\tilde{I}}) \; ; \; \int_{F_0} (\Delta u_{3,2})^2 \, dF_0 \le C_2(F_0)(\tilde{I}-\tilde{\tilde{I}}) \tag{6.4}$$

and for the deflection

$$\int_{F_0} (\Delta u_3)^2 \, dF_0 \le C_3(F_0)(\tilde{I}-\tilde{\tilde{I}}) \; . \tag{6.5}$$

Then from (6.3 - 6.5) we deduce that the error connected with the inserted trial functions belongs to some SOBOLEV-space

$$\Delta u_3 \in W_{2,2}(F_0) \tag{6.6}$$

SOBOLEV's imbedding theorem [16] gives

$$\sup_{F_0} |\Delta u_3| \le C(F_0)(\tilde{I}-\tilde{\tilde{I}})^{1/2} \; .$$

It turns out that the maximum error for the plate deflection tends to zero as the square root of the complementary bound difference. The obtained results remain valid for linear plate theory. A generalization for geometrically nonlinear shallow shells is possible.

There is some similarity. Thus the very attractive problem arises, to generalize some methods, used in the linear KIRCHHOFF plate theory, and to obtain pointwise error bounds for all field quantities in the geometrically nonlinear finite displacement plate theory. This still open problem will be explained in forthcoming papers.

References

[1] WEBER,C., Eingrenzung von Verschiebungen mit Hilfe der Minimalsätze, Z. angew. Math. Mech. 22, S. 130, 1942

[2] RIEDER, G., Über punktweise Eingrenzung in der Elastizitätstheorie, Vortrag auf dem Mathematikerkongress "Equadiff II", Bratislava 1966

[3] STUMPF, H., Eingrenzungsverfahren in der Elastomechanik, Forschungsberichte des Landes Nordrhein-Westfalen, Nr. 2116, Köln 1970

[4] PRAGER, W., Synge J. L., Approximations in Elasticity Based on the Concept of Function Space, Quaterly of Applied Math. 5, S. 241, 1947

[5] DIAZ, J. B., Greenberg H. J., Upper and Lower Bounds for the Solution of the First Boundary Value Problem of Elasticity, Quart. Appl. Math. 6, S. 326, 1948

[6] TH. v. KÁRMÁN, Festigkeitsprobleme im Maschinenbau, Enz. d. math. Wiss. BD IV, pp 348 - 352, 1910

[7] NOBLE, B., Complementary Variational Principle for Boundary Value Problems I: Basic Principles with an Application to Ordinary Differential Equations, MRC Technical Summary Report ## 473, November 1964

[8] SEWELL, M. J., The Governing Equations and Extremum Principles of Elasticity and Plasticity Generated from a Single Functional, Part I, J. Struct. Mech 2(1), pp 1 - 32, 1973, Part II, J. Struct. Mech. 2(2), pp 135 - 158, 1973

[9] RALL, L. B., On Complementary Variational Principles, J. Math. Anal. Appl. 14, 174 - 184, 1966

[10] KNIGHTLY, G. H., An Existence Theorem for the von Kármán Equations, Arch. Rat. Mech. Anal., vol. 36, pp 65 - 78, 1970

[11] DUVAUT, G. et LIONS, J. L., Problèmes unilatéraux dans la théorie de la flexion forte des plaques, J. d. Méc., vol. 13, No. 1, pp 51 - 74, Mars 1974

[12] LABISCH, F., STUMPF, H., Pauschale Fehlerschranken in der nichtlinearen Plattentheorie, ZAMM, GAMM-Sonderheft 1974

[13] LABISCH, F., Zur Fehlerabschätzung in der nichtlinearen Kármánschen Plattentheorie, ZAMM, GAMM-Sonderheft 1975

[14] STUMPF, H., Dual extremum principles and error bounds in the theory of plates with large deflections, Archives of Mechanics (Archiwum Mechaniki Stosowanej) 27, 3, 485 - 496 (1975)

[15] AGMON, S., Lectures on Elliptic Boundary Value Problems, Princeton, Van Nostrand 1965

[16] SOBOLEV, S. L., Applications of Functional Analysis in Mathematical Physics, A.M.S. 1963 translated from the Russian 1950 edition

CRITERES DE VALIDITE DE LA THEORIE NON-LINEAIRE DES COQUES ELASTIQUES

P. LADEVEZE
Département de Mécanique
Université P et M Curie Paris/FRANCE

Des estimations de la validité des hypothèses sous-jacentes aux versions usuelles de la théorie non-linéaire des coques élastiques sont établies. Les résultats sont du même type que ceux obtenus par Koiter et Danielson [1970] en théorie linéaire. De par leur signification physique et aussi leur finesse, ils constituent des critères d'utilisation de ces modèles bidimensionnels considérés comme approchés du modèle tridimensionnel de l'élasticité non-linéaire classique.

La théorie non-linéaire des coques élastiques comme d'ailleurs la théorie linéaire est le fruit d'hypothèses mixtes qui exploitent le fait que dans la majorité des conditions de chargement des coques le champ de contraintes normales à la surface moyenne est négligeable. En fait, il n'existe pas une théorie non-linéaire mais plusieurs que nous avons regroupées sous une même formulation utilisant un opérateur Λ variant de l'une à l'autre et caractérisant la relation de comportement bidimensionnelle . La théorie correspondant à la valeur Λ de cet opérateur est notée: Λ- version / La solution $(\tilde{u}, (\tilde{\mathbb{N}}, \tilde{\mathbb{M}}))$ d'une Λ- version est constituée d'un champ $\hat{u}$ de déplacement et d'un champ $(\tilde{\mathbb{N}}, \tilde{\mathbb{M}})$ de contrainte généralisée définis sur la surface moyenne Σ_m . La solution $(\tilde{U}_0, \tilde{C}_0)$ du modèle de référence, c'est-à-dire de l'élasticité non-linéaire classique, est composée d'un champ de déplacement et d'un champ de contrainte (de Piola Lagrange) définis sur le domaine tridimensionnel Ω occupé par le milieu dans son état naturel . Ces solutions étant de nature différente ne peuvent être comparées.

Notre méthode consiste à comparer $(\tilde{U}_0, \tilde{C}_0)$ à un champ $(\hat{U}, \hat{C})$ de même nature mathématique et représentatif de la solution coque, en ce sens qu'il appartiendra à l'ensemble $\tilde{V} \times \tilde{W}$ des éléments (U, C) constitués d'un champ de déplacement U (cinématiquement) admissible et prenant la valeur $\tilde{u}$ sur Σ_m , et d'un champ de contrainte C (statiquement) admissible dont la résultante et le moment sont égaux à $\tilde{\mathbb{N}}$ et $\tilde{\mathbb{M}}$.

Cet élément $(\hat{U}, \hat{C})$ est élaboré avec le souci qu'il vérifie "au mieux" la relation de comportement tridimensionnelle . Il est à noter que dans cette étape essentielle, $(\hat{U}, \hat{C})$ est construit explicitement en fonction uniquement des données et de la solution coque.

L'élément $(\hat{U}, \hat{C})$ apparaît ainsi comme une solution approchée du modèle de référence, et le problème de l'estimation de la validité des hypothèses de la Λ- version de la théorie non-linéaire des coques élastiques est ramené à l'étude du degré d'approximation de cette solution approchée.

Cette question est d'abord envisagée dans le cadre général où $(\hat{U}, \hat{C})$ est un couple admissible quelconque (n'appartenant pas nécessairement à $\tilde{V} \times \tilde{W}$). Deux types d'erreurs sont distingués : erreur en relation de comportement $\hat{C} - C(\hat{U})$ où $C(\hat{U})$ est le champ de contrainte associé à $\hat{U}$ par la relation de comportement de référence, et erreur en solution $(\hat{U} - \tilde{U}_0 \ , \ \hat{C} - \tilde{C}_0)$.

Des liens entre ces deux erreurs sont établis en supposant que pour tout $\lambda \in [0,1]$ les configurations $\lambda \tilde{U}_0$ sont stables au sens du critère habituellement utilisé. Il en résulte des estimations de l'erreur en solution qui n'utilisent que la solution approchée $(\hat{U}, \hat{C})$.

Ces résultats sont ensuite appliqués au couple $(\hat{U}, \hat{C})$ représentatif de la solution de la Λ- version. En introduisant des grandeurs caractéristiques de cette solution, des données, de Λ et de la géométrie, nous obtenons des estimations des erreurs relatives. En outre, les versions usuelles de la théorie linéaire des coques élastiques étant des Λ- versions particulières de la théorie non-linéaire, les résultats obtenus fournissent aussi des critères de validité pour ces théories linéaires considérées comme approchées du modèle tridimensionnel de l'élasticité non-linéaire classique.

Le détail des démonstrations pourra être trouvé dans [5]

1 - Résultats de comparaison

1.1 Modèle de milieu hyperelastique considéré

Sur une partie $\partial_1\Omega$ de la frontière de Ω, le déplacement est imposé. En outre, le milieu est soumis à une charge donnée f qui est supposée indépendante du déplacement ; le cas contraire devrait pouvoir être traité par notre méthode: a priori il n'introduit que des complications d'ordre technique.

Pour tout espace vectoriel X (sur $\mathbb{R}$), L(X) désigne l'espace vectoriel des endomorphismes de X, $L_S(X)$ celui des endomorphismes symétriques et $L_S^+(X)$ l'ensemble des endomorphismes définis positifs. La transposition euclidienne relative à

l'espace euclidien orienté à trois dimensions E_3 est notée par un trait : $\overline{\cdot}$;
M désigne le point courant de Ω

Pour préciser le modèle de référence, nous introduisons deux espaces vectoriels $\mathcal{S} \overset{déf}{=} \{C \mid C \in [L(E_3)]^{\Omega}, C \in L^{4/3}(\Omega)\}$ et $\mathcal{E} \overset{déf}{=} \{E \mid E \in [L(E_3)]^{\Omega}, E \in L^4(\Omega)\}$ mis en dualité séparante par la forme bilinéaire $(C,E) \longmapsto \int_\Omega Tr[\overline{C}E]\,d\Omega$.

$$\mathcal{S} \times \mathcal{E} \longrightarrow \mathbb{R}$$

$U_0 + \mathcal{U}$ (*) est l'espace affine des champs de déplacement admissibles et $J \overset{déf}{=} \{\frac{\partial U}{\partial M} | U \in \mathcal{U}\}$ un sous-espace de $\mathcal{E}$. L'espace affine des champs de contrainte admissibles (de Piola-Lagrange) est un sous-espace de $\mathcal{S}$ noté $C_0 + J^0$ où J^0 est l'orthogonal de J.

Le problème de l'équilibre du milieu hyperelastique considéré peut être formulé ainsi :

Trouver un couple $(\tilde{U}_0, \tilde{C}_0)$ tel que :

(1.1) $\tilde{U}_0 \in U_0 + \mathcal{U}$ (équations de liaison)

(1.2) $\tilde{C}_0 \in C_0 + J^0$ (équations d'équilibre)

(1.3) $\tilde{C}_0 = C(\tilde{U}_0)$ (relation de comportement)

avec $C(U) = (I + \frac{\partial U}{\partial M}) K_0 . D(U)$

où K_0 est le champ d'opérateur de Hooke caractéristique du matériau $(K_0(M) \in L_S^+(L_S(E_3)))$
I l'endomorphisme identité et $D(U) \overset{déf}{=} \frac{1}{2} (\frac{\partial U}{\partial M} + \overline{\frac{\partial U}{\partial M}} + \overline{\frac{\partial U}{\partial M}} \frac{\partial U}{\partial M})$.

Ce modèle admet une énergie de déformation que nous écrirons
$E_D(\frac{\partial U}{\partial M}) = \frac{1}{2} \int_\Omega Tr[D(U) K_0 D(U)]\,d\Omega$; il est à signaler que l'existence des solutions de (1) n'est démontrée que dans des cas particuliers .

1.2 Erreur en relation de comportement et erreur en solution

A tout couple admissible $(\hat{U}, \hat{C})$, qui par définition appartient à $(U_0 + \mathcal{U}) \times (C_0 + J^0)$ et donc vérifie (1.1) et (1.2), on peut associer la quantité $\hat{C} - C(\hat{U})$ qui caractérise le degré de vérification de la relation (1.3). Considérant $(\hat{U}, \hat{C})$ comme solution approchée, nous appelons erreur en relation de comportement cette quantité $\hat{C} - C(\hat{U})$

Soit $\psi : C \longmapsto \int_\Omega Tr[\overline{C} K_0^{-1} C]\,d\Omega$ (*) ; $[\psi(\hat{C} - C(\hat{U}))]^{\frac{1}{2}}$

$$\mathcal{S} \longrightarrow \mathbb{R}^+$$

(*) On impose à $U_0 + \mathcal{U}$ d'appartenir à l'espace de Sobolev $W_4^1(\Omega)$

(*) K_0 est ici un prolongement du champ d'opérateur de Hooke ; $K_0(M) \in L_S^+(L(E_3))$.

est une mesure particulière de cette erreur, appelée erreur quadratique en relation de comportement.

Il est aisé d'interpréter physiquement cette notion qui se généralise à bien d'autres types de milieux. En effet, on peut penser intuitivement que dans le domaine "avant flambage", si d'une part l'erreur en relation de comportement est "petite" et si d'autre part les configurations $\hat{U}$ et $\tilde{U}_o$ sont "stables", le champ $\hat{U}$ est "proche" de $\tilde{U}_o$; de toute façon $(\hat{U}, \hat{C})$ est solution de (1) si l'erreur en relation de comportement est nulle. Cependant les relations intéressantes entre ce type d'erreur et l'erreur en solution $(\hat{U} - \tilde{U}_o, \hat{C} - \tilde{C}_o)$ sont limitées par les moyens mathématiques actuels ; en particulier les seules mesures de l'erreur en solution que l'on peut escompter sont relatives au cadre de régularité naturel du problème (1).

Cette question est envisagée ici en supposant que la solution $(\tilde{U}_o, \tilde{C}_o)$ de (1) retenue vérifie la condition de stabilité suivante :

$$N_o \left| \begin{array}{l} (\forall S \in J) \\ (\forall \lambda \in [0\ 1]) \end{array} \right. \quad D^2(E_D)\left(\lambda \frac{\partial \tilde{U}_o}{\partial M}\right)(S)(S) \geq 0$$

condition qui signifie que les configurations $\lambda\tilde{U}_o$ sont stables au sens du critère habituellement utilisé pour tout λ compris entre 0 et 1 . $D^2(E_D)(A)$ désigne la différentielle seconde de E_D en $A \in \mathcal{E}$.

On pose :

$$(\forall A,B \in (L_S(E_3))^{\Omega} \cap L^2(\Omega)) \qquad <A,B> \stackrel{\text{déf}}{=} \int_{\Omega} \mathrm{Tr}\,[\bar{A}\,K_o\,B]\,d\Omega$$

$$\begin{pmatrix} \forall S \in \mathcal{E} \\ \forall \lambda \in \mathbb{R}^+ \end{pmatrix} \quad M(A, \lambda, S) \stackrel{\text{déf}}{=} \frac{2 - 3\lambda}{4} < \bar{S}S, \bar{S}S > + \frac{3}{4}(1 - \lambda) < \bar{S}\hat{A} + \bar{\hat{A}}S, \bar{S}S > +$$

$$+ D^2(E_D)(\lambda\hat{A})(S)(S) - \frac{1}{2\lambda} D^2(E_D)(\lambda\hat{A})(S)(S)$$

$$\text{avec} \quad \hat{A} = \frac{\partial \hat{U}}{\partial M}$$

La condition N_o et les relations (1) vérifiées par $(\tilde{U}_o, \tilde{C}_o)$ entrainent que $\hat{S} \stackrel{\text{déf}}{=} \frac{\partial U_o}{\partial M} - \frac{\partial \hat{U}}{\partial M}$ est solution de l'inégalité

$$(2) \quad \left| \ (\forall \lambda \in [0\ 1]) \quad M(\hat{A}, \lambda, \hat{S}) \leq \int_{\Omega} \mathrm{Tr}\,\left[\overline{\hat{C} - C(\hat{U})}\,\hat{S}\right] d\Omega \right.$$

Si nous établissons la proposition

$$(P) \quad \left| \ \exists \alpha > 0 \text{ tel que } (\forall S \in \mathcal{E}) \quad \alpha \|S\|^2_{L^2(\Omega)} \leq \sup_{\lambda \in [0\ 1]} M(\hat{A}, \lambda, S) \right.$$

nous déduisons de (2) $||\hat{S}||_{L^2(\Omega)} \leq \frac{1}{\alpha} ||\hat{C} - C(\hat{U})||_{L^2(\Omega)}$ c'est-à-dire une estimation du type cherché.

Pour préciser le domaine de validité de la proposition (P), on introduit

$$\gamma(\hat{A}) \overset{\text{déf}}{=} \left[\frac{\varepsilon(A)}{\varepsilon_c} \right]^{1/2} \quad \text{où} \quad \varepsilon(\hat{A}) = \sup \left(\left|\left| \frac{\hat{\bar{A}} - \hat{A}}{2} \right|\right|^2_{L^\infty(\Omega)}, \left|\left| \frac{\hat{A} + \hat{\bar{A}}}{2} \right|\right|_{L^\infty(\Omega)} \right)$$

et $\varepsilon_c = \sup \{\delta | \delta \in \mathbb{R}^+ ; (\forall \delta' \in [0\ \delta], \forall S \in J)\ L(S,\delta') \geq 0\}$ avec

$$L(S,\delta') = < \frac{S + \bar{S}}{2}, \frac{S + \bar{S}}{2} > - \delta' \int_\Omega \operatorname{Tr} \left[\frac{\bar{S} - S}{2} K_0 \frac{S - \bar{S}}{2} \right] d\Omega$$

Pour les structures métalliques. ε_c , qui peut être appelée déformation critique, a pour ordre de grandeur en général 10^{-3} .

Nous avons établi ([5]) que pour $g(\gamma(\hat{A}))$ positif, g étant l'application $\gamma \longmapsto g(\gamma) = \frac{10}{46} - \frac{\gamma}{4} - \frac{7}{4}\gamma^2$, la proposition (P) est vraie et que l'on a :

$\mathbb{R}^+ \longrightarrow \mathbb{R}$

$$(3) \quad \left| \begin{aligned} & 2\, E_{DL}(\hat{S}) \overset{\text{déf}}{=} < \frac{\hat{S} + \hat{\bar{S}}}{2}, \frac{\hat{S} + \hat{\bar{S}}}{2} > \leq \frac{1}{[g(\hat{\gamma})]^2} \hat{e} \\ & \frac{1}{16} < \hat{\bar{S}}\hat{S}, \hat{\bar{S}}\hat{S} > \leq \frac{g(\hat{\gamma}) + \frac{\hat{\gamma} + 3\hat{\gamma}^2}{4}}{[g(\hat{\gamma})]^2} \hat{e} \\ & E_D(\hat{S}) \leq \frac{2}{[g(\hat{\gamma})]^2} \hat{e} \\ & \text{avec} \quad \hat{\gamma} = \gamma(\hat{A}) \\ & \qquad \hat{e} = \inf \psi(C - C(\hat{U})) \\ & \qquad\quad \underline{C \in C_0 + J_0} \\ & \qquad\quad C - C(\hat{U}) = C - C(\hat{U}) \end{aligned} \right.$$

Ce résultat peut être étendu aux cas où $g(\gamma(\hat{A})) \leq 0$. En effet, il suffit en utilisant (3) de comparer les champs de déplacement $\tilde{U}_0$ et $\hat{U}_r \overset{\text{déf}}{=} U_0 + r(\hat{U} - U_0)$, le réel positif r étant choisi de manière à vérifier $g(\gamma(\hat{U}_r)) > 0$, ce qui est toujours possible dans le cas où $A_0 = 0$.

2 - Critères de validité de la théorie non-linéaire des coques élastiques.

2.1 Λ - version.

La coque est définie géométriquement par la surface de référence

Σm et par l'épaisseur $2h(m)$, m désignant le point courant de Σm . La configuration de référence Ω est alors

$$\Omega = \{M \mid M = m + N(m)Z, \quad m \in \Sigma m, -h(m) < Z < h(m)\}$$

où $N(m)$ est le vecteur normal unitaire en m de la surface Σm qui est supposée régulière, orientée et connexe.

On pose :

$$\left|\begin{array}{l} \Sigma_+ = \{M \mid M = m + h(m)\, N(m),\ m \in \Sigma m\} \\ \Sigma_- = \{M \mid M = m - h(m)\, N(m)\ ,\ m \in \Sigma m\} \\ \Sigma_L = \partial\Omega - (\Sigma_+ \cup \Sigma_-) \ : \text{surface latérale ou bord de la coque} \\ \Sigma_{1L} = \partial_1\Omega \cap \Sigma_L \qquad\qquad \Sigma_{2L} = \Sigma_L - \Sigma_{1L} \\ Cm = \partial\ \Sigma m \\ \pi(m) : \text{opérateur de projection orthogonale sur le plan tangent à } \Sigma m \text{ en m.} \\ S = \pi + Z\,\dfrac{\partial N}{\partial m} \qquad\qquad k = \det S \end{array}\right.$$

Dans le cas où (m,Z) est un système de coordonnées de Ω et où le déplacement n'est imposé que sur une partie éventuellement vide du bord définie par un arc C_1m de Cm, on peut construire des théories bidimensionnelles définies sur Σm, appelées théories des coques, qui remplacent avantageusement les modèles de référence tridimensionnels. Nous ne considérons ici que les versions usuelles de la théorie des coques correspondant au modèle de référence de l'élasticité non-linéaire (1) que l'on trouvera par exemple dans [Koiter et Simmonds 3], et qui sont suivant notre terminologie des Λ-versions particulières de la théorie non-linéaire des coques élastiques.

Par Λ-version, nous entendons le problème aux limites suivant :

Trouver $(\tilde{u}, (\tilde{\mathbb{N}}, \tilde{\mathbb{M}}))$ où $\tilde{u}$ est un champ de déplacement défini sur Σm et $(\tilde{\mathbb{N}}, \tilde{\mathbb{M}})$ un champ de contrainte généralisée également défini sur Σm tel que

$$\tilde{u} = \tilde{U}_{|\Sigma m}$$

$$(\tilde{\mathbb{N}}, \tilde{\mathbb{M}}) = \int_{-h}^{h} (\pi\, \tilde{C}'\pi\, ,\ \pi\tilde{C}'\pi\ Z)\ S^{-1}\ k\ dZ$$

$$\left|\begin{array}{lll} \text{où} & (\tilde{U},\tilde{C}') & \text{est une solution des équations (4) ci-dessous :} \\ & (4\text{-}1) & \tilde{U} \in U_0 + \mathcal{U}_{KL} \\ & (4\text{-}2) & (\forall\ U \in \mathcal{U}_{KL}) \qquad \displaystyle\int_\Omega \mathrm{Tr}\ [\bar{\tilde{C}}'\ \frac{\partial U}{\partial M}]\,d\Omega = f(U) \\ & (4\text{-}3) & \tilde{C}' = \Lambda\,(\dfrac{\partial \tilde{U}}{\partial M}) \end{array}\right.$$

$\mathcal{U}_{KL}$ est le sous-espace des champs de déplacement de Kirchoff-Love de trace nulle sur Σ_{1L} c'est-à-dire l'ensemble

$$\mathcal{U}_{KL} = \{U \mid U(m,Z) = u(m) - Z\,[\,\overline{\frac{\partial u}{\partial m}}\,N](m),\ u \in E_3^{\Sigma m},\ U \in \mathcal{U},\ \bar{N}u \in W_4^2(\Sigma m)\,\}$$

Λ est un opérateur arbitraire mais donné qui ne vérifie que la condition : $\bar{N}\,\Lambda(\frac{\partial U}{\partial M}) = 0$ pour tout $U \in U_0 + \mathcal{U}_{KL}$. Cette condition qui n'est pas indispensable est vérifiée par les versions classiques.

Généralement, par équations de la théorie non-linéaire des coques, on entend les équations vérifiées par $\tilde{u}, \tilde{\mathbb{N}}, \tilde{\mathbb{M}}$ obtenues en éliminant les inconnues intermédiaires $\tilde{U}, \tilde{C}'$. Les versions usuelles sont déduites, par des hypothèses simplificatrices portant sur l'expression en fonction de Z de la densité d'énergie et sur les termes non-linéaires, de la Λ_0 version définie par

$$\Lambda_0(A) = (I + A)K'_0 \,.\, (\frac{A + \bar{A} + \bar{A}A}{2})$$

K'_0 est le champ d'opérateur.

$$K'_0(M) : \quad D \longmapsto [\pi K_0(\pi D \pi)\pi - \pi K_0(N\bar{N})\pi \times \frac{\bar{N}K_0(\pi D \pi)N}{\bar{N}K_0(N\bar{N})N}]\ (M)$$

$$L_S(E_3) \longrightarrow L_S(E_3)$$

Signalons également que l'existence et "l'unicité" du problème (4) sont des questions encore ouvertes excepté pour la version de Koiter. En effet, pour cette théorie l'opérateur intégro-différentiel est pseudomonotone et de ce fait non sans difficulté du fait en particulier de la courbure de Σm, les méthodes classiques s'appliquent [4].

2-2 <u>Construction d'une solution tridimensionnelle admissible</u>

Le problème de l'estimation des hypothèses $\mathcal{H}^{\Lambda}$ qui permettent de déduire la Λ-version du modèle (1), passe par l'élaboration d'un élément appartenant à l'ensemble $\tilde{V} \times \tilde{W}$ des solutions tridimensionnelles admissibles de la Λ-version qui est défini par :

$$\tilde{V} = \{U \mid U \in U_0 + \mathcal{U}, \qquad U_{|\Sigma m} = \tilde{u}\}$$

$$\tilde{W} = \{C \mid C \in C_0 + J^0\,,\ \int_{-h}^{h} (\pi C \pi, \pi C \pi Z)\, S^{-1} k dZ = (\tilde{\mathbb{N}}, \tilde{\mathbb{M}})\}$$

L'élément retenu dans $\tilde{V} \times \tilde{W}$, que nous noterons $(\hat{U}, \hat{C})$, est choisi de manière à minimiser (au moins de façon approchée) l'erreur quadratique en relation de comportement. Cette construction est menée sous les hypothèses simplificatrices suivantes :

| matériau homogène isotrope, charge volumique nulle, champ de déplacement U_0 de type Kirchoff-Love, et épaisseur $2h(m)$ constante .

Le cas général peut être traité comme nous l'avons montré en théorie linéaire [5], mais au prix de difficultés techniques assez sérieuses.

Examinons d'abord la construction de $\hat{C}$; il doit vérifier en particulier les équations d'équilibre (1-2) que nous avons écrites en utilisant les coordonnées m et Z de M sous la forme :

$$\left|\begin{array}{ll}
(5\text{-}1) & (\bar{N}\bar{C}\,\pi k)_{/Z} + \bar{N}C\pi k\dfrac{\partial N}{\partial m}S^{-1} + \underset{m}{\operatorname{div}}\,(S^{-1}k\,\pi\bar{\bar{C}}\,\pi).\,\pi = 0 \\
(5\text{-}2) & (\bar{N}\,C\,N\,k)_{/Z} + \underset{m}{\operatorname{div}}\,(S^{-1}\pi\,\bar{C}\,N\,k) - \operatorname{Tr}\left[S^{-1}k\,\pi\bar{C}\,\pi\,\dfrac{\partial N}{\partial m}\right] = 0 \\
(5\text{-}3) & [\bar{N}\bar{C}\,k\,\pi - \varepsilon\,\bar{F}\,\pi\,k]_{|\Sigma_+ \sqcup \Sigma_-} = 0 \\
(5\text{-}4) & [\bar{N}\,C\,N\,k - \varepsilon\,\bar{F}\,N\,k|_{|\Sigma_+ \sqcup \Sigma_-} = 0 \\
(5\text{-}5) & [\bar{n}\,\bar{C}\,k - \bar{G}\,k\,]_{|\Sigma_{2L}} = 0
\end{array}\right.$$

avec : n : normale extérieure unitaire à Σ_L

$$\left|\begin{array}{ll}
f : (F,G) & \\
\varepsilon_{|\Sigma_+} = +1 & \varepsilon_{|\Sigma_-} = -1
\end{array}\right.$$

La vérification de la relation en contrainte généralisée associée à $\widetilde{W}$ ainsi que les résultats de comparaison (3) nous ont conduit à imposer à $\hat{C}$ les conditions supplémentaires

$$(6)\quad \left|\begin{array}{l}
\pi\,C\,\pi = \overline{\pi\,C'\pi} \\
C - \tilde{C}' = \overline{C - \tilde{C}'}
\end{array}\right.$$

Ces conditions ainsi que (5.1-2-3-4) constituent un système d'équations différentielles en Z avec conditions pour $Z = \pm h$ possédant une solution unique $\hat{C}$ que nous avons construite. Ce $\hat{C}$ ne vérifie pas en général exactement les conditions au bord (5.5), mais il les vérifie "en moyenne" au sens suivant :

$$(7)\quad \left|\ (\forall\, U \in \mathcal{U}_{KL}) \int_{\Sigma_{2L}} (\bar{n}_2\,\bar{\bar{C}} - \bar{G})\,U\,d\Sigma_L = 0\,.\right.$$

Quant au champ $\hat{U}$ retenu, il est égal à $\tilde{U} + \tilde{\tilde{U}}$ où $\tilde{\tilde{U}}$ est une solution approchée du système non-linéaire (8) :

$$(8)\quad \left|\begin{array}{l}
\tilde{\tilde{U}}_{|\Sigma m} = 0 \\
\bar{N}\,K_0\,D(\tilde{U} + \tilde{\tilde{U}}) = \dfrac{1}{2}\,(\bar{N}\bar{\bar{C}}(I + \dfrac{\partial}{\partial M}(\tilde{U} + \bar{\tilde{\tilde{U}}}))^{-1} + \overline{\bar{N}(I + \dfrac{\partial}{\partial M}(\tilde{U} + \bar{\tilde{\tilde{U}}}))^{-1}\hat{C}})
\end{array}\right.$$

Si on a bien $\hat{U}_{|\Sigma m} = \tilde{u}$, $\hat{U}$ ne vérifie qu'en première approximation les liaisons imposées sur le bord, au sens suivant :

$$\left|\begin{array}{l}
\hat{U}_{|C_{1m}} = U_{o|C_{1m}} \\
\left[\dfrac{\partial\bar{N}\hat{U}}{\partial m}\,n\right]_{|C_{1m}} = \left[\dfrac{\partial\bar{N}U_o}{\partial m}\,n\right]_{|C_{1m}}
\end{array}\right.$$

En définitive, l'élément $(\hat{U},\hat{C})$ que nous venons de préciser vérifie les conditions sous-jacentes à $\tilde{V} \times \tilde{W}$ excepté les conditions au bord qu'il ne vérifie qu'en première approximation. Cependant, l'indétermination due à la minceur de la coque de ces conditions au bord rend loisible l'hypothèse suivante qui est très satisfaisante dans la plupart des problèmes concrets :

| les liaisons cinématiques sont schématisées par $\hat{U}_{|\Sigma_{1L}}$
| la charge surfacique sur Σ_{2L} est schématisée par $[\hat{C}n]_{|\Sigma_{2L}}$

Cette hypothèse entraine que $(\hat{U},\hat{C}) \in \hat{V} \times \hat{W}$. Signalons que dans [5], nous avons envisagé une voie qui permet d'éviter cette hypothèse de schématisation des données au bord.

2-3 Estimation des erreurs.

Pour chaque Λ-version, nous définissons en supposant que la solution $(\tilde{u}, (\tilde{\mathbb{N}}, \tilde{\mathbb{M}}))$ est suffisamment régulière les grandeurs R, η, L, q, $d_1(\Lambda)$, $d_2(\Lambda)$:

. $\tilde{U} = \tilde{u} - Z \overline{\frac{\partial \tilde{u}}{\partial m}} N$ $\qquad \tilde{C}' = \Lambda(\frac{\partial \tilde{U}}{\partial M})$ $\qquad \tilde{A} = \frac{\partial \tilde{U}}{\partial M}$ $\qquad \tilde{D} = D(\tilde{U})$

. $\| \quad \| \overset{\text{déf}}{=} \| \quad \|_{L^2(\Omega)}$ $\qquad \| \quad \|_{\Sigma m} = \| \quad \|_{L^2(\Sigma m)}$

. $\frac{h}{R} = \| h \frac{\partial N}{\partial m} \|_{L^\infty(\Sigma m)}$

. $\tilde{\varepsilon} = \varepsilon(\tilde{U})$ $\qquad \hat{\varepsilon} = \varepsilon(\hat{U})$

. $\tilde{\gamma} = \gamma(\tilde{U})$ $\qquad \hat{\gamma} = \gamma(\hat{U})$

. $\frac{1}{2} \| \bar{N} \,\bar{\tilde{A}}\tilde{A}\, \pi \| = \eta \, \| \pi \, \tilde{D} \, \pi \|$ $\qquad \eta$: coefficient de non-linéarité

. $\frac{h}{L} = \inf \{\alpha \mid \| \frac{\partial}{\partial m} B \|_V \; h \leq \alpha \, \|B\|_V$ pour les B,V ci-dessous}

B		V
$\tilde{C}'$, $h \frac{\partial}{\partial m} \tilde{C}'$	$\bar{\tilde{A}}\tilde{A}$, $h \frac{\partial}{\partial m} \bar{\tilde{A}}\tilde{A}$	$L^2(\Omega)$
$\tilde{A}$, $h \frac{\partial \tilde{A}}{\partial m}$	$\frac{1}{2}(\tilde{A}+\bar{\tilde{A}})$, $h \frac{\partial}{\partial m} . \frac{1}{2}(\tilde{A}+\bar{\tilde{A}})$	$L^\infty(\Omega)$
F		$L^2(\Sigma m)$

L : longueur d'onde caractéristique de la solution de la Λ-version

$$. q = \sup_{m \in \Sigma m} \left\{ \frac{|\pi F_{|\Sigma^+}| + |\pi F_{|\Sigma^-}|}{|[Sk\ \pi\ F\varepsilon]^+_-|} , \frac{|\bar{N}F_{|\Sigma^+}| + |\bar{N}F_{|\Sigma^-}|}{|[K\ \bar{N}\ F\ \varepsilon]^+_-|} \right.$$

$$\text{ou } [\quad]^+_- = [\quad]^+ - [\quad]_-$$

q : coefficient caractéristique de la répartition de la charge entre Σ_+ et Σ_-

$$. d_1(\Lambda) = \frac{1}{\sqrt{2}\ ||\Lambda_0(A)||}\ ||\frac{1}{2}\ \overline{\{(\Lambda - \Lambda_0)(\tilde{A}) + (\Lambda - \Lambda_0)(\tilde{A})\}}||$$

$$d_2(\Lambda) = \frac{1}{\sqrt{2}\ ||\Lambda_0(A)||}\ ||\frac{1}{2}\ \{(\Lambda - \Lambda_0)(\tilde{A}) - \overline{(\Lambda - \Lambda_0)(\tilde{A})\}}||$$

a- Erreur en relation de comportement

Nous obtenons que

$$\left[\frac{\psi(\hat{C} - C(\hat{U}))}{\psi(C(\hat{U}))} \right]^{\frac{1}{2}} \leq c_1$$

et que l'erreur quadratique en relation de comportement de la Λ-version, E_1 vérifie :

$$E_1 \overset{\text{déf}}{=} \left[\inf_{(U,C) \in \tilde{V} \times \tilde{W}} \psi(C - C(U)) \right]^{\frac{1}{2}} \leq \varepsilon_1 \left[\psi(C(\hat{U})) \right]^{\frac{1}{2}}$$

$$\text{avec } \varepsilon_1 = \left[\frac{3\lambda + 2\mu}{2\mu} \right]^{\frac{1}{2}} \left\{ \frac{\lambda + 2\mu}{2\mu} \left[16q^2 \left(\frac{h}{R} + \frac{h^2}{L^2}\right)^2 + 4n^2 \left(\frac{h}{R} + \frac{h}{L}\right)^2 \right]^{\frac{1}{2}} + d_1(\Lambda) + d_2(\Lambda) \right\}$$

λ, μ sont les coefficients de Lamé .

b - Erreur en solution.

Nous supposons que la solution retenue $(\tilde{U}_0 , \tilde{C}_0)$ du problème (1) vérifie la condition de stabilité N_0 . D'autre part, nous nous plaçons dans le cas où $g(\hat{\gamma})$ est positif . On démontre que :

$$\left[\frac{E_{DL}(\hat{S})}{E_{DL}(\hat{A})} \right]^{\frac{1}{2}} \leq \varepsilon_2$$

$$\left[\frac{E_D(\hat{S})}{E_D(\hat{A})} \right]^{\frac{1}{2}} \leq \varepsilon_3$$

$$\frac{1}{8} < \hat{\bar{S}}\hat{S} , \hat{\bar{S}}\hat{S} > \leq \varepsilon_4^2 \quad E_D(\hat{A})$$

* Nous supposons évidemment que $(\frac{h}{R} , \frac{h}{L} , \tilde{\varepsilon} , d_1(\Lambda) , d_2(\Lambda))$ sont inférieurs à 1. D'autre part, pour alléger les calculs nous n'avons pas tenu compte des termes où apparaissent les dérivées des courbures de Σm qui sont en général négligeables .

avec $\varepsilon_2^2 = \frac{\varepsilon_3^2}{4} = 2\,\varepsilon_4^2 = \frac{1}{[g(\hat{\gamma})]^2} \quad \frac{3\lambda + 2\mu}{2\mu} \quad [\, d_1^2(\Lambda) + \frac{d_2^2}{\varepsilon C}(\Lambda) \; +$

$$+ \left(\frac{\lambda + 2\mu}{2\mu} \right)^2 \left[16q^2 \left(\frac{h}{R} + \frac{h^2}{L^2}\right)^2 + 4(\eta^2 + 2\tilde{\gamma}) \left(\frac{h}{R} + \frac{h}{L}\right)^2 \right] \,]$$

c- Commentaires sur les résultats

Dans les expressions de ε_1 ,..., ε_4 , il ne s'introduit que des grandeurs caractéristiques des données et de la solution de la -version. Pour la version de Koiter, $d_1(\Lambda)$ et $\frac{d_2^2(\Lambda)}{\varepsilon_c}$ sont de l'ordre de $(\tilde{\varepsilon} + \frac{h}{R})$. Notons également que pour la version la plus classique de la théorie linéaire, $d_1(\Lambda)$ et $\frac{d_2(\Lambda)}{\varepsilon_c^{1/2}}$ sont de l'ordre de $(\frac{h}{R} + \tilde{\gamma})$.

Pour conclure cette étude, il nous semble intéressant de poser le problème du raffinement des estimations de la 1ere partie pour étudier finement la validité des hypothèses $\mathcal{H}^{\Lambda}$ dans un voisinage de la charge critique.

BIBLIOGRAPHIE :

[1] WT. KOITER : On the foundations of the linear theory of thin shells. Pro. Kon. Ned. Ak. Wet B 73 n° 3 (1970)

[2] DA. DANIELSON : Improved error estimates in the linear theory of thin elastic shells. Pro. Kon. Ned. Ak. Wet B 74 (1970) p. 294-300

[3] WT. KOITER. JG. SIMMONDS : Foundations of shells theory. Report n° 473 (1972) Laboratory of Engineering Mechanics Mekelweg 2. Delft

[4] P. ROUGEE : Equilibre des coques élastiques minces inhomogènes en théorie non-linéaire . Thèse Paris VI (1969)

[5] P. LADEVEZE : Comparaison de modèles de milieux continus Thèse Paris VI (1975)

FUNCTIONAL ANALYSIS APPROACH FOR THE DERIVATION OF HYBRID VARIATIONAL FUNCTIONALS

L.G.Napolitano
Institute of Aerodynamics
University of Naples, It.

0.- INTRODUCTION

For the application of finite element methods, based on variational principles, to the solution of problems from the field of fluid-dynamics (or, more generally, of momentum, mass and energy transfers) the authors has advocated [1] a unified functional approach as opposed to ad hoc extensions, for each new problem being considered, of techniques mainly developed in the field of structural mechanics. A first step in this direction, taken in [1] , has made it possible to show that generalized "virtual work principles" and such well known theorems of elasticity and structural mechanics as the theorems of Betti, Castigliano, Clapeyron, Pasternak hold for all formally self-adjoint elliptic problems irrespective of the particular fields of continuum mechanics they belong to. As a further contribution along this line, the author has presented in [2] a unified derivation of the so called hybrid variational method, originally introduced in structural mechanics problems and here and there subjected to ad-hoc reformulations for other classes of problems [3] . The therminology "hybrid variational functional"is here taken in a broadened sense and it is ment to indicate multi-field variational functionals whose fields have different tensorial orders. In [2] only b.v.problems associated with the Laplace operator were dealt with It was therein shown that the suggested unified Hilbert space approach afford also the derivation of new two and three-fields hybrid variational functionals which may form the basis for the development of new approximating solution techniques. The open litterature presents,in addition, examples of hybrid variational functionals only for fourth order elliptic differential operators.

The more general case of b.v. problems associated with arbitrary formally self-adjoint elliptic linear differential operators is considered in the present paper which offers a unified derivation of hybrid variational functionals for these operators. The equivalence between the solution of certain classes of boundary value problems and the minimization of functionals defined over appropriate Banach spaces is well known [4] , [5] . Equally well established is the theory of duality (in the sense of Fenchel and Rockfallar) in convex analysis and its several applications in variational, optimization, numerical problems [5] , [6] . However, hybrid variational formulations of b.v. problems cannot be derived from such known approaches and new ones must be deviced. The approach proposed here generalizes the one introduced in [2] and can be broadly described as follows: The original boundary value problem in embedded in a larger class of b.v. problems pertaining to physical quantities of tensorial order higher than that of the original problem. The new b.v.

problem is interpreted as defining an isomorphism between two appropriate Hilbert spaces: the constraints space $\mathcal{C}$ and the solution space H, the unique solution of the original b.v. problem being thus characterized as the unique element of H satisfying a prescribed set of constraints. Then variational formulations are deduced by: i) defining subsets of H satisfying an arbitrary number of such constraints (thus generalizing the notions of "equilibrating" and "compatible" solutions); ii) characterizing the solution vector as the unique element common to two such subsets; iii) finding the appropriate functional, defined over these two subsets, which is rendered stationary (in particular minimum) by the exact solution. The subject paper will present detailed developments up to step ii) above. Once this step has been achieved the further formal development is completely independent of the order of the differential operator considered. Thus the pertinent essential results proven in [2] will only be briefly reported here for completeness sake. The explicit formulation of all possible multi-field hybrid functionals depends on the explicit expression of the original differential operator. This further development will not be pursued here and only remarks applying to the general case will be offered.

To render the development formally rigorous a basic assumption, related to the hypothesis of existence and uniqueness of the solution of the original b.v. problem, needs to be made. This assumption is closely related to the notion of V-ellipticity of sesquilinear forms entering the variational theory of b.v. problems (cfr.[4]) as well as to the general problem of the characterization (or "optimum" choice) of the "data" and "solution" spaces, as formulated and analysed in [4] . These aspects of the problem will not be dealt with here.

1.- THE BOUNDARY VALUE PROBLEM

Let Ω be a bounded open subset of $\underline{R}^n$ with boundary $\Gamma = \Gamma_1 \cup \Gamma_2$, Γ_i being open disjoint subsets of Γ and let $V(\Omega) \subset K(\Omega)$ be two appropriate closed subspaces of the Hilbert space $L^2(\Omega)$.

The boundary value problem to be considered here is formally stated as follows:

$$\text{in } \Omega \quad L v = \bar{f} \; ; \quad \text{on } \Gamma_1 \quad P_j v = \bar{p}_{1j} \; ; \quad \text{on } \Gamma_2 \quad N_j v = \bar{n}_{2j} \tag{1.1}$$

Here: $(0 \leq j \leq m-1)$; $v \in V$ is a real scalar or vector valued function, $\bar{f} \in K'(\Omega)$, the dual of K; L is a formally self-adjoint elliptic differential operator of order $2m$; P_j (resp. N_j) is a boundary "principal" (resp. "natural") differential operator of order j (resp. of order 2m-1-j) defined, in a manner to be specified shortly, in terms of the Green's formula corresponding to the scalar product (v', Lv) in L^2.

There is no essential difficulty in dealing with a more general case where the couple of boundary conditions $P_j v, N_j v$ are each prescribed on different complementary oper subsets Γ_j, $\Gamma - \Gamma_j$ of the boundary Γ.

Since we are going to assume a well defined factorization of L the sets $\{P_j\}$ and $\{N_j\}$ are uniquely characterized and, consequently, so is the class of possible sets of boundary conditions [4]. Other types of boundary conditions could be obtained, as known [4], by relating them to Green's formulae corresponding to the L^2 scalar product $(\Lambda v', Lv)$ with Λ a suitable operator. This development will not be pursued here. We assume that L can be factorized as L=DG, with D and G differential operators of order (m). The quantity $\xi = Gv$ is considered as a physical quantity of tensorial order higher than v: the differential equation $Lv = D\xi = \bar{f}$ amounts to prescribing the value of its "source" $D\xi$ in Ω and the boundary conditions, if the problem admits of an unique solution, amount to prescribing the additional data on the boundary needed to characterize ξ uniquely. We then embed the set $\{\xi\}$ into a larger set $\{\zeta\}$ of the same "physical" quantities by adding to ξ another "part" ω and look for a boundary value problem for ω which would uniquely define ω in terms of its "source" and its boundary values.

The boundary value problem appropriate for the present purpose is formally stated as follows. Let: $\zeta = \xi + \omega = Gv + R\eta \;;\; D\omega = 0$ with $\xi, \omega, \zeta \in \mathcal{U}(\Omega)$, an appropriate Hilbert space, and R is a differential operator of order m such that DR is the null operator. Then the b.v. problem reads:

$$\text{in } \Omega \qquad ; \qquad \text{on } \Gamma$$
$$RR\eta = q \qquad ; \qquad C_j R\eta = 0 \qquad B_k\eta = g_k \tag{1.2}$$

where the boundary operators C_j $(0 \le j \le m-1)$ are defined by $N_j = C_jG$ and the number and type of boundary differential operators B_k are such that the problem is well posed.

$RR\eta = R\omega$ can be interpreted as the "source" of ω. In general, however, while $D\zeta = D\xi$, $R\omega \neq R\zeta$ since $RG \neq \phi$. In some cases it may turn to be either convenient or appropriate to impose the further restriction $D\eta = 0$ on η. When such is the case $RR\eta = L\eta$. The restriction $C_j\omega = 0$ is imposed in order to guarantee that the "natural" boundary values of ζ coincide with those of ξ.

A general proof of the existence of the operator R will be reported elsewhere and we shall limitourselves here to furnish some specific examples.

Thus, for m=1, take: $G = -\underline{\nabla}(\;\;)$; $D = \underline{\nabla}\cdot(\;\;)$; $R = \underline{\nabla}\wedge(\;\;)$ so that: $DG = -\nabla^2(\;\;)$; $DR = \phi$; $RR = -GD + DG$ and hence $RR\eta = L\eta$ if $D\eta = 0$. Physically v and η are the scalar and vector potentials. For m=2, if v is a transversal vector $\underline{h}$ $(\underline{\nabla}\cdot\underline{h} = 0)$, take $G = -2[\underline{\nabla}\,\underline{\nabla}\wedge(\,)]^S$ $D = \underline{\nabla}\wedge[\underline{\nabla}\cdot(\,)]$; $R = \underline{\nabla}\wedge[\underline{\nabla}\wedge(\,)]$ so that $DR = \phi$; $DG = RR = -\nabla^2 R$ and, upon the transversality of $\underline{h}$, $DGu = \nabla^4 h$. Hence, once again, $RR\eta = L\eta$ if $D\eta = 0$. Another example for m=2 is given by the case in which v is a scalar. Take $G = \underline{\nabla}\,\underline{\nabla}$; $D = \underline{\nabla}\cdot[\underline{\nabla}\cdot\;]$ $R = \underline{\nabla}\wedge[\underline{\nabla}\wedge(\,)]$ so that now both DR and RG are null operators, $DG = \nabla^4 = RR$ if $D\eta = 0$, and $D\zeta = D\xi$; $R\zeta = R\omega$.

Upon the self-adjointness of L, formal integration by parts yields (Green's formula)

$$a(v,v')=\int_{\Omega} Gv\,Gv'dx = \int_{\Omega} v\,Lv'dx - \sum_{j=0}^{m-1}\int_{\Gamma} P_j v\,N_j v'd\sigma$$

This formula defines the previously introduced boundary operators P_j, N_j and indicates the symmetric bilinear form a(u,v) as a basis for the inner product in $\mathcal{U}(\Omega)$. If $\langle , \rangle$ and $(,)$ denote, respectively, inner product in $\mathcal{U}$ and L^2 one obtains

$$\langle \zeta,\zeta' \rangle = (v,f') + \langle \omega,\omega' \rangle - \sum_{j=0}^{m-1}\int_{\Gamma} P_j v\,N_j v'd\sigma \qquad (1.3)$$

since DR=$\emptyset$ and $C_j\omega=0$. The scalar product $\langle \omega,\omega' \rangle$ could be further expanded and would yield additional boundary integrals. Such development is not needed for the present purposes but may well be worth pursuing.

2.- BASIC ASSUMPTION AND REFORMULATION OF THE B.V. PROBLEM.-

We now view the r.h.s. of eqs.(1.1) and (1.2) as prescribed values of a set of "constraints" for the element $\zeta=\xi+\omega$ and assume that each set of specific values determines ζ uniquely.

More precisely, since in the subject case we shall not need any further decomposition of the subset $\{\omega\}$, we shall consider only 2(m+1) types of constraints: the "source" constraint f, the 2m boundary constraints on Γ_1 and Γ_2 and the "transersality" constraint (t) which combines (in a cartesian product) the "source" constraint q and the boundary constraints g_k. Since ξ can be characterized on Γ_i either through its "principal" values p_j or its "natural" values n_j we introduce, in the pertinent cartesian product space, the "constraints equivalence classes" and denote as c = $[C_i]$ = = $[f, b_{1j}, b_{2j}, t]$ $(0 \le j \le m-1)$ the element of an equivalence class. Elements belong to the same equivalence class when the values of the constraints lead to the same element [2].

We now make the basic assumption that there exist proper subsets $\mathcal{M}_{C_i}$ of the pertinent L^2 spaces, closed with respect to appropriate norms and such that the linear transformation $F: \mathcal{C}=\mathcal{M}_{C_0}\otimes\mathcal{M}_{C_1}\otimes\cdots\otimes\mathcal{M}_{C_{2m+1}} \to \mathcal{U}(\Omega)$ defined by problems (1.1) and (1.2) is closed and bounded, closure in the image space being with respect to the previously defined scalar product in $\mathcal{U}$. It follows that: $\mathcal{D}(F)=\mathcal{C}$; $\mathcal{R}(F)$=H, a closed subspace of U; F^{-1}=A exists, is linear and bounded, with $\mathcal{D}(A)$=H; $\mathcal{R}(A)=\mathcal{C}$. The unique element $\bar{\xi}$ of H such that: $A\bar{\xi}=\bar{c}=[f, \bar{p}_{1j}, \bar{n}_{2j}, 0]$ $(0\le j\le m-1)$ will be called the generalized solution of problem (1.1). The original problem can thus be formally stated as follows: find the unique element of the Hilbert space H=F($\mathcal{C}$) such that $\bar{\xi}=F(\bar{c})$

The basic assumption entails, in most cases, that the derivatives be taken in the sense of distributions, over Ω or Γ , and that the boundary conditions may have to be taken in a "variational" sense unless Ω and Γ are sufficiently regular and the data spaces appropriate for the "trace theorems" [4] to be applicable.

3.- DIRECT SUM DECOMPOSITION OF H.-

The Hilbert space H is decomposed into a direct sum of 2(m+1) mutually orthogonal closed subspaces (basic subspaces) $\Sigma_i = F(\mathcal{C}_i)$ where the subsets $\mathcal{C}_i$ of $\mathcal{C}$ are defined by:

$$\mathcal{C}_o = \{c \in \mathcal{C} : p_{1j} = p_{2j} = 0 ; (0 \le j \le m-1) ; t = 0\}$$

$$\mathcal{C}_i = \{c \in \mathcal{C} : f = 0 ; p_{1j} = p_{2j} = 0 \; (j \ne i-1) ; n_{2,i-1} = 0\}$$

$$\mathcal{C}_{i+m} = \{c \in \mathcal{C} : f = 0 ; p_{1j} = p_{2j} = 0 \; (j \ne i-1) ; p_{1,i-1} = 0\}$$

$$\mathcal{C}_{2m+1} = \{c \in \mathcal{C} : f = 0 ; n_{1j} = n_{2j} = 0 \; (0 \le j \le m-1)\}$$

Given the assumed properties of the transformation F the only thing which needs to be proved the orthogonality of the subspaces and this follows from Green's formula (1.3).

From the definition it also follows that $(\sigma_i \in \Sigma_i)$: $\sigma_i = G v_i \; (0 \le i \le 2m)$; $D\sigma_i = 0 \; (1 \le i \le 2m+1)$ and that each restriction of the inner product $<,>$ to one of the Σ_i corresponds to one of the terms of the r.h.s. of equation (1.3).

Notice, finally, that with an obvious notation for the restriction of $\mathcal{C}$ and $(1 \le j \le m)$:

$$\sigma_o = G v_o = F(f) \quad ; \quad \sigma_j = G v_j = F(b_{1,j})$$
$$\sigma_{2m+1} = F(t) \quad ; \quad \sigma_{j+m+1} = G v_{j+m+1} = F(b_{2,j}) \tag{3.1}$$

where $b_{ij} (i=1,2)$ can be the value of either a principal or a natural boundary condition. Each of the equations (3.1) represents the solution of a corresponding b.v. problem. The basic assumption can thus be recast into uniqueness statements for each of these b.v.problems.

The decomposition of the solution element $\bar{s}$ leads to a corresponding decomposition of the original b.v.problem according to the expressions given by eqs.(3.1). The last m of the (2m+1) b.v.problems thus obtained are "coupled" insofar as the $b_{2,j}$ contain the data $\bar{n}_{2,j}$ prescribed on Γ_2 as well as those of all the other v_i (with the only eception of v_{j+1}). This type of coupling depends on the choice made for the decomposition of H. Other choices are possible but no-one will lead to complete uncoupling. This type of coupling however does not constitute any drawback expecially when approximate solutions are sought (see [2] for a discussion of these points in the case m=1).

The decomposition of the space H makes it possible to carry out a systematic and exhaustive analysis of the adjoint operators that can be associated with the original b.v. problem. This will not be pursued here. It will only be mentioned that, as a plausible extrapolation of the results of [2], one may expects as many as (2m+1) couples of adjoint operators and that, as in [2], each relevant couple may be associated to a particular classical or hybrid functional.

4.- LINEAR VARIETIES.-

Identify the element (k) of the index set $J = \{0,1,2,\ldots,2m+1\}$ with the k-th constraint satisfied by the element ζ and let the two subsets J_α, J_β, of cardinality α and β respectively, characterize two arbitrary subsets of constraints. Associate to the basic subspaces $\Sigma_i (0 \le k \le 2m+1)$ projection operators $\mathcal{P}_k$ $[\Sigma_i = \mathcal{P}_i H]$ let $\mathcal{P}_\alpha = \Sigma \mathcal{P}_i \ (i \in J_\alpha)$; $\mathcal{Y}_\alpha = \mathcal{P}_\alpha H$; $\mathcal{P}_{\bar\alpha} = I - \mathcal{P}_\alpha$; $\bar{\mathcal{Y}}_\alpha = \mathcal{P}_{\bar\alpha} H$, with I the indetity operator and define the linear variety $\mathcal{V}_\alpha$ "parallel" to $\mathcal{Y}_\alpha$ as:

$$\{\zeta_\alpha\} = \mathcal{V}_\alpha = \mathcal{P}_{\bar\alpha}\bar\xi + \mathcal{P}_\alpha H = \mathcal{P}_{\bar\alpha}\bar\xi + \mathcal{Y}_\alpha \tag{4.1}$$

where $\bar\xi$ is the exact solution of the origianl b.v.problem. The variety $\mathcal{V}_{\bar\alpha}$ associated with the index set $\bar J_\alpha = (I - J_\alpha)$ will be called dual of $\mathcal{V}_\alpha$.

Upon the properties of the projection operators one readily proves that:

i) $\mathcal{V}_\alpha$ is the subset of vectors $\zeta \in H$ whose projection on the subspace $\bar{\mathcal{Y}}_\alpha$ equals that of the exact solution. When $\mathcal{P}_{\bar\alpha}\bar\xi$ satisfies the same boundary constraints as $\bar\xi$ (this implies some restrictions on the subset J_α, see [2]) then $\mathcal{V}_\alpha$ is the subset of vectors $\zeta \in H$ satisfying the same set $\bar J_\alpha = (I - J_\alpha)$ of boundary conditions as the exact solution

ii) The exact solution $\bar\xi$ is the only element common to two dual linear varieties

iii) More generally, given two subsets $J_\alpha, J_\beta \subset J$ with $J_\alpha \cup J_\beta \subseteq J$ the exact solution $\bar\xi$ is the unique element common to the two linear varieties $\mathcal{V}_\alpha$ and $\mathcal{V}_\beta$ iff: $J_\alpha \cap J_\beta = \phi$. The two varieties V_α and V_β constitute the generalization of the notion of equilibrated and compatible solutions which correspond to a single specific choice of dual varieties $(J_\alpha \cup J_\beta = J)$.

5.- THE VARIATIONAL FUNCTIONALS CONCLUDING REMARKS.-

Once the possible linear varieties have been characterized, the analysis proceeds in a manner completely analogous to that presented in [2] We report here only the relevant results for completeness sake.

Consider the variational functionals:

$$2K(\zeta_\alpha - \zeta_\beta) = \langle \zeta_\alpha - \zeta_\beta \rangle^2 - \langle \zeta_\beta - \bar{\zeta} \rangle^2$$
$$\zeta_\alpha \in \mathcal{V}_\alpha \; ; \; \zeta_\beta \in \mathcal{V}_\beta$$

As shown in [2] they attain the stationary null value for value ζ^*_α and ζ^*_β such that:

$$\zeta^*_\alpha = \bar{\zeta} \qquad \mathcal{P}_\alpha \, \zeta^*_\beta = \mathcal{P}_\varepsilon \bar{\zeta}$$

The projection of ζ^*_β over $\bar{\mathcal{Y}}_\alpha$ remains undetermined, i.e. ζ^*_β is an arbitrary element of the variety $\mathcal{V}_\gamma$, associazed with the index set $J_\gamma = J_\beta \cap (J - J_\alpha)$, "parallel" to the subspace $\mathcal{Y}_\beta \cap \mathcal{Y}_\alpha$. The following max-mini property holds:

$$0 = \min_{\substack{\zeta_\alpha \in \mathcal{V}_\alpha \\ \zeta_\beta \in \mathcal{V}_\beta}} K = \max_{\substack{\zeta_\beta \\ \in \mathcal{P}_\alpha \zeta_\beta}} \left[\min_{\zeta_\alpha \in \mathcal{V}_\alpha} K \right]$$

with the minimum over V_α being attained for ζ_α an arbitrary element of the linear variety parallel to the subspace $\mathcal{Y}_\alpha \cap \mathcal{Y}_\beta$.

Multifield variational functionals are obtained by suitable choices of the index subsets J_α, J_β. For m=1 one, two and three field functionals can be ontained [2], the fields being scalar, vectorial or both (hybrid functional). It is apparent that in the present general case functionals involving up to (2m+1) fields could be obtained and that, for m=2, the fields could be scalar, vectorial or tensorial. The actual formulation of these variational functionals requires that the operazor L be given explicitely. The discussion of even the simplest m=2 case would take us beyond the limit of the present analysis and will be presented in future works.

We conclude by pointing out that the classical one field variational formulation is recoveres by taking ζ_β as a fixed element of the variety $\mathcal{V}_\beta$ associated with an index subset J_β such that $J_\beta \cap J_\alpha = \emptyset$. The class of such functionals contains however more than one functional since it is not necessary to take,as in the classical case, $J_\beta \cup J_\alpha = J$. [This correspond to the already mentioned fact that the linear varieties generalize on more than one account the notion of equilibrated and compatible solutions]. Primal and dual one field functionals are obtained by interchanging the roles of the varieties $\mathcal{V}_\alpha$ and $\mathcal{V}_\beta$. The notion of duality for multi-field functionals is more complex and will not be considered here.

REFERENCES

1.- L.G.NAPOLITANO: "Unified functional approach to the solution of formally adjoint problems of continuum mechanics" - Letters in Applied and Engineering Sciences - Vol.I - pp/465-479, Pergamon Press Inc.

2.- L.G.NAPOLITANO: "On the functional analysis derivation and generalization of hybrid methods" - II Meeting AIMETA (Naples October 16-19, 1974)
L.G.NAPOLITANO: "Functional Analysis derivation and generalization of hybrid variational methods", IAR No.218; Dec.1973 - Final Scientific Report, Grant AFOSR-73-2508

3.- ZIENKIEVICZ, O.C.: "The finite element methods in continuum mechanics" McGraw Hill, 1971

4.- J.L.LIONS, E.MAGENES: "Problêmes aux limites non homogênes et applications" - Vol.I, Dunod, Paris 1968

5.- EKELAND, I., TEMAN, R.: "Analyse convexe et problêmes variationnels", Dunod, Gauthier-Villars, 1974

6.- LUENBERGER, D.G.: "Optimization by vector space methods" - J.Wiley & Sons, New York, 1969

ACKNOWLEDGEMENT

This work has been sponsored in part by the Air Force Office of Scientific Research through the European Office of Aerospace Research, OAR, United States Air Force, Under Grant No.AFOSR-74-2704.-

STABILITY OF EQUILIBRIUM IN ELASTIC-PLASTIC SOLIDS

Nguyen Quoc Son and Dragos Radenkovic
Laboratoire de Mécanique des Solides
Ecole Polytechnique, Paris

INTRODUCTION

The study of the stability of an equilibrium state under dead loading finds its place in the more general context of problems concerning the evolution of a material system. Namely, in order to appreciate the stability, we are bound to follow the motion, or at least to estimate the work done, due to an arbitrary perturbation of the initial equilibrium state. From this point of view, Liapunov's direct method proves to be a precious tool for dealing with our problem. It has been applied with success in the theory of elastic stability [1] , [8] , [7] in order to establish the criteria, i.e. sufficient conditions which ensure the stability of an equilibrium. Our task here is to extend the domain of its application to a fairly large class of non-elastic solids.

In this paper, first, a general formulation of the problem of evolution is given, in order to fix the terminology and the notations. After that, generalized standard materials, plastic or visco-plastic, are introduced and an a priori estimation of the internal work is proposed, by examining the optimal strain paths for any given strain increment in a material element. Finally on the basis of a definition of stability with corresponding measures of perturbation and displacement, which in general are not small, but finite, a Liapunov's functional is constructed. A condition of positivity of this functional expresses the required criterion of stability. A well known result of Hill [5] , [15] , [6] is thus precisely formulated and generalized to finite deformations.

The discussion is presented in a purely mechanical frame, a more general formulation which should take into account thermal effects can be developed on the same basis.

1.- INITIAL BOUNDARY VALUES PROBLEM

1.1.- Evolution problem.

Consider a body occupying in the space referred to Cartesian coordinates a bounded volume V° in the reference state V°, the same shorthend symbol being used both for volume and for the corresponding configuration. Let V be the volume of the body in the current state ; equilibrium states V^{eq} will have to be defined in the following.

Let Π be the symmetric Kirchhoff's stress tensor, ε Green's strain tensor and $u(a,t)$ the displacement vector of a material point whose initial position is a ; u is assumed to be regular . The equations of motion with respect to the reference state can be expressed in the form of the virtual work principle :

$$\int_{V^\circ} \Pi_{ij}\,\delta\varepsilon_{ij}\,dV + \int_{V^\circ} \rho\,\ddot{u}_i\,\delta u_i\,dV = \int_{V^\circ} \rho F_i\,\delta u_i\,dV + \int_{S^\circ_T} T_i\,\delta u_i\,dS + \int_{S^\circ_u} (\delta_{ik} + u_{i,k})\,\Pi_{jk} n_j\,\delta u_i\,dS \tag{1}$$

δu being an arbitrary field of virtual displacements ; F is here the specific body force, T the surface traction with respect to the reference state given on the part S°_T of the surface S° of the body, S°_u being the complementary part of S° on which the displacements u are given ; virtual strain field $\delta\varepsilon$, associated with δu is defined by :

$$2\,\delta\varepsilon_{ij} = \delta u_{i,j} + \delta u_{j,i} + \delta u_{k,i}\,u_{k,j} + \delta u_{k,j}\,u_{k,i} \tag{2}$$

The energy balance at any instat t :

$$\mathcal{P}^i_t + \frac{d}{dt} K_t = \mathcal{P}^e_t \tag{3}$$

is obtained from (1) by taking for δu the real velocities $\dot{u}$ and putting :

$\mathcal{P}^i_t = \int_{V^\circ} \Pi_{ij}\,\dot{\varepsilon}_{ij}\,dV$ - work rate of internal forces

$K_t = \int_{V^\circ} \rho\,\dot{u}_i\,\dot{u}_i\,dV$ - kinetic energy

$\mathcal{P}^e_t = \int_{V^\circ} \rho\,F_i\,\dot{u}_i\,dV + \int_{S^\circ_T} T_i\dot{u}_i\,dS + \int_{S^\circ_u} (\delta_{ik} + u_{i,k})\,\Pi_{jk}\,n_j\dot{u}_i\,dS$ - work rate of external forces.

By introducing constitutive law $\Pi = \mathcal{H}\{\underset{-\infty}{\overset{t}{\varepsilon}}\}$, $\mathcal{H}$ being in the general case a functional, the evolution of the body is determined when initial values and boundary values at any instant t are given.

If we restrict ourselves to materials such that $\Pi(t)$ depends only on the material internal state $E(t)$ which is a functional of $E(0)$ and the history of the deformation during the time interval $(0, t)$:

$$E(t) = G\left(E(0) , \{ \overset{t}{\underset{0}{\varepsilon}} \} \right)$$

then clearly , the initial values of the evolution problem are defined by $E(0)$, $u(0)$, $\dot{u}(0)$.

1.2.- Equilibrium under dead loading

The body is said to be in an equilibrium state V^{eq} under the dead loading $F(a)$, $T(a)$, $u(a)$ if there exist time independent stress and displacement fields Π^{eq} and u^{eq} respectively , related by the constitutive law, which verify equilibrium equations derived from (1) , the inertia term being suppressed . To a given dead loading may correspond either one or several equilibrium states or none at all .

Note that for a rate independent material the relation between Π^{eq} and u^{eq} depends on the previous history ; for a rate dependent material a quasi-static evolution occurs in general under dead loading as a succession of states in external equilibrium . However , following previous definition we consider as equilibrium state V^{eq} only the stationary equilibrium states .

Suppose an equilibrium state , corresponding to a given dead loading to have been determinated . In order to examine its stability in the sense of Liapounov we shall have to consider the motions starting from arbitrary initial values $E(0)$, $u(0)$, $\dot{u}(0)$ obtained from E^{eq} , u^{eq} , $\dot{u}^{eq} = 0$ by not necessary small perturbations which however must be compatible . This means that $E(0)$, $u(0)$, $\dot{u}(0)$ must be defined by physically admissible transformations from the equilibrium state V^{eq} , respecting imposed displacements on S_u° .

It is convenient to consider that a perturbation is defined by a compatible strain path or strain history $\{ \overset{0}{\underset{-T}{\varepsilon}} \}$ starting at arbitrary time $-T$ from the equilibrium value at each material point of V° and resulting from perturbing forces acting on S_T° and V° during the interval $(-T , 0)$.

Liapounov functional will be constructed by estimating the work supplied by such a perturbation .

Splitting the work rate P_t^e of the external forces into two parts : P_t^ℓ due to the given dead loading and P_t^p due to the perturbing forces :

$$(4) \qquad P_t^e = P_t^\ell + P_t^p$$

we obtain from the energy balance at any $t \in (-T, +\infty)$ the work supplied :

$$\mathcal{T}^p_t = \int_{-T}^{t} P^i_\tau \, d\tau + K_t - \int_{-T}^{t} P^\ell_\tau \, d\tau$$

Explicitly written , this gives :

$$\mathcal{T}^p_t = \int_{-T}^{t} \int_{V^\circ} \Pi_{ij}(a,\tau)\, \dot{\varepsilon}_{ij}(a,\tau)\, dV\, d\tau + \int_{V^\circ} \rho\, \dot{u}_i(a,t)\, \dot{u}_i(a,t)\, dV$$

(5)

$$- \int_{-T}^{t} \left\{ \int_{V^\circ} \rho\, F_i(a)\dot{u}_i(a,\tau)\, dV + \int_{S^\circ_T} T_i(a)\dot{u}_i(a,\tau)\, dS \right\} d\tau$$

The equation (5) can be simplified taking into account that for an equilibrium state , the equation (1) is satisfied . We obtain then :

$$\mathcal{T}^p_t = \int_{-T}^{t} \int_{V^\circ} (\Pi_{ij}(a,\tau) - \Pi^{eq}_{ij}(a))\, \dot{\varepsilon}_{ij}(a,\tau)\, dV\, d\tau + K_t$$

(6)

$$+ \frac{1}{2} \int_{V^\circ} \Pi^{eq}_{ij}(a)\, (u_{k,i}(a,t) - u^{eq}_{k,i}(a))(u_{k,j}(a,t) - u^{eq}_{k,j}(a))\, dV$$

For an elastic material with strain energy potential $\Phi(\varepsilon)$ such that $\Pi = \partial\Phi / \partial\varepsilon$, the first term in (6) can be integrated with respect to time thus becoming :

$$\int_{V^\circ} \{ \Phi(\varepsilon(a,t)) - \Phi(\varepsilon^{eq}(a)) - \frac{\partial\Phi}{\partial\varepsilon}(\varepsilon^{eq}(a))(\varepsilon_{ij}(a,t) - \varepsilon^{eq}_{ij}(a))\} dV$$

So , in this particular case , the expression (6) for $\mathcal{T}^p_t$ depends only on the actual strain state . In the general case , i.e. for all non-elastic materials the work $\mathcal{T}^p_t$ depends essentially on the strain path $\{ \overset{t}{\underset{-T}{\varepsilon}} \}$ through the first term of the expression (6).

However , a lower bound for $\mathcal{T}^p_t$ depending only on the actual strain state can be given for certain specific non-elastic materials . This is done by examining the value of the work supplied to an element of the body , which is given by the functional :

$$V_t = \int_{-T}^{t} (\Pi_{ij}(\tau) - \Pi^{eq}_{ij})\, \dot{\varepsilon}_{ij}(\tau)\, d\tau$$

for the class of generalized standard materials .

2.- OPTIMAL STRAIN PATHS FOR GENERALIZED STANDARD MATERIALS

2.1.- Generalized standard materials.

The definition of a general standard material, either elastieplastic or elastic-visco-plastic, given by one of the authors Nguyen Q.S. [13] , based on the hypothesis of normal dissipativity, makes it possible to bring the problems for a general hardening material into the well known mathematical frame of perfect plasticity. In this way some results concerning the existence and the unicity of the solutions (stress and strain-fields, parameters of hardening) of the evolution problem for such materials have been obtained in the case of small deformation [13] , [10]. We shall use it here in order to estimate the functional V_t .

A simplified presentation of standard materials, which is still valid with in the frame of a Lagrangian description for small strains and finite rotations, will be given in what follows.

Total strain ε is split into elastic and plastic parts :

$$\varepsilon_{ij} = \varepsilon^e_{ij} + \varepsilon^p_{ij} \tag{7}$$

the approximation (7) being acceptable in most problems of stability. The internal state of the material is now defined by ε^e and a finite number of parameters α^m , (m = 1 , n) , which describe the hardening in particular. Let the isothermal free energy be denoted by $\Phi(\varepsilon^e, \alpha)$ and the intensive or force parameter associated with α by A ; the force parameter associated with ε^e is obviously Π . We then have :

$$\tag{8} \begin{cases} \Pi = \Pi(\varepsilon^e, \alpha) = \dfrac{\partial \Phi}{\partial \varepsilon^e} \\ A = A(\varepsilon^e, \alpha) = \dfrac{\partial \Phi}{\partial \alpha} \end{cases}$$

The evolution of the plastic strain ε^p and of the internal parameters α is given by :

$$\tag{9} \begin{cases} \dot{\varepsilon}^p = \lambda \dfrac{\partial F}{\partial \Pi} + \dfrac{\partial \Omega}{\partial \Pi} \\ -\dot{\alpha} = \lambda \dfrac{\partial F}{\partial A} + \dfrac{\partial \Omega}{\partial A} \end{cases} \qquad \begin{matrix} \lambda = o \text{ if } F(\Pi, A) < 0 \\ \lambda \geqslant o \text{ if } F(\Pi, A) = 0 \end{matrix}$$

The inequality $F(\Pi, A) \leqslant 0$ defines, in the $\Pi x A$ space a convex domain C ; Ω is a convex function in the same space, constant on a convex subdomain C^V containing the origin O . When the elastic domain C^V is defined by the inequality $F^V(\Pi, A) \leq 0$, Ω is usually taken in the form $\Omega \equiv \frac{1}{2\eta}\langle F^V(\Pi, A)\rangle^2$ i.e. Ω is the square of the distance of (Π, A) from C^V, η being the coefficient of viscosity.

The relations (8) introduce the hypothesis of a generalized elastic potential $\Phi(\varepsilon^e, \alpha)$, which will be supposed convex in what follows.

The relations (9) introduce a potential of dissipation $\Psi_C(\Pi, A) + \Omega(\Pi, A)$ which is a convex function ; Ψ_C is here the indicator function of the convex C. These relations express the hypothesis of normal dissipativity [2][4] in the frame of plasticity and visco-plasticity.

2.2.- Optimal strain paths - Let us now consider the functional :

$$V\{\,{}_{t_1}^{t_2}\varepsilon\} = \int_{t_1}^{t_2} (\Pi_{ij}(t) - \Pi_{ij}(t_1))\,\dot{\varepsilon}_{ij}(t)\,dt$$

The initial state $\Pi(t_1)$ and $A(t_1)$ being given, the problem relevant to our study is the following:

Find the strain paths minimizing the functional V among all paths $\{\,{}_{t_1}^{t_2}\varepsilon\}$ which satisfy the conditions

(i) $\varepsilon(t_2) = \varepsilon^2$, given in advance

(ii) $\varepsilon(t)$ and $\Pi(t)$ associated by constitutive law.

Obviously the analogous problem concerning optimal stress paths can be formulated.

2.2.1.- Plastic materials ($\Omega \equiv 0$). Let us introduce the generalized variables $\Sigma = (\Pi, A)$, $\mathbb{E} = (\varepsilon, 0)$, $\mathbb{E}^e = (\varepsilon^e, \alpha)$, $\mathbb{E}^p = (\varepsilon^p, -\alpha)$; in this way the analysis for a hardening material is simplified, as the corresponding constitutive equations reduce to those known from the theory of perfect plasticity.

Let $\Sigma^* \in C$ and put $\Delta\mathbb{E} = \mathbb{E}(t_2) - \mathbb{E}(t_1)$, $\Delta\Sigma = \Sigma(t_2) - \Sigma(t_1)$, $\Delta\Sigma^* = \Sigma^* - \Sigma(t_1)$

V can be written in the form :

$$V\{\,{}_{t_1}^{t_2}\varepsilon\} = \int_{t_1}^{t_2} (\Sigma(t) - \Sigma^*)\dot{\mathbb{E}}\,dt + \int_{t_1}^{t_2} (\Sigma^* - \Sigma(t_1))\,\dot{\mathbb{E}}\,dt$$

or, rearanging the terms :

$$V\{\,{}_{t_1}^{t_2}\varepsilon\} = \int_{t1}^{t_2} (\Sigma(t) - \Sigma^*)\dot{\mathbb{E}}^p dt + \Delta\Sigma^*\Delta\mathbb{E} + \Phi(\mathbb{E}_1^e + \Delta\mathbb{E}^e) - \Phi(\mathbb{E}_1^e) - (\Sigma_1 + \Delta\Sigma^*)\Delta\mathbb{E}^e$$

Then consider the convex function :

$$\Theta(\Delta E^e) = \Phi(E_1^e + \Delta E^e) - \Phi(E_1^e) - \frac{\partial \Phi}{\partial E_1^e} \Delta E^e$$

and its Legendre-Fenchel transform :

$$\Theta^*(\Delta S) = \operatorname*{Max}_{\Delta E^e} \quad \Delta S \ \Delta E^e - \Theta(\Delta E^e)$$

We have :

$$\Theta(\Delta E^e) + \Theta^*(\Delta \Sigma^*) \geq \Delta \Sigma^* . \Delta E^e$$

Taking into account this inequality and the fact that the first term in the above expression for V is positive due to the normality, we obtain a bound, valid $\forall$ $\Sigma^* \in C$, $\forall \ \{ \overset{t_2}{\underset{t_1}{\varepsilon}} \}$:

$$V \{ \overset{t_2}{\underset{t_1}{\varepsilon}} \} \geq \Delta \Sigma^* \Delta E - \Theta^*(\Delta \Sigma^*)$$

Now a particular value Σ^* can be chosen, among all $\Sigma^* \in C$, maximizing the right hand side. Thus a function $B(\Delta E)$ is introduced, defined by :

$$(10) \qquad B(\Delta E) = \operatorname*{Max}_{\Sigma^* \in C} \quad \Delta E . \Delta \Sigma^* - \Theta^*(\Delta \Sigma^*) = \Delta E . \Delta \tilde{\Sigma} - \Theta^*(\Delta \tilde{\Sigma})$$

In fact, $B(\Delta E)$ is the Legendre-Fenchel transform of the function :

$$\Theta^*(\Delta S) + \Psi_C(\Sigma_1 + \Delta S)$$

Let us note that $B(\Delta E) > 0$ if $\Delta \varepsilon \not\equiv 0$.

We have thus obtained a lemma which plays a fundamental role in the study of the stability of non-elastic materials.

<u>LEMMA</u> : For any strain path $\{ \overset{t_2}{\underset{t_1}{\varepsilon}} \}$ corresponding to a given increment $\Delta \varepsilon$

$$(11) \qquad V \{ \overset{t_2}{\underset{t_1}{\varepsilon}} \} \geq B(\Delta E)$$

the function $B(\Delta E)$ being defined by (10).

In the case of perfect plasticity, optimal strain paths can be effectively determined. Namely, the preceding decomposition shows that an optimal path consists in this case of two parts : first the stress $\tilde{\Pi}$ is reached in a <u>purely elastic</u> deformation ; then ΔE is completed by a <u>purely plastic</u> strain, obviously under the condition that the later is plastically admissible i.e. that $\Delta E - \Delta \tilde{E}^e$ is normal to C at $\tilde{\Pi}$. Now from (10) it is seen that $\tilde{\Pi}$ is defined by the variational inequality :

$$\left(\Delta E - \frac{\partial \Theta^*}{\partial \Delta \tilde{\Pi}} \right) \left(\Delta \tilde{\Pi} - \Delta \Pi^* \right) \leq 0 \qquad \forall \ \Pi_1 + \Delta \Pi^* \in C$$

As $\Delta\tilde{E}^e = \frac{\partial \Theta^*}{\partial \Delta\tilde{\Pi}}$ and as $\Delta\tilde{\Pi} - \Delta\Pi^* = \tilde{\Pi} - \Pi^*$ we have $\forall\ \Pi^* \in C$ the inequality :

$$(\Delta E - \Delta\tilde{E}^e)(\tilde{\Pi} - \Pi^*) \geqslant 0$$

which completes the proof.

In the case of hardening materials, optimal paths depend essentially on the form of the elastic domain C in $\Pi \times A$ so that a general solution can not be given.

2.2.2.- Visco-plastic materials. - In the absence of instantaneous plasticity the equations (9) reduce to :

$$\begin{cases} \dot{\varepsilon}^p = \frac{\partial \Omega}{\partial \Pi} \\ -\dot{\alpha} = \frac{\partial \Omega}{\partial A} \end{cases}$$

Following the definition of Ω given in §2.1 it is convenient to associate to each visco-plastic material a corresponding plastic material with the same elastic domain C^V . Consider again the inequality :

$$V \{ \varepsilon_{t_1}^{t_2} \} \geq \int_{t_1}^{t_2} (\Sigma(t) - \Sigma)\, \dot{E}^p\, dt + B(\Delta E)$$

The first term of the right-hand member is still non-negative ; indeed $\frac{\partial \Omega}{\partial \Sigma^*} = 0$ if $\Sigma^* \in C^V$, so that Ω being convex :

$$\int_{t_1}^{t_2} (\Sigma - \Sigma^*)\, \dot{E}^p\, dt = \int_{t_1}^{t_2} (\Sigma - \Sigma^*)(\dot{E}^p - \dot{E}^{p*})\, dt$$

Thus, optimal p aths are the same as for the associated plastic material, the representative point moving infinitely slowly along the path if the interval $t_2 - t_1$ is not fixed in advance.

3.- STABILITY OF EQUILIBRIUM IN THE PLASTIC DOMAIN

3.1.- A sufficient condition of stability.

An equilibrium state V^{eq} of an elasti-plastic body at $t = -T$ is described by the set of data $u^{eq} = u(-T)$, $\dot{u}^{eq} = 0$, $E^e(-T)$, $\alpha\ (-T)$, the last two terms being needed here in order to characterize the internal state of the body. We shall say that such a state is stable if a perturbation of the initial data during $(-T, 0)$, which may be finite, produces only limited additional displacements for every $t > 0$.

In order to give a precise definition, let us consider the set H of compatible additional displacement fields Δv, $\Delta v = 0$ on S_u°. Let us consider a measure $|\Delta v|$ of the distance of displacement to the equilibrium configuration. If we take as the measure of the initial perturbation $d^\circ = \mathrm{Max}\ (T_o^p , |\Delta u_o|)$ and the measure of the additional displacement $d_t = |\Delta u_t|$, a definition of stability can be formulated as follows :

(12) V^{eq} is stable if and only if for every $\varepsilon > 0$, there exists $\delta > 0$ such that all motions which satisfy $d^\circ \leqslant \delta$ also satisfy $d_t \leqslant \varepsilon$ for all positive values of time t.

A sufficient condition of stability can now be formulated by using Liapunov's direct method :

(13) V^{eq} is stable if the functional :

$$B(\Delta v) = \int_{V^\circ} \{ B(\Delta E) + \frac{1}{2} \Pi_{ij}^{eq} \Delta v_{k,i} \Delta v_{k,j} \} dV$$

defined on H is positive definite in the sense that there exist a number $\Delta > 0$ and a function $g(r) : R \rightarrow R$ strictly increasing, $g(0) = 0$, such that :

$$B(\Delta v) \geqslant g(|\Delta v|) \ ; \forall \Delta v \in H, \ |\Delta v| \leqslant \Delta$$

The proof is readily obtained by taking T_t^p as the Liapunov functional. Namely we have by definition $\dot{T}_t^p = 0$ for all positive values of t

and $T_t^p \geqslant B(\Delta u_t)$

the first term in (6) being estimated following (11).

Whatever $\varepsilon \in]0, \Delta[$, let δ be determined as $\delta = \mathrm{Min}(\varepsilon, g(\varepsilon))$ the condition $d^\circ \leqslant \delta$ implies for every $t > 0$ $T_t^p = T_o^p \leqslant \delta$. Suppose now that $d_t > \varepsilon$ at certain t ; then as $\delta \leqslant \varepsilon$ the distance to the equilibrium configuration d_t attains at t_1 an intermediate value $\varepsilon_1 \in]\varepsilon, \Delta[$ because of the continuity. The proposed criterion, which can be applied at t_1 thus gives :

$$d_{t_1} \leqslant g^{-1}(B(\Delta u_{t_1})) \leqslant g^{-1}(T_{t_1}^p) \leqslant g^{-1}(\delta) \leqslant \varepsilon$$

which is impossible. It follows that $d_t \leqslant \varepsilon$ for all positive values of t.

Note that the proposed criterion (13), valid for plastic and viscoplastic standard materials, is obtained under the sole hypothesis of normal dissipativity, whatever be the form of the potentials.

3.2.- Infinitesimal perturbations

The set H may be difficult to explore for three-dimensional structures. But if we introduce the assumption of small perturbed motions i.e. if only small transformations near the equilibrium state 2 are investigated, the criterion (13) is very simple. Indeed, all additional quantities Δu , $\Delta u_{k,i}$,ΔE , $\Delta\Sigma$ are infinitesimal we have , taking V^{eq} as the reference state :

$$\Delta\varepsilon_{ij} = \Delta e_{ij} = \frac{1}{2}(\Delta u_{i,j} + \Delta u_{j,i})$$

$$\Theta(\Delta E^e) = \Theta_{lim}(\Delta E^e) = \frac{1}{2}\Delta E^e \, L \, \Delta E^e , \quad L = \left.\frac{\partial^2 \Phi}{\partial E^e \, \partial E^e}\right|_{eq}$$

$$B(\Delta E) = B_{lim}(\Delta E) = \underset{\Delta S \in C}{\text{Max}} \; \Delta E \, \Delta S - \frac{1}{2}\Delta S \, L^{-1} \Delta S$$

Where C is the tangent cone to C at Σ^{eq} :

$$C = \begin{cases} \Pi \times A \quad \text{space if } F(\Sigma^{eq}) < 0 \\ \{ \Delta S \mid \frac{\partial F}{\partial \Sigma^{eq}} \Delta S \leq 0 \} \quad \text{if } F(\Sigma^{eq}) = 0 \end{cases}$$

If the fields $L_{...}$ and Π^{eq}_{ij} are elements of $L^{\infty}(V°)$, the functional

$$B_{lim}(\Delta v) = \int_{V°} \{ B_{lim}(\Delta E) + \frac{1}{2}\Pi^{eq}_{ij} \, \Delta v_{k,i} \, \Delta v_{k,j} \} \, dV$$ is defined on the space K :

$$K = \{ \Delta v \mid \Delta v \in H^1(V°) \cap C°(\bar{V}°) \; ; \; \Delta v = 0 \text{ on } S^{°}_{u} \}$$

Observing that $B_{lim}(\lambda\Delta v) = \lambda^2 B_{lim}(\Delta v)$, the criterion (13) is now written in the form (14) with $| \Delta v | = ||\Delta v||_{H^1(V°)}$:

(14) $\begin{cases} V^{eq} \text{ is stable if the functional } B_{lim}(\Delta v) \text{ is coercive on K i.e.} \\ \text{there exists a constant } c > 0 \text{ such that :} \\ \qquad B_{lim}(\Delta v) \geq c \, . || \Delta v ||^2_{H^1(V°)} \end{cases}$

$B_{lim}(\Delta v)$ is the functional introduced earlier by Hill [5] who proposed as a condition of stability $B_{lim}(\Delta v) > 0$ for every $\Delta v \not\equiv 0$, kinematically admissible .

Thus , for elastic plastic standard materials we have interpreted Hill' s criterion as a sufficient condition of stability in the Liapunov' s sense , corresponding to a partially " linearized " form of the evolution problem considered. In the " linearized " problem , namely , the elastic domain C is represented in the vicinity of the equilibrium state V^{eq} by the tangent cone C and in the energy term $B(\Delta E)$ only the linear part of the strains appear .

CONCLUSIONS

A suitable definition of a general standard material has allowed us to consider the evolution problem in its general non-linearized form, in order to formulate a criterion of stability valid for finite perturbations of the equilibrium under dead loading. It should be relatively easy to complete this study by taking into account thermal effects in the now well defined thermodynamical frame of standard materials which exhibit normal dissipativity.

For practical use it is worthwhile to give a mechanical interpretation of the estimation obtained. The function B is in fact defined in such a way that plastically admissible stress $\Sigma = \Sigma^1 + \frac{\partial B}{\partial E}$ corresponds to a given final state E , which is a well known property of a Hencky material. It follows that the lower bound of the work supplied is given by a kind of deformation theory, i.e. essentially as if the behaviour were "elastic". Therefore, in the applications it is sufficient to consider, as a first approximation, the stability of corresponding elastic structures with a convex strain potential appropriately defined on the basis of experimental loading curves.

REFERENCES

1 ERICKSON J.L., A thermokinetic view of elastic stability theory. Int. J. Solids & Structures, 1966, pp. 573-580.

2 GERMAIN P., Cours de Mécanique des Milieux Continus. Masson & Cie, Paris, 1973.

3 GREEN A.E. & NAGHDI P.M., A general theory of an elastic plastic continuum. Arch. Rat. Mech. An., 1965, pp. 251-281.

4 GYARMATI I., Non equilibrium thermodynamics. Springer-Verlag, 1970.

5 HILL R., A general theory of uniqueness and stability in elastic plastic solids. J. Mech. Phys. Solids, 1958, pp. 236-249.

6 HUTCHINSON J.W., Plastic buckling. Advances in Applied Mechanics, vol.14, 1974.

7 KNOPS R.J. & WILKES E.W., Theory of elastic stability. Handbuch der Physik III, 1973, pp. 125-302.

8 KOITER W.T., On the thermodynamic background of elastic stability theory. Report n°360, Dept. Mech. Eng., Tech. Univ. Delft, 1967.

9 MANDEL J., Plasticité Classique et Viscoplasticité. Lecture Note, CISM, Udine, 1971.

10 MOREAU J.J., On unilateral constraints, friction and Plasticity. Lecture note, CIME, Bressanone, 1973.

11 MURPHY L.M. & LEE L.H.N., Inelastic buckling process of axially compressed cylindrical shells subject to edge constraints. Int. J. Solids & Structures, 1971, pp. 1153-1170.

12 NAGHDI P.M. & TRAPP J.A., On the general theory of stability for elastic bodies. Arch. Rat. Mech. Analys., 1973, pp. 165-191.

13 NGUYEN Q.S., Contribution à la theorie macroscopique de l'élastoplasticité avec écrouissage. Thèse, Paris, 1973.

14 NGUYEN Q.S. & HALPHEN B., Sur les lois de comportement élasto-visco-plastiques à potentiel généralisé. C. R. Ac. Sc., 277, Paris, pp. 319-322.

15 SEWELL M.J., A general theory of elastic and inelastic plate failure (I,II). J. Mech. Phys. Solids, 1963, pp. 377-393. J. Mech. Phys. Solids, 1964, pp. 279-297.

16 PONTER A.R.S. & MARTIN J.B., Some extremal properties and energy theorems for inelastic materials and their relationship to the deformation theory of plasticity. J. Mech. Phys. Solids, 1972, pp. 281-300.

SOLUTIONS IN THE LARGE FOR CERTAIN NONLINEAR HYPERBOLIC SYSTEMS ARISING IN SHOCK-WAVE THEORY

Takaaki Nishida and Joel A. Smoller[1]

1. Introduction

We are concerned with systems of partial differential equations which serve as models for the full set of gasdynamic equations. Specifically, we consider systems of the form

(1) $$v_t - u_x = 0, \quad u_t + p(v)_x = 0,$$

where $p(v) = k^2 v^{-\gamma}$, $k = \text{const.} > 0$, and γ, the adiabatic gas constant, is of the form $\gamma = 1 + 2\varepsilon$, $\varepsilon \geq 0$. Here v is the specific volume (reciprocal of the density ρ), u is the velocity and p denotes the pressure. It is generally believed that the system (1) is a fairly good approximation to the equations of one-dimensional isothermal (or isentropic) gas flow.

For the system (1) we shall discuss the following three problems:

(a) The pure Cauchy problem. Here we specify initial data

(2) $$(v(x,0), u(x,0)) = (v_0(x), u_0(x)), \quad -\infty < x < \infty.$$

The physical model we have in mind is that of a "shock tube". Thus, we imagine a very long thin tube containing a gas. We know the density and velocity at time $t = 0$, and we are required to find these quantities at any time $t > 0$.

(b) The piston problem. For this problem, we consider the system (1) in a quarter space $x \geq 0$, $t \geq 0$. In addition to specifying the initial data (2) on $x \geq 0$, we also specify the piston velocity:

(3) $$u(0,t) = u_1(t), \quad t \geq 0.$$

(c) The double-piston problem. Here we take as our domain the region $0 \leq x \leq 1$, $t \geq 0$. We specify the initial data (2) on $0 \leq x \leq 1$, and we also specify the two piston velocities: (3) and

(4) $$u(1,t) = u_2(t), \quad t \geq 0.$$

For all three of these problems, the data functions are assumed to be bounded functions having finite total variation. We set

$$V_1 = \text{T.V.}\{v_0\} + \text{T.V.}\{u_0\}, \quad V_2 = V_1 + \text{T.V.}\{u_1\}, \quad V_3 = V_2 + \text{T.V.}\{u_2\}.$$

Our goal is to prove existence theorems for the problems (a), (b) and (c) where the V_i's are arbitrarily large.[2]

---

1. Research supported in part by AFOSR Contract Number AROSR-71-2122 and LASL contract number W-7405-Eng-36; this paper was written while J.A.S. was visiting the Los Alamos Scientific Laboratory. J.A.S. wishes to thank the many people at this institution (and in particular, Burton Wendroff) for their kind hospitality.

2. For the problem (a) where the oscillation multiplied by $1+V_1$ is small, see [2]; in [3] these results were extended to problem (b), but for small V_2.

The results which we discuss here are of the following form. First, there is a constant c such that if $\varepsilon V_1 < c$, then problem (a) has a solution defined for all $t > 0$, while if $\varepsilon V_2 < c$, problem (b) has a global solution. For the double piston problem, we need $\varepsilon V_3 < c$, and in addition, we must place a condition on the piston velocities which preclude the possibility of the pistons coming together or going infinitely far apart.

2. Background

In this section we shall review some of the standard notions from shock-wave theory, together with the celebrated Glimm difference scheme, [2].

It is well-known that solutions of (1) generally are not smooth (or even continuous!) for all $t > 0$. This is due to the nonlinear terms which force the characteristic speeds to depend on the solution. Thus one considers "weak" (or distribution) solutions of (1). For example, for the Cauchy problem (a), we say that a pair of bounded, measurable functions $v(x,t)$, $u(x,t)$ is a (weak) solution of (a) if the following two identities hold:

$$\iint_{t\geq 0} v\phi_t - u\phi_x + \int_{t=0} v_0\phi = 0\,, \qquad \iint_{t\geq 0} u\psi_t + p(v)\psi_x + \int_{t=0} u_0\psi = 0,$$

for all smooth functions ϕ,ψ which have compact support in $t \geq 0$.

Our definition of a solution implies that the following "jump conditions" must hold across any smooth curve of discontinuity $x = x(t)$; namely

$$\sigma[v] = -[u], \quad \sigma[u] = [p(v)]. \tag{5}$$

Here $\sigma = dx/dt$ is the "shock speed" at the point in question, and $[f]$ denotes the difference in the quantity f across the discontinuity curve. Such a curve of discontinuity is called a back-shock wave (S_1) if

$$-\sqrt{-p'(v_\ell)} > \sigma > -\sqrt{-p'(v_r)}, \tag{6_-}$$

while if

$$\sqrt{-p'(v_\ell)} > \sigma > \sqrt{-p'(v_r)}, \tag{6_+}$$

it is called a front-shock wave (S_2). The quantities $v_\ell = v(x-0,t)$, and $v_r = v(x+0,t)$ denote, respectively, the values of v on the left and right sides of the shock, while $\pm\sqrt{-p'(v)}$ denote the characteristic (or sound) speeds. Thus conditions (6_+) and (6_-) imply that the shock speed is intermediate to the characteristic speeds on both sides of the shock (see figure 1).

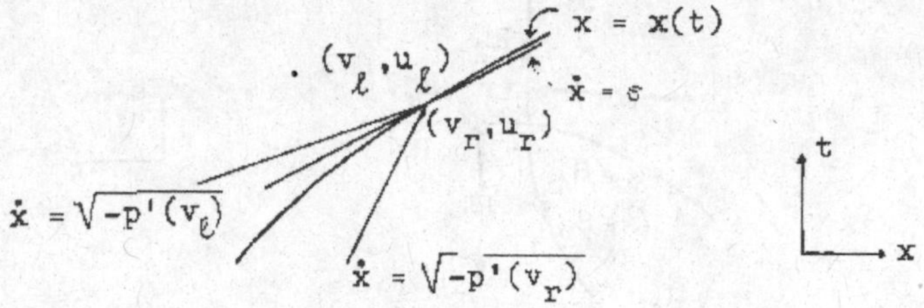

Figure 1

It can be shown that conditions (6), for the full gas dynamics equations, are equivalent to the fact that the entropy increases across shocks. For this reason conditions (6) are often called the "entropy inequalities".

Given a state (v_ℓ, u_ℓ), the equations (5) define two curves in v-u space which represent those states which can be connected to (v_ℓ, u_ℓ) by a shock wave. We call these curves "shock curves". Our results rely on a careful analysis of the global geometry of these curves.

In order to define another important class of solutions, we turn now to the notion of Riemann invariants. First, we write the system (1) in vector form:

$$(1) \qquad U_t + F(U)_x = 0,$$

where $U = (v,u)$, $F(U) = (-u, p(v))$. An easy calculation shows that the functions

$$r(v,u) = u - \int^v \sqrt{-p'(\theta)}\, d\theta, \qquad s(v,u) = u + \int^v \sqrt{-p'(\theta)}\, d\theta,$$

have the property that their gradients are left eigenvectors of $F'(U)$; these functions are called "Riemann invariants". In terms of the Riemann invariants we can discuss the special class of solutions called centered rarefaction waves. These are continuous solutions which are functions of x/t such that one of the Riemann invariants is constant in the region, while the other increases monotonically as a function of x/t. Thus, a backward (resp. forward) rarefaction wave R_1 (resp. R_2) is a continuous solution of x/t such that s (resp. r) is constant in the region and r (resp. s) increases.

Now in complete analogy to the shock-wave case, one can show that for a given state (v_ℓ, u_ℓ), there are two 1-parameter family of states which can be connected to (v_ℓ, u_ℓ) by a rarefaction wave on the right; they are called the forward and backward rarefaction-wave curves. We depict them, together with the shock curves, in figure 2, where $(r_\ell, s_\ell) = (r(v_\ell, u_\ell), s(v_\ell, u_\ell))$.

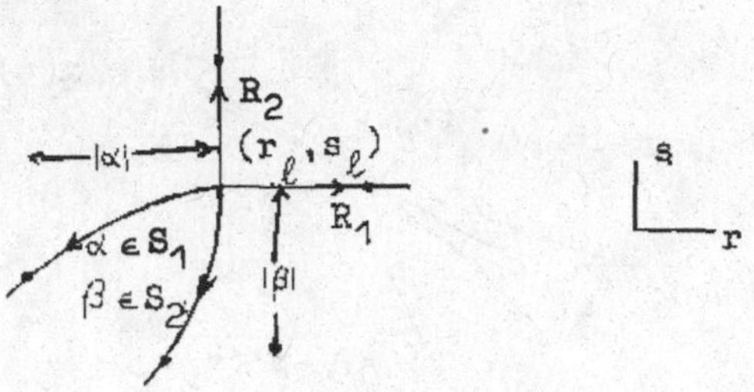

Figure 2

In terms of the r-s coordinates, we can define the "strength" of a back (resp. front) shock $\alpha \in S_1$ (resp. $\beta \in S_2$) by $|\alpha|$ = change in r (resp. $|\beta|$ = change in s) across the shock (cf. figure 2).

These particular solutions, shocks and rarefaction waves, are the building blocks out of which more general solutions are constructed. To see why this is true, we must mention the so-called Riemann problem for (1). This problem consists of solving (1) with initial data of the form

$$(7) \qquad (v(x,0),u(x,0)) = \begin{cases} (v_-,u_-), & x < 0 \\ (v_+,u_+), & x > 0; \end{cases}$$

that is, the data consists of two constant states. A complete discussion of this problem can be found in [7]; briefly, the solution of (1), (7) consists of at most three constant states separated by shock and rarefaction waves.[3]

We turn now to Glimm's difference scheme. The idea here is to assume that on the line $t = n\Delta t$ (n a non-negative integer) we have a piecewise constant approximate solution. To see how to achieve this, we divide the upper-half-plane $t \geq 0$ into a grid $x = m\Delta x$, $t = n\Delta t$, $m \in Z$, $n \in Z^+$, m+n even, and let $\{\alpha_n\}$ be a random sequence of equi-distributed numbers in (-1,1). Let $a_m^n = (m\Delta x + \alpha_n\Delta x, n\Delta t)$. We assume that on each interval $k\Delta x \leq x < (k+2)\Delta x$, the solution is constant. In order to obtain a piecewise constant solution on the line $t = (n+1)\Delta t$, we solve Riemann problems; i.e., we solve (1) with initial data

$$(v(x,n\Delta t),u(x,n\Delta t)) = \begin{cases} (v_m^n, u_m^n) & m-2 \leq x \leq m \\ (v_{m+2}^n, u_{m+2}^n) & m \leq x \leq m+2, \end{cases}$$

3. One has to avoid the vacuum; i.e., one must avoid $\rho = 0$. This will be the case if $s(v_-,u_-) - r(v_+,u_+) > 2(u_+-u_-) - \frac{2k\sqrt{\gamma}}{\varepsilon}$. We shall assume throughout this paper that this condition holds.

where (v_m^n, u_m^n) is the value of the approximate solution on $t = n\Delta t$, $m-2 \le x \le m$. The mesh ratios are chosen in such a way that the resulting waves do not interact with each other (cf. figure 3). In order to obtain a piecewise constant solution on

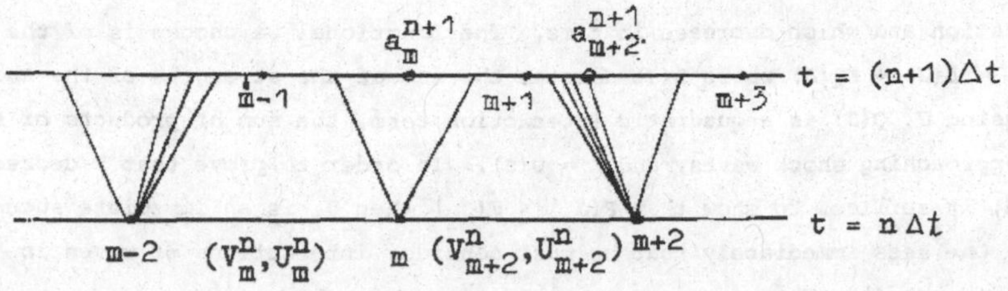

Figure 3

the line $t = (n+1)\Delta t$, we define the new approximate solution on $t = (n+1)\Delta t$, $m-1 \le x \le m+1$ to be the value of the Riemann problem solution at the point $(a_m^{n+1}, (n+1)\Delta t)$. This defines the difference scheme.

In order to prove that the approximate solutions converge to a solution, it is necessary to obtain estimates on the total variation of the approximations. This is done by considering the solution on piecewise linear (or smooth) curves, called J-curves. A J-curve is a piecewise linear curve composed of straight line segments connecting the mesh point a_m^n to $a_{m\pm1}^{n+1}$. The J-curve J_2 is called an immediate successor to the J-curve J_1 if they agree on all but two mesh points, and J_2 doesn't lie below J_1. In order to bound the total variation, it suffices to construct functionals which dominate the total variation and which decrease on J-curves (in the sense of the above defined partial order). These functionals are defined on J-curves and they measure the strength of shock waves crossing J.

3. The Cauchy Problem

In [4], Nishida proved that the Cauchy problem for (1) with the special choice of pressure $p(v) = v^{-1}$, has a solution. The key observation made by Nishida was that for this system, the shock curves satisfy a remarkable geometric property. Namely, if (r_1,s_1) and (r_2,s_2) are any two points in the r-s plane and S_1 and S_2 are backward (resp. forward) shock curves starting at (r_1,s_1) and (r_2,s_2) respectively, then S_1 and S_2 are congruent; i.e., one is simply a translate of the other! This property enabled Nishida to study interactions of waves, and to estimate strengths of outgoing waves in terms of strengths of the incoming interacting waves. These estimates allowed him to apply Glimm's method to prove global esistence of a solution.

For the case where $p(v) = v^{-(1+2\varepsilon)}$, $\varepsilon > 0$, the above geometric property is no longer valid; i.e., S_1 and S_2 are no longer congruent. However, if the data is bounded, then for small $\varepsilon > 0$, S_1 and S_2 are "almost" congruent, and things still work. This was the observation made in [5]; we proceed to describe this now.

As we stated above, one needs to construct functionals which dominate the total variation and which decrease in time. The functional we choose is of the form $F(J) = L(J) + KQ(J)$ where $L(J)$ denotes the sum of the strengths of the shock waves crossing J, $Q(J)$ is a quadratic interaction term, the sum of products of strengths of approaching shock waves, and $K = O(\varepsilon)$. In order to prove that F decreases in time, it suffices to show that $F(J_2) \leq F(J_1)$ when J_2 is an immediate successor to J_1. One sees immediately that we must consider interactions of waves in "diamonds" (cf. figure 4). That is

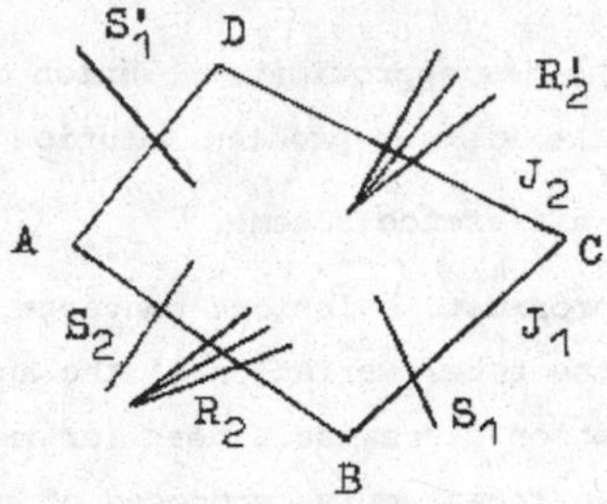

Figure 4

J_1 and J_2 agree outside of the "diamond" ABCD, AD and DC are part of J_2, while AB and BC are part of J_1. Waves (S_2, R_2, S_1) enter the diamond, interact in the diamond, and then the new waves (R_2', S_1') leave the diamond. The goal is to estimate the strengths of the outgoing waves in terms of the incoming waves.

<u>Theorem</u>. Let $v_0(x)$, $u_0(x)$ be bounded functions, $0 < \underline{v} \leq v_0(x) \leq \bar{v}$, each having finite total variation. There is a constant c, depending only on $\bar{v}$ and $\underline{v}$ such that if $\varepsilon TV(v_0,u_0) < c$, then the problem (1), (2) has a globally defined solution. This solution has bounded total variation on each line t = const. > 0.

We remark that an alternate proof of this theorem was given by DiPerna in [1]. We shall discuss DiPerna's result in §5.

4. <u>The Piston Problem</u>

In order to solve the piston problem, we need a lemma, the simplest mixed problem. This lemma plays the role of the solution of the Riemann problem for the Cauchy problem.

<u>Lemma</u>. Consider the system (1) with constant data $(v(x,0),u(x,0),u(0,t)) = (v_+,u_+,u_-)$, $v_+ > 0$. This problem has a piecewise continuous solution satisfying the inequalities $r(x,t) \geq r(v_+,u_+) = r_+$, $s(x,t) \leq \max\,[s_+ = s(v_+,u_+),\ 2u_- - r_+]$, $\Delta s \leq 2\max\,[0,u_- - u_+]$, where Δs is the variation of s across a forward shock.

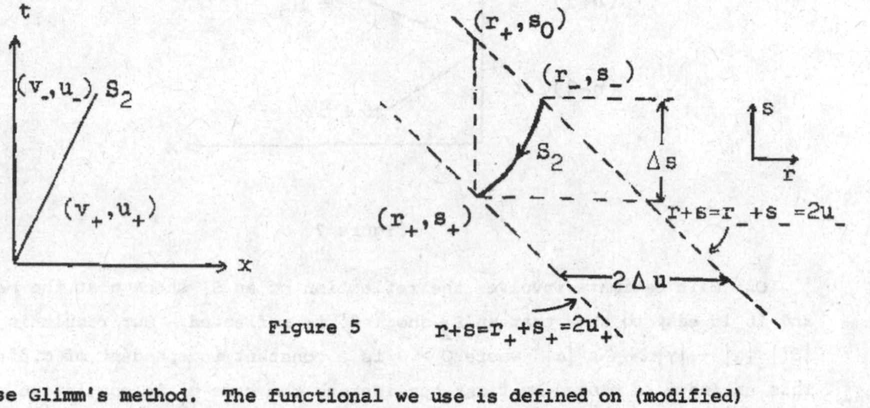

Figure 5

We now use Glimm's method. The functional we use is defined on (modified) J-curves. These curves consist of any space-like curve connecting the mesh points a_k^n in $x \geq a_1^n > 0$ where a_1^n is the first mesh point on $t = n\Delta t$. Then a_1^n is connected to either a_0^{n+1} or a_0^{n-1} by a straight line, and then $t \geq a_0^{n+1}$ or $t \geq a_0^{n-1}$, respectively (see figure 6)

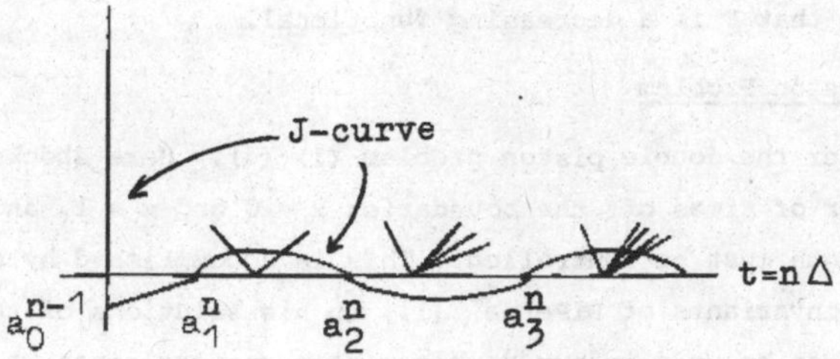

Figure 6

The functional we use is given by $F(J) = L(J) + KQ(J)$ where $K = O(\varepsilon)$. Here

$$L(J) = \sum_J \{|\alpha_k| + |\beta_\ell| + |\gamma_j|\}$$

where α_k is an S_1 crossing J, β_ℓ is an S_2 crossing J and $|\gamma_j| = 2\max\{0,\ u_1(a_0^{j+1}) - u_1(a_0^{j})\}$ for all integers j such that $(0,\ j\Delta t \pm 1/2\,\Delta t)$ lies on J. Finally

$$Q(J) = \sum_J \{|\alpha_k||\beta_\ell| + |\alpha_k||\alpha_\ell| + 1/2|\alpha_k|^2 + |\alpha_k||\gamma_j|\}$$

where in the above sum we only include the term $|\alpha_k||\beta_\ell|$ if α_k and β_ℓ approach each other, and we include the term $|\alpha_k||\alpha_\ell|$ if $k < \ell$.

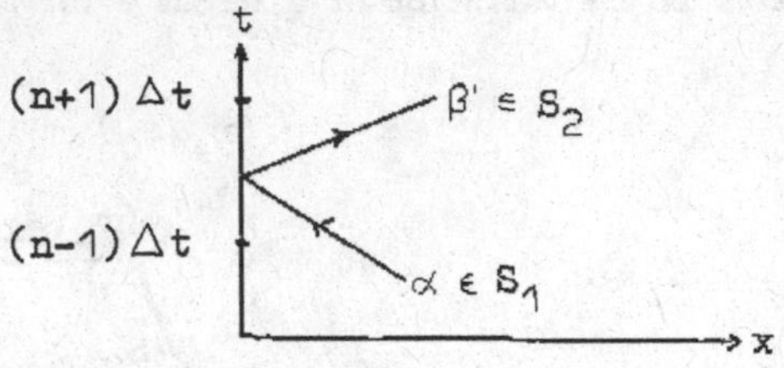

Figure 7

Our main estimate involves the reflection of an S_1 shock α at the boundary $x = 0$, and it is easy to show that an S_2 shock β' is reflected. Our result is that $|\beta'| \leq |\alpha| + |\gamma| + C\varepsilon\, |\alpha|^2$ where $C > 0$ is a constant independent of α, β', and ε. This estimate is proved by first considering the case u_1 is continuous at the point $t = n\Delta t$, $x = 0$. For this problem we show that there exist $\alpha' \in S_1$, $\beta \in S_2$ such that $|\alpha| = |\beta|$, $|\alpha'| = |\beta'|$ and $\alpha + \beta \to \alpha' + \beta'$; i.e. the interaction of α with β yields α' and β'. Thus, we have reduced the reflection problem to a pure Cauchy problem.

It is worthwhile to observe that the reflected shock β' increases in strength, but never again interacts with the boundary $x = 0$; thus the increase in strength is controlled. All other interactions are the same as in the pure Cauchy problem, and allow us to prove that F is a decreasing functional.

5. The Double Piston Problem

We now consider the double piston problem (1)-(4). Here shocks get reflected an infinite number of times off the boundaries $x = 0$ and $x = 1$, and the resulting increase in strength must be controlled. This is accomplished by employing the generalized Riemann invariants of DiPerna, [1], in his solutions of the Cauchy problem (1), (2). These can be used in our double-piston context since we have already shown that shock reflection can be reduced to shock interaction for the Cauchy problem. In terms of DiPerna's coordinates, $|\beta'| \leq |\alpha| + |\gamma|$; i.e., the strength of the reflected shock does not increase, modulo boundary terms. Thus, one would think that everything would go through. However, this is not enough as we show by the following example.

Let $p(v) = v^{-1}$ and consider the problem (1)-(4) with constant data

$$(v_0(x), u_0(x), u_1(t), u_2(t)) = (v_0, 0, 1, 0),\ v_o > 0.$$

For this problem, an S_2 shock shoots out of the origin and impinges on the boundary $x = 1$; it then gets reflected as an S_1 shock which impinges on $x = 0$, and gets reflected, etc. Using the jump conditions (5) we can show that $v \to 0$ and $p(v) \to +\infty$ in a finite time. Hence one cannot obtain a solution to this problem which is defined for all positive time.

To analyze this example, we change to Eulerian coordinates; the equations then become

$$\rho_t + (\rho u)_q = 0, \quad (\rho u)_t + (\rho u^2 + p(\rho))_q = 0.$$

Here

$$x = \int_0^q \rho(s,t)ds > 0, \; q_t = u, \; q_x = \rho^{-1}, \; u_x = v_t,$$

and

$$q(x,t) = q(x,0) + \int_0^t u(x,t)dt.$$

The piston path is given by $x = 0$, so that in q coordinates it is $q(0,t) = t$. On the other hand, the wall is given by $q(1,t) = v_0$; thus the piston collides with the wall at $t = v_0$ (see figure 8)

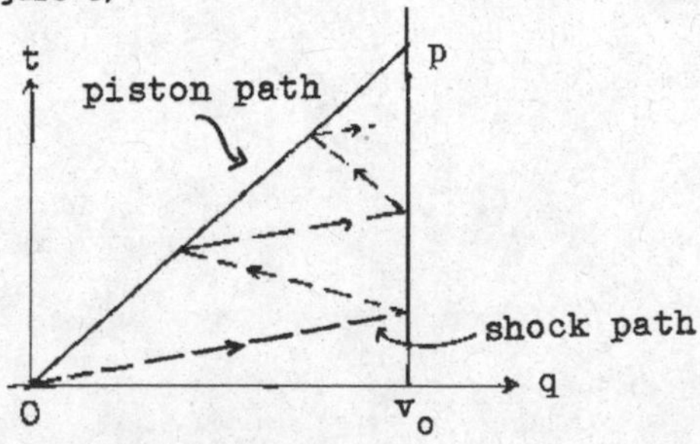

Figure 8

In order to rule out this umphysical situation, we require that $q(0,t) < q(1,t)$ for all $t \geq 0$. Moreover, we also do not want the pistons to move infinitely far apart, for otherwise the density ρ tends to zero as $t\to\infty$. Thus is it necessary to bound $q(1,t) - q(0,t)$ from above. Hence if we set

$$Q(t) = \int_0^1 v_0(x)dx + \int_0^t [u_2(s)-u_1(s)]ds,$$

then in order to get a global existence theorem it is necessary to assume that there exist constants Q_1, Q_2 such that

(8) $$0 < Q_1 \leq Q(t) \leq Q_2 < \infty, \quad t \geq 0.$$

<u>Theorem</u>. Let the data $v_0(x)$, $u_0(x)$, $u_1(t)$, $u_2(t)$ be bounded functions having bounded total variation, with $0 < \underline{v} \leq v(x)$. Suppose that (8) holds. Then there is a constant $c > 0$ such that if $\varepsilon TV\{v_0,u_0,u_1,u_2\} < c$, the double-piston problem (1)-(4) has a global solution.

We shall briefly sketch the proof. The idea is to use Glimm's method, together with DiPerna's functional (see [1]), now modified, however, to include boundary interactions. But the generalized Riemann invariants of DiPerna are not globally defined. Thus, in order to get our approximate solutions to lie in the region where these generalized Riemann invariants are defined, we must first solve our problem locally in time; i.e., we first fix εV_3 to be sufficiently small. We thus get a solution defined in a region $0 \leq t \leq t_0$, $0 \leq x \leq 1$. This solution is L^1-continuous in t and thus satisfies (8) in this region. Then using this "a-priori bound", we take $(v(x,t_0), u(x,t_0))$ as new "initial" data. We then solve the problem locally and repeatedly in regions $nt_0 \leq t \leq (n+1)t_0$, $0 \leq x \leq 1$, $n = 1,2,\ldots$, to get a global solution.

References

1. DiPerna, R., Existence in the large for quasilinear hyperbolic conservation laws, Arch. Rat. Mech. Anal., Vol. 52, (1973), pp. 244-257.
2. Glimm, J., Solutions in the large for nonlinear hyperbolic systems, Comm. Pure Appl. Math., Vol. 18, (1965), pp. 697-715.
3. Kassin, J., Ph.D. thesis, New York University, 1965.
4. Nishida, T., Global solution for an initial boundary value problem of a quasi-linear hyperbolic system, Proc. Japan, Acad. Vol. 44, (1968), pp. 642-648.
5. Nishida, T., and Smoller, J.A., Solutions in the large for some nonlinear hyperbolic conservation laws, Comm. Pure Appl. Math., Vol. 26, (1973), pp. 183-200.
6. Nishida, T., and Smoller, J.A., Mixed problems for nonlinear conservation laws, to appear.
7. Smoller, J.A., On the solution of the Riemann problem with general data for an extended class of hyperbolic systems, Mich. Math. J., Vol. 16, (1969), 201-210.

Kyoto University
Kyoto, Japan, and
The University of Michigan
Ann Arbor, Michigan 48104

CAUCHY PROBLEM IN A SCALE OF BANACH SPACES AND ITS APPLICATION TO THE SHALLOW WATER THEORY JUSTIFICATION

L.V.Ovsjannikov
Siberian Branch of the USSR Academy of Sciences
Institute of Hydrodynamics
Novosibirsk 630090 USSR

SUMMARY. An exposition is given of the abstract theorem concerning Cauchy problem solution in a Scale of Banach Spaces. Its basis is the concept of quasidifferential operator. The second part is devoted to a strict justification of the shallow water theory in a class of analytical functions. This is approached by means of the abstract theorem on the example of the plane problem for unsteady periodical waves.

INTRODUCTION. The idea of Cauchy problem consideration in a Scale of Banach Spaces occured in connection with one old hydrodynamical problem. The question is about non-stationary liquid motion with a free boundary. Particularly, Cauchy-Poisson's problem of surface waves is the well known one in this field. Despite very many approximate theories, the list of which may be found in the Stocker's book [1], no precise results were obtained in this problem up to the recent time. Therefore an attempt was natural to establish for this problem a theorem of Cauchy-Kovalevskaja type about the analytical solution existence and uniqueness for the small time interval provided that analytical initial data were given. The development of an appropriate analytical technique was obstructed by the nonlocal character of this problem. To this end the method of estimates was elaborated by means of Scales of Banach Spaces of analytical functions. This method is a natural logical development of the Cauchy's majorant method. It is analogous to the method being applied by Leray and Ohya in the theory of nonlinear hyperbolic problems [2]. At the first time the author formulated an appropriate basic theorem by means of "singular" operator concept [3]. Nevertheless, further it was cleared up that the "singular" operator concept, sufficiently well taking into account nonlocality, is badly working in nonlinear case. That is why in the subsequent author's work [4] it was replaced by the "quasidifferential" operator concept. Just on this base it succeeded to obtain an abstract equivalent of Cauchy-Kovalevskaja theorem which is good for applications to hydrodynamical free boundary problems.

The first precise results as to the non-stationary free boundary problem were obtained by Nalimov [5] and the author [6]. Recently Nalimov made a following significant step in this hydrodynamical problem-proved the existence and uniqueness theorem of the plane Cauchy-Poisson's problem in the function classes of finite smoothness [7]. By the way, he obtained a strict justification of the linear wave theory. This claimed to develop rather new technique founded on the theory of pseudodifferential operators.

The getting of precise results in the wave theory enables to consider a question of justification of different approximate theories. It was already mentioned about the linear theory. The "shallow water" approximation is also widely used in hydrodynamics leading to nonlinear equations of the gasdynamics equation type. The systematic deduction of these equations firstly was done by Friedrichs [8]. But the strict justification of the "shallow water" theory remained still desirable because the Friedrichs'construction was the formal one. Such a justification was elaborated by the author [9] (firstly called in 1973 on the XI Polish Symposium on Advanced Problems and Methods in Fluid Mechanics). This result related to the plane problem with periodic waves.

General concepts connected with Scales of Banach Spaces are communicated in the first part of the paper. Some examples of Scales of Banach Spaces are considered. The definition of quasidifferential operator and the formulation of the basic theorem related to a Cauchy problem solution are given. This theorem is followed by the solution estimate important for applications. The second part is devoted to the exposition of the analytical technique related to the plane Cauchy-Poisson's problem of waves above the plane bottom. The deduction of the "shallow water" approximation is given by means of modelling with respect to a small parameter. The basic estimate of the solution region existence is made more precise by means of a new norm. The scetch of the strict justification of the "shallow water" theory using some results from the first part completes the paper.

CAUCHY PROBLEM

SCALES OF BANACH SPACES. Let B_ρ be a Banach space which is put into correspondence to each value of the real parameter $\rho > 0$. The norm of the element $u \in B_\rho$ is designated as $\|u\|_\rho$. The union $S = \bigcup_{0<\rho} B_\rho$ is called Scale of Banach Spaces (SBS for short) if

$$0 < \rho' < \rho \implies B_{\rho'} \supset B_\rho \,, \quad \|u\|_{\rho'} \leqslant \|u\|_\rho \,. \qquad (1)$$

It is possible to show a number of ways for constructing SBS of analytical functions. One of them is described in the following notations. To the vector $x \in R^m$ having a coordinate representation $x = (x^1, \ldots, x^m)$ the covector of differentiation $D = (D_1, \ldots, D_m)$ is compared where $D_i = \partial/\partial x^i \quad (i = 1, \ldots, m)$. For a multiindex $\alpha = (\alpha_1, \ldots, \alpha_m)$ it is laid $|\alpha| = \alpha_1 + \ldots + \alpha_m$, $\alpha! = \alpha_1! \ldots \alpha_m!$, $D^\alpha = D_1^{\alpha_1} \ldots D_m^{\alpha_m}$. An open set $\omega \subset R^m$ is fixed. The space B_ρ is determined as a set of all functions $u : \omega \to R$ for which the norm is finite:

$$\|u\|_\rho = \sum_{n=0}^{\infty} \rho^n \sum_{|\alpha|=n} \frac{1}{\alpha!} \sup_{x \in \omega} |D^\alpha u(x)| \,. \qquad (2)$$

It is possible to use Fourier transformation for constructing SBS of functions determined on the whole R^m. Here B_ρ is determined as a set of all functions $u : R^m \to R$ for which the norm is finite:

$$\|u\|_\rho = \int_{R^m} e^{|\xi|\rho} |\hat{u}(\xi)| d\xi \,. \qquad (3)$$

Fourier series may be used to construct SBS of periodic functions. For example, in one-dimensional case of 2π-periodic functions $u : R \to R$ the sequence of Fourier coefficients

$$u_n = \frac{1}{2\pi} \int_0^{2\pi} e^{-inx} u(x) dx$$

determines the norm in B_ρ :

$$\|u\|_\rho = \sum_{n=-\infty}^{+\infty} (1 + |n|) e^{|n|\rho} |u_n| \,. \qquad (4)$$

It is clear that the union $S = \bigcup_{0<\rho} B_\rho$ is a Scale of Banach Spaces for any norm (2)-(4). Each function belonging to any of B_ρ is analytical and the region of convergence of a Taylor series which represents it at any point contains an open sphere of the radius ρ

The norms (2)-(4) posses following properties important for applications:

1°. Each space B_ρ is the Banach algebra relative to multiplication of functions, and the estimate is valid

$$\|uv\|_\rho \leqslant \|u\|_\rho \cdot \|v\|_\rho \;;$$

2°. For a fixed $u \in S$ the norm $\|u\|_\rho$ is the analytical function of the parameter ρ, and all its derivatives are non-negative;

3°. Triangle inequality $\|u+v\|_\rho \le \|u\|_\rho + \|v\|_\rho$ in the space B_ρ admits term-by-term differentiation:

$$\frac{\partial}{\partial\rho}\|u+v\|_\rho \le \frac{\partial}{\partial\rho}\|u\|_\rho + \frac{\partial}{\partial\rho}\|v\|_\rho \ ;$$

4°. The estimate of the first order derivatives is valid:

$$\|D_i u\|_\rho \le \frac{\partial}{\partial\rho}\|u\|_\rho \ .$$

The usual design of direct product of Banach spaces is used to construct SBS of vector-functions. If $u = (u^1, \ldots, u^s)$ and $u^j \in B_\rho$ $(j=1,\ldots,s)$ then it is assumed:

$$\|u\|_\rho = \|u^1\|_\rho + \ldots + \|u^s\|_\rho \ .$$

Thereby all the properties 2°, 3°, 4° are conserved.

Note. The following property of the norm is valid in SBS of analytical functions:

$$u \in B_\rho \, , \ \frac{\partial}{\partial\rho}\|u\|_\rho = \infty \, , \ \rho < \sigma \implies u \notin B_\sigma \ . \qquad (5)$$

This means that for $u \in B_\rho$ and any $\rho' < \rho$ always will be $\frac{\partial}{\partial\rho'}\|u\|_{\rho'} < \infty$. In arbitrary SBS the property (5) is valid if $\|u\|_\rho$ as the function of parameter ρ is convex downward.

NONLINEAR CAUCHY PROBLEM. Let a Scale of Banach Spaces $S = \bigcup_{0<\rho} B_\rho$ and a mapping $f : S \times R \to S$ are given. The problem is considered of finding out a mapping $u : R \to S$ satisfying for $t \in R$ the differential equation and in $t = 0$ the initial condition

$$\frac{\partial u}{\partial t} = f(u,t) \, , \quad u(0) = \theta \, . \qquad (6)$$

The problem (6) has proved to be correct if the right-hand part of the equation is so-called quasidifferential operator (the analog of the function satisfying Lipschitz condition).

Open spheres $O_{r,\rho}$ with the center in zero of the spaces B_ρ and with radius r and the union of closure of these spheres $V(r,\rho_0) = \bigcup_{0<\rho\le\rho_0} \overline{O}_{r,\rho}$ are considered.

DEFINITION. Mapping $g : V(r,\rho_0) \to S$ is called quasidifferential operator if $g(\theta) \in B_{\rho_0}$ and if such a number $Q > 0$ exists that for arbitrary $\rho < \rho_0$ and any $u, v \in \overline{O}_{r,\rho}$ the inequality is valid

$$\|g(u)-g(v)\|_\rho \leq Q\left(1+\frac{\partial}{\partial\rho}\right)\left[\left(1+\|u\|_\rho+\|v\|_\rho\right)\|u-v\|_\rho\right]. \qquad (7)$$

It is worth noting that the image $g(\bar{O}_{r,\rho})$, generally speaking, is not contained in B_ρ. Nevertheless, in SBS with property (5) always $g(\bar{O}_{r,\rho}) \subset B_{\rho'}$ for any $\rho' < \rho$.

The problem (6) correctness is established under the following suppositions concerning Scale of Banach Spaces S and mapping f :
(a) For any fixed $u \in S$ the norm $\|u\|_\rho$ as a function of parameter ρ is twice differentiable and convex downward (i.e. $\frac{\partial^2}{\partial\rho^2}\|u\|_\rho \geq 0$);
(b) The property 3° is valid in the Scale of Banach Spaces S ;
(c) There exist positive numbers r, ρ_0, t_0 such that for any fixed $t \in [0,t_0]$ the mapping $f_t : V(r,\rho_0) \rightarrow S$ is quasidifferential operator with a constant Q independent on t ;
(d) For any fixed $u \in O_{r,\rho} \subset V(r,\rho_0)$ the mapping $f_u : [0,t_0] \rightarrow B_{\rho'}$ is continuous for any $\rho' < \rho$.

THEOREM 1. The conditions (a)-(d) being satisfied, the problem (6) has the unique solution

$$u : [0, (\rho_0-\rho)/\varkappa] \longrightarrow B_\rho \qquad (\rho < \rho_0)$$

and the estimate is valid

$$\|u(t)\|_\rho \leq \int_\rho^{\rho+\varkappa t} e^{\sigma-\rho} \sup_{t\in[0,t_0]} \|f(\theta,t)\|_\sigma \, d\sigma \qquad (8)$$

where the number $\varkappa$ is defined as follows:

$$\varkappa = \max\left\{Q(1+4r), \ \frac{1}{r}\int_0^{\rho_0} e^{\sigma} \sup_{t\in[0,t_0]} \|f(\theta,t)\|_\sigma \, d\sigma \ , \ \frac{\rho_0}{t_0}\right\}$$

This theorem is proved by the method of successive approximations [4]. It is interesting to note that though in the theorem 1 there are no analyticity conditions, the generalized form of Cauchy-Kovalevskaja theorem for differential equations follows from it when the right-hand part of the equations is subjected to the continuity condition with respect to t only. In order to obtain the last result it is sufficient to use SBS with the norm (2) and to note that any quasilinear first order differential operator is quasidifferential in this SBS owing to the properties 1° and 4°.

SHALLOW WATER THEORY

CAUCHY-POISSON'S PROBLEM. The plane problem is considered of non-stationary wave theory for non-viscous, incompressible fluid located above horizontal bottom and being under gravity forces. It is supposed that the liquid motion is vortexless and that the surface pressure is equal to zero. The Cartesian coordinate system (x,y) is fixed so that the bottom line is $y=0$ and the free surface line depending on time t is $y=\gamma(x,t)>0$. Then full description of the motion is given by means of function γ and the velocity potential $P=P(x,y,t)$ determined in the domain $\omega^t=\{(x,y)\mid x\in R,\ 0\leq y\leq\gamma(x,t)\}$.
In these notations the starting equations are

$$P_{xx}+P_{yy}=0 \qquad \left((x,y)\in\omega^t\right) \qquad (9)$$

$$\left.\begin{aligned} \gamma_t+\gamma_x P_x-P_y&=0\\ P_t+\tfrac{1}{2}\left(P_x^2+P_y^2\right)+y&=0\end{aligned}\right\} \qquad \left(y=\gamma(x,t)\right) \qquad (10)$$

$$P_y=0 \qquad (y=0) \qquad (11)$$

In the Cauchy-Poisson's problem it is required to obtain the functions γ and P satisfying the equations (9)-(11) and the initial conditions

$$\gamma=\gamma_0(x)>0\,,\quad P=P_0(x,y) \qquad (t=0) \qquad (12)$$

with a function P_0 which is harmonic in the domain ω^0.

Let P be a solution of (9) satisfying the condition (11) and let $P(x,\gamma(x,t),t)=\Phi(x,t)$. The linear operator N is introduced acting on functions Φ in accordance with formula

$$\Phi\longrightarrow N\Phi=\left(P_y-\gamma_x P_x\right)\Big|_{y=\gamma(x,t)} \qquad (13)$$

Operator N is a singular integro-differential one; it depends non-linearly on the function γ. By means of it the system (10) may be rewritten in the form:

$$\gamma_t = N\Phi\ ,\ \ \Phi_t = -\gamma - \frac{1}{2}\cdot\frac{\Phi_x^2 - 2\gamma_x\Phi_x N\Phi - (N\Phi)^2}{1+\gamma_x^2} \qquad (14)$$

The initial conditions for this system pass from (12):

$$\gamma(x,0) = \gamma_0(x),\ \ \Phi(x,0) = \Phi_0(x) \qquad (15)$$

The problem obtained (14), (15) for the pair (γ, Φ) is equivalent to the starting Cauchy-Poisson's problem.

SHALLOW WATER EQUATIONS. For the shallow water theory construction the Cauchy problem is considered for the system (14) when the initial conditions contain a parameter ε

$$\gamma(x,0) = \varepsilon\gamma_*(x),\ \ \Phi(x,0) = \varepsilon^{\frac{1}{2}}\Phi_*(x)\ . \qquad (16)$$

Shallow water equations are the result of the following operations: dilatation of all the variables

$$x \to x,\ y \to \varepsilon y,\ t \to \varepsilon^{-\frac{1}{2}} t,\ P \to \varepsilon^{\frac{1}{2}} P,\ \gamma \to \varepsilon\gamma,\ \Phi \to \varepsilon^{\frac{1}{2}}\Phi \qquad (17)$$

calculation of the operator N in the smallest order with respect to ε , substitution into (14) and formal limit transition when $\varepsilon \to 0$ This leads to

$$\gamma_t + (\gamma\Phi_x)_x = 0\ ,\qquad \Phi_t + \tfrac{1}{2}\Phi_x^2 + \gamma = 0\ . \qquad (18)$$

These equations are equivalent to the isentropic gas dynamics equations for polytropic gas with the specific ratio = 2. The initial conditions following from (16) and (17) for the system (18) are

$$\gamma(x,0) = \gamma_*(x),\ \ \Phi(x,0) = \Phi_*(x)\ . \qquad (19)$$

Note. They say that the transformation of the variables (17) determines a modelling process of the starting problem. It is easy to verify that the same equations (18) may be obtained as a result of the one-parameter family of modelling processes with the parameter κ

$$x \to \varepsilon^{2\kappa} x, \quad y \to \varepsilon^{1+2\kappa} y, \quad t \to \varepsilon^{-\frac{1}{2}+\kappa} t, \quad P \to \varepsilon^{\frac{1}{2}+3\kappa} P,$$
$$\gamma \to \varepsilon^{1+2\kappa} \gamma, \quad \Phi \to \varepsilon^{\frac{1}{2}+3\kappa} \Phi .$$

In particular the modelling process corresponding to $\kappa = -\frac{1}{2}$ is often called "long waves approximation".

The justification of the shallow water theory is meant in this paper in the sense of proving the following statements:
A. The solution of the system (14) with initial data (16) exists for any sufficiently small values $\varepsilon > 0$; let it be the pair of functions $\gamma(x,t,\varepsilon)$, $\Phi(x,t,\varepsilon)$. Moreover, the solution of the problem (18), (19) exists; let it be $\gamma_*(x,t)$, $\Phi_*(x,t)$.
B. The limit equalities are valid in an appropriate metric

$$\lim_{\varepsilon \to 0} \varepsilon^{-1} \gamma(x, \varepsilon^{-\frac{1}{2}} t, \varepsilon) = \gamma_*(x,t), \quad \lim_{\varepsilon \to 0} \varepsilon^{-\frac{1}{2}} \Phi(x, \varepsilon^{-\frac{1}{2}} t, \varepsilon) = \Phi_*(x,t). \quad (20)$$

The necessary elements of such a justification are the theory of solvability of the problem (14), (16) and the solution estimates receiving which are uniform with respect to ε . It will be shown below how this may be done in the case of periodic waves.

CONFORMAL MAPPING. The most difficult part of the planned programme is connected with the fact that it is too complicate to realise the operator N . This difficulty is overcome in this paper by means of the special transformations of the system (14).

The strip $\Pi_\delta = \{ \zeta \mid 0 < \eta < \delta \}$ is taken in the plane of the complex variable $\zeta = \xi + i\eta$. The conformal mapping $g: \Pi_\delta \to \omega^t$ is considered depending on the parameters t, δ and normalized so that it transforms axis $\eta = 0$ onto axis $y = 0$ and the line $\eta = \delta$ onto the boundary line $y = \gamma(x,t)$. The system (14) is transformed by means of this mapping into the equivalent one on the straight line $\eta = \delta$. Thereafter obtained equations are subjected to "quasilinearization" by means of differentiation with respect to ξ . If to put $x(\xi,t,\delta) = \mathrm{Re}\, g(\xi + i\delta, t, \delta)$ then for functions

$$\bar{u}(\xi,t,\delta) = \frac{\partial}{\partial \xi} x(\xi,t,\delta), \qquad \bar{v}(\xi,t,\delta) = \frac{\partial}{\partial \xi} \Phi(x(\xi,t,\delta),t)$$

the system of equations will arise which contains a linear operator

A_δ instead of N. It is defined as follows: if the function $\alpha + i\beta$ is analytical in Π_δ, continuous in $\overline{\Pi}_\delta$ and $\beta(\xi, 0) = 0$ then $A_\delta \alpha(\xi,\delta) = \beta(\xi,\delta)$

The modelling process analogous to (17) is applied to the obtained system with parameter δ

$$\xi \to \xi, \quad \eta \to \delta\eta, \quad t \to \delta^{-\frac{1}{2}} t, \quad \bar{u} \to \bar{u}, \quad \bar{v} \to \delta^{\frac{1}{2}} \bar{v}.$$

As a result of this modelling for functions

$$u(\xi, t, \delta) = \bar{u}(\xi, \delta^{-\frac{1}{2}} t, \delta), \quad v(\xi, t, \delta) = \delta^{-\frac{1}{2}} \bar{v}(\xi, \delta^{-\frac{1}{2}} t, \delta). \tag{21}$$

the system of equations is formed

$$\left.\begin{aligned} u_t &= \frac{\partial}{\partial \xi}\left[H A_\delta u \cdot A_\delta v - u B'_\delta (H A_\delta v)\right], \\ v_t &= -\frac{1}{\delta} A_\delta u + \frac{\partial}{\partial \xi}\left[\tfrac{1}{2} H\left((A_\delta v)^2 - v^2\right) - v B'_\delta (H A_\delta v)\right], \\ &\qquad H = \left[u^2 + (A_\delta u)^2\right]^{-1}, \\ u(\xi, 0, \delta) &= u_*(\xi, \delta), \quad v(\xi, 0, \delta) = v_*(\xi, \delta). \end{aligned}\right\} \tag{22}$$

where B'_δ is an appropriate inverse operator to A_δ. In accordance with (21) the initial data in (22) are expressed through the initial values of functions $\bar{u}$, $\bar{v}$ by analogy with (16)

$$\bar{u}(\xi, 0, \delta) = u_*(\xi, \delta), \quad \bar{v}(\xi, 0, \delta) = \delta^{\frac{1}{2}} v_*(\xi, \delta).$$

Further on the problem (22) is considered in a class of functions 2π-periodical relative to the variable ξ. Auxiliary operators A_δ, B'_δ in this class are determined by their action on separate harmonics $e^{ik\xi}$ $(k = 0, \pm 1, \pm 2, \ldots)$ as follows:

$$A_\delta e^{ik\xi} = i e^{ik\xi} \operatorname{th}(k\delta), \quad B'_\delta e^{ik\xi} = -i e^{ik\xi} \operatorname{cth}(k\delta), \quad B'_\delta(1) = 0. \tag{23}$$

Shallow water theory follows from (22) by means of formal limit transition when $\delta \to 0$ taking into account relations valid by virtue of (23)

$$\lim \frac{1}{\delta} A_\delta w = \frac{\partial w}{\partial \xi}, \quad \lim \delta B'_\delta w = B_0 w \quad \left(B_0 e^{ik\xi} = -\frac{i}{k} e^{ik\xi}\right). \tag{24}$$

The problem arises for limit functions $U = u(\xi, t, 0)$, $V = v(\xi, t, 0)$:

$$\left.\begin{aligned} U_t &= -\frac{\partial}{\partial\xi}\left[U B_o\left(V U^{-2}\right)\right], \\ V_t &= -\frac{\partial}{\partial\xi}\left[U + \tfrac{1}{2}V^2 U^{-2} + V B_o\left(V U^{-2}\right)\right], \\ & U(\xi,0) = U_o(\xi), \quad V(\xi,0) = V_o(\xi). \end{aligned}\right\} \qquad (25)$$

EXISTENCE THEOREMS. The problem (22) is considered in the Scale of Banach Spaces $S = \bigcup_{0<\rho} B_\rho$ with the norm (4). Existence of the solution is proved by the reference to the theorem 1 after having been stated that the right-hand part in (22) is a quasidifferential operator.

For this approach the operators A_δ, B'_δ estimates and the function u estimate from below are necessary. The last follows from the supposition that the bottom does not dry in the considered wave motion of fluid. This is equivalent to the existence of a constant $b > 0$ such that the representation is valid $u_* = b + u_1$ where the function u_1 has a mean over period value equal to zero. The constant b is supposed here to be independent on δ. It determines the wave period equal to $2\pi\delta$ in the physical plane. Sufficient estimates for proving the existence theorem are

$$\left.\begin{aligned} \|A_\delta u\|_\rho \leq \|u\|_\rho, \quad \|\tfrac{1}{\delta} A_\delta u\|_\rho \leq \frac{\partial}{\partial\rho}\|u\|_\rho, \\ \|B'_\delta(w A_\delta v)\|_\rho \leq \|w\|_\rho \cdot \|v\|_\rho. \end{aligned}\right\} \qquad (26)$$

THEOREM 2. Let the initial data $u_* = b + u_1$, v_* of the problem (22) belong to the space B_{ρ_0} and let $\|u_1\|_{\rho_0} + \|v_*\|_{\rho_0} < \frac{1}{8} b$. Then there exists such a number $\varkappa > 0$ that the problem (22) has a unique solution $u, v \in B_\rho$ for all values (ρ, t) from the region

$$0 \leq t, \quad 0 < \rho, \quad \rho + \varkappa t \leq \rho_0. \qquad (27)$$

An analogous theorem is valid for the problem (25). It follows immediately from the fact that estimates (26) and the constant b are independent on δ. With these theorems the part of shallow water theory justification is accomplished corresponding to the statement A .

SHALLOW WATER THEORY JUSTIFICATION. Precise formulation of the statement B and the equality (20) as relates to the system (22) and its limit form (25) is given in the following theorem.

THEOREM 3. If suppositions to the theorem 2 are valid and if

$$\lim_{\delta\to 0}\left(\|u_*(\xi,\delta)-U_0(\xi)\|_{\rho_0}+\|v_*(\xi,\delta)-V_0(\xi)\|_{\rho_0}\right)=0 \tag{28}$$

then for all (ρ,t) from a region of (27) type also

$$\lim_{\delta\to 0}\left(\|u(\xi,t,\delta)-U(\xi,t)\|_{\rho}+\|v(\xi,t,\delta)-V(\xi,t)\|_{\rho}\right)=0 . \tag{29}$$

The differences are considered for the proof

$$\tilde{u}=u-U-u_*+U_0\,;\qquad \tilde{v}=v-V-v_*+V_0\,.$$

A Cauchy problem is obtained for these differences from (22) and (25) with zero initial data. Right-hand parts of this problem are linear combinations of some addends proportional to three types of values: linear operators L_δ with arguments $\tilde{u}$, $\tilde{v}$, differences u_*-U_0 , v_*-V_0 and values of the type

$$A_\delta w\,,\quad \left(\tfrac{1}{\delta}A_\delta-\tfrac{\partial}{\partial\xi}\right)w\,,\quad \left(\delta B_\delta'-B_0\right)w$$

with different functions w and with coefficients being uniformly (relative to δ) bounded together with their derivatives by the B_ρ-norm in the region (27). Herewith the estimate of the form (8) is used for solutions of the problems (22) and (25). It is established that the linear operators L_δ are quasidifferential ones and allow the estimate

$$\|L_\delta w\|_\rho \le C_1\left(1+\frac{\partial}{\partial\rho}\right)\|w\|_\rho$$

and that the last values have, in addition to (26), the estimates

$$\left\|\left(\frac{1}{\delta}A_\delta-\frac{\partial}{\partial\xi}\right)w\right\|_\rho \le \delta C_2\frac{\partial^2}{\partial\rho^2}\|w\|_\rho,\quad \left\|\left(\delta B_\delta'-B_0\right)w\right\|_\rho \le \delta^2 C_3\frac{\partial}{\partial\rho}\|w\|_\rho,$$

where positive constants C_1, C_2 , C_3 are independent on δ . The solution $(\tilde{u},\tilde{v})$ estimate follows from these estimates and from the estimate of the solution (8) in the theorem 1. It shows that the equality (28) implies the equality (29).

This completes the verification of the statement $\mathcal{B}$. Therefore the shallow water theory justification is finished in the framework of the suppositions made.

FINAL REMARK. The theory given above leaves a field for further investigations. The following progress in the shallow water theory justification may be connected, for example, with the consideration of non-periodical waves and the investigation of three-dimensional problem.

REFERENCES

[1] Stoker J.J. Water Waves. The Mathematical Theory with Applications (Interscience Publishers, Inc., New York, 1957).

[2] Leray J. et Ohya Y. Equations et Systemes Non-Lineares, Hyperboliques Non-Stricts. Math. Ann. 170 (1967), 167-205.

[3] Ovsjannikov L.V. Singular Operator in a Scale of Banach Spaces. Dokl. Akad. Nauk SSSR 163 (1965), 819-822.

[4] Ovsjannikov L.V. A Non-linear Cauchy Problem in a Scale of Banach Spaces. Dokl. Akad. Nauk SSSR 200 (1971), 789-792.

[5] Nalimov V.I. A priori Estimates of the Solution of Elliptic Equation Problem. Dokl. Akad. Nauk SSSR 189 (1969), 45-48.

[6] Ovsjannikov L.V. On the Bubble Upflow. (Russian) Some Problems of Mathematics and Mechanics. Izd. Akad. Nauk SSSR (1970), 209-222.

[7] Nalimov V.I. Cauchy-Poisson Problem. (Russian) Continuum Dynamics (Institute of Hydrodynamics, Siberian Branch USSR Acad. Sci.) 18 (1974), 104-210.

[8] Friedrichs K.O. On the Derivation of the Shallow Water Theory. Appendix to "The Formation of Breakers and Bores" by J.J.Stoker. Comm. Pure Appl. Math. 1 (1948), 81-85.

[9] Ovsjannikov L.V. To the Shallow Water Theory Foundation. Arch. Mech. (Arch. Mech. Stos.) 26 (1974), 407-422.

PERTURBATION RESULTS AND THEIR APPLICATIONS TO PROBLEMS IN STRUCTURAL DYNAMICS

A.J.Pritchard
Dept. of Engineering
University of Warwick

W.T.F. Blakeley
Dept. of Mathematics
The Polytechnic,
Wolverhampton

§1 Introduction

All dynamical systems are subject to perturbations. If the basic mathematical model is of the form

(1.1) $\dot{x} = -Ax \qquad x(0) = x_0$

where A may be a non-linear unbounded operator then these perturbations may arise through the initial state, dynamic loads or by perturbation of the operator A. Thus a more realistic model may be

(1.2) $\dot{x} = -(A+B)x + f \qquad x(0) = \bar{x}.$

In this paper we develop a perturbation theory to obtain estimates for bounds on the perturbed motion and criteria for stability when the exact characteristics of the perturbing operator and dynamical loads are not known. The results are illustrated by application to shallow arch, elastic column and panel flutter problems.

§2 Perturbation Results

In this section we obtain some new perturbation results which extend the theorems of Kato[1,2], Nelson and Gustafson[3] and Okazawa[4,5] for linear and non-linear m-accretive operators.

Kato[1] has shown that if -A, -B both generate linear contraction semi-groups on a Banach space X with D(B) ⊃ D(A) and if B is relatively bounded with respect to A with A bound less than ½, that is

(2.1) $\|Bx\| \leq a\|x\| + b\|Ax\|, \quad a>0, \quad 0<b<\tfrac{1}{2}, \quad x\in D(A)$

then -(A+B) generates a contraction semi-group on X. Nelson and Gustafson [3] modified this result to show that for the same conditions on A but with B accretive and b<1, then -(A+B) will generate a contraction semi-group. It is well known [4] that -A generating a contraction semi-group is equivalent to A being m-accretive and densely defined. If X is reflexive and A m-accretive then A is automatically densely defined and for such spaces Okazawa[4] has extended the Nelson and Gustafson result to b=1.

For Hilbert spaces Okazawa[5] has shown that if A is m-accretive, B accretive, D(B) ⊃ D(A) and there are non-negative constants a and b≤1 such that

(2.2) $0 \leq \mathrm{Re}\langle Ax,Bx\rangle + a\|x\|^2 + b\|Ax\|^2, \quad x\in D(A)$

then A+B is m-accretive if b<1 and the closure of A+B is m-accretive if b=1.

We extend these linear results in the following theorems.

Theorem 2.1 Let X be a reflexive Banach space and A be m-accretive, B accretive with D(B) ⊃ D(A). If for any integer n≥0 there exists an a≥0 such that

(2.3) $$\|Bx\| \leq a\|x\| + \left\|\left(2 - \frac{1}{2^{(2^n-1)}}\right)Ax + \left(1 - \frac{1}{2^{(2^n-1)}}\right)Bx\right\|, \quad x\in D(A)$$

then A+B is m-accretive. In the limit as n→∞ this condition becomes

(2.4) $\|Bx\| \leq a\|x\| + \|(2A+B)x\|, \quad x\in D(A), \quad a\geq 0.$

Proof:- Let A+B = A + βB + (1-β)B where 0≤β≤1.
Since B is accretive and D(B) ⊃ D(A) then βB is accretive and D(βB)⊃D(A) and so we may apply Okazawa's result[4] to conclude that A+βB is m-accretive if there exists a ≥0 such that

(2.5) $\|\beta Bx\| \leq a\|x\| + \|Ax\| \quad x\in D(A).$

If (2.5) is valid then since D((1-β)B) ⊃ D(A+βB) we may again apply Okazawa's theorem to the perturbation (1-β)B of A+βB to conclude that A+B is m-accretive if there exists a'≥0 such that

(2.6) $\|(1-\beta)Bx\| \leq a'\|x\| + \|(A+\beta B)x\|, \quad x\in D(A)$

We now choose β so that (2.6) implies (2.5).
Clearly if (2.6) is valid then

$$(1-\beta)\|Bx\| \leq a'\|x\| + \|Ax\| + \beta\|Bx\|$$

or $$(1-2\beta)\|Bx\| \leq a'\|x\| + \|Ax\|$$

and this implies (1.5) if a' = a and β = 1/3. So we have shown that A+B will be m-accretive if there exists a≥0 such that

(2.7) $\|Bx\| \leq \tfrac{3}{2}a\|x\| + \|(\tfrac{3}{2}A+\tfrac{1}{2}B)x\| \quad x\in D(A).$

We note that if B satisfies

(2.8) $\|Bx\| \leq a\|x\| + \|Ax\|$

then $\|Bx\| \leq \tfrac{3}{2}a\|x\| + \tfrac{3}{2}\|Ax\| - \tfrac{1}{2}\|Bx\|$

hence $\|Bx\| \leq \tfrac{3}{2}a\|x\| + \|(\tfrac{3}{2}A+\tfrac{1}{2}B)x\|$

Thus any B which satisfies the Okazawa hypothesis satisfies (2.7) however B = ρA, 1<ρ≤3 satisfies (2.7) but not (2.8).

To prove (2.3) for n=2 we let A+B = A+γB+(1-γ)B where 0≤γ≤1 and derive two conditions similar to (2.5) and (2.6) but using (2.7) instead of Okazawa's condition (2.8). Then by setting $\gamma=\frac{1}{5}$ we obtain the single condition

(2.9) $\|Bx\| \leq \|(\tfrac{15}{8}A+\tfrac{7}{8}B)x\| + a\|x\|, \quad x\in D(A).$

Noting that (2.7) and (2.9) are just (2.3) with $n = 1,2$ respectively the general result is obtained by an induction argument.

One of the disadvantages of the above results is the assumption that B is accretive, we relax this in the following theorem.

Theorem 2.2. Let A be m-accretive and A+B be accretive on a reflexive B-space X with $D(B) \supset D(A)$, if there exists an $a>0$ and an α, $0 \leq \alpha < 1$, such that

$$(2.10) \quad \|(A+B)x\| \leq \|(\tfrac{2-\alpha}{\alpha}A+B)x\| + a\|x\| \qquad x \in D(A)$$

then $A+\alpha B$ is m-accretive.

Proof:- By setting $A+\alpha B = (1-\alpha)A + \alpha(A+B)$ for $0 \leq \alpha < 1$ and applying Theorem 2.1 (2.4) we find $A+\alpha B$ is m-accretive if

$$\|\alpha(A+B)x\| \leq \|2(1-\alpha)Ax + \alpha(A+B)x\| + a\|x\| \qquad x \in D(A)$$

which is (2.10)

For Hilbert spaces we have a similar result to (2.2)

Theorem 2.3. Let A be m-accretive and A+B accretive on a Hilbert space H with $D(B) \supset D(A)$. If there exists $a' \geq 0$, $b>1$ such that

$$(2.11) \quad 0 \leq \mathrm{Re}\langle Ax,Bx\rangle + a'\|x\|^2 + b\|Ax\|^2, \qquad x \in D(A)$$

then $A + \frac{1}{b}B$ is m-accretive.

Proof:- Now $\|(\frac{2-\alpha}{\alpha}A+B)x\|^2 - \|(A+B)x\|^2 = \frac{4(1-\alpha)}{\alpha}\left[\frac{1}{\alpha}\|Ax\|^2 + \langle Ax,Bx\rangle\right]$.

Let $\alpha = \frac{1}{b}$ then by (2.11)

$$\|(A+B)x\|^2 \leq a\|x\|^2 + \|(\tfrac{2-\alpha}{\alpha})Ax + Bx\|^2$$

where $a = 4a'(1-\alpha)/\alpha$. Hence by Theorem 2.2 $A+\frac{1}{b}B$ is m-accretive.

We can obtain similar results for the theorems of Kato[1] and Nelson and Gustafson[3].

Corollary Let A be m-accretive and $\frac{1}{b}A+\tilde{B}$, $b>1$ accretive on a Hilbert space H with $D(\tilde{B}) \supset D(A)$. If there exists a constant $a \geq 0$ such that

$$0 \leq \mathrm{Re}\langle Ax,\tilde{B}x\rangle + a\|x\|^2 + \|Ax\|^2 \qquad x \in D(A)$$

then $A+\tilde{B}$ is m-accretive.

Proof: Set $B = b\tilde{B}$ in Theorem 2.3.

For non-linear operators we will generalize the following two perturbation results of Kato[2] by essentially the same methods used in the linear case.

(i) Let A and B be m-accretive operators, possibly non-linear and multiple-valued. Let B be locally A-bounded so that $D(B) \supset D(A)$ and for each $x \in X$ there are a neighbourhood U of x and constants a and b such that

$$(2.12) \quad |||Bx||| \leq a + b|||Ax||| \quad \text{for } x \in D(A) \cap U,$$

where $|||Ax||| = \inf_{s\in Ax} \|s\|$. If b<1 then A+B is m-accretive.

(ii) Let A be m-accretive and B single-valued and accretive. Let B satisfy the conditions of (i) above. Furthermore assume that for each $x_0 \in D(A)$ there are a neighbourhood $\cup$ of x_0 and constants a' and b' such that

$$(2.13) \quad \|Bx-By\| \leq a'\|x-y\| + b'|||Ax-Ay||| \quad \text{for } x,y\in D(A)\cap\cup.$$

Then A+B is m-accretive if b<1 and b'<1.

Theorem 2.5 With the same conditions on the operators A and B as in (i) above but with (2.12) replaced by

$$(2.14) \quad \|Bx\| \leq a\|x\| + b\left|\left|\left|\left(2 - \frac{1}{2^{(2^n-1)}}\right)Ax + \left(1 - \frac{1}{2^{(2^n-1)}}\right)Bx\right|\right|\right| \quad x\in D(A)\cap\cup$$

for any integer $n\geq 0$, $a>0$, $b<1$ then A+B is m-accretive. In the limit as $n\to\infty$ this condition becomes

$$(2.15) \quad \|Bx\| \leq a\|x\| + b|||(2A+B)x||| \quad x\in D(A)\cap\cup.$$

The proof of this result is similar to the linear case, the essential difference will be illustrated in the proof of the next theorem.

Theorem 2.6 With the same conditions on the operators A and B as in (ii) above but with (2.12) replaced by (2.14) and (2.13) replaced by

$$(2.16) \quad \|Bx - By\| \leq a'\|x - y\| + b'\left|\left|\left|\left(2 - \frac{1}{2^{(2^n-1)}}\right)(Ax - Ay) + \left(1 - \frac{1}{2^{(2^n-1)}}\right)(Bx - By)\right|\right|\right|,$$

then A+B is m-accretive if b<1, b'<1.

Proof: Consider $A+B = A+\beta B+(1-\beta)B$, $0<\beta<1$, as in the proof of Theorem 2.1. Then applying Kato's result (ii) we find that A+B is m-accretive if there exist non-negative constants a, a_1, a_2, a_3, b, b_1, b_2, b_3 with each b<1 such that for $x,y\in D(A)\cap\cup$

$$(2.17) \quad \|\beta Bx\| \leq a + b|||Ax|||$$

$$(2.18) \quad \|\beta(Bx-By)\| \leq a_1\|x-y\| + b_1|||Ax-Ay|||$$

$$(2.19) \quad \|(1-\beta)Bx\| \leq a_2 + b_2|||(A+\beta B)x|||$$

$$(2.20) \quad \|(1-\beta)(Bx-By)\| \leq a_3\|x-y\| + b_3|||(A+\beta B)x - (A+\beta B)y|||$$

If (2.19) holds then we have

$$(1-\beta-b_2\beta)\|Bx\| \leq a_2 + b_2|||Ax|||$$

By setting $a = \frac{\beta a_2}{1-\beta(1+b_2)}$ and $b = \frac{\beta b_2}{1-\beta(1+b_2)}$ we can see that (2.17) holds if $\frac{\beta b_2}{1-\beta(1+b_2)} < 1$, that is if $\beta < \frac{1}{1+2b_2}$. Since $b_2<1$ we can choose $\beta = \frac{1}{3}$

In a similar way with $\beta = \frac{1}{3}$ we can show that (2.20) implies (2.18) so that (2.17) - (2.20) may be replaced by

(2.21) $\|Bx\| \leq \frac{3}{2}a_2 + b_2 |||(\frac{3}{2}A+\frac{1}{2}B)x|||$ $\qquad x\in D(A)\cap U$

and

(2.22) $\|Bx-By\| \leq \frac{3}{2}a_3\|x-y\| + b_3|||\frac{3}{2}(Ax-Ay) + \frac{1}{2}(Bx-By)|||$ $\qquad x,y\in D(A)\cap U$

The rest of the proof follows as in the linear case.

We can again relax the accretiveness condition on B as in Theorem 2.2 to obtain the following theorems

<u>Theorem 2.7</u> Let A be single-valued and m-accretive and A+B single-valued and accretive with $D(B) \supset D(A)$. If there exist non-negative constants a, a_1, b, b_1 with $b<1$, $b_1<1$ and an α, $0\leq\alpha<1$ such that

(2.23) $\|(A+B)x\| \leq a + b\|\frac{2-\alpha}{\alpha}Ax + Bx\|$ $\qquad x\in d(A)\cap U$

(2.24) $\|Ax-Ay + Bx-By\| \leq a_1\|x-y\| + b_1\|(\frac{2-\alpha}{\alpha})(Ax-Ay) + Bx-By\|$ $\qquad x,y\in D(A)\cap U$

then $A+\alpha B$ is m-accretive.

<u>Theorem 2.8</u> Let A be single-valued and m-accretive, A+B be single-valued and accretive on a Hilbert space H with $D(B) \supset D(A) \supset 0$. If $A0 = B0 = 0$ and there exist a, $b>1$ such that

(2.25) $\|Bx-By\|^2 \leq a^2\|x-y\|^2 + b^2\|Ax-Ay\|^2$ $\qquad x,y\in D(A)\cap U$

then $A + \frac{2}{1+b^2}B$ is m-accretive.

<u>Proof</u>: If

(2.26) $(1-b_1^2)\|Bx-By\|^2 \leq a_1^2\|x-y\|^2 + (\frac{1}{b_1^2} - 1)\|Ax-Ay\|^2$

and $\quad b_1^2(2-\alpha)/\alpha = 1$ then

$$\|Ax-Ay + Bx-By\|^2 \leq a_1^2\|x-y\|^2 + b_1^2\|(\frac{2-\alpha}{\alpha})(Ax-Ay) + Bx-By\|^2$$

But (2.26) is valid if we set $b^2 = 1/b_1^2$, $a^2 = a_1^2/(1-b_1^2)$.

Moreover since $A0 = B0 = 0$ we have

$$\|Bx\|^2 \leq a^2\|x\|^2 + b^2\|Ax\|^2 \qquad x\in D(A)\cap U$$

and we may use a similar argument to show that

$$\|(A+B)x\| \leq a_1\|x\| + b_1\|\frac{2-\alpha}{\alpha}Ax + Bx\| \qquad x\in D(A)\cap U$$

Thus we have satisfied the conditions of Theorem 2.7 and hence $A+\alpha B = A+\frac{2}{1+b^2}B$ is m-accretive.

<u>Corollary</u> With the same conditions as in Theorem 2.8 but with $(1-\epsilon)A+\widetilde{B}$ accretive instead of A+B accretive and (2.25) replaced by

$$\|\widetilde{B}x-\widetilde{B}y\|^2 \leq a^2\|x-y\|^2 + (1-\epsilon^2)\|Ax-Ay\|^2 \qquad x,y\in D(A)\cap U$$

then $A+\widetilde{B}$ is m-accretive.

<u>Proof</u>: Set $B = \frac{(1+b^2)}{2}\widetilde{B}$ and $2/(1+b^2) = 1-\epsilon$ in Theorem 2.8.

§3 <u>Perturbation of Semi-groups</u>

In this section we will consider inhomogeneous linear evolution equations of the form

(3.1) $\dot{x}(t) = Ax(t) + B(t)x(t) + f(t), \qquad x(0) = x_0$

The operator B(t) will in general be unbounded. However for the applications considered in the next section we so choose the underlying space so that $B \in B_\infty([0,T],\mathcal{L}(X))$, the space of $\mathcal{L}(X)$ valued operators which are essentially bounded in [0,T], so there exists K such that

(3.2) $\quad \|B(t)\|_{\mathcal{L}(X)} \leq K \qquad$ a.e. on [0,T].

This can easily be generalized to the case where $\|B(t)\|_{\mathcal{L}(X)} \leq K(t)$ [6] and for the generalization to unbounded operators B see [7].

We assume A is the infinitesimal generator of a strongly continuous semi-group T_t on X with

(3.3) $\quad \|T_t\| \leq Me^{-\omega t}, \qquad \omega>0.$

Hence

(3.4) $\quad x(t) = T_t x_0$

is the solution of the evolution equation

(3.5) $\quad \dot{x} = Ax \qquad x(0) = x_0 \in D(A)$

and $\|x(t)\| \leq Me^{-\omega t}\|x_0\|$. Thus the equilibrium point x=0 will be asymptotically stable in the large in the sense of Liapunov[8] for the system (3.5).

Before giving the conditions on f(t) we must interpret what is meant by a solution of (3.1), to do this we associate with the operator A+B(.) an evolution operator U(.,.) on $\Delta(\tau) = \{(t,s)\ 0\leq s<t\leq T\}$ defined by

(3.6) $\quad U(t,s)x = T_{t-s}x + \int_s^t T_{t-\rho}B(\rho)U(\rho,s)x\, d\rho.$

We have the following theorem.

<u>Theorem 3.1</u> There exists a unique solution of (3.6) with the following properties.

(i) U(t,.) is strongly continuous on [o,t] and U(.,s) is strongly contintuous on [s,T],

(ii) $U(t,\rho)U(\rho,s) = U(t,s), \quad U(t,t) = I \quad T\geq t\geq\rho\geq s\geq 0,$

(iii) $\|U(t,s)\| \leq Me^{-(\omega-MK)(t-s)}$

where we have assumed (3.3) i.e. $\|T_t\| \leq Me^{-\omega t}$,

(iv) U(t,s) is a quasi-evolution operator [6] in the sense

$\frac{\partial U}{\partial s}(t,s) = -U(t,s)[A+B(s)]x, \qquad x\in D(A)$

<u>Proof</u>: We construct U(t,s) by the usual iterative scheme

$U^0(t,s) = T_{t-s}$.

$U^n(t,s)x = \int_s^t T_{t-\rho}B(\rho)U^{n-1}(\rho,s)x\, d\rho$

where the integral in the above is a well-defined Bochner integral since

$B(\ ) \in B_\infty[[0,T],\mathcal{L}(X)]$. Then exactly as in [6] $U(t,s)$ defined by

$$U(t,s) = \sum_{n=0}^{\infty} U^n(t,s)$$

can be shown to satisfy (i), (ii) and (iv). To show (iii) we have

$$\|U(t,s)x\| \leq Me^{-\omega(t-s)}\|x\| + \int_s^t Me^{-\omega(t-\rho)}K\|U(\rho,s)\|d\rho$$

hence

$$\|U(t,s)x\|e^{\omega(t-s)} \leq M\|x\| + \int_s^t Me^{-\omega(s-\rho)}K\|U(\rho,s)\|d\rho$$

and by Gronwall's inequality

$$\|U(t,s)x\|e^{\omega(t-s)} \leq Me^{MK(t-s)}\|x\|$$

or $\quad \|U(t,s)\| \leq Me^{-(\omega-MK)(t-s)}$.

We define a "mild" solution of (3.1) to be

$$x(t) = U(t,0)x_0 + \int_0^t U(t,\rho)f(\rho)d\rho \tag{3.7}$$

so that if for example $f \in L_2[0,T,X]$, we have $x(.) \in C[[0,T],X]$. However in general we are not able to differentiate (3.7) to obtain (3.1). It is easy to show that (3.7) is equivalent to

$$x(t) = T_t x_0 + \int_0^t T_{t-\rho}B(\rho)x(\rho)d\rho + \int_0^t T_{t-\rho}f(\rho)d\rho. \tag{3.8}$$

Using (iii) of Theorem 3.1 and (3.7) gives

$$\|x(t)\| \leq Me^{-(\omega-MK)t}\|x_0\| + \int_0^t Me^{-(\omega-MK)(t-\rho)}\|f(\rho)\|d\rho. \tag{3.9}$$

From this expression we can estimate the effect of the perturbation operator B(.) and the forcing term f(.). However in applications it may be easier to estimate these effects via a Liapunov functional [9] constructed from the unperturbed homogeneous system (3.5) which can also be used to obtain the estimate (3.3) for the semi-group T_t.

We suppose X is a Hilbert space and are given an operator $P(t) \in \mathcal{L}(X)$ for $t \in [0,T]$ such that

$$\frac{d}{dt}\langle x,P(t)y\rangle + \langle P(t)x,Ay\rangle + \langle Ax,P(t)y\rangle + \langle x,Wy\rangle = 0, \quad x,y \in D(A), \tag{3.10}$$
$$P(T) = G,$$

where $\langle\ ,\ \rangle$ is the inner product on X, $W>0$, $G\geq 0$, $\langle x, P(t)y\rangle \in C^1[0,T]$ for $x,y \in D(A)$ and

$$\inf_{t\in[0,T]} \langle x,P(t)x\rangle \geq p\|x\|^2, \qquad k>0 \tag{3.11}$$

If we consider the Liapunov functional

$$V(t) = \langle x(t),P(t)x(t)\rangle$$

then formally for the system (3.5)

$$(3.12)\quad \dot V(t) = \langle \dot x(t), P(t)x(t)\rangle + \langle x(t), \dot P(t)x(t)\rangle + \langle x(t), P(t)\dot x(t)\rangle = -\langle x(t), Wx(t)\rangle$$

and for the system (3.1) we have

$$(3.13)\quad \dot V(t) = -\langle x(t), Wx(t)\rangle + \langle x(t), P(t)B(t)x(t)\rangle + \langle B(t)P(t)x(t), x(t)\rangle + \langle x(t), P(t)f(t)\rangle + \langle P(t)f(t), x(t)\rangle$$

Before we use these results it is necessary to justify the formal differentiation. Since the strict solution of $\dot x = Ax$ is $x(t) = T_t x_0$ for $x_0 \in D(A)$ it is not difficult to prove (3.12). However since we have only a mild solution (3.7) or (3.8) of (3.1) the derivation of (3.13) is not so immediate, we need the following theorem.

<u>Theorem 3.2.</u> If P(t) satisfies (3.10) then

$$(3.14)\quad P(t)x = T^*_{T-t} G T_{T-t} x + \int_t^T T^*_{s-t} W T_{s-t} x \, ds$$

and if x(t) is given by (3.8)

$$(3.15)\quad \langle x(t), P(t)x(t)\rangle = \langle x(T), Gx(T)\rangle + \int_t^T \Big[-\langle x(s), P(s)B(s)x(s)\rangle - \langle x(s), P(s)f(s)\rangle - \langle P(s)B(s)x(s), x(s)\rangle - \langle P(s)f(s), x(s)\rangle + \langle x(s), Wx(s)\rangle \Big] ds$$

<u>Proof:</u> Proof of (3.15) is obtained by substituting (3.8) directly into (3.14) (For more details see [6]).

To prove (3.14) we note that T^*_t is the dual semi-group of T_t and since X is a Hilbert space T^*_t is strongly continuous. Clearly P(t) given by (3.14) satisfies (3.10) so the only problem is uniqueness. Let Q(t) be another solution then R(t) = P(t)-Q(t) must satisfy

$$\frac{d}{dt}\langle x, R(t)y\rangle + \langle Ax, R(t)y\rangle + \langle R(t)x, Ay\rangle = 0, \quad R(T) = 0$$

If we let $D(t) = T^*_{t-s} R(t) T_{t-s}$ for any $s \in [0,T)$ we obtain

$$\frac{d}{dt}\langle x, D(t)y\rangle = 0 \qquad x, y \in D(A)$$

Hence $\langle x, D(t)y\rangle = 0$ since D(T) = 0, x,y D(A) and so $\langle T_{t-s}x, R(t)T_{t-s}y\rangle = 0$. But D(A) is dense in X so that on letting $s \to t$ we obtain P(t) = Q(t) and the solution is unique.

Differentiating (3.15) we obtain (3.13) for almost all $t \in [0,T]$. From the conditions on P(t) and W there exist positive constants p, w such that $W > wI$, $\bar p > \sup_{t\in[0,T]} \|P(t)\|$

Hence using (3.11) and (3.12) gives

$$p\|x(t)\|^2 \le V \le \bar p\|x(t)\|^2$$

and

$$\dot V \le -w\|x(t)\|^2 \le -\frac{w}{\bar p}V$$

so that

$$\|x(t)\| \le \sqrt{\frac{\bar p}{p}}\, e^{-wt/2\bar p}\, \|x_0\|$$

and $\qquad \|T_t\| \leq \sqrt{\frac{\bar{p}}{p}}\, e^{-wt/2\bar{p}}$

Using (3.13), if $w>2\bar{p}K$ we obtain

$$\dot{V} \leq -\left(\frac{w-2\bar{p}K}{\bar{p}}\right) V(t) + \frac{2V^{\frac{1}{2}}}{p^{\frac{1}{2}}}\,\bar{p}\|f(t)\| \quad \text{a.e.}$$

Setting $V(t) = U^2(t)$ we find

$$U(t) \leq e^{-\lambda t}U(0) + \frac{\bar{p}}{p^{\frac{1}{2}}}\int_0^t e^{-\lambda(t-s)}\|f(s)\|\,ds$$

where $\lambda = \dfrac{w-2\bar{p}K}{2\bar{p}}$

Hence $\|x(t)\| \leq \left(\frac{\bar{p}}{p}\right)^{\frac{1}{2}} e^{-\lambda t}\|x_0\| + \frac{\bar{p}}{p}\int_0^t e^{-\lambda(t-s)}\|f(s)\|\,ds$

which is a similar estimate to (3.9).

For many applications it is convenient to take the operator P to be independent of t so that

$$Px = \int_0^\infty T_t{}^*WT_t x\,dt$$

which is well defined if $\|T_t\| \leq Me^{-\omega t}$ $(\omega>0)$. In this case we can simplify the analysis by introducing a new Hilbert space X_0 with inner product $\langle x,y\rangle_{X_0} = \langle x,Py\rangle_X$

(3.16) Then $V(t) = \|x(t)\|^2_{X_0}$ and $p = \bar{p} = 1$.

From (3.12)

$$\dot{V}(t) = -\langle x(t),Wx(t)\rangle_X = -\frac{\langle x(t),Wx(t)\rangle_X}{\langle x(t),Px(t)\rangle_X}\,V$$

$$\leq -\|W^{-\frac{1}{2}}PW^{-\frac{1}{2}}\|^{-1}_{\mathcal{L}(X)}\,V = -2\mu V \text{ (say).}$$

Thus $\|x(t)\|^2_{X_0} \leq e^{-2\mu t}\|x(0)\|^2_{X_0}$

and $\|T_t\|_{\mathcal{L}(X_0)} \leq e^{-\mu t}$.

The estimate equivalent to (3.9) will now have the form

(3.17) $\quad \|x(t)\|_{X_0} \leq e^{-(\mu-K)t}\|x(0)\|_{X_0} + \int_0^t e^{-(\mu-K)(t-\rho)}\|f(\rho)\|_{X_0}\,d\rho$

where $\operatorname{ess\,sup}_{t\in[0,T]} \|B(t)\|_{\mathcal{L}(X_0)} \leq K$.

Since the space X_0 and the constants μ,K are essentially determined by the operator W a variety of estimates of the form (3.17) can be obtained. This could lead to an optimisation problem if suitable criteria could be formulated. We note that the dependence on K in the estimate (3.17) is such that it cannot be improved by splitting the operator B into $\beta B + (1-\beta)B$ as in §2.

§4 Applications

Example 1. Shallow Arch Problem

The non-dimensional non-linear equation of motion of a shallow arch [9] is assumed to be

$$(4.1)\quad y_{tt} + 2\xi y_t + (y-y_o)_{xxxx} + 2y_{xx}\int_0^1\left[(y_o)_x^2 - y_x^2\right]dx = 0 \qquad 0\leq x\leq 1,\quad t\geq 0.$$

$y = y_{xx} = 0$ at $x = 0,1$ where $y_o(x)$ denotes the equilibrium configuration. For simplicity we will assume that $y_o(x) = 0$.

Set $y_t = v$, $w = [y,v]^T$ then (4.1) takes the form

$$(4.2)\qquad \dot{w} + Aw + Bw = 0$$

where formally $Aw = \left[-v, 2\xi v + y_{xxxx}\right]^T$, $Bw = \left[0, -2y_{xx}\int_0^1 y_x^2 dx\right]^T$

and $v = y = y_{xx} = V_{xx} = 0$ at $x = 0,1$.

Consider the real Hilbert space H with inner product

$$\langle w_1, w_2\rangle = \int_0^1 [y_{1xx}y_{2xx} + v_1v_2 + \xi(v_1y_2 + v_1y_2) + 2\xi^2 y_1y_2]dx$$

and let $D(A) = \{w\in H : Aw\in H,\ v = y = v_{xx} = y_{xx} = 0 \text{ at } x = 0\}$.

Then it is easy to show that A is m-accretive. However B is not accretive. Nevertheless it is possible to show that for $\cup : \|w\| < k$ where k is sufficiently small there exists an ε such that $(1-\varepsilon)A + B$ is accretive. Moreover

$$\|Bw_1 - Bw_2\|^2 \leq a^2\|w_1 - w_2\|^2 \qquad w_1, w_2\in D(A)\cap\cup.$$

Hence we may apply the corollary to theorem 2.8 to conclude that A+B is m-accretive. Thus -(A+B) generates a non-linear contraction semi-group T_t and the solution of (4.2) is given by

$$w(t) = T_t w_o, \qquad w_o\in D(A)\cap\cup \quad\text{and}\quad \|w(t)\| \leq \|w_o\|.$$

We can obtain a better estimate by considering the Liapunov functional $V = \frac{1}{2}\|w\|^2$ so

$$\dot{V} = -\langle w, (A+B)w\rangle = -\int_0^1 2vy_{xx}\left(\int_0^1 y_x^2 dx\right)dx + \xi\int_0^1 (v^2 + y_{xx}^2)dx + 2\xi\left(\int_0^1 y_x^2 dx\right)^2$$

from which it is easy to show that if

$$\|w_o\| < (1-\alpha)\sqrt{\frac{\xi(\pi^4+\xi^2)}{\pi^2}}$$

then $\|w(t)\| \leq e^{-\alpha\xi[1-\xi/(\xi^2+\pi^4)^{1/2}]t}\|w_o\|$, $w_o\in D(A)\cap\cup$

hence the origin is asymptotically stable in the sense of Liapunov.

Example 2. Forced Elastic Column

The non-dimensional linearized equation of motion of a fixed elastic column [9] is taken to be

$$(4.4)\quad y_{tt} + 2\xi y_t + y_{xxxx} + p(t)y_{xx} = q(x,t),\quad 0\leq x\leq 1,\quad t>0,\quad y = y_{xx} = 0 \text{ at } x=0,1.$$

If we consider the same Hilbert space as in Example 1 and assume that

$$\operatorname*{ess\,sup}_{t\in[0,T]} |p(t)| < p \text{ and } q\in L_2[0,T;L_2[0,1]]$$

we may write (4.4) as

$$\dot{w}+Aw+Bw = Q$$

where Aw is as in example 1, $B(t)w = [0,p(t)y_{xx}]^T$, $Q = [0,q(x,t)]^T$.

Thus $\|B(t)w\|^2 \leq p^2\int_0^1 y_{xx}^2 dx \leq \frac{p^2\pi^4}{\pi^4+\xi^2} \|w\|^2 = K^2\|w\|^2$

and $\|Q(\cdot,t)\|^2 = \int_0^1 q^2(x,t)dx.$

By considering the Liapunov functional $V = \|w\|^2$ (3.16) for the unperturbed system we have

$$\dot{V}(t) \leq -2\xi\int_0^1 (v^2+y_{xx}^2)dx \leq -2\xi\left[1-\frac{\xi}{(\xi^2+\pi^4)^{\frac{1}{2}}}\right]V = -2\mu V$$

Thus A generates a semi-group T_t such that $\|T_t\| \leq e^{-\mu t}$.

Hence for the perturbed system (3.17) gives

$$\|w(t)\| \leq e^{-(\mu-K)t}\|w_0\| + \int_0^t e^{-(\mu-K)(t-\rho)}\|Q(\cdot,\rho)\|d\rho$$

We note that it is easy to find a constant λ such that

$$\|w(t)\| \geq \lambda \sup_{x\in[0,1]} |y(x,t)|$$

and hence the above estimate will give a bound on the maximum displacement.

<u>Example 3. Panel Flutter</u>

The non-dimensional equations for panel flutter in a supersonic stream are [10]

(4.5) $\mu y_{tt} + y_t + dy_{xxxx} - fy_{xx} + My_x = 0,$

$y = y_{xx} = 0$ at $y = 0,1$, where M is the Mach number.

We set $w = [y,v]^T$ then (4.5) can be written as $\dot{w} + Aw + Bw = 0$

where $Aw = \left[-v,\frac{1}{\mu}v + \frac{d}{\mu}y_{xxxx} - \frac{f}{\mu}y_{xx}\right]^T$, $Bw = \left[0,\frac{M}{\mu}y_x\right]^T$.

Consider the Hilbert space normed by

$$V = \|w\|^2 = \int_0^1 \left[\mu v^2 + dy_{xx}^2 + fy_x^2 + vy + \frac{1}{2\mu}y^2\right]dx$$

then for the system with Bx = 0 we have $\dot{V} \leq -2\lambda V$ where

$$2\lambda = \frac{1}{\mu}\left(1-\frac{1}{\sqrt{4\mu(d\pi^4+f\pi^2+1/4\mu)}}\right)$$

and $\|Bw\|^2 = \mu\int_0^1\left(\frac{M}{\mu}y_x\right)^2 dx \leq K\|w\|^2$

where $K^{\frac{1}{2}} = M\pi/\sqrt{(d\pi^4+f\pi^2+1/4\mu)\mu}$

Thus from (3.17) the condition for stability is that

$$\sqrt{d\pi^4 + f\pi^2 + \frac{1}{4\mu}} - \frac{1}{2\sqrt{\mu}} > 2M\pi\mu^{\frac{1}{2}}.$$

The results of these examples could be improved upon by taking different Liapunov functionals.

REFERENCES

[1] T.Kato. "Perturbation Theory for Linear Operators", Springer (1966)

[2] T.Kato. "Accretive Operators and nonlinear evolution equations in Banach spaces". Proc. Symp. Pure Math. Amer. Math Soc. 18 Part 1 (1970) 138-161.

[3] K.Gustafson. "A Perturbation Lemma", Bull. Amer. Math. Soc. 72 (1966) 334-338.

[4] N.Okazawa. "A Perturbation Theorem for Linear Contraction Semi-groups on Reflexive Banach Spaces", Proc. Japan Acad. 47 (1971) 947-949.

[5] N.Okazawa. "Perturbations of Linear m-Accretive Operators", Proc. Amer. Math. Soc. 37 (1973) 169-174.

[6] R.Curtain and A.J.Pritchard, "The Infinite Dimensional Riccati Equation for Systems Defined by Evolution Operators", SIAM J. of Control (to appear).

[7] W.Blakeley and A.J.Pritchard. University of Warwick Control Theory Centre Report No.40 (to appear).

[8] J.La Salle and S.Lefschetz, "Stability by Liapunov's Direct method", Academic Press, (1961)

[9] R.H.Plaut and E.F.Infante, "Bounds on Motions of some Lumped and Continuous Dynamic Systems", Amer. Soc. Mech. Eng. Paper No. 71-APMW-3.

[10] P.C.Parks, "A Stability Criterion for Panel Flutter via the Second Method of Liapunov", AIAA J, (1966) 175-177.

ON THE PHYSICAL INTERPRETATION OF CERTAIN INNER PRODUCTS AS A GUIDE TO THE APPLICATION OF FUNCTIONAL ANALYSIS

G. Rieder
Institut für Technische Mechanik
Rheinisch-Westfälische Technische Hochschule
D-51 Aachen

1. Introduction: Abstraction means to extract the common property from otherwise possibly entirely different, real or irreal things. The best known results of such an abstraction are the cardinal numbers. So the number ten is the common property of ten nuts, ten jet planes, ten lions and ten unicorns, the ontological difference between a jet plane and a unicorn being of no relevance at all to the number ten.

But abstraction is no privilege of mathematics. The concept of energy, for example, undoubtedly belongs to physics. But it is not less abstract than an operator defined on an abstract space composed of abstract sets of numbers. Nobody has ever seen, heard of touched it; nevertheless it is recognized to-day to be a fundamental means of subsistence for mankind. And the danger to run short of it has made clear even to the public, that it is a measurable quantity, like sugar, wheat, coal, oil, hydrogen etc. But it does not exist in the same sense as these materials; really it is only a common measure for them, based on the gauge of mechanical work that can be gained form them by some irreversible process and depending on the process envisaged. So you get an entirely different energy quantity for the same mass of hydrogen by burning on earth or by reacting on the sun after the BETHE-WEIZSÄCKER-cycle. Note that energy is not measured by numbers, but by numbers times units ("dimension" in the physical sense [1]), and that its unit, the product of the length unit with the force unit, reflects the gauge on which this abstract entity is measured.

Now turn to functional analysis, known to be one of the most abstract branches of mathematics; only norms and scalar products, if there are any, have kept some perceptuality, as they are real numbers and so stick on a comparatively low level of abstraction. For the rest, we have abstract operators defined and acting on some set of abstract elements. As to the nature of the elements, there seems to be only one common agreement, namely that they are numbers or structures of numbers, if by this term we understand vectors, tensors, matrices, functions etc., that is, number and function n-tuples with certain invariance properties and transformation laws. But just this last agreement seems to me a prejudice, based on the opinion that mathematics must always have to do with numbers and only with numbers and concepts derived from them.

But in reality, without contesting that numbers will come with mathematical treatment sooner or later, we have to admit that the tools of mathematical logic may work on things that have no more to do with numbers than oxen before being counted. And this not only in the branch of mathematics, called the applied one, but also in physics and, to lesser extent, in chemistry, biology, economics, political sciences etc.

It is just the most abstract parts of mathematics that lend itself best to such non-numerical applications as it is always the most abstract concept that can be filled with the most different contents. And so we can drop the assumption that the elements of an abstract space must be number structures and try to find out if something useful is achieved if we regard as its elements, say, the possible states of a given physical system.

It should be stressed that we mean the states themselves, not their translation into number structures with the help of some system of units. In fact, this translation involves two distinct steps of abstraction. First we have to decide by which quantity we want to represent each state (e.g., whether we want to represent the state of an electrical OHM resistor network by branch currents, voltages, node sources and mesh circular EMFs, or node potentials and mesh circular currents). Each of these sets of quantities constitutes a "representation space" to the original "state space", and they can be grouped into pairs that are "dual" in a physical sense (e.g. for the electrical network yielding the power by pairing). And it is only in the second step, by measuring the choosen quantities by an appropriate unit that, after dropping the units and retaining the numbers, we arrive at a function space in the usual sense.

It is clear that we have to go through such a function space to find quantitative results. But what can be resolved beforehand, in the state space and then in the representation spaces? Can we use functional analysis immediately as a tool of physics and to see how far it bears before we apply the tools of classical and functional analysis in the usual sense? It is hoped that this will better clarify the physical content of the problem and its method of solution which may, for the physicist and engineer, otherwise be obscured by the elaborate and complicated reasoning of analysis. Not that this were unnecessary. We only try to defer it as far as possible.

For a norm we need a physical quantity. It should be a real scalar, as real scalars are real numbers with a unit and so have a one-to-one correspondence to real numbers. For the electric network, it is the root of the power, for linear elastic systems the root of the double elastic energy.

2. Fixed linear elastic systems. Linear elastic systems, be it finite or infinite networks of springs or continuous bodies, are defined by a linear connection between static and geometric quantities and the existence of an elastic energy, which means that work done in deforming an elastic system can always be retrieved in allowing

the relaxing system to do work on some other body. From linearity, it follows that the double elastic energy of a state $\mathfrak{f}$ of a given body must be a positive quadratic form which we symbolize by $\{\mathfrak{f},\mathfrak{f}\}$. It follows, moreover, that for the linear superposition of two states $\hat{\mathfrak{f}}$ and $\hat{\hat{\mathfrak{f}}}$, denoted by $\hat{\mathfrak{f}}+\hat{\hat{\mathfrak{f}}}$, we have distributivity and homogeneity, and thus

$$\{\hat{\mathfrak{f}}+\hat{\hat{\mathfrak{f}}},\hat{\mathfrak{f}}+\hat{\hat{\mathfrak{f}}}\} = \{\hat{\mathfrak{f}},\hat{\mathfrak{f}}\}+\{\hat{\hat{\mathfrak{f}}},\hat{\hat{\mathfrak{f}}}\}+\{\hat{\mathfrak{f}},\hat{\hat{\mathfrak{f}}}\}+\{\hat{\hat{\mathfrak{f}}},\hat{\mathfrak{f}}\} \tag{2.1}$$

The existence of a unique double energy implies commutativity

$$\{\hat{\mathfrak{f}},\hat{\hat{\mathfrak{f}}}\} = \{\hat{\hat{\mathfrak{f}}},\hat{\mathfrak{f}}\} \tag{2.2}$$

and so we arrive at the conclusion that the "interaction energy" $\{\hat{\mathfrak{f}},\hat{\hat{\mathfrak{f}}}\}$ has all necessary properties for a scalar product defining a HILBERT space $\mathscr{H}$, the space of states of the given elastic system.

How to generate now a non-zero state $\mathfrak{f}$? The first, classical, way is to let a system of forces p act on the system which generates a system of displacements u, both representing the same state $\mathfrak{f}'$. We write the double work done by p in the symbolic form

$$(p,u) = \{\mathfrak{f}',\mathfrak{f}'\} \tag{2.3}$$

For an elastic continuum, this is shorthand for the BETTI-MAXWELL integral theorem; for discrete structures [2], we have corresponding sum theorems, and combinations are also possible. We call any state $\mathfrak{f}'$ generated in this way a *load stress state*.

The other method is cutting, plastically deforming or filling in resp. removing of material, and then rewelding. After this process, first considered for certain singular states by VOLTERRA [3], also applicable in the continuum limit and to thermal stresses [4], the elastic system may be no more compatible in the relaxed state with its intrinsic EUKLIDean geometry or with its rigid supports. Therefore there will remain a non-zero state $\mathfrak{f}''$ without any forces acting on the system. We call it a ***self stress state***.

If we symbolize by η a measure of this geometric incompatibility and by χ the stress function generating the zero-force equilibrium stress field of the state $\mathfrak{f}''$, we have for a self stress state $\mathfrak{f}''$

$$(\chi,\eta) = \{\mathfrak{f}'',\mathfrak{f}''\} \tag{2.4}$$

For an elastic continuum, η symbolizes the incompatibility tensor and the misfit on the rigidly supported parts S_u of its surface S; for an elastic network (frame), the ensemble of VOLTERRA type discontinuities (BURGERS vectors and rotation discontinuities) in each mesh and misfits in supports. Well known examples of stress functions χ are AIRYs function for slices in plane deformation, PRANDTL's function for DE ST. VENANT torsion and shear bending, MINDLIN-SCHAEFER's vector function for slabs in bending and, for the three-dimensional elastic body, BELTRAMI-GWYTHER-FINZI's tensor function with its specializations by MORERA, MAXWELL, BLOCH, SCHAEFER and others [5] . In elastic networks, we have "mesh-dynames" of motor type, whose connection to AIRY's stress function in the plane case was already known to MAXWELL. Note that for representing self stress, χ must correspond to zero force on free parts of the boundary resp.

on free nodes. The set of all u of an elastic system constitutes a representation space for all f', and the set of all χ a representation space for all f''. Only both representation spaces together will represent any possible state

$$f = f' + f'' \tag{2.5}$$

of the system. The same applies to the sets of all p and all η, forming representation spaces dual to the above mentioned ones with respect to the scalar product (2.2).

The spaces $\mathcal{H}'$ of all load stress states f' and $\mathcal{H}''$ of all self stress spaces f'' are subspaces of the space $\mathcal{H}$ of all stress states f

$$\mathcal{H}' \oplus \mathcal{H}'' = \mathcal{H} \tag{2.6}$$

This follows from COLONNETTI's theorem

$$\{f,f\} = \{f',f'\} + \{f'',f''\} \tag{2.7}$$

or, equivalently after (2.1,2)

$$\{f', f''\} = 0 \tag{2.8}$$

That is, load stress and self stress states are orthogonal. This follows without any analysis from the fact that a linear system will react on additional forces independently of prevailing stresses. Therefore, to a linear elastic system with self stress one has to apply the same forces to get the same additional displacements as in zero stress. With (2.3), this implies (2.7).

On (2.8) rests the hypercircle method of PRAGER & SYNGE 1947 for bounding a state f with the help of two approximate states $f^{\sim}$ and $f^{\approx}$ complying with the geometric resp. static conditions of the problem [6, 7]. Fig. 1 gives a symbolic representation in two-dimensional projection [2] .

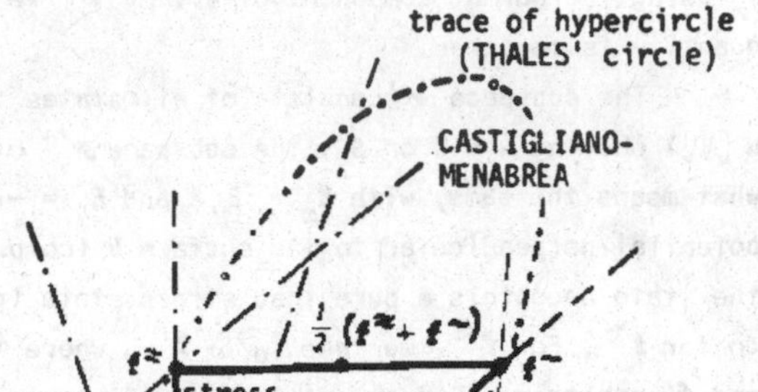

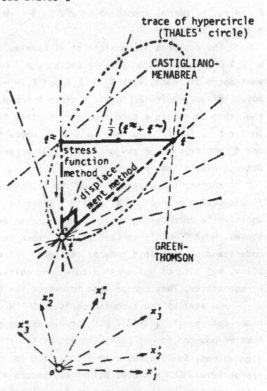

Figure 1. The hypercircle method. Dashed lines span $\mathcal{H}'$ and dash-dotted lines $\mathcal{H}''$ or subsets parallel to them.

Clearly the method can be extended to any linear system with a positive measure such as energy or power. Examples are generalized elastic continua or combinations with electric, magnetic or exchange fields or electric networks [2,8]. It should reflect the structure of the physical laws [9,10].

Example: Soap film B with constant surface tension G $[\mathrm{N{\cdot}m^{-1}}]$(conducting foil B

with constant surface conductivity G $[\Omega^{-1}]$) rigidly fixed in zero displacement (kept on zero potential) on part S_1 of its boundary and elastically fixed with C $[\mathrm{N{\cdot}m^{-2}}]$ for vertical displacement (connected to zero potential over a transverse line conductivity C $[\Omega^{-1}\mathrm{m}^{-1}]$) on the rest S_2 of the boundary and under the action of pressure Gf $[\mathrm{N{\cdot}m^{-2}}]$ (of a surface source density Gf $[\mathrm{A{\cdot}m^{-2}}]$). Be $\boldsymbol{E}$ [1] the gradient of the soap film surface ($\boldsymbol{E}[\mathrm{V{\cdot}m^{-1}}]$ the electric field), then the elastic interaction energy (electric interaction power) of two states $\hat{\mathbf{f}}$ and $\hat{\hat{\mathbf{f}}}$ is

$$\{\hat{\mathbf{f}},\hat{\hat{\mathbf{f}}}\} = G\int_B \boldsymbol{E}^2 \mathrm{d}B + \frac{G^2}{C}\int_{S_1} E_n^2 \,\mathrm{d}S \qquad [\mathrm{N{\cdot}m}]([\mathrm{W}]) \tag{2.9}$$

E_n being the normal component of $\boldsymbol{E}$. By [ ] we denote the unit in which the preceding quantity is measured.

The subspace $\mathscr{H}'$ consists of all states $\mathbf{f}'$ with a displacement u [m] (potential u [V]) obeying $u = 0$ on S_1, the subspace $\mathscr{H}''$ of all states $\mathbf{f}''$ obeying $G\nabla\cdot\boldsymbol{E} = 0$ or, what means the same, with $E_x = \partial_y A$ and $E_y = -\partial_x A$, A [m] ([V]) constituing a vector potential perpendicular to the surface B (compare also PRANDTL's stress function). As the state sought is a pure load stress state (pure potential state), any state $\mathbf{f}'$ will do for $\mathbf{f}^{\sim}$. For $\mathbf{f}^{\approx}$, we take $\mathbf{f}_0^{\approx} + \mathbf{f}''$, where $\mathbf{f}_0^{\approx}$ is some fixed state with $-\nabla\cdot\boldsymbol{E}_0 = F$ and $\mathbf{f}''$ arbitrary. For any state $\mathbf{f}'$ (2.9) apparently takes the form (2.3) of [11] . Other examples see, e.g., [7].

Another useful property of the HILBERT space is the possibility of defining a finite scalar product of a "regular" state of finite norm with a "singular" one with an infinite norm belonging to it only as the improper limit of a sequence of regular states. Most "GREEN's states" in continuous bodies are singular in this sense, if we understand by them in a generalized sense not only the state generated by a force in a point, but also of different dipoles and multipoles and corresponding singular self stress states. They can be used to extend the hypercircle method to pointwise bounds.

In statics the singular parts of GREEN's states can be eliminated, yielding good numerical results [12-17]. But for eigenvalue problems, the norm of the GREEN's state must be bounded to get pointwise bounds for the eigenfunctions from known bounds on the eigenvalues. So it takes more trouble and is impossible for infinite norm of the appropriate GREEN's state. But it may become finite by application of the operator, which can also be physically defined by the RAYLEIGH quotient of two different norms, one of which represents for stationary oscillations the maximum elastic energy and the other one the equal maximum of the cinetic energy [16-18].

For practical applications on continuous systems combination with conforming finite elements [19] seems most promising [14]. By renouncing the exact RITZ approximation for the complementary extremal problem and constructing, e.g., a state $\mathbf{f}^{\sim}$ in calculating a displacement by piecewise integration of the incompatible strain field of a CASTIGLIANO approximation $\mathbf{f}^{\approx}$ and averaging out the incompatibilities at the corners of each element, it is being tried to save computing labour [20] .

Remember also the classical VOIGT and PRANDTL-REUSS bounds on the average elastic constants of a polycrystal [21].

3. <u>States of different bodies.</u> Instead of comparing different states in the same body we can also compare states in different bodies which we identify by some arbitrary prescription.

We use here strain resp. stress identification, which means that two states in different bodies are said to be identical if their strain resp. stress is the same.

Now be there a given "real body" B^+ that may be non-uniform in its elastic properties, with prescribed forces and sources of self stress (e.g. dislocations or non-uniform heating). Then we take a fictive "basic body" B^O of the same shape in which the solution for such problems is known to exist; it may be uniform and isotropic. We want to show constructively the existence of a solution in the sense of the energy norm of the real or the basic body.

Any self stress problem can be replaced by a fictive load stress problem by introducing fictive forces and stresses, and vice versa by introducting fictive self stress sources and elastic strains, allowing two ways of solution.

We define an operator $\mathscr{A}'(\mathscr{A}'')$ on the space $\mathscr{H}_0'(\mathscr{H}_0'')$ of load (self) stress states of the basic body B^O by the following definition of $\mathfrak{k}_2' = \mathscr{A}'\mathfrak{k}_1'$ ($\mathfrak{k}_2'' = \mathscr{A}''\mathfrak{k}_1''$): Apply the prescribed forces p_1 (self stress sources η_1) to the body B^O, producing there a certain displacement u_1 (stress function χ_1). Force this displacement (stress function) on the body B^+ ; applying the forces p_2 (self stress sources η_2) required in B^+ to B^O will produce there the state $\mathfrak{k}_2'$ ($\mathfrak{k}_2''$). In other words, we have constructed the linear operator $\mathscr{A}'$ ($\mathscr{A}''$) by strain (stress) identification between the bodies B^O and B^+. After [22] a basic body B^O can always be constructed such that the operator $\mathscr{B}' = \mathscr{E}' - \mathscr{A}'$ ($\mathscr{B}'' = \mathscr{E}'' - \mathscr{A}''$) is contracting to zero if only the energy norm $\{\mathfrak{k},\mathfrak{k}\}_+^{\frac{1}{2}}$ in the real body B^+ is a full norm, which means that the tensor of elastic constants and its inverse exist everywhere in B^+. And this implies that the operator equation $\mathscr{A}'\mathfrak{k}' = \mathfrak{g}'$ ($\mathscr{A}''\mathfrak{k}'' = \mathfrak{g}''$) has a solution in the energy norm sense. But the displacement of $\mathfrak{k}'$ (stress function of $\mathfrak{k}''$) is the sought displacement in B^+ under the action of the forces producing $\mathfrak{g}'$ in B^O (stress function in B^+ under the action of the self stress sources producing $\mathfrak{g}''$ in B^O), and this completes the proof. To determine the class of functions or distributions representing the state sought we need, of course, mathematical analysis. But it seems sound practice first to make sure that a solution exists in a physical sense, and then go on to settle the other questions.

Comparing the double energy of states $\{\mathfrak{k},\mathfrak{k}\}_+$ and $\{\mathfrak{k},\mathfrak{k}\}_0$ identical by strain (stress) in B^+ and B^O leads to the definition of a RAYLEIGH quotient $\{\mathfrak{k},\mathfrak{k}\}_+/\{\mathfrak{k},\mathfrak{k}\}_0$ and by

$$(3.1) \qquad \{\hat{\mathfrak{k}}',\mathscr{A}'\hat{\mathfrak{k}}'\}_0 = \{\hat{\mathfrak{k}}',\hat{\mathfrak{k}}'\}_+ \quad (\{\hat{\mathfrak{k}}'',\mathscr{A}''\hat{\mathfrak{k}}''\}_0 = \{\hat{\mathfrak{k}}'',\hat{\mathfrak{k}}''\}_+)$$

to the conclusion that these operators must be symmetric on $\mathscr{H}'(\mathscr{H}'')$, their spectrum bounded by the extreme ratio of the local elastic energy densities on $\mathscr{H}$ within (0,1) by suitable choice of B^O . But these bounds may not be limits for $\mathfrak{k}$ being confined to $\mathscr{H}'$ or $\mathscr{H}''$ or a subset of them.

This must be remembered if $\{\mathfrak{k},\mathfrak{k}\}_+$ is only a half norm, i.e. if the elastic energy density may be zero for non-zero strain (stress) in B^+. Then the tools of analysis must be used at an earlier stage.

Take, e.g., for B^o a uniform infinite elastic medium, and for B^+ a finite part of it, bounded by the surface S. Equivalently, B^+ might be considered as infinite, with the tensor of elastic constants (its inverse) disappearing outside S. Now $\{f',f'\}_+^{\frac{1}{2}}$ ($\{f'',f''\}_+^{\frac{1}{2}}$) is a half norm, as it disappears for any strain (stress) outside S . Therefore we limit ourselves to states $f'\in\mathscr{H}_0'(S)$ ($f''\in\mathscr{H}_0''(S)$) caused by surface forces on S (displacement discontinuities in S). It can be shown, then, that the spectrum of the operator $\mathscr{A}_S'(\mathscr{A}_S'')$ (a) is bounded within (0,1) (b) has exactly two accumulation points in the plane symmetric to $\frac{1}{2}$ and three in space, all within (0,1) for physically realistic (isotropic or anisotropic) material (c) has a zero space of no more dimensions than B^+ has rigid movement components.

For summarizing results, see [22-25] and compare with [26,27]. In [26], also fictitions materials occur, and the well known LAURICELLA-SHERMAN equations can thus also be re-derived by them [28,29].

Theory predicts that those integral equations will have the best numerical properties where argument and image state are represented by quantities of the same physical nature, but this works not so simply for mixed problems requiring simultaneous application of static and geometric singularities, the corresponding operators not being symmetric, as by the difference of strain and stress identification

(3.2) $$\mathscr{H}_0'\ni f'\notin \mathscr{H}_+' \qquad (\mathscr{H}_0''\ni f''\notin \mathscr{H}_+'')$$

An operator adjoined in $\mathscr{H}_0$ has been established in [22] for $\{f,f\}_+^{\frac{1}{2}}$ being a full norm. Mixed boundary problems can be studied with the help of "transposition operators" mapping one representation space onto another one (e.g. displacements on S onto the corresponding force distributions on S in B^o and the inverse) and "comparison operators" dividing up force resp. dislocation distributions on S in B^o to B^+ and its complement B^-, without writing down the integral operators explicitly [30]. It turns out that the adjoined operator follows by interchanging the inner (outer) static boundary conditions with the outer (inner) geometric ones, which is in remarkable agreement with known results on mixed boundary problems in [26]. For simply supported plates, see [30]

Another interesting question is the connection of the COSSERAT spectrum of a boundary problem [32] with the spectrum of the corresponding operator $\mathscr{A}$. The author's conjecture is, that a point of the COSSERAT spectrum is hit every time when, on changing POISSON's constant, one of the eigenvalues of $\mathscr{A}$ becomes zero.

4. <u>A Sequence of comparison bodies.</u> The SAPONDZHYAN-BABUŠKA paradox says, that the solution for a sequence of surface loaded simply supported polygonal plates (slabs) approaching a smooth boundary will not tend to the smooth boundary solution for a real material, but only to the solution for the physically irreal POISSON's ratio $\nu = 1$ owing to the singularities occuring in the corners [33,34]. This is illustrated very

clearly by the numerical experiments of RAO & RAJAIAH [35] on loaded circular plates, using sums of regular displacement functions for approximation (fig.2). On augmenting the number of functions M on each segment (better approximation of corner singularities), the approximation curve sags down to the $\nu = 1$ solution, whilst on augmenting the number of sides N for constant M (neglecting the increasing corner singularity) the real solution is approached. And HANUŠKA's sequence [36], accounting for the singularity exactly throughout, tends straightforwardly to the $\nu = 1$ solution.

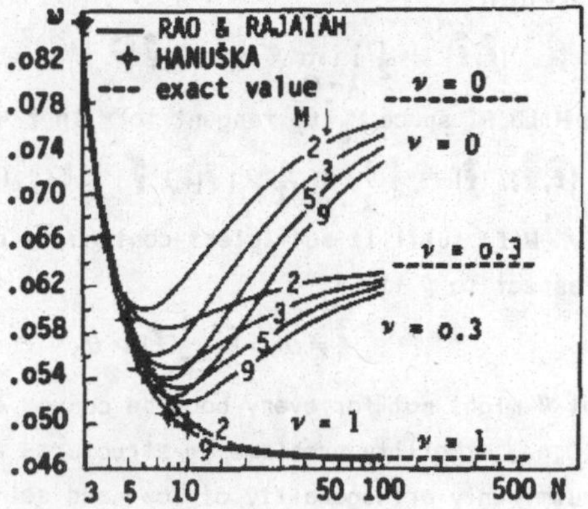

Figure 2: Numerical experiments on plate paradox (normed central deflection w).

The numerical experiment can be extended to a general proof by constructing approximations $\mathfrak{f}^{\sim}$ and $\mathfrak{f}^{\approx}$ after §2 for the sequence of polygonal plates which must, following HANUŠKA, contain the exact singularities. Only the proof of existence of converging sequences of upper and lower hypercircle bounds is not difficult in principle, but rather laborious. Rather crude comparison states, on the other hand, will do to confirm BABUŠKA's result that dropping BERNOULLI's hypothesis (REISSNER type plate theory) suffices to achieve convergence to the real solution (see preliminary report [38]). So the approximation problem, e.g., with finite elements [37], does not show up in procedures that neglect the singularity altogether.

5. <u>The scalar product in the non-linear case.</u> Complementary extremal principles having been studied by others, the present author will confine himself some remarks on extending the notion of the scalar product to the non-linear case. His proposal is to replace the state space by a set $\mathscr{W}$ of states $\mathfrak{f}$ and to assign to each $\mathfrak{f}$ a positive scalar quantity (double energy, power etc.) $W(\mathfrak{f})$. If $W(\mathfrak{f}) = 0$, $\mathfrak{f}$ is said to be a zero state. Addition, or rather several sorts of addition, can be defined with the help of representation spaces. By the symbol

(5.1) $$\mathfrak{f} = \hat{\mathfrak{f}} \overset{\xi}{(+)} \hat{\hat{\mathfrak{f}}}$$

we mean that the not yet specified representations $\hat{\xi}$ and $\hat{\hat{\xi}}$ are added to constitute the new state $\mathfrak{f} \in \mathscr{W}$ in the given body.

To define scalar products we remember that for a linear space $\mathscr{H}$

(5.2) $$\|\hat{\mathfrak{f}} + \lambda \hat{\hat{\mathfrak{f}}}\|^2 = \|\hat{\mathfrak{f}}\|^2 + 2\lambda \{\hat{\mathfrak{f}}, \hat{\hat{\mathfrak{f}}}\} + \lambda^2 \|\hat{\hat{\mathfrak{f}}}\|^2$$

and therefore

(5.3) $\qquad \{\hat{\mathfrak{f}},\hat{\hat{\mathfrak{f}}}\} = \frac{1}{2}\lim_{\lambda\to 0}\frac{\partial}{\partial\lambda}\|\hat{\mathfrak{f}}+\lambda\,\hat{\hat{\mathfrak{f}}}\|^2 \,,\; \|\hat{\hat{\mathfrak{f}}}\|^2 = \frac{1}{2}\frac{\partial^2}{\partial\lambda^2}\|\hat{\mathfrak{f}}+\lambda\,\hat{\hat{\mathfrak{f}}}\|^2$

Thus a HILBERT space $\mathscr{H}_\xi(\mathfrak{f})$ tangent to $\mathscr{W}$ in $\mathfrak{f}$ with respect to ξ can be defined by

(5.4) $\{\hat{\mathfrak{f}},\hat{\hat{\mathfrak{f}}}\}_\xi(\hat{\mathfrak{f}}) = \frac{1}{2}\lim_{\lambda\to 0}\frac{\partial}{\partial\lambda}W(\hat{\mathfrak{f}}\overset{\xi}{(+)}\lambda\,\hat{\hat{\mathfrak{f}}})\,,\; \|\hat{\hat{\mathfrak{f}}}\|^2_\xi(\hat{\mathfrak{f}}) = \frac{1}{2}\lim_{\lambda\to 0}\frac{\partial^2}{\partial\lambda^2}W(\hat{\mathfrak{f}}\overset{\xi}{(+)}\lambda\,\hat{\hat{\mathfrak{f}}})$

if only $W(\mathfrak{f})$ fulfills sufficient continuity conditions. $\mathscr{W}$ is said to be convex in $\hat{\mathfrak{f}}$ with respect to ξ if

(5.5) $$\hat{\hat{\mathfrak{f}}} \neq 0 := \|\hat{\hat{\mathfrak{f}}}\|^2_\xi(\hat{\mathfrak{f}}) > 0$$

The set $\mathscr{W}$ might not for every body be convex everywhere with respect to every ξ ; take, e.g., stability problems in structures and the PEIERLS dislocation [3] in a continuum. Only orthogonality of load and self stress states subsists in the limited, non-commutative sense

(5.6) $$(\sigma'',\varepsilon') = 0$$

at least for small deformations.

For the hypercircle method, it is proposed to compute bounds, e.g., in $\mathscr{H}_\varepsilon(\mathfrak{f}^{\sim})$ or $\mathscr{H}_\sigma(\mathfrak{f}^{\approx})$. If $\mathscr{W}$ is conex with respect to the ξ used and bounds are known for the difference of scalar products and norms in the different tangent spaces $\mathscr{H}_\xi(\mathfrak{f})$ within a ball around $\mathfrak{f}^{\sim}$ resp. $\mathfrak{f}^{\approx}$ in $\mathscr{H}_\varepsilon(\mathfrak{f}^{\sim})$ resp. $\mathscr{H}_\sigma(\mathfrak{f}^{\approx})$ [39] they may be extended to practically useful bounds and even made to converge if the different tangent spaces approach each other on contracting the ball on $\mathfrak{f}^{\sim}$ resp. $\mathfrak{f}^{\approx}$.

To find pointwise bounds, we can use the scalar product with the appropriate GREEN's state in a tangent space. As the singularities cancel out in the static case [12-16] only bounds for the regular part confined to some ball in $\mathscr{H}_\varepsilon(\mathfrak{f}^{\sim})$ resp. $\mathscr{H}_\sigma(\mathfrak{f}^{\approx})$ are needed. Moreover, by the well known relation

(5.7) $$\sigma_i = \frac{\partial\Phi(\varepsilon)}{\partial\varepsilon_i}\,,\quad W = \int_B 2\Phi(\varepsilon)\,dB$$

$\Phi(\varepsilon)$ being the local energy density expressed by ε , and ε_i as well as the integral understood in a generalized sense, applicable to continuous and discrete structures, we can prove

(5.8) $$\mathscr{H}_\sigma(\mathfrak{f}) = \mathscr{H}_\varepsilon(\mathfrak{f})$$

For, inserting (5.7) into (5.4), we have with summation convention in i and j ,

(5.9)
$$\frac{1}{2}\lim_{\lambda\to 0}\frac{\partial}{\partial\lambda}W(\hat{\mathfrak{f}}\overset{\varepsilon}{(+)}\lambda\hat{\hat{\mathfrak{f}}}) = \lim_{\lambda\to 0}\frac{\partial}{\partial\lambda}\int_B 2\Phi(\hat{\varepsilon}+\lambda\hat{\hat{\varepsilon}})\,dB$$

$$= \lim_{\lambda\to 0}\int_B(\sigma_i(\hat{\varepsilon})+\lambda\sigma_i(\hat{\hat{\varepsilon}}))\,\hat{\hat{\varepsilon}}_i\,dB = \int\hat{\sigma}_i\hat{\hat{\varepsilon}}_i\,dB$$

$$\frac{1}{2}\lim_{\lambda\to 0}\frac{\partial}{\partial\lambda}W(\hat{\mathfrak{f}}\overset{\varepsilon}{(+)}\lambda\hat{\hat{\mathfrak{f}}}) = \int_B\hat{\hat{\sigma}}_i\hat{\varepsilon}_i\,dB$$

If, now, the relation

(5.10) $$d\sigma_i(\varepsilon) = C_{ij}(\varepsilon)\,d\varepsilon_j \qquad C_{ij}(\varepsilon) = \frac{\partial \sigma_i(\varepsilon)}{\partial \varepsilon_j} = \frac{\partial^2 \Phi(\varepsilon)}{\partial \varepsilon_i\, \partial \varepsilon_j} = \frac{\partial \sigma_j(\varepsilon)}{\partial \varepsilon_i} = C_{ji}(\varepsilon)$$

can be inverted, then the same result comes by using $\Psi(\sigma)$ instead of $\Phi(\varepsilon)$, and (5.8) is proved. Finally, we find for small deformations, inserting the corresponding differential resp. integral expressions instead of ε resp. σ into (5.7-9)

(5.11) $$\mathfrak{f} = \mathfrak{f}' \in \mathscr{W}' := \mathscr{H}'_u(\mathfrak{f}') = \mathscr{H}'_\varepsilon(\mathfrak{f}') = \mathscr{H}'_p(\mathfrak{f}') \;\cdot\; \mathfrak{f} = \mathfrak{f}'' \in \mathscr{W}'' := \mathscr{H}''_\chi(\mathfrak{f}'') = \mathscr{H}''_\sigma(\mathfrak{f}'') = \mathscr{H}''_\eta(\mathfrak{f}'')$$

so that we can write

(5.12) $$\mathscr{H}_\varepsilon(\mathfrak{f}) = \mathscr{H}_\sigma(\mathfrak{f}) = \mathscr{H}'_u(\mathfrak{f}') \oplus \mathscr{H}''_\chi(\mathfrak{f}'') = \mathscr{H}'_p(\mathfrak{f}') + \mathscr{H}''_\eta(\mathfrak{f}'') = \mathscr{H}(\mathfrak{f})$$

at least in the geometrically linear case.

6. Conclusion. It was not the author's aim to produce new concepts, really the reader will find most of the preceding notions in the literature on mathematical physics, more or less hidden in sumptuous mathematical clothing. He just wanted to strip the mechanical core of its mathematical wrapping and so to show that functional analysis in mechanics is not bound to create a new branch of "pure numerical analysis" [40]. To take another picture: just as in tunnelling from both sides of the mountain to meet in the middle the same problem can be attacked from the mathematical and from the mechanical side. Both ways must meet in the correct solution, but, in the author's impression, the mechanical side is often greatly undermanned.

REFERENCES

1. QUADE, W.: Über die algebraische Struktur des Größenkalküls in der Physik. Abh. Braunschweig. Wiss. Ges. 13 (1961), pp. 24-72.
2. RIEDER, G.: On MAXWELL's, KLEIN's and WIEGHARDTS's Stress Functions for Discontinuous Systems. Reports DCAMM, Technical University of Denmark, to appear.
3. LARDNER, R.W.: Mathematical Theory of Dislocations and Fracture. University of Toronto Press 1974.
4. KRÖNER, E.: Kontinuumstheorie der Versetzungen und Eigenspannungen. Springer-Verlag 1958.
5. RIEDER, G.: Die Randbedingungen für den Spannungsfunktionentensor an ebenen und gekrümmten Oberflächen. Österr. Ing.-Archiv 18 (1964), pp. 208-243.
6. PRAGER, W. and SYNGE, J.L.: Approximations in Elasticity based on the concept of Function Space. Quarterly Appl. Math. 5 (1947), pp. 241-269.
7. VELTE, W.: Komplementäre Extremalprobleme bei nichtlinearen Randwertaufgaben. Reprint Nr. 7, Math. Inst. Universität Würzburg 1975.
8. RIEDER, G.: On the Decomposition of the HILBERT Space of the Fields in a Linear Continuum into Two Orthogonal Subspaces. In "Trends in Elasticity and Thermoelasticity", Witold Nowacki Anniversary Volume, ed. E. CZARNOTA-BOJARSKI, M. SOKOLOWSKI and H. ZORSKI, Wolters-Noordhoff Publ., Groningen 1971.
9. MINAGAWA, S.: On the Variational Principles for Deformation from the Standpoint of the Strain Space and the Stress Space. RAAG Research Notes, Univ. of Tokyo, Ser. 3, No. 53 (1962).
10. TONTI, E.: On the Mathematical Structure of a Large Class of Physical Theories.

Rend. Accad. Naz. Lincei, Class.Sci.fis.mat.nat. (8) 52 (1972), pp. 48-56.

11. VELTE. W.: Komplementäre Extremalprobleme bei nichtlinearen Randwertaufgaben. In "Numerische Lösung nichtlinearer partieller Differential- und Integro-Differentialgleichungen", ed. R. ANSORGE and W. TÖRNIG, Lecture Notes in Mathematics Nr. 267, Springer-Verlag 1972, pp. 323-335.

12. WEBER, C.: Eingrenzungen von Verschiebungen und Zerrungen mit Hilfe der Minimalsätze. Z. Angew. Math. Mech. 22 (1942), pp. 126-130, pp. 130-136.

13. STUMPF, H.: Eingrenzungsverfahren in der Elastomechanik. Westdeutscher Verlag, Köln und Opladen 1970, Nr. 2116, Forschungsberichte NRW.

14. SCHOMBURG, U.: Lokale Eingrenzung bei Randwertproblemen der Elastizitätstheorie mit der Methode der finiten Elemente. Thesis, Aachen 1972.

--: The Finite Element Method and Local Bounds for Boundary Value Problems of Elastic Structures.Preprint, 2nd Int. Conf. on "Structural Mechanics in Reactor Technology" in Berlin 1973, Vol. V, M 2/9.

ANTES, H.: Punktweise Eingrenzung bei flachen Schalen. To appear.

15. YOUNG, R.C. and MOTE, C.D.Jr.: Solution of Mixed Boundary Value Problems with Local Error Bound by the Finite Element Method. Computer Methods in Appl.Mech. Engng. 2 (1973), pp. 159-183.

--: Local Error Bounds in Mathematical Physics by Finite Element Methods. In "Variational Methods in Engineering", ed. C.A. BREBBIA and H. TOTTENHAM, University of Southampton Press 1973, Vol. I, pp. 2/92-108.

16. RIEDER, G.: On Bounds in Linear Elasticity and Potential Theory. In "Continuum Mechanics and Related Problems of Analysis", to the 80th Birthday of N.I. MUSKHELISHVILI, ed. L.I. SEDOV and G.K. MIKHAILOV, "Nauka" Publ.House,Moscow 1972,pp.409-417.

--: Eingrenzungen in der Elastizitäts- und Potentialtheorie. Z.Angew.Math.Mech. 52 (1972), pp. T340-T347.

17. STUMPF,H.: Lower Bounds to the Frequencies of Continuous Elastic Systems. Same Vol. as 1st title [16], pp. 667-676.

18. BIEHL, F.-J.: Bounding of the Electric Field in Wave Guides. Work in progress, Aachen 1975.

19. ZIENKIEWICZ, O.C.: The Finite Element Method in Engineering Science. McGraw Hill, London 1971, German edition VEB Fachbuchverlag, Leipzig and Carl Hanser Verlag, München 1975.

20. SCHOMBURG, U. and HABALI, S.: Bounds by Hypercircle with Finite Element Method, Simplified Calculation. Work in progress, Aachen 1975.

21. DEDERICHS, P.H. and ZELLER, R.: Elastische Konstanten von Vielkristallen. Berichte der Kernforschungsanlage Jülich, Jül-877 - FF, Juli 1972.

22. RIEDER, G.: Iterationsverfahren und Operatorgleichungen in der Elastizitätstheorie. Abhandlungen Braunschweig.Wiss. Ges. 14/2 (1962), pp. 109-343.

23. HEISE, U.: Formulierung und Ordnung einiger Integralverfahren für Probleme der ebenen und räumlichen Elastostatik unter besonderer Berücksichtigung mechanischer Gesichtspunkte. Habilitation Thesis, Aachen 1975.

24. RIEDER, G., HEISE, U., PAHNKE, U., ANTES, H., GLAHN, H. and KOMPIŠ, V.: Berechnung von elastischen Spannungen in beliebig krummlinig berandeten Scheiben und Platten. Westdeutscher Verlag, Köln and Opladen, in preparation.

25. RIEDER, G.: Über Eingrenzungsverfahren und Integralgleichungsmethoden für elastische Scheiben, Platten und verwandte Probleme. Wiss. Z. Hochschule für Architektur und Bauwesen Weimar 19 (1972), pp. 217-222.

26. KUPRADZE, V.D., GEGELIA, T.G., BASHELEISHVILI, M.O. and BURCHULADZE, T.V.: Three-dimensional Problems in the Mathematical Theory of Elasticity (in Russian). University of Tbilisi Publishing House 1968. New edition to appear in "Nauka" Publishing House, Moscow 1975 (Novye Knigi No.2-1975,(65)). See also W. NOWACKI in Math. Rev. 38 (1969), ##2985.

27. MAZYA, V.G. and SAPOZHNIKOVA, V.G.: Remark on the Regularization of a System of the Isotropic Theory of Elasticity (in Russian). Vestnik Leningradskogo Universiteta, Ser.mat.mech.astron. 1964:7, pp. 165-167.

28. RIEDER, G.: On KUPRADZE's generalized Stress, its Physical Meaning and its Application to certain Integral Operators of Plane Elasticity. Paper dedicated to H.PARKUS, Vienna 1974. To be published.

29. KOMPIŠ, V.: Integralgleichungsverfahren zur Lösung der ersten Randwertaufgabe der ebenen Elastizitätstheorie. Thesis, Aachen 1970.

--: Integralgleichungen in Festigkeitsproblemen. Paper to be read on the Schiffstechnisches Symposium Rostock 1975. To be published by the University of Rostock.

30. RIEDER, G.: Adjungierte Integralgleichungen in der Elastizitätstheorie. (See [29]).

31. GLAHN, H.: Eine Integralgleichung zur Berechnung gelenkig gelagerter Platten bei krummem Rand. Ingenieur-Archiv 44 (1975), pp. 189-198.

32. MIKHLIN, S.G.: The Spectrum of a Pencil of Operators in the Theory of Elasticity (in Russian). Uspekhi Mat. Nauk 28:3(171) (1973), pp.43-82 (For English translation see Russian Math. Surveys, London).

33. SAPONDZHYAN, O.M.: Bending of a Simply Supported Polygonal Plate (in Russian). Izv. Akad. Nauk Arm. SSR 5 (1952), pp. 29-46.

34. BABUŠKA, I.: Stabilität des Definitionsgebietes mit Rücksicht auf grundlegende Probleme der Theorie der partiellen Differentialgleichungen auch im Zusammenhang mit der Elastizitätstheorie (in Russian, German summary) I.-II., Čechoslovackij matematičeskij žurnal 11/86 (1961), pp. 76-105, pp. 165-203.

--: Die Stabilität mit Rücksicht auf das Definitionsgebiet und die Frage der Formulierung des Plattenproblems /Vorbericht/. Aplikace matematiky 7 (1962), pp. 463-470.

35. RAO, A.K., RAJAIAH, K.: Polygon-circle Paradox of Simply Supported Thin Plates Under Uniform Pressure. AIAA Journal 6 (1968), pp. 155-156.

36. HANUŠKA, A.: On the Validity of the Solution of Simply Supported Uniformly Loaded Polygonal Plates. USTARCH-SAV Bratislava 1969, unpublished.

37. CHERNUKA, M.W., COWPER, G.R., LINDBERG, G.M. and OLSON, M.D.: Finite Element Analysis of Plates with Curved Edges. Int. J. Numerical Methods in Engineering 4 (1972), pp. 49-65.

38. RIEDER, G.: On the Plate Paradox of SAPONDZYAN and BABUŠKA. Mech. Res. Comm. 1 (1974), pp. 51-53.

39. GHIZZETTI, A.(ed.): Theory and Applications of Monotone Operators. Edizione "Odersi", Gubbio 1969.

40. FOX, L. Why do I need Functional Analysis when I have got on very well so far without it? The Inst. Math. Appl. Bull. 10 (1974), pp. 112-113.

The author thanks Mrs. Zastrow, Mr. Osthoff and Mr. Hirsch for carefully preparing the manuscript. Thanks are also due to the Landesamt für Forschung Nordrhein-Westfalen for supporting part of the work cited.

BRANCHING AND STABILITY FOR NONLINEAR SHELLS

D. Sather
University of Colorado
Boulder, Colorado 80302 USA

1. Introduction. The general nonlinear theory of stability for thin, elastic shells due to Koiter (e.g., see [10]) can be described in its simplest setting as follows. The relevant displacements of the shell are assumed to be of the form $w = \xi u + v$, where u is the "buckling mode", ξ is the (unknown) amplitude of u , and v is in some sense orthogonal to u . For a fixed value of ξ , the minimum value of the potential energy $E = E(w)$ then determines v as a function of ξ , $\xi < \xi_0$, which in turn determines E as a function of ξ only. The extremum values of $E(\xi u + v(\xi))$ are then obtained and the sign of the second derivative of E with respect to ξ determines whether the corresponding equilibrium state is stable or unstable. Clearly, such an approach can be formulated in terms of some of the standard methods of nonlinear functional analysis.

A second useful approach to studying buckling problems for shells can be described roughly as follows. By employing the modern theory of partial differential equations and some standard theorems of functional analysis, the coupled differential equations for a number of well-known shell buckling problems can be written as a pair of uncoupled nonlinear operator equations of the form

$$(1.1a) \qquad f = Kw - \frac{1}{2} B(w,w)$$

$$(1.1b) \qquad w = -Kf + B(w,f) - \lambda Aw \qquad f,w \in \mathcal{H} .$$

Here $\mathcal{H}$ is an appropriate real Hilbert space with inner product $(\cdot,\cdot)$, w is typically a measure of the normal displacement of the shell from its initial configuration, f is some sort of stress or shear function, λ is a measure of the load parameter, and the linear operators $A:\mathcal{H} \to \mathcal{H}$ and $K:\mathcal{H} \to \mathcal{H}$, and the symmetric bilinear operator $B:\mathcal{H} \times \mathcal{H} \to \mathcal{H}$ are all completely continuous; the operator K is usually related to a geometrical property of the shell such as the curvature. For example, for various boundary conditions, the von Kármán-Donnell equations (e.g., see [4;22]) and the general equations for shallow shells in [11 ,§11] as well as some of the more special equations for axisymmetric deformations of shells of revolution (e.g., see [2;14;15]) can all be written in the form (1.1). If we now substitute f from (1.1a) into (1.1b) we obtain a single operator equation for the determination of the buckled states of the shell, namely,

(1.2) $\quad w - \lambda Aw + K^2w + F(w) + G(w) = 0 \qquad w \in \mathscr{H}$

where $F(w) = \frac{1}{2}KB(w,w) + B(w,Kw)$ and $G(w) = \frac{1}{2}B(w,B(w,w))$ are homogeneous operators of degree two and three, respectively. A constructive approach to finding small solution branches of equations such as (1.2) near the buckling load λ_1 is provided by the Lyapunov-Schmidt method which, as in the approach of Koiter, consists of projecting (1.2) into a pair of subspaces of $\mathscr{H}$, solving a nonlinear problem in the infinite-dimensional subspace, and then solving a second well-set problem in the appropriate finite-dimensional subspace. We shall describe the Lyapunov-Schmidt method in more detail in Section 2 (see also [13;16;18;21]).

In the present paper we consider an equation which is an important special case of (1.2), namely,

(*) $\quad w - \lambda Aw + \sigma^2A^2w + \sigma Q(w) + C(w) = 0 \qquad w \in \mathscr{H}$

where λ is typically a load parameter, σ is typically a geometric parameter of the shell, $A:\mathscr{H} \to \mathscr{H}$ is a positive, compact, self-adjoint operator, and $Q:\mathscr{H} \to \mathscr{H}$ and $C:\mathscr{H} \to \mathscr{H}$ are continuous, homogeneous operators of degree two and three, respectively, which are the (strong) gradients of the functionals $q(w) = \frac{1}{3}(Q(w),w)$ and $c(w) = \frac{1}{4}(C(w),w)$. That is, there exists a functional $r:\mathscr{H} \times \mathscr{H} \to R^1$ such that, for w and h in $\mathscr{H}$,

(1.3) $\quad q(w + h) - q(w) = (Q(w),h) + r(w,h)$

and $r(w,h)/\|h\| \to 0$ as $\|h\| \to 0$; a similar relationship holds for the functional c and the operator C. The assumption that Q and C are the gradients of the particular functionals q and c is a consequence of the homogeneity of Q and C. An equation of the form (*) can be derived, for example, in buckling problems for uniformly compressed complete spherical shells and spherical caps, and axially compressed cylindrical panels and shells (e.g., see [9;19] and also Sec. 5).

The outline of the paper is as follows. Since the "potential energy" associated with (*) is a functional E of the form

(1.4) $\quad E(w) = \frac{1}{2}(L_\lambda w,w) + \frac{\sigma}{3}(Q(w),w) + \frac{1}{4}(C(w),w) \qquad w \in \mathscr{H}$

where $L_\lambda = I - \lambda A + \sigma^2A^2$, we begin in Section 2 by showing that, for shell buckling problems whose potential energy E is given by (1.4), the stability method of Koiter is equivalent to the Lyapunov-Schmidt method for determining nontrivial solution branches of (*) together with some sort of a stability analysis of the resultant solution branches. In Section 3 we describe some general branching results which are available for the study of (*) , and in Section 4 we discuss the stability of some of the solution branches obtained in Section 3; the stability theorems are established by means of elementary results of nonlinear functional analysis and are consistent with the usual concept of "linearized stability". In Section 5 we outline some applications of the results in Section 3 and Section 4 to some specific buckling problems;

other applications to buckling problems for (complete) cylinders and spheres will appear in some forthcoming joint work with Professor George H. Knightly. In all of these applications, branching and stability results are obtained which are not only more precise but also in many cases more complete than results obtained previously by various formal methods of classical analysis alone. For example, in the problem of an axially-compressed, simply-supported, "narrow" cylindrical panel, we obtain a description of all stable and unstable buckled states near the buckling load which depend continuously on the load parameter λ ; in addition, the buckled states obtained exhibit a number of important features such as the asymptotic forms of the buckled states near the buckling load, the exchange of stabilities at the buckling load, and stable "outward" buckling at the ends of the panel.

2. The branching equations. In this section we describe the Lyapunov-Schmidt method for (*) and derive the "branching equations" associated with (*) .

We begin by describing the linear theory for (*) . We will say that λ_0 is an eigenvalue of the linear, self-adjoint operator $L_\lambda = I - \lambda A + \sigma^2 A^2$ if $L_\lambda u = 0$ for some $u \in \mathcal{H}$, $u \neq 0$. The eigenvalues of L_λ are related to the characteristic values of A in the following way. Since

(2.1) $$L_\lambda = (I - \mu^+ A)(I - \mu^- A) = (I - \mu^- A)(I - \mu^+ A)$$

where $\mu^\pm = \frac{1}{2}[\lambda \pm (\lambda^2 - 4\sigma^2)^{\frac{1}{2}}]$, it follows that λ_0 is an eigenvalue of L_λ if and only if at least one of the two numbers μ^+ or μ^- is a characteristic value of A (i.e., $w - \mu^\pm Aw = 0$ for some $w \in \mathcal{H}$, $w \neq 0$), that is, if and only if $\lambda_0 = \mu_0 + \sigma^2 \mu_0^{-1}$ for some characteristic value μ_0 of A . Moreover, if λ_0 is an eigenvalue of L_λ then the corresponding eigenvectors of A (i.e., the eigenvectors for μ^+, μ^- or both) are also eigenvectors of L_{λ_0} and span the null space $\mathcal{N} \equiv \mathcal{N}(L_{\lambda_0})$ of L_{λ_0} . In particular then the relationship $\lambda = \mu + \sigma^2 \mu^{-1}$ shows that $\lambda_0 > 0$ and that the multiplicity of λ_0 (i.e., the dimension of $\mathcal{N}$) is that of either μ^+ or μ^- (whichever one is a characteristic value of A) unless μ^+ and μ^- are both distinct characteristic values of A , in which case the multiplicity of λ_0 is the sum of the multiplicities of μ^+ and μ^- . The relationship $\lambda = \mu + \sigma^2 \mu^{-1}$ also shows that the multiplicity of λ_0 is the sum of the multiplicities of μ^+ and μ^- if and only if σ^2 is the product of the distinct characteristic values μ^+ and μ^- ; hence, the multiplicity of the smallest (positive) eigenvalue $\lambda_1 = \lambda_1(\sigma)$ of L_λ is the sum of the multiplicities of μ^+ and μ^- if and only if $\sigma^2 = \mu^+\mu^-$ and μ^+ and μ^- are successive characteristic values of A .

The Lyapunov-Schmidt method for (*) can be described as follows. For fixed $\sigma > 0$, let $\lambda_0 = \lambda_0(\sigma)$ be an eigenvalue of L_{λ_0} and let $\mathcal{N} \equiv \mathcal{N}(L_{\lambda_0})$ denote the null space of L_{λ_0} ; we assume that $\mathcal{N}$ is n-dimensional. Then $\mathcal{H}$ can be written as the direct sum $\mathscr{W} = \mathcal{N} \oplus \mathcal{N}^{\perp}$ where $\mathcal{N}^{\perp}$ is the orthogonal complement of $\mathcal{N}$ in $\mathscr{W}$. Let P denote the orthogonal projection of $\mathscr{W}$ onto $\mathcal{N}$. The operator I - P is a projection of $\mathscr{W}$ onto $\mathcal{N}^{\perp}$ and an element $w \in \mathscr{W}$ can be written as $w = u + v$ where $u = Pw \in \mathcal{N}$ and $v = (I - P)w \in \mathcal{N}^{\perp}$. Since $\mathcal{N}$ is spanned by the eigenvectors of μ^{+}, μ^{-} or both μ^{+} and μ^{-}, A maps $\mathcal{N}$ into itself so that AP = PA (e.g., see [7, p. 278]). Hence, finding solutions of (*) is equivalent to finding solutions of the two equations obtained by projecting (*) into $\mathcal{N}^{\perp}$ and $\mathcal{N}$, namely,

(2.2a) $$L_{\lambda}v + (I - P)[\sigma Q(u + v) + C(u + v)] = 0$$

(2.2b) $$L_{\lambda}u + P[\sigma Q(u + v) + C(u + v)] = 0$$

Let $\{z_1,\ldots,z_n\}$ be any basis for $\mathcal{N}$ such that $(Az_i,z_j) = \delta_{ij}$ $(i,j = 1,\ldots,n)$ where $\delta_{ij} = 0$ if $i \neq j$ and $\delta_{ij} = 1$ if $i = j$. Then $u \in \mathcal{N}$ can be written as

$$u = \sum_{j=1}^{n} \xi_j z_j \qquad \xi = (\xi_1,\ldots,\xi_n) \in R^n$$

and the system (2.2) is equivalent to the system

(2.3a) $$L_{\lambda_0} v - \tau A + (I - P)[\sigma Q(\sum_{j=1}^{n} \xi_j z_j + v) + C(\sum_{j=1}^{n} \xi_j z_j + v)] = 0$$

(2.3b) $$-\tau\xi_i + (\sigma Q(\sum_{j=1}^{n} \xi_j z_j + v) + C(\sum_{j=1}^{n} \xi_j z_j + v), z_i) = 0 \qquad (i = 1,\ldots,n)$$

where $\tau = \lambda - \lambda_0$. The Lyapunov-Schmidt method for solving equation (*) is to first solve (2.3a) for $v = v(\xi,\tau)$ in $\mathcal{N}^{\perp}$ and then, if possible, to solve (2.3b) for $\xi = \xi(\tau)$ in R^n.

A standard argument using the contraction mapping principle (e.g., see [21, p. 19]) shows that, for $\xi < \rho_0$ and $\tau < \tau_0$ (ρ_0 and τ_0 sufficiently small), there exists a unique solution $v = v(\xi,\tau)$ of (2.3a) which is analytic in ξ and τ. Moreover, since Q is homogeneous of degree two, v is "higher order" in the sense that there exists a constant c depending only on ρ_0 and τ_0 such that

(2.4) $$\|v(\xi,\tau)\|^2 \leqslant c|\xi|^2$$

for $|\xi| < \rho_0$ and $|\tau| < \tau_0$ (e.g., see [18, p. 230]). Thus, finding solutions $w = w(\lambda)$ of equation (*) in $\mathscr{W}$ near $w = 0$ is equivalent to finding sufficiently small solutions $\xi = \xi(\tau)$ in R^n of the system of analytic equations

(2.5) $$-\tau\xi_i + \sigma(Q(\sum_{j=1}^{n} \xi_j z_j), z_i) + h^i(\xi,\tau) = 0 \qquad (i = 1,\ldots,n)$$

where

$$h^i(\xi,\tau) = \sigma(Q(\sum \xi_j z_j + v(\xi,\tau)) - Q(\sum \xi_j z_j), z_i) + (C(\sum \xi_j z_j + v(\xi,\tau)), z_i) \quad .$$

The equations in (2.5) are called the "branching equations" associated with (*).

<u>Remark 2.1</u>. Since Q is Frechet differentiable near w = 0 and C is

homogeneous of degree three, it follows easily from (2.4) that the analytic functions h^i are "higher order" in the sense that if $|\tau| < \tau_0$ then $h^i(\xi,\tau)/|\xi|^2 \to 0$ as $|\xi| \to 0$.

The form of the branching equations in (2.5) and the homogeneity of Q suggest for $\tau > 0$ the substitution

(2.6) $\xi = \tau\beta \qquad \beta \in \mathbb{R}^n$.

(For negative τ one sets $\xi = |\tau|\beta$ and proceeds in an analogous way). After dividing by τ^2 the system becomes

(2.7) $$F^i(\beta,\tau) \equiv -\beta_i + \sigma(Q(\sum_{j=1}^{n} \beta_j z_j), z_i) + \tau^{-2} h^i(\tau\beta,\tau) = 0 \qquad (i = 1,\ldots,n) .$$

Using then Remark 2.1 and formally setting $\tau = 0$ in (2.7), we obtain the closely related system

(2.8) $$f^i(\beta) \equiv -\beta_i + \sigma(Q(\sum_{j=1}^{n} \beta_j z_j), z_i) = 0 \qquad (i = 1,\ldots,n) .$$

It is clear that every solution $\beta = \beta(\tau)$ of (2.7) generates a solution $\xi = \xi(\tau)$ of the branching equations in (2.5) by means of the substitution (2.6). On the other hand, if β^* is a nontrivial solution of (2.8) from which we can obtain a solution of (2.7) by an argument involving some sort of an implicit function theorem, then the substitution (2.6) yields a nontrivial solution of (2.5) which, in turn, generates a nontrivial solution of (*) . Thus, the problem of finding nontrivial solutions of (*) near $w = 0$ reduces to the problem of finding nontrivial solutions of (2.8) together with the development of suitable implicit function theorems for the system (2.7). Such an approach can be used to establish the branching results in Section 3.

Let us next examine the formal connection between the Lyapunov-Schmidt method for (*) and the stability method of Koiter. Let E be the potential energy functional given by (1.4), and suppose that, for fixed $\lambda \in \mathbb{R}^1$ and fixed $u = \sum_{j=1}^{n} \xi_j z_j$ in η , $E(u + v)$ has a critical point at v . Then, for all $h \in \eta^\perp$,

(2.9) $$\lim_{t\to 0} t^{-1}[E(u + v + th) - E(u + v)] = 0 .$$

Hence, by making use of (1.3) and a similar relationship for the functional c , we see that

(2.10) $$(L_\lambda(u + v) + Q(u + v) + C(u + v),h) = 0$$

for all $h \in \eta^\perp$. Since A maps η into η it follows that L_λ maps η into η so that (2.10) is equivalent to (2.2a). Thus, the first step in the stability method of Koiter is equivalent to solving (2.2a) for v as a function of u and λ . Let us also suppose that, for fixed $\lambda \in \mathbb{R}^1$, $E(\sum \xi_j z_j + v(\xi,\lambda))$ has a critical point at $\xi \in \mathbb{R}^n$. In order to calculate $\partial E/\partial\xi_i$ we again make use of (1.3) to obtain

$$(2.11)\qquad \lim_{t\to 0} t^{-1}[q(\sum \xi_j z_j + tz_i + v(\xi + t\theta)) - q(\sum \xi_j z_j + v(\xi))] = (Q(\sum \xi_j z_j + v(\xi)), z_i + \frac{\partial v}{\partial \xi_i})$$

where θ is a unit vector in R^n with all components zero except the ith component. Carrying out similar calculations for the other terms in E we see that $(i = 1,\ldots,n)$

$$(2.12)\qquad 0 = \lim_{t\to 0} t^{-1}[E(\sum(\xi_j + t\theta_j)z_j + v(\xi + t\theta) - E(\sum \xi_j z_j + v(\xi))]$$

$$= (L_\lambda(\sum \xi_j z_j + v(\xi)) + \sigma Q(\sum \xi_j z_j + v(\xi)) + C(\sum \xi_j z_j + v(\xi)), z_i + \frac{\partial v}{\partial \xi_i}) \quad .$$

However, $L_\lambda v$ and $\partial v/\partial \xi_i$ belong to $\eta^\perp$, and v is a solution of (2.2a) so that (2.12) reduces to (2.5). Thus, for shell buckling problems whose potential energy is given by (1.4), the stability method of Koiter is equivalent to the Lyapunov-Schmidt method for determining small solutions of (*) together with some sort of a stability analysis of the resultant nontrivial solution branches of the branching equations (2.5).

3. The branching results. The first branching result is similar to results in [13;16] and is stated for the convenience of the reader; the essential ideas of the proof may be bound in [16]. Throughout this section the Jacobian $\partial(f^1,\ldots,f^n)/\partial(\beta_1,\ldots,\beta_n)$ is denoted by $j(f,\cdot)$.

Theorem 1. Suppose that $\eta \equiv \eta(L_{\lambda_0})$ is n-dimensional. Suppose that the functional $(Q(u),u)$ restricted to the ellipse $\mathcal{E} = \{u \in \eta : (Au,u)\} = 1$ in η has a positive relative maximum at u^* . Let the basis $\{z_1,\ldots,z_n\}$ for η be chosen so that $z_1 = u^*$ and suppose that $j(f,\beta^*) \neq 0$ where $\beta^* = (\theta^{-1},0,\ldots,0)$ and $\theta = (Q(u^*),u^*)$. Then there exists a positive constant δ such that, for $0 < |\lambda - \lambda_0| < \delta$, the equation (*) has a branch of nontrivial solutions of the form

$$(3.1)\qquad w^*(\lambda) = (\sigma\theta)^{-1}(\lambda - \lambda_0)u^* + U^*$$

where U^* is an analytic function of $\tau = \lambda - \lambda_0$ which satisfies $\lim_{\tau\to 0} \tau^{-1}U^*(\tau) = 0$

The proof consists of verifying that $\sigma^{-1}\beta^*$ is a nontrivial solution of (2.8) and applying the ordinary implicit function theorem in R^n .

If $(Q(u),u)$ restricted to $\mathcal{E}$ has a positive relative minimum we have the following much stronger result.

Theorem 2. Suppose that $\eta \equiv \eta(L_{\lambda_0})$ is n-dimensional, and suppose that $(Q(u),u)$ restricted to $\mathcal{E}$ in η has a positive relative minimum at u^* . Then, for $0 < |\lambda - \lambda_0| < \delta$, there exists a branch of nontrivial solutions of (*)

which has the form (3.1) with $\theta = (Q(u^*),u^*)$.

The proof consists of showing that, for a positive relative minimum, one necessarily has $j(f,\beta^*) \neq 0$, where $\beta^* = (\theta^{-1},0,\ldots,0)$ and the basis $\{z_1,\ldots,z_n\}$ for $\mathcal{N}$ is chosen so that $z_1 = u^*$; a complete proof will appear in [19] (see also Theorem 3.4 in [6]).

The condition in Theorem 1 and Theorem 2 that the extremum value be positive is essentially a normalization. If $(Q(u),u)$ restricted to $\mathcal{E}$ has a positive relative extremum at u^* then it also has a negative relative extremum at $-u^*$ which generates a solution branch of (*) . However, since $(Q(u^*),u^*)u^* = -(Q(-u^*),-u^*)u^*$, the solutions of (*) corresponding to u^* and $-u^*$ may not be distinct.

The following general result is our main branching theorem for equation (*) . It is a consequence of a known "curve selection lemma" in the theory of real analytic sets (see [5]) together with some results on calculating the topological index of certain types of fixed points in R^n (see [6;17]); a proof will appear in [19]).

<u>Theorem 3</u>. Suppose that $\mathcal{N} \equiv \mathcal{N}(L_{\lambda_0})$ is n-dimensional and suppose that the functional $(Q(u),u)$ restricted to $\mathcal{E}$ in $\mathcal{N}$ has isolated, positive relative extrema at the m points $u_1^*,\ldots,u_m^*$. Then there exists a positive constant δ such that, for $0 < |\lambda - \lambda_0| < \delta$, the equation (*) has at least m nontrivial solution branches which are analytic in some fractional power of $\tau = \lambda - \lambda_0$.

If $(Q(u),u)$ vanishes identically on $\mathcal{E}$ then, of course, the above theorems do not apply and one must consider also the "higher order" terms in the branching equations. We state here only the following result for the special case $\sigma = 0$ when the panel becomes a thin, flat rectangular plate; in the special case $\sigma = 0$, note that L_λ is replaced by $I - \lambda A, \lambda_0$ is replaced by a characteristic value μ_0 of A , and $\mathcal{E}$ is replaced by the unit sphere $\mathcal{S}$ in $\mathcal{N}$.

<u>Theorem 4</u>. Suppose that $\mathcal{N} \equiv \mathcal{N}(I - \mu_0 A)$ is n-dimensional. Suppose that $(C(u),u)$ restricted to $\mathcal{S}$ in $\mathcal{N}$ has a positive relative extremum at u^* . Let the basis $\{z_1,\ldots,z_n\}$ for $\mathcal{N}$ be chosen so that $z_1 = u^*$ and suppose that $j(f,\beta^*) \neq 0$ where $\beta^* = (\gamma , 0,\ldots,0)$ and $\gamma^{-2} = (C(u^*),u^*)$. Then there exists a positive constant δ such that, for $\mu_0 < \lambda < \mu_0 + \delta$, the equation (*) has a branch of nontrivial solutions of the form

$$w^*(\lambda) = \gamma\eta^{\frac{1}{2}}u^* + W(\eta) \qquad \eta = (\lambda/\mu_0) - 1 \quad , \tag{3.2}$$

where W is an analytic function of $\eta^{\frac{1}{2}}$ which satisfies $\lim_{\eta\to 0^+} \eta^{-\frac{1}{2}}W(\eta) = 0$.

4. <u>The stability results</u>. In this section we consider the stability properties of some of the solution branches of (*) constructed in Section 3.

We begin with a discussion of what we will mean by "stability" (see also the related discussion in [13,§4]). For fixed $\sigma > 0$, let $\lambda_1 = \lambda_1(\sigma)$ denote the smallest (positive) eigenvalue of L_λ, and let w^* be either the trivial solution $w = 0$ or a nontrivial solution branch of (*) constructed as in Theorem 1 or Theorem 2 with $\lambda_0 = \lambda_1$. Then the "derived" operator associated with w^* is of the form

$$(4.1) \qquad D(w^*,\lambda) = L_{\lambda_1} - (\lambda - \lambda_1)A + \sigma Q'(w^*) + C'(w^*) \qquad \lambda \in \Lambda$$

where $\Lambda = \{\lambda \in R^1 : 0 < |\lambda - \lambda_1| < \delta\}$, and $Q'(w)$ and $C'(w)$ denote the Fréchet derivatives at w of Q and C. Since w^* is of the form $w^* = (\theta\sigma)^{-1}\tau u^* + U^*$, where $\tau = \lambda - \lambda_1$ and $\|U^*(\tau)\| = o(\tau)$ as $\tau \to 0$, and since $Q'(w^*)$ and $C'(w^*)$ are symmetric operators (e.g., see [20, p. 56]), the operator $D(w^*,\lambda)$ is a symmetric perturbation of a self-adjoint operator and is therefore itself self-adjoint (e.g., see [7,p. 278]) and has only real eigenvalues. Hence, in shell buckling problems where the spectrum of $D(w^*,\lambda)$ is discrete, the signs of the eigenvalues of $D(w^*,\lambda)$ determine whether w^* is stable or unstable; namely, for fixed $\lambda \in \Lambda$, if all of the eigenvalues of $D(w^*,\lambda)$ are positive then w^* is stable at λ, whereas if $D(w^*,\lambda)$ has at least one negative eigenvalue then w^* is unstable at λ. (In particular then one can show that the trivial solution $w = 0$ is stable for $\lambda < \lambda_1$ and unstable for $\lambda > \lambda_1$). Using the above definition of stability we have, for example, the following stability result.

Theorem 5. Suppose that $\mathcal{N} \equiv \mathcal{N}(L_{\lambda_1})$ is n-dimensional where, for fixed $\sigma > 0$, $\lambda_1 = \lambda_1(\sigma)$ denotes the smallest (positive) eigenvalue of L_λ. If the functional $(Q(u),u)$ restricted to $\mathcal{E}$ in $\mathcal{N}$ has a positive relative minimum at u^* and if w^* is the nontrivial solution branch of (*) constructed as in Theorem 2 with $\lambda_0 = \lambda_1$, then $w^* = w^*(\lambda)$ is stable for $\lambda_1 < \lambda < \lambda_1 + \delta$ and unstable for $\lambda_1 - \delta < \lambda < \lambda_1$. If $(Q(u),u)$ restricted to $\mathcal{E}$ has a positive relative maximum at u^* and if, in addition, the Jacobian hypothesis of Theorem 1 is satisfied, then the nontrivial solution branch $w^* = w^*(\lambda)$ constructed as in Theorem 1 with $\lambda_0 = \lambda_1$ is unstable for $0 < |\lambda - \lambda_1| < \delta$.

A stability result for the special case $\sigma = 0$ is given by

Theorem 6. Suppose that $\mathcal{N} \equiv \mathcal{N}(I - \mu_1 A)$ is n-dimensional where μ_1 denotes the smallest (positive) characteristic value of A. Suppose that the hypotheses of Theorem 4 with $\mu_0 = \mu_1$ are satisfied and let $w^* = w^*(\lambda)$ be the resultant nontrivial solution branch of (*). If $(C(u),u)$ restricted to $\mathcal{S}$ in $\mathcal{N}$ has a positive relative minimum [respectively, maximum] at u^* then w^* is stable [respectively, unstable] for $\mu_1 < \lambda < \mu_1 + \delta$.

The proofs of the above results will appear in [19]; the methods used are closely related to those in [13;16], however, they depend heavily upon the fact that Q and C are homogeneous gradient operators.

5. Applications to buckling problems. In this section we outline some specific applications of the results in Section 3 and Section 4.

As a first application we consider the problem of determining the buckled states of a thin, flat elastic plate that is rectangular in shape, simply-supported along its edges, and subjected to a constant compressive thrust applied normal to its two short edges. We assume that the deformations of the plate are described by the von Karman equations (see [1])

$$\text{(5.1a)}\qquad \Delta^2 f = -\tfrac{1}{2}[w,w] \qquad \text{in } \Omega$$

$$\text{(5.1b)}\qquad \Delta^2 w = [w,f] - \lambda w_{xx}$$

together with the simply-supported boundary conditions

$$\text{(5.2)}\qquad w = f = \Delta w = \Delta f = 0 \qquad \text{on } \partial\Omega .$$

Here $\Omega = \{(x,y): 0 < x < d \;,\; 0 < y < 1\}$ corresponds to the middle plane of the undeflected plate, $\partial\Omega$ denotes its boundary, Δ denotes the Laplacian in x,y coordinates, and

$$\text{(5.3)}\qquad [u,v] = u_{xx}v_{yy} + u_{yy}v_{xx} - 2u_{xy}v_{xy} .$$

The function $w = w(x,y)$ is a measure of the deflection of the middle plane of the plate out of the x,y plane, $f = f(x,y)$ is an "excess" stress function and the parameter λ is a measure of the compressive load.

Let $\mathscr{H}$ be the real Hilbert space defined as the closure, in the norm $\|\cdot\|_{2,2}$ of the Sobolev space $W_{2,2}(\Omega)$, of the set of smooth functions defined on $\bar{\Omega}$ and vanishing on $\partial\Omega$; since the bilinear form $(u,v) = \int_\Omega \Delta u \Delta v dxdy$ is coercive over $\mathscr{H}$, it defines an equivalent norm for $\mathscr{H}$ which we denote by $\|\ \|$. Then, finding classical solutions of (5.1) which satisfy (5.2) is (essentially) equivalent to finding (generalized) solutions in $\mathscr{H}$ of a single operator equation of the form (*) with $\sigma = 0$ (see [3;8]). Some general branching results for the rectangular plate are described in [8] (see also Theorem 4).

Let us consider in more detail the special case when $d = \sqrt{2}$. Then the smallest (positive) characteristic value $\mu_1 = 9\pi^2/2$ of A has multiplicity two and we have the following description of the buckled states of the plate near the buckling load λ_1 (see [8]): There exists a constant δ such that, for $\mu_1 < \lambda < \mu_1 + \delta$, there are <u>exactly</u> eight buckled states which depend continuously on the load parameter λ; the eight buckled states are of the form

$$\text{(5.4)}\qquad \pm w(\lambda) = \pm[\eta^{\frac{1}{2}}\theta_i^{-\frac{1}{2}}u_i + U_i(\lambda)] \qquad (i = 1,2,3,4)$$

where $\eta = (\lambda/\mu_1) - 1$, $\theta_i = (C(u_i),u_i)$, the U_i are analytic functions of $\eta^{\frac{1}{2}}$ which satisfy $\lim_{\eta\to 0^+} \eta^{-\frac{1}{2}}U_i = 0$, and, if

$$\text{(5.5)}\qquad u_{mn} = c_{mn} \sin\frac{m\pi x}{d} \sin n\pi y \qquad (m,n = 1,2,\dots)$$

with $c_{mn} = 2d^{\frac{3}{2}}/\pi^2(m^2 + d^2n^2)$, then $u_1 = u_{21}$, $u_2 = u_{11}$ and u_3, u_4 are linear combinations of u_1 and u_2. Moreover, since the functional $(C(u),u)$ restricted to the unit sphere $\mathscr{S}$ in $\mathscr{N}(I - \mu_1 A)$ has extrema at $\pm u_i (i = 1,2,3,4)$, and since u_1 and u_2 correspond to minima with

(5.6) $$\min_{u \in \mathscr{S}} (C(u),u) = \theta_1 < \theta_2 < \theta_3 = \theta_4 = \max_{u \in \mathscr{S}} (C(u),u) ,$$

it follows from Theorem 6 that the four solution branches $\pm w_1, \pm w_2$ are stable for $\mu_1 < \lambda < \mu_1 + \delta$ and the four solution branches $\pm w_3$, $\pm w_4$ are unstable for $\mu_1 < \lambda < \mu_1 + \delta$. Finally, although all four solution branches $\pm w_1$, $\pm w_2$ are stable, it also follows from (5.6) that the two solutions $\pm w_1$ have less potential energy than the two solutions $\pm w_2$.

As a second application we consider the problem of determining the buckled states of an axially-compressed cylindrical panel; a detailed analysis of the problem will appear in some forthcoming joint work with Professor George H. Knightly. We assume that the deformations of the panel are described by the von Karman-Donnell equations

(5.7a) $$\Delta^2 f = -\frac{1}{2}[w,w] + \sigma w_{xx}$$

(5.7b) $$\Delta^2 w = [w,f] - \lambda w_{xx} - \sigma f_{xx} \qquad \text{in } \Gamma$$

together with the boundary conditions

(5.8) $$w = f = \Delta w = \Delta f = 0 \qquad \text{on } \partial\Gamma$$

(see [12] for the description of a more realistic set of boundary conditions). Here $\Gamma = \{(x,s): 0 < x < a , 0 < s < b < 2\pi R\}$, where x is proportional to the axial coordinate and s is proportional to the circumfirential coordinate of the panel, R is the (constant) radius of the panel and $\sigma = 1/R$, and Δ and $[\cdot,\cdot]$ in (5.3) are now defined with respect to the x,s coordinates. The function $w = w(x,s)$ is a measure of the radial displacement of the panel and λ is a measure of the uniform compressive load.

Let $\mathscr{H}$ be the closed subspace of $W_{2,2}(\Gamma)$ obtained by the closure, in the norm $\|u\|^2 = \int_\Gamma (\Delta u)^2 dxds$, of smooth functions defined on $\bar{\Gamma}$ and vanishing on $\partial\Gamma$. Then, finding classical solutions of (5.7) which satisfy (5.8) is (essentially) equivalent to finding (generalized) solutions in $\mathscr{H}$ of a single operator equation of the form (*) with $\sigma = 1/R$ (e.g., see [4;8]). Since the relevant operators Q and C are also gradient operators the results of Section 3 and Section 4 can be applied in a number of cases to establish the existence and stability of buckled states of the panel.

Let us now consider the special case of a "narrow" cylindrical panel (i.e., a panel with sufficiently small curvature) and describe the buckled states in more detail. The panel being considered is called narrow if $\sigma \leqslant 4\pi^2/b^2$; such a definition of a narrow cylindrical panel is in agreement with that used previously in [12]. We further restrict a and b so that $a^2 = m(m+1)b^2$ where m is a positive integer; if a^2/b^2 is not the product of successive integers then, for a narrow panel, $\dim (L_{\lambda_1}) = 1$ and the methods used to carry out a branching and stability analysis are fairly standard. If μ_1 denotes the smallest (positive) characteristic value of A and if $a^2 = m(m+1)b^2$ then

$$\mu_1 = \frac{\pi^2}{b^2}\left(4 + \frac{1}{m(m+1)}\right) . \tag{5.9}$$

Hence, for narrow panels with $a^2 = m(m+1)b^2$, $\sigma \leqslant (4\pi^2/b^2) < \mu_1$ and it follows that $\dim(L_{\lambda_1}) = \dim(I - \mu_1 A) = 2$, where $\lambda_1 = \mu_1 + \sigma^2\mu_1^{-1}$ denotes as usual the smallest (positive) eigenvalue of L_λ . It then easily follows that $\{u_m, u_{m+1}\}$ is an orthonormal basis for $\mathcal{N}(L_{\lambda_1})$, where

$$u_k \equiv u_{k1} = d_k \sin\frac{k\pi x}{a} \sin\frac{\pi s}{b} \tag{5.10}$$

and d_k is an appropriate constant $(k = 1,2,\ldots)$. For the special case being considered, we have the following description of the buckled states of the panel near the buckling load λ_1 . There exists a constant δ such that, for $0 < |\lambda - \lambda_1| < \delta$, there are exactly three buckled states which depend continuously on the load parameter λ . The three buckled states are of the following form: (1) If m is even then $w_e = a_m(\tau/\sigma)u_{m+1} + U(\tau)$ and $w_e^{\pm} = b_m(\tau/\sigma)(u_{m+1} \pm c_m u_m) + W^{\pm}(\tau)$, where a_m , b_m and c_m are constants depending on m , $\tau = (\lambda - \lambda_1)/\mu_1$, and U , $W^{\pm}$ are analytic functions of order $o(\tau)$ as $\tau \to 0$, (2) If m is odd then $w_o = \alpha_m(\tau/\sigma)u_m + U$ and $w_o^{\pm} = \beta_m(\tau/\sigma)(u_m \pm \gamma_m u_{m+1}) + W^{\pm}(\tau)$ where α_m , β_m and γ_m are constants depending on m , and U , $W^{\pm}$ are analytic functions of order $o(\tau)$ as $\tau \to 0$. Moreover, since w_e and w_o are generated by minima of the functional $(Q(u),u)$ restricted to $\mathcal{E}$, they are stable for $\lambda_1 < \lambda < \lambda_1 + \delta$ and unstable for $\lambda_1 - \delta < \lambda < \lambda_1$ (see Theorem 5) so that, for m either even or odd, there is an exchange of stabilities at $\lambda = \lambda_1$; note also that, for m either even or odd, the stable buckled state of the panel for $\lambda_1 < \lambda < \lambda_1 + \delta$ is always the one in which the panel buckles "outwards" at the ends. Since $w_e^{\pm}$ and $w_o^{\pm}$ are generated by maxima of $(Q(u),u)$ restricted to $\mathcal{E}$, the other buckled states of the panel are all unstable for $0 < |\lambda - \lambda_1| < \delta$. Moreover, since c_m and γ_m both tend to 1 as $m \to \infty$, for panels with a much larger than b , the unstable buckled states in the special case being considered are essentially of the form $w_e^{\pm}$ = constant $\tau(u_{m+1} \pm u_m)$, if m is even, and $w_o^{\pm}$ = constant $\tau(u_m \pm u_{m+1})$ if m is odd. These four unstable buckled states clearly represent situations in which essentially all of the buckling takes place at one of the ends of the panel; note that even in these situations, for m either even or odd, the buckling is still "outwards" at the end at which it occurs.

In summary, a "simply-supported" cylindrical panel exhibits more or less the stable behavior of a plate rather than the unstable behavior of a (complete) cylinder provided that it is a narrow panel (compare the somewhat different stability results in [12, p. 5] for a different set of boundary conditions). On the other hand, if $b = 1$ and $m = 1$, then the panel has a single stable solution for $\lambda_1 < \lambda < \lambda_1 + \delta$, namely w_o , whereas the above buckling problem for a plate with sides 1 and $\sqrt{2}$ has four stable buckled states for $\mu_1 < \lambda < \mu_1 + \delta$, namely $\pm w_1$ and $\pm w_2$;

thus, although a narrow, simply supported cylindrical panel exhibits some of the stability properties of a plate, it is not true that the stable buckled states of the two problems are "more or less the same" even when σ is very small.

References

[1] L. Bauer & E.L. Reiss, "Nonlinear buckling of rectangular plates," S.I.A.M. J. 13(1965), 603-626.

[2] L. Bauer, H.B. Keller & E.L. Reiss, "Axisymmetric buckling of hollow spheres and hemispheres," Comm. Pure Appl. Math. 23(1970), 529-568.

[3] M.S. Berger & P.C. Fife, "On von Kármán's equations and the buckling of a thin elastic plate", Bull. Amer. Math. Soc. 72(1966), 1006-1011.

[4] M.S. Berger, "On the existence of equilibrium states of thin elastic shells (I)," Indiana Math. J. 20(1971), 591-602.

[5] R. Böhme, "Die Lösung der Verzweigungsgleichungen für Nichtlineare Eigenwert-problems", Math. Z. 127(1972), 105-126.

[6] A. Grundmann, "Der topologische Abbildungsgrad homogener Polynomoperatoren", Dissertation, Universität Stuttgart, 1974.

[7] T. Kato, "Perturbation Theory for Linear Operators", Springer-Verlag, New York, 1966.

[8] G.H. Knightly & D. Sather, "Nonlinear buckled states of rectangular plates", Arch. Rational Mech. Anal. 54(1974), 356-372.

[9] G.H. Knightly & D. Sather, "Nonlinear axisymmetric buckled states of shallow spherical caps", S.I.A.M. J. Math. Anal. 6(1975).

[10] W.T. Koiter, "Elastic stability and post-buckling behavior", Nonlinear Problems edited by R. Langer, University of Wisconsin Press, Madison, 1963.

[11] W.T. Koiter, "On the nonlinear theory of thin elastic shells I", Proc. Kon. Ned. Akad. Wet. 69(1966), 1-54.

[12] W.T. Koiter, "Buckling and post-buckling behavior of a cylindrical panel under axial compression, NLR Report S476, Amsterdam, 1956, and also Reports and Transactions, National Aero. Res. Inst. 20.

[13] J.B. McLeod & D.H. Sattinger, "Loss of stability and bifurcation at a double eigenvalue", J. Funct. Anal. 14(1973), 62-84.

[14] E.L. Reiss, "Bifurcation buckling of spherical caps", Comm. Pure Appl. Math. 18(1965), 65-82.

[15] E. Reissner, "On the axisymmetrical deformation of thin shells of revolution," Proc. of Symposia in Appl. Math. 3(1950), 27-52.

[16] D. Sather, "Branching of solutions of an equation in Hilbert space," Arch. Rational Mech. Anal. 36(1970), 47-64.

[17] D. Sather, "Nonlinear gradient operators and the method of Lyapunov-Schmidt," Arch. Rational Mech. Anal. 43(1971), 222-244.

[18] D. Sather, "Branching of solutions of nonlinear equations," Proceedings of the Seminar on Nonlinear Eigenvalue Problems, Santa Fe, New Mexico, 1971, Rocky Mtn. J. Math. 3(1973), 203-250.

[19] D. Sather, "Branching and stability for nonlinear shells", to appear.

[20] M.M. Vainberg, "Variational Methods for the Study of Nonlinear Operators," Holden-Day, Inc., San Francisco, 1964.

[21] M.M. Vainberg & V.A. Trenogin, "The methods of Lyapunov and Schmidt in the theory of nonlinear equations and their further development," Russian Math. Surveys 17, No. 2(1962), 1-60.

[22] V.Z. Vlasov, "General Theory of Shells and Its Applications in Engineering," National Aeronautics and Space Administration, Washington, D.C., 1964.

On a free surface problem

D. H. Sattinger

University of Minnesota, U.S.A.

1. Much of the extant mathematical literature on free surface phenomena excludes the effects of vicosity and surface tension. In many applications such an approximation is no doubt perfectly valid, while in other situations such effects are of principle importance. One such example is the phenomenon of "climbing" of a viscous, non-Newtonian fluid. Experiments of such have been carried out by Beavers, Fosdick, and Joseph at the University of Minnesota [1], [2]. In these experiments a rod inserted in a viscous fluid rotates at a constant angular speed ω. In the case of a Newtonian fluid the surface dips in the vicinity of the rod, while in the non-Newtonian case the fluid may climb up the rod as it spins.

In order to develop a quantitative theory which would explain the effects of viscosity, Fosdick and Joseph developed a perturbation scheme which allowed for the perturbation of the domain caused by the displacement of the fluid. They developed the solutions in a power series in ω, the angular velocity of the rod, both for the Newtonian case and for the non-Newtonian case of a simple fluid. A rigorous proof of the convergence of their perturbation series for the Newtonian case, under the assumption of a zero wetting angle, is given in [3] and will be discussed in this summary. The non-Newtonian case is considerably more difficult, for in this case the stress tensor depends not only on the current fluid velocities, but on the past history of the flow as well. There thus remains open a large class of interesting, though perhaps very difficult, mathematical problems relating to non-Newtonian fluids.

2. The flow of a Newtonian fluid is governed by the Navier-Stokes equations, which in tensor notation take the form

$$g^{jk}(T_{ij})_{,k} - \Phi_{,i} = u^j \, u_{i,j}$$

$$g^{ij}u_{i,j} = 0 \qquad ;$$

u_i being the covariant velocities, g_{ij} the metric tensor, the gravitational potential, an T_{ij} the stress tensor:

$$T_{ij} = -pg_{ij} + 2\mu D_{ij} \quad ,$$

$$D_{ij} = \frac{1}{2}(u_{i,j} + u_{j,i}) \quad .$$

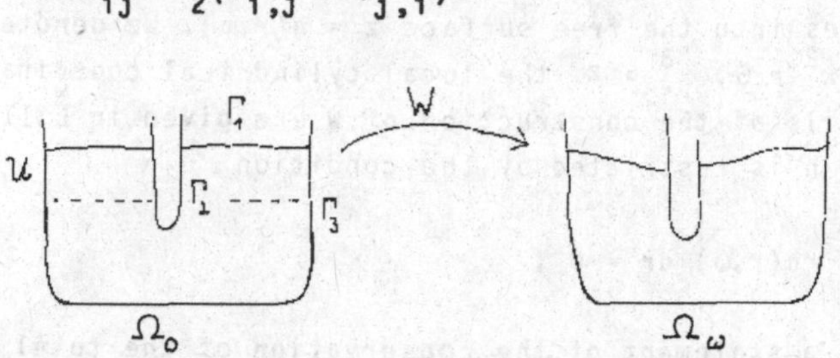

These equations are to hold in the domain Ω_ω. The boundary conditions are

$\vec{u} = 0$ on the container wall

$\vec{u}\cdot\hat{k} = \vec{u}\cdot\hat{r} = 0$, $\vec{u}\cdot\hat{\Theta} = \omega$ on the side of the rod

and

$$\left.\begin{array}{l} \vec{u}\cdot\vec{n} = 0 \\ T\cdot\vec{n} = -p_a\vec{n} + \sigma J\vec{n} \end{array}\right\} \quad \text{on the free surface } z = h(r)$$

where

$\vec{n}$ = normal vector to free surface

p_a = atmosphere pressure

J = mean curvature

$= \frac{1}{r}\left(\frac{rh'}{\sqrt{1+h'^2}}\right)'$

σ = coefficient of surface tension

In tensor notation the conditions on the free surface take the form

$$u_i n^i = 0\ ,\quad T_{ij} n^j = -p_a n^i + \sigma J n^i$$

3. Since the domain is unknown and varies with ω we proceed as follows. Let W be a volume-preserving mapping from Ω_0 to Ω_ω, which, in the region U at the top of Ω_0, has the form

$$W(r,\Theta,z) = (r\cos\Theta,\ r\sin\Theta,\ z + h(r,\omega))$$

where $h(r,\omega)$ is the (unknown) free surface. Then the boundary $z = 0$ goes into the free surface $z = h(r,\omega)$. We denote by $x^1 = r$, $x^2 = \Theta$, $x^3 = z$ the local cylindrical coordinate in U. The details of the construction of W are given in [3]. The function h is restricted by the condition

$$\int_a^R rh(r,\omega)\,dr = 0\ ,$$

which is a statement of the conservation of the total volume occupied by the fluid. In the region U the metric tensor g is given by

$$g_{ij} = \frac{\partial W}{\partial x^i}\cdot\frac{\partial W}{\partial x^j} = \begin{pmatrix} 1+h'^2 & 0 & h' \\ 0 & r^2 & 0 \\ h' & 0 & 1 \end{pmatrix}$$

If $\vec{v}$ is a vector field in Ω_ω the contravariant components of $\vec{v}$ in Ω_0 are given by

$$v^i = g^{ij}\left(\frac{\partial W}{\partial x^j}\cdot\vec{v}\right)$$

By virtue of the transformation W the equations and boundary conditions are pulled back to the domain Ω_0. Joseph and Fosdick [1] construct a formal perturbation series for the unknown functions $\{v^i,p,h\}$. We prove the convergence of the perturbation series by an application of the implicit function theorem on a Banach space.

4. Implicit Function Theorem. Let $\mathcal{E}$, $\mathcal{F}$ be complex Banach spaces, C the complex numbers, and $F: \mathcal{E} \times C \to \mathcal{F}$ a Frechet differentiable (hence analytic) mapping in a neighbourhood of the origin of $\mathcal{E} \times C$. Suppose $F(0,0) = 0$ and $F_x(0,0)$ is an isomorphism from $\mathcal{E}$ to $\mathcal{F}$. Then there exists an analytic $\mathcal{E}$ valued function $x(\varepsilon)$ such that $F(x(\varepsilon),\varepsilon) \equiv 0$ for sufficiently small ε.

As an example of the application of this theorem to a nonlinear boundary value problem, consider the problem

$$\Delta u + f(u) = \varepsilon h \qquad \zeta(u)\big|_{\partial D} = \varepsilon g$$

on a smoothly bounded domain D. Assume that f and ζ are analytic in a neighbourhood of the origin; that $f(0) = f'(0) = 0$ and $\zeta(0) = 0$, $\zeta'(0) \neq 0$; and that $h \in C_\alpha(\overline{D})$, $g \in C_{2+\alpha}(\partial D)$. We set

$$\mathcal{E} = C_{2+\alpha}(\overline{D}) \;, \qquad \mathcal{F} = C_\alpha(\overline{D}) \times C_{2+\alpha}(\partial D)$$

$$F(u,\varepsilon) = (\Delta u + f(u) - \varepsilon h,\ \zeta(u)\big|_{\partial D} - \varepsilon g)$$

Then F is a Frechet differentiable mapping from $\mathcal{E}$ to $\mathcal{F}$, with Frechet derivative

$$F'_u(0,0)v = (\Delta v,\ \zeta'(0)v\big|_{\partial D})$$

The verification that $F'_u(0,0)$ is an isomorphism from $\mathcal{E}$ to $\mathcal{F}$ consists in showing that the linear boundary value problem

$$\Delta v = h \qquad \zeta'(0)v\big|_{\partial D} = g$$

is solvable for $h \in C_\alpha(\overline{D})$ and $g \in C_{2+\alpha}(\partial D)$ and that the solution v satisfies the estimate

$$|v|_{2+\alpha} \leq c(|h|_\alpha + |g|^{\partial D}_{2+\alpha})$$

for some constant c independent of h and g. This estimate, however, is precisely the content of the classical Schauder estimates for elliptic boundary value problems.

5. Our free surface problem may be treated in a similar manner, but there is an added difficulty: the domain Ω_0 is not smooth but has a ridge at the intersection of Γ_1 and Γ_2 and of Γ_1 and Γ_3 ($z = 0$, $r = a, R$). The classical Schauder estimates do not hold in such a domain; however, let us consider another simple example:

$$\begin{cases} \Delta u = f \quad x \geq 0,\ y \geq 0 \\ u(x,0) = g_1(x) \quad u(0,y) = g_2(y) \end{cases}$$

We assume that $f \in C_\alpha(R^+ x R^+)$ and that $g_1, g_2 \in C_{2+\alpha}(R^+)$. In general, the solution u may not be smooth at the corner but may have logarithmic singularities in its second derivatives. If u is regular then clearly we must have

$$\boxed{\begin{aligned} & g_1(0+) = g_2(0+) \\ & g_1''(0+) + g_2''(0+) = f(0,0) \end{aligned}}$$

It turns out that these conditions are also sufficient for u to be regular ($C_{2+\alpha}(\overline{R^+ x R^+})$). (see [4]). We shall call such conditions "compatibility conditions" or " consistency conditions" for the inhomogeneous data at the corner. It turns out that there are also a set of consistency conditions for our fluid mechanics problem.

6. In our fluid mechanics problem we may associate with the nonlinear boundary value problem a nonlinear mapping $F(v,p,h,\omega)$. The detailed form of F, which is somewhat complicated, is given explicitly in [3]. The space $\mathcal{E}$ consists of triples $\{v,p,h\}$, where v is a vector field and p and h are functions satisfying certain regularity and consistency conditions to be given below. The space $\mathfrak{F}$ consists of functions $\{f,g^1,g^2,H\}$ where f is a vector field on Ω_0, and g', g^2, and H are functions on Γ_1. These functions

also satisfy a certain set of regularity and consistency conditions to be given below. We now write down the Frechet derivative of F in the local coordinates which are valid in the region U :

$$\Delta v^1 - \frac{v'}{r^2} - \frac{\partial p}{\partial r} = f^1$$

$$\Delta v^2 + \frac{2}{r}\frac{\partial v^2}{\partial r} = f^2$$

$$\Delta v^3 - \frac{\partial p}{\partial z} = f^3 \qquad (6.1)$$

$$\frac{1}{r}\frac{\partial}{\partial r}(rv^1) + \frac{\partial v^3}{\partial z} = 0$$

$$\frac{\partial v^1}{\partial z} = g^1(r) \quad \frac{\partial v^2}{\partial z} = g^2(r) \quad v^3 = 0 \quad \text{on } \Gamma_1 \qquad (6.2)$$

$$v^1 = v^3 = 0 \ , \quad v^2 \equiv \text{const.} \qquad \text{on } \Gamma_2 \cap U \qquad (6.3)$$

$$v^1 = v^2 = v^3 \qquad \text{on } \Gamma_3 \qquad (6.4)$$

$$gh - \frac{\sigma}{r}(rh')' + p + 2\frac{\partial v^3}{\partial z} = H \qquad \text{on } \Gamma_1 \qquad (6.5)$$

We now write down the explicit forms of the spaces $\mathcal{E}$ and $\mathcal{F}$, including the consistency conditions :

$$\underline{\mathcal{E} = \{(v,p,h)\}}$$

1. $v^i_{,i} = 0$, $v^i \in C_{2+\alpha}(\overline{\Omega_0})$

2. $h \in C_{3+\alpha}(\Gamma_1)$, $p \in C_{1+\alpha}(\overline{\Omega_0})$

3. $\int_a^R rh(r,\omega)\, dr = 0$

4. $h' = 0$, $\frac{\partial v^1}{\partial z} = \frac{\partial v^2}{\partial z} = \frac{\partial^2 v^1}{\partial r\, z} = 0$, $\frac{\partial p}{\partial z} = 0$ at $r = a,R$, $z = 0$

(Conditions 4. are the consistency conditions)

$\mathcal{F} = (f, g^1, g^2, H)$

1. $f^i \in C_\alpha(\overline{\Omega_0})$, $g', g^2, H \in C_{1+\alpha}(\Gamma_1)$

2. $f^3 = 0$, $g' = g^2 = \frac{dg'}{dr} = 0$ at $r = a, R$, $z = 0$.

It can be shown, by a detailed investigation of the structure of the nonlinear equations, that these consistency conditions are invariant under the nonlinear transformation F : that is, if (v,p,h) lie in $\mathcal{E}$ then $(f,g^1,g^2,H) = F(v,p,h,\omega)$ lie in $\mathcal{F}$. Furthermore the transformation F is easily seen to be analytic in a neighborhood of the origin of $\mathcal{E} \times C$. Finally, one can show that the Frechet derivative of F is an isomorphism between $\mathcal{E}$ and $\mathcal{F}$. The details are given in [3]. One can thus apply the implicit function theorem on the restricted subspaces $\mathcal{E}$ and $\mathcal{F}$, obtaining a regular solution to the fluid problem.

7. Remarks.

i) Condition 3. on h may be satisfied as follows. Equations (6.1) - (6.4) are first resolved - they comprise a linear elliptic boundary problem. The pressure p is not unique but determined only up to an additive constant. This constant may be adjusted in (6.5) so as to ensure that h satisfies condition 3.

ii) The assumption of zero wetting angle ($h' = 0$) considerably simplifies the analysis. It is not clear whether the solution would be regular at the corner ($r=a,R;z=0$) if this condition were dropped. We have not investigated the convergence of the series in this case.

iii) The problem of the non-Newtonian fluid remains open. The analysis is complicated by the fact that the stress tensor depends on the past history of the flow.

Bibliography

[1] D.D. Joseph, and R.L. Fosdick, "The Free Surface on a Liquid Between Cylinders Rotating at Different Speeds, Part I," Arch.Rat.Mech. Anal. 49 (1973) pp. 321-380.

[2] D.D.Joseph, G.S. Beavers, and R.L. Fosdick, " The Free Surface on a Liquid Between Cylinders Rotating at Different Speeds, Part II," Arch.Rat.Mech.Anal. 49 (1973), pp. 381-401.

[3] D.H. Sattinger, " On the Free Surface of a Viscous Fluid Motion," Preprint

[4] E.A. Volkov, "Differentiability Properties of Solutions of Boundary Value Problems for the Laplace and Poisson Equations on a Rectangle," Proceedings of the Steklov Institute of Mathematics 77 (1965), pp. 101-126.

THEORETICAL CONSTRUCTIONS OF SELECTION OF ACTUAL EVENTS FROM THE VIRTUAL ONES

L.I. SEDOV
Moscow State University
Moscow V-234/USSR

The mathematical methods of research used in natural sciences are based on introducing certain independent and dependent characteristic elements for which such mathematical operations are determined which allow to formulate in their terms the problems to be solved by means of theoretical (mathematical) or experimental methods.

The initial step of any investigation is to get an idea about the sense of phenomena under investigation to list the characteristic conceptions introduced and to specify the relations following from definitions or specified by the conditions at which the problem is posed. These conditions are taken from the experiments or from the initial postulates of the theory. The other words, one must explicitly fix (in the list of the characteristic parameters used) the system of independently given determining parameters, and the characteristic parameters, functions and laws obtained by means of models on which the investigation of the problem considered is based.

To construct models and closed systems of mathematical relationships it is necessary to use certain basic, largely universal, relationships. These are usually treated as laws suitable for various models and phenomena, the other hand they can and must serve as a field for introduction of different hypotheses and laws to be checked in the course of study of physical phenomena.

For example, proceeding from the convention that the space is Euclidean and that time is obsolute, we can, after choosing an inertial coordinate system and defining the concept of material point, construct the analytical mechanics of the interacting material particle system on the base of the Newton equation

$$m\bar{a} = \bar{F} \tag{1}$$

where m is the particle mass, $\bar{a}$ is its acceleration with respect to the inertial coordinate system, and $\bar{F}$ is the acting force, which either is determined from (1) or specified by means of postulates, i.e. the laws previously verified on the base of experiments with an obligatory use of equation (1).

As it is known the basic relation (1) can be replaced in the general case or for important classes of phenomena by other equivalent statements, and in particular, by variational "principles".

It is useful to note that for system motions, when "gyroscopic" forces for each point are absent, the basic equation (1) is equivalent to the corresponding energy equation.

But one of the most important re-statement of the system of equation of the form (1) in rational mechanics is one scalar equation expressing the Lagrangian principle of virtual displacements or one scalar variational equation in the Lagrangian form.

Mechanics and physics today are in a process of merger. Especially clearly one can see it in the macroscopic theories of continuum mechanics. The thermodynamic and electromagnetic characteristics of the internal state of bodies and fields are gaining a major importance in many phenomena along with such purely mechanical characteristics as vectors of displacement of medium points and their derivates of different orders with respect to time and coordinates. It becomes impossible to describe effectively the macroscopic relations and models of continuum media with internal degrees of freedom by means of equation (1) alone.

The problem arises to obtain the basic relations of a more general nature than equation (1), which would involve as their particular cases, the corresponding basic laws already proved and shown to be fruitful for the study of material media and fields in mechanics, physics and chemistry.

The literature is now flooded by papers devoted to construction of macroscopic models of material media and fields with internal degrees of freedom. In some cases various "conservation laws" are taken for the basic initial relations. This technique, however, is effective only for a limited number of internal degrees of freedom and it is applicable only to infinitesimal particles or volumes, e.g. in the general theory of relativity and in the case of fields in general. To study the body and field motions with strong discontinuities one has to formulate physically meaningful integral relations. The laws of plastic deformation and kinetic for irreversible phenomena are established on the base of some hypotheses and principles of the irreversible thermodynamics, which are not connected with the conservation laws of the classical type.

Sometimes the basic relations are written down by means of formal mathematical generalizations. In particular, in this way the systems of moment equations are obtained from Boltzmann equation or from similar relations. (People who use these theories often loose the feeling of the obligatory character of such complicated generalizations, and this leads to serious difficulties in obtaining effective results.)

Such methods should not be rejected, however, as they can give and do give useful results in determination of the thermodynamic functions for macroscopic systems and in establishing the laws of energy dissipation.

It is symptomatic that some eminent authors of new models are joined together in closed groups-institutionalized or in statu nascendi - which are acting independently. This tendency is promoted by the complexity of the new models as well as by the fact that often it is more difficult to understand the gist of the other author model than to construct a new one describing similar phenomena within the framework of "one's own ideas". A weak connection of such theories with the applications is another factor favouring this isolation tendency.

To construct a model means to establish a system of closed relations suitable for describing vast classes of motions and processes, which are, as a rule,non-uniform and unsteady ones. Under such general formulation, any preliminary empiric foundation can be only partial ; historical experience, intuition and common sense can help to formulate the minimally necessary mathematical and physical assumptions. The authors of these theories are encouraged and inspired in their inventive activity by the numerous examples when laws established for the limited range of phenomena proved to be applicable far beyond their original scope.

As the complicated properties and mechanisms of interactions of the adjoint particles in the newly made models are hard to grasp, their authors cannot help using in their designs regular experimentally checked methods. It is clear that it is reasonable to take into account the approximate correspondence of schematic model and reality and to ensure because of this fact a maximal mathematical simplicity which allows effectively to carry out qualitative investigations and quantitative calculations.

It is obvious that the need arises to discuss and develop some common stand - points for the designers of new models. In these problems (concerned with complex auxiliary mathematical theories unknown to the broad community of scientists, engineers and experimentators) some unification and agreement between different points of views is even more essential than in international establishing uniform systems of measurement units for geometrical or physical quantities. The essence of the matter is made even more complicated by the fact that the use of various individual, local and special models (as well as ad hoc measurement units) is acceptable in principle and sometimes really justified. Yet this is hardly a sufficient argument against the need for coordination in the methods of model establishing and in standardization of the good examples of particular models.

For many years mechanicians and physicists have been making use of the following universal equations : the law of the conservation of energy involving the balance of all kinds of energies, the equation of the principle of virtual displacements and the equation of the Lagrangian variational principle.

The principle of virtual displacements is usually formulated or derived from mechanic of equations to describe purely mechanical phenomena. The Lagrangian variational principle also is derived for continuous mechanical and electrodynamic phenomena. The extension of these relations to involve the case of internal degrees of freedom can be accepted as appropriate postulates by means of specifying coefficients for the variations of the unknown functions. Like the definition of force in the equation of Newton's second law, the specifying these coefficients as a whole means specifying the medium or field model.

It must be emphasized that if the infinitesimal virtual variations of the unknown functions coincide with the small increments of these functions in the corresponding actual processes, one can readily show that for a small particle as well as for the body as a whole the equation of energy, the equation of the principle of virtual displacements and the Lagrangian variational equation coincide exactly.

In particular cases when real infinitesimal increments of the unknown functions can be arbitrary (for this a linear independence is sufficient) for different processes in which the coefficients of the increments of these functions do not depend on these increments, it follows that from the equation of energy as well as from the equation of the principle of virtual displacements and from the variational equation one can derive immediately (provided there are no additional connections) the equations of motion, equations of state and other physico-chemical process equations for additional variable physical parameters.

However, in the general case, the equation of the principle of virtual displacements and the variational equation are of a more universal nature than the energy equation where the "gyroscopic terms" are absent. (In the equation of virtual displacement principle and in the variational equation these terms are present.)

Below we consider a theory for the construction of continuum medium field models by means of a basic equation of the following form :

$$\delta \int_V \Lambda \, d\tau + \delta W^* + \delta W = 0 \qquad (2)$$

Here $d\tau$ is an element of the arbitrary four-dimensional space-time volume V of the medium or the field.

In this equation, the density of the Lagrangian function Λ and the linear functional with respect to the variations of the unknown functions, δW^*, represented by an integral with respect to the volume V and by a three-dimensional surface integral of a linear expression with respect to variations and their coordinate and time derivatives over the three-dimensional surface $\Sigma + S_\pm$ (Σ - the surface bounding the volume V, and $S_\pm$ two-sided surfaces at which the unknown functions can have strong discontinuities) are given.

The functional δW is determined by Λ and δW^* and it is an integral over the surface Σ.

Arguments of the scalar Λ and of the coefficients of the variations in δW^* are expressed in terms of unknown functions and of their coordinate and time derivatives of various orders.

In Newton's mechanics it is often (but not always) possible to admit :

$$\Lambda = \frac{\rho v^2}{2} - \rho U \qquad (3)$$

where ρ is the mass density, v is the velocity of point particles of the medium, and U is the internal energy of particle per mass unit (a thermodynamic characteristic of the medium depending on the mechanical and thermodynamic parameters characterizing the state of the infinitesimal particles of the medium). If an electromagnetic field is involved in the system under consideration it is necessary to add to the right side of (3) other well known terms.

The physico-thermodynamic interpretation of δw^* in some cases is reduced to the following.

The integrand for δw^* is a full input of the internal energy to an infinitesimal particle due to the variation of the determining parameters. It involves the work of the external forces, the influxes of external heat $dQ^{(e)}$ and the additional energy input dQ^{**} of a nonthermal nature.

Since the volume V and the Σ are arbitrary, determination of the coefficients in the linear formula by the variations and by their derivatives with respect to unknown functions in the integrand of the surface integral for δw is none other than determination of the state equations in the model under discussion. With this determination of the state equations, when the determining parameters include the entropy S or the temperature T , it is necessary to use the second law of thermodynamics in the form

$$\int_V \rho\, T ds d\tau = dQ^{(e)} + dQ' \qquad (4)$$

to eliminate $dQ^{(e)}$ from δw^*.

In equation (4), $dQ' \geqslant 0$ is non-compensated heat. To specify the quantity dQ' means to establish the energy dissipation laws for the model being introduced.

Therefore, in introducing the quantities Λ and δw^*, in order to fix model one can be guided by various physical and mathematical considerations, concerned with the thermodynamic properties of matters through the density of the internal energy U and through the dissipation laws determining the dQ'. The quantity dQ^{**} must be also introduced from the admitted physical mechanisms of energy exchange of the given small particle with the adjoining particles and with the external bodies and fields.

Proceeding from a physical consideration one can postulate equation (2) in its initial form (with $dQ^{(e)}$ not only for processes continuous in space and time, but also for processes with strong discontinuities at some surfaces $S_\pm$ inside the volume of the medium under consideration. In this case one can introduce into the quantity δw^* an additional integral over $S_\pm$ characterizing the concentrated energy inputs at discontinuities along $S_\pm$, proceeding from the physical consideration about such energy inputs.

Equation (2) is a natural generalization and development of the Lagrangian variational principle of analytical mechanics.

The proposed generalization and development of this principle is based on the following essential points :

1.- extension of the variational equation (2) to continuum media with mechanical and physico-chemical processes involving the internal freedom degrees which are characterized by macroscopic parameters having in general a nonmechanical nature ;

2.- application of equation (2) to describe irreversible phenomena. The Eulerian equations system includes not only general equations of conservation laws, but also kinetic equations of chemical reactions and phase transformations, and also the Maxwellian equations wherever the electrodynamic phenomena are taken into account ;

3.- equation (2) is considered for arbitrary volumes and for arbitrary continuous variations of the unknown functions, including the variations that do not equal to zero at the boundary Σ . This makes it possible to formulate the thermodynamic state equations which determine the characteristics of mechanical, chemical and electrodynamic interaction in bodies and in the field, and, for the metric properties of the space, this allows to find its geometrical characteristics ;

4.- after specifying the Lagrangian function Λ and the functional δw^* expressed in terms of the external energy influx to the particle, it becomes possible, on the base of the integral form of equation (2), to establish the conditions at the strong discontinuity surfaces - the jumps of the characteristic determining parameters inside the medium. These conditions can be used, in particular, to formulate the initial and boundary conditions ;

5.- as the base for all sorts of approximate and numerical methods of solution of particular problems one can use the integral equation (2), instead of the differential equations with additional initial and boundary conditions on jumps

and

6.- equation (2) can be used to construct special models of thin bodies - plates, shells, rods, of fluid motions in films or shallow channels (in thin fluid layers), for calculating composite structures, etc. In such cases one can single out certain directions, say, the direction of the plate thickness or the direction of the rod cross- section. One can then specify with respect to these directions (e.g. along the fine small thickness) the distributions of the state characteristics under study, expressing them in a simplified form by specific particular functions along the initial normal to the plate which include "internal parameters" that may depend on the coordinates along the plate and must be variable and determined by means of equation (2) in the course of problem solution.

An application of a similar method in its various modifications to the three - dimensional case can be a direct or indirect generalization of Bubnov's method.

After similar transformation of unknown functions one can make some integrations in equation (2) and the variational equation (2) reduces the problem to considering complicated model with internal freedom degrees but with fewer dimensions ; for example, in the case of plates and shells we have a two-dimensional model ; in the case of rods - an one-dimensional one, etc.

In connection with the basic equation (2) it is essential to pay attention to the following points.

Fixing the system of Eulerian equations including the equations of momentum, of moments of momentum, of energy, of entropy production, kinetic equations, etc. which can be not independent, does not determine the density of the Lagrangian function Λ . It is obvious that the system of Eulerian equations is not changed when we add non-zero "divergent terms" to the Lagrangian function, but in this case the expression of δW is changed and hence it has an influence on the form of the state equations. Consequently, a specification of a closed system of Eulerian equations, i.e. of all equations valid for different processes, is not sufficient to the state equations.

One often come across a statement, in textbooks and research papers that the conditions at strong discontinuities can be obtained if a closed system of differential equations describing some phenomena within the choosen model framework is given.

This is certainly wrong ! Indeed, first of all the discontinuity motions cannot be regarded in general case as the limiting situations of continuous motions within the framework of the same model, and secondly, when deriving the conditions at strong discontinuities it is necessary to use certain integral relationships which are not unique for the given system of differential equations.

For continuous motions many different systems of integral relations can correspond to a given system of differential equations and are fully equivalent to it.

For discontinuous motions, different systems of integral equations give different conditions contradicting each other.

Therefore it is clear that for establishing the conditions at strong discontinuities one must postulate as an empiric law an appropriate system of integral relations that would be valid for continuous motion as well as for the motions with strong discontinuities.

Experience shows that one must take as the integral relation,applicable to continuous as well as to discontinuous motions, the law of energy conservation expressed in an integral form. Note, for example, the equation of the heat flow in an integral form or the integral equation of the entropy conservation for adiabatic processes lead to incorrect conditions at a strong discontinuity.

In connection with this it is clear that the variational equation (2) in an integral form can be applied to discontinuous processes and to deriving conditions at jumps when expression of the density of the Lagrangian function Λ and of the functional δw^* are correctly defined (according to empiric data). In particular, when deriving the conditions at the discontinuities it is necessary to retain in the expression of δw^* all the terms characterizing energy flows and to introduce new concentrated energy flows to the discontinuity surfaces in the same form as in the energy equation. The function Λ also must be fixed (and this determines the divergent term), proceeding from physical considerations, which, in the final analysis, are justified by the agreement of the model with experiments in which strong discontinuities of state characteristics are observed.

Thus, it is obvious that, using the base equation (2) and the physical meanings of its terms, one can establish direct contacts with thermodynamics, electrodynamics and chemistry. For many further specific details and conclusions concerned with the use of equation (2) for reconstructing, on its base all known models of material media and fields, as well as for constructing new models, see the author's book "Mechanics of Continuum Media", v.1, 2nd ed., 1973 (in Russian) and also relevant research papers listed with brief summaries in the paper : L.I.Sedov, "Theoretical Models" (in Russian). (This paper was published in Proceedings of the Macedonian Academy of Sciences, Scople, Yugoslavia, 1975).

STEADILY ROTATING CHAINS

C.A. Stuart
Battelle Institute
Advanced Studies Center
1227 Carouge-Geneva
Switzerland

Introduction

We give a comparative study of two problems concerning the steady rotations of a chain, which we henceforth assume to be uniform, inextensible and perfectly flexible. The first problem (K) was originally treated by Kolodner [1] using classical, but delicate, analysis. The second problem (M) is a model for a spinning process and seems intractible by classical methods. In [2], both problems are treated, taking as a starting point the recent global results in bifurcation theory (e.g. [3]). Indeed, once Rabinowitz' theorem, [3], has been used to establish the bifurcation of unbounded branches of solutions, most of Kolodner's results follow from an easy application of comparison theorems. Although technically more complicated, the same approach can be successfully applied to Problem M, [2].

Statement of problems

We consider a chain with ends A and B. The end A is fixed and we introduce rectangular co-ordinates (x_1,x_2,x_3) with origin at A and with gravity acting in the direction of the positive x_3-axis. Arc-length measured along the chain from A is denoted by s. Let $\underline{u}(s,t)$ and $T(s,t)$ denote the position and tension at the point s at time t. We seek only motions in which the chain is stationary with respect to a plane which rotates with constant angular velocity ω about the x_3-axis. That is, T is independent of t and there exist functions v and w of s such that

$$\underline{u}(s,t) = (v(s)\cos \omega t, v(s)\sin \omega t, w(s)) .$$

Then we have $$v(0) = w(0) = 0 , \tag{1}$$

$$v'(s)^2 + w'(s)^2 = 1 , \quad (2)$$

$$-\rho\omega^2 v(s) = (T(s)v'(s))' , \quad (3)$$

$$-\rho g = (T(s)w'(s))' , \quad (4)$$

for $0 < s < \ell$, where ℓ is the length of the chain and ρ its density.

Problem K . Given $\rho > 0$ and $\ell > 0$, consider the system (1-4) subject to the additional boundary condition

$$T(\ell) = 0 . \quad (5)$$

This represents a chain of length ℓ with lower end free. For each $\omega \geq 0$, there is a trivial configuration: $v \equiv 0$, $w(s) = s$, $T(s) = \rho g(\ell-s)$.

Problem M . Given $\rho > 0$, $L > 0$ and $\alpha \geq 0$, consider the system (1-4) subject to the additional boundary conditions:

$$v(\ell) = 0 , \quad w(\ell) = L , \quad (6)$$

$$w'(\ell) > 0 \quad \text{and} \quad T(\ell) = \alpha . \quad (7)$$

The conditions (6) mean that the end B is fixed at a distance L vertically below A and the tension at B is α . As we shall see, $w'(\ell) > 0$ implies that B is the lowest point on the chain. Let us emphasise that the length, ℓ , of the chain is not prescribed. For any α , $\omega \geq 0$ and ρ , $L > 0$, we again have a trivial configuration:

$$\ell = L , \quad v \equiv 0 , \quad w(s) = s \quad \text{and} \quad T(s) = \alpha + \rho g(L-s) .$$

By choosing appropriate variables, both problems can be reduced to non-linear eigenvalue problems. Indeed, Kolodner shows that (K) reduces to the system:

$$-u''(s) = \lambda u(s)\{u(s)^2+s^2\}^{-\frac{1}{2}} \quad 0 < s < 1 \quad (8)$$

$$u(0) = u'(1) = 0 \quad (9)$$

where $\lambda = \omega^2\ell/g$, s is normalised arc-length measured from B and $u(s) = T(s)v'(s)/\ell$.

For (M) , arc-length is a less convenient independent variable and we prefer to use x_3 . From (4) we see that, provided $T(\ell) \geq 0$ and $w'(\ell) > 0$ or $T(\ell) = 0$ and $w(\ell) \geq w(0)$, $T(s) > 0$ and $w'(s) > 0$ for $0 < s < \ell$. Thus w is a homeomorphism of $[0,\ell]$ onto $[0,w(\ell)]$, and we introduce the new variables: $L = w(\ell)$, $z = L^{-1}w(s)$, $\mu = \omega^2 L/g$, $x(z) = L^{-1}v(s)$, $\tilde{T}(z) = (\rho g L)^{-1}T(s)$. Then (1-4) become:

$$x(0) = 0 \tag{10}$$

$$-\{(1+x'(z)^2)^{-\frac{1}{2}}x'(z)\tilde{T}(z)\}' = \mu x(z)(1+x'(z)^2)^{\frac{1}{2}} \tag{11}$$

$$-\{(1+x'(z)^2)^{-\frac{1}{2}}\tilde{T}(z)\}' = (1+x'(z)^2)^{\frac{1}{2}} \tag{12}$$

for $0 < z < 1$.

From (12) we have that

$$\tilde{T}(z) = (1+x'(z)^2)^{\frac{1}{2}}\left\{T(1)(1+x'(1)^2)^{-\frac{1}{2}} + \int_z^1 (1+x'(t)^2)^{\frac{1}{2}}dt\right\} . \tag{13}$$

In terms of z - x variables (M) becomes:

$$-\left\{\left[\beta(1+x'(1)^2)^{-\frac{1}{2}} + \int_z^1 (1+x'(t)^2)^{\frac{1}{2}}dt\right]x'(z)\right\}' = \mu x(z)(1+x'(z)^2)^{\frac{1}{2}} ,\quad 0 < z < 1 \tag{14}$$

with $x(0) = x(1) = 0$ and $|x'(1)| < \infty$ (15)

where $\beta = (\rho g L)^{-1}\alpha$, and Problem K becomes:

$$-\left\{\int_z^1 (1+x'(t)^2)^{\frac{1}{2}}dt\ x'(z)\right\}' = \mu x(z)(1+x'(z)^2)^{\frac{1}{2}} ,\quad 0 < z < 1 \tag{16}$$

with $x(0) = 0$, $|x(1)| < \infty$ and $\int_0^1 (1+x'(t)^2)^{\frac{1}{2}}dt = \ell/L$ (17)

Properties of solutions

We use the usual notation, [2], for the spaces of continuously differentiable real-valued functions on $[0,1]$. The norms in $C([0,1])$ and $C^1([0,1])$ are denoted by $\| \ \|$ and $\| \ \|_1$. Let $X = \{x \in C^1([0,1]) \cap C^2((0,1)) : x(0) = 0\}$. Then in the $z - x$ variables, any solution of (M) or (K) is an element of X which satisfies (14). But, for any $x \in X$ which satisfies (14) with $\lambda, \beta \geq 0$, it is easy to see that the zeros of x interlace the zeros of x' and that, between two consecutive zeros, x is first convex and then concave. Also multiplying (14) by $x'(z)(1+x'(z)^2)^{-\frac{1}{2}}$ and integrating from z to 1 yields:

$$\tilde{T}(z) = 1 - z + \beta + \tfrac{1}{2}\mu\{x(1)^2 - x(z)^2\} \quad \text{for } 0 < z < 1 . \tag{18}$$

If x is a solution of (K), then $\beta = 0$ and it follows from (13) and (18) that $x(z)^2 < x(1)^2$ for $0 \leq z < 1$. That is, the maximal displacement is at the end B. If x is a solution of (M) then $x(1) = 0$ and so $\tilde{T}(0) = 1 + \beta$. Thus prescribing a tension α at B is equivalent to prescribing a tension $\alpha + \rho g L$ at A. Furthermore, it follows from (18) that, for a solution of (M), the length of the chain does not exceed $L(1+2\alpha/\rho g L)^{\frac{1}{2}}$ and the displacement from the vertical axis does not exceed $(\alpha L/2\rho g)^{\frac{1}{2}}$. These bounds do not depend on λ and, in particular, we see that $x \equiv 0$ if $\beta = 0$.

It is also shown in [2] that, for $x \in X$ which satisfies (14), we have

$x(z_1)^2 < x(z_2)^2$ provided that $z_1 < z_2$ and $x'(z_1) = x'(z_2) = 0$
and $x'(z_1)^2 < x'(z_2)^2$ provided that $z_1 < z_2$ and $x(z_1) = x(z_2) = 0$.
Thus, for any solution of (M), we see that the maximum displacement occurs at the lowest turning value.

Solutions of (8-9) are also elements of X and clearly u is concave between any two consecutive zeros. Since $u'(s) = -\lambda v(s)/\ell$, it follows from the results about x that $|u'|$ takes its maximum value at B. Multiplying (8) by $u'(s)(u(s)^2+s^2)^{\frac{1}{2}}$ and integrating from s to 1, we find that $u(s)^2 < u(1)^2$ for $0 \leq s < 1$. Thus the horizontal component of the tension is greatest at A.

Spectrum of Problem K

Let E denote the real Banach space $\{u \in C^1([0,1]) : u(0) = 0\}$ with the norm $\| \|_1$. Let Gh be the unique solution of

$$-u'' = h \ , \quad u(0) = u'(1) = 0$$

and let $Lu(s) = s^{-1}u(s)$. Then $G : C([0,1]) \to E$ is a compact linear operator and $L : E \to C([0,1])$ is a bounded linear operator. Setting

$$\begin{aligned} F(u)(s) &= s^{-1}u(s)\{(s^{-2}u(s)^2+1)^{-\frac{1}{2}}-1\} & 0 < s \leqslant 1 \\ &= u'(0)\{(u'(0)^2+1)^{-\frac{1}{2}}-1\} & s = 0 , \end{aligned}$$

we have that $F : E \to C([0,1])$ is bounded and continuous and $\|u\|_1^{-1}\|F(u)\| \to 0$ as $\|u\|_1 \to 0$. Thus (8-9) is equivalent to

$$u = \lambda GLu + \lambda GF(u) \quad \text{for } (u,\lambda) \in E \times \mathbb{R} , \tag{19}$$

and we set $\mathcal{S} = \{(u,\lambda) \in E \times \mathbb{R} : u$ satisfies (19) and $u \neq 0\}$. The set of positive integers is denoted by $\mathcal{N}$ and, for $n \in \mathcal{N}$, $S_n = \{u \in C^1([0,1]) : u$ has exactly $n-1$ zeros in $(0,1)$ and all the zeros in $[0,1]$ are simple$\}$. Then $\mathcal{S} \subset \bigcup_{n=1}^{\infty} S_n \times \mathbb{R}$.

The linearisation of (19) at the trivial solution, $(0,\lambda) \in E \times \mathbb{R}$, is equivalent to:

$$-u''(s) = \lambda s^{-1}u(s) \ , \quad 0 < s < 1 \ \text{ with } u(0) = u'(1) = 0 . \tag{20}$$

The n-th eigenvalue of (20) is $(\frac{1}{2}\sigma_n)^2$ where σ_n is the n-th zero of the Bessel function J_o. The corresponding eigenfunction, $s^{\frac{1}{2}}J_1(\sigma_n s^{\frac{1}{2}})$, is in S_n.

For $n \in \mathcal{N}$, $\mathcal{S} \cup \{(0,(\frac{1}{2}\sigma_n)^2)\}$ is a metric space with the $E \times \mathbb{R}$ metric and we denote by $\mathcal{C}_n$ the component (i.e. maximal connected subset) containing $(0,(\frac{1}{2}\sigma_n)^2)$. As in [3], we have $\mathcal{C}_n \setminus \{(0,(\frac{1}{2}\sigma_n)^2)\} \subset S_n \times \mathbb{R}$ and $\mathcal{C}_n$ is unbounded. An easy lemma sharpens this result.

<u>Lemma 1</u>. Suppose $(u,\lambda)\in \mathcal{S}$ with $u\in S_n$. Then

$$(\tfrac{1}{2}\sigma_n)^2 < \lambda < \min\{(\tfrac{1}{2}\sigma_n)^2(1+u'(0)^2)^{\frac{1}{2}},(\tfrac{1}{2}n\pi)^2(1+u(1)^2)^{\frac{1}{2}}\}\ , \qquad (21)$$

and

$$|u'(0)| = \|u'\| \leqslant \lambda \qquad (22)$$

<u>Proof</u>: Consider the "regular singular" Sturm-Liouville problem:

$$-v''(s) = \gamma s^{-1}v(s)\{s^{-1}u(s)^2+1\}^{-\frac{1}{2}}\ , \quad 0 < s < 1 \qquad (23)$$

$$v(0) = v'(1) = 0\ . \qquad (24)$$

Then $u\in S_n$ is a solution of (23-24) and consequently $\gamma = \lambda$ is the n-th eigenvalue. However

$$\{\|u'\|^2+1\}^{-\frac{1}{2}} \leqslant \{s^{-2}u(s)^2+1\}^{-\frac{1}{2}} \leqslant 1$$

and

$$\{\|u\|^2+1\}^{-\frac{1}{2}} \leqslant \{u(s)^2+s^2\}^{-\frac{1}{2}} \quad \text{for } 0 < s < 1\ .$$

But, as shown above, $\|u'\| = |u'(0)|$ and $\|u\| = |u(1)|$. Hence, (21) follows from Theorem 7 on page 411 of [4]. For (22), we have

$$u'(s) = -\int_s^1 u''(t)dt = \lambda\int_s^1 u(t)\{u(t)^2+t^2\}^{-\frac{1}{2}}dt,$$

and so $|u'(s)| \leqslant \lambda(1-s)$ for $0 < s < 1$.

To describe the spectrum of (K), let

$$\sigma_K(n) = \{\lambda : \exists (u,\lambda)\in \mathcal{S} \text{ with } u\in S_n\}\ .$$

<u>Corollary 2</u>. For each $n\in \mathcal{N}$, we have

$$\{|u(1)| : (u,\lambda)\in \mathcal{C}_n\} = \{|u'(0)| : (u,\lambda)\in \mathcal{C}_n\} = [0,\infty)$$

and

$$\{\lambda : (u,\lambda)\in \mathcal{C}_n\} = [(\tfrac{1}{2}\sigma_n)^2,\infty)\ .$$

Indeed,

$$\sigma_K(n) = ((\tfrac{1}{2}\sigma_n)^2,\infty)\ .$$

<u>Proof</u>: This follows from the unboundedness of $\mathcal{C}_n$ and (21-22).

Note that as $\lambda \to \infty$, $|u(1)| \to \infty$ and so the horizontal component of the tension at A approaches infinity. Indeed, using (13), we have that $T(0) = \rho g L \ell / w'(0)$ and so the slope of the chain at A converges to the horizontal as $\lambda \to \infty$.

Kolodner proves that $\mathcal{C}_n$ is a curve parametrised by λ . For $n = 1$, a simple proof is available using the comparison theorem. The map $I - \lambda GL - \lambda GF : E \times \mathbb{R} \to E$ is Fréchet differentiable at every point (u,λ) of $E \times \mathbb{R}$ and its derivative with respect to u has a bounded inverse provided that λ is not an eigenvalue of the problem:

$$-v''(s) = \gamma s^{-1} v(s)\{s^{-2}u(s)^2+1\}^{-3/2} \;,\; 0 < s < 1 \quad \text{with} \quad v(0) = v'(1) = 0 \qquad (25)$$

But, if $(u,\lambda) \in \mathcal{C}_1$, λ is the first eigenvalue of (23-24). Since

$$\{s^{-2}u(s)^2+1\}^{-3/2} \leq \{s^{-2}u(s)^2+1\}^{-\frac{1}{2}} \quad \text{for} \quad 0 < s < 1$$

the comparison theorem shows that λ cannot be an eigenvalue of (25). By the implicit function theorem it now follows that $\mathcal{C}_1$ is a curve parametrised by λ .

Spectrum of Problem M

Let E_o denote the Banach space $\{x \in E : x(1) = 0\}$. We need only consider $\beta > 0$ and we denote by $H(\beta)h$ the unique solution of:

$$-\{(1-z+\beta)x'(z)\}' = h(z) \quad \text{with} \quad x(0) = x(1) = 0 \ .$$

Then, for $\beta > 0$, $H(\beta) : C([0,1]) \to E_o$ is a compact linear operator and, rearranging (14), we find that (M) is equivalent to,

$$x = \mu H(\beta)x + H(\beta)R(\beta,\mu,x) \quad \text{for} \quad (x,\mu) \;\; E_o \times \mathbb{R} \;, \qquad (26)$$

where $R(\beta,.,.) : E_o \times \mathbb{R} \to C([0,1])$ is bounded continuous and $\|x\|_1^{-1}\|R(\beta,\mu,x)\| \to 0$ as $\|x\|_1 \to 0$ uniformly for μ in bounded intervals. Let $\mathcal{J}(\beta) = \{(x,\mu) \in E_o \times \mathbb{R} : x$ satisfies (26) and $x \neq 0\}$. Then $\mathcal{J}(\beta) \subset \bigcup_{n=1}^{\infty} S_n \times \mathbb{R}$.

The linearisation of (26) at the trivial solution $(0,\mu)$ is equivalent to

$$-\{(1-z+\beta)x'(z)\}' = \mu x(z) \;,\quad 0 < z < 1 \;\text{ with }\; x(0) = x(1) = 0 \;. \quad (27)$$

The general solution of (27) is

$$AJ_0(2\mu^{\frac{1}{2}}(1-z+\beta)^{\frac{1}{2}}) + BY_0(2\mu^{\frac{1}{2}}(1-z+\beta)^{\frac{1}{2}})$$

where J_0 and Y_0 are Bessel functions of the first and second kind. Hence the eigenvalues, $\mu_n(\beta)$, of (27) are given by the zeros of

$$J_0(2\mu^{\frac{1}{2}}(1+\beta)^{\frac{1}{2}})Y_0(2\mu^{\frac{1}{2}}\beta^{\frac{1}{2}}) - J_0(2\mu^{\frac{1}{2}}\beta^{\frac{1}{2}})Y_0(2\mu^{\frac{1}{2}}(1+\beta)^{\frac{1}{2}})$$

and the corresponding eigenfunction,

$$J_0(2\mu_n(\beta)^{\frac{1}{2}}(1-z+\beta)^{\frac{1}{2}})Y_0(2\mu_n(\beta)^{\frac{1}{2}}\beta^{\frac{1}{2}}) - J_0(2\mu_n(\beta)^{\frac{1}{2}}\beta^{\frac{1}{2}})Y_0(2\mu_n(\beta)^{\frac{1}{2}}(1-z+\beta)^{\frac{1}{2}})$$

is in S_n . For fixed n , $\mu_n(\beta)$ is an increasing function of β and

$$\mu_n(\beta) \to \infty \quad \text{as} \quad \beta \to \infty$$
$$\mu_n(\beta) \to (\tfrac{1}{2}\sigma_n)^2 \quad \text{as} \quad \beta \to 0 \;.$$

For $n \in \mathcal{N}$, let $\mathcal{D}_n(\beta)$ denote the component of $\mathcal{J}(\beta) \cup \{(0,\mu_n(\beta))\}$ containing $(0,\mu_n(\beta))$. As in [3], $\mathcal{D}_n(\beta)\backslash\{(0,\mu_n(\beta))\} \subset S_n \times \mathbb{R}$ and $\mathcal{D}_n(\beta)$ is unbounded. This conclusion can again be sharpened.

Lemma 3 . Suppose that $(x,\mu) \in \mathcal{J}(\beta)$ and $x \in S_n$. Then

$$(1+2\beta)^{-\frac{1}{2}}(\tfrac{1}{2}\sigma_n)^2 < \mu < \mu_n(\beta) \;.$$

A proof is given in [2] and we thus have found an interval containing $\{\mu : (x,\mu) \in \mathcal{D}_n(\beta)\}$. We now seek an interval contained in $\{\mu : (x,\mu) \in \mathcal{D}_n(\beta)\}$. For this we set

$$\Sigma(\beta,n) = \{\mu : \exists\, \{(x_m,\gamma_m)\} \subset \mathcal{D}_n(\beta) \text{ such that } \gamma_m \to \mu$$
$$\text{and } \|x_m'\| \to \infty \text{ as } m \to \infty\} \;.$$

The following information is obtained in [2].

Lemma 4 . For each $\beta > 0$ and $n \in \mathcal{N}$, $\Sigma(\beta,n) \neq \phi$. Suppose that $\mu \in \Sigma(\beta,n)$ and that $\{(x_m,\gamma_m)\} \subset \mathcal{D}_n(\beta)$ is such that $\gamma_m \to \mu$ and

$\|x_m'\| \to \infty$ as $m \to \infty$. Then $|x_m'(1)| \to \infty$ and there exists $v \in C'([0,1))$ such that $x_m \to v$ in $C^1([0,b])$ for each $b < 1$. Furthermore $v' \in L^1(0,1)$ and

$$\int_0^1 (1+v'(z)^2)^{\frac{1}{2}}dz = \lim_{m\to\infty} \int_0^1 (1+x_m'(z)^2)^{\frac{1}{2}}dz \geq \{1 + 8\beta/(4+\mu_n(\beta))\}^{\frac{1}{2}} \equiv \gamma_n(\beta) .$$

The comparison theorem now gives an upper bound for $\sum(\beta,n)$.

<u>Lemma 5</u>. For $\beta > 0$ and $n \in \mathcal{N}$,

$$(1+2\beta)^{-\frac{1}{2}}(\tfrac{1}{2}\sigma_n)^2 \leq \inf \textstyle\sum(\beta,n) \leq \sup \textstyle\sum(\beta,n) \leq \mu_n^*(\beta)$$

where $\mu_n^*(\beta) = \gamma_n(\beta)^{-1}\mu_n(\beta\gamma_n(\beta)^{-1})$.

To describe the spectrum of (M), let

$$\sigma_M(\beta,n) = \{\mu : \exists(x,\mu) \in \mathcal{J}(\beta) \text{ with } x \in S_n\} .$$

<u>Corollary 6</u>. For $\beta > 0$ and $n \in \mathcal{N}$,

$$((1+2\beta)^{-\frac{1}{2}}(\tfrac{1}{2}\sigma_n)^2, \mu_n(\beta)) \subset \sigma_M(\beta,n) \subset (\mu_n^*(\beta), \mu_n(\beta)) .$$

Unlike the case (K), we have not been able to determine the spectrum exactly. It would be interesting to know what happens if the condition $|x'(1)| < \infty$ in (15) is removed.

<u>Comments</u>

In many ways the Problems K and M exhibit complementary behaviour. In (K) the bifurcations are to the right and in (M) they are to the left. Also $\sigma_K(n)$ is unbounded and $\sigma_K(n+1) \subset \sigma_K(n)$, whereas $\sigma_M(\beta,n)$ is bounded and, for β small and $m \neq n$,

$$\sigma_M(\beta,m) \cap \sigma_M(\beta,n) = \phi .$$

For (K) the tension at A tends to infinity as ω approaches infinity. For (M) the tension at all points is bounded by $\alpha + \rho L$, independent of ω. For (K) the slope of the chain at A approaches the horizontal as $\omega \rightarrow \infty$, whereas for (M) the slope remains bounded away from the horizontal at all points except B where it does approach the horizontal.

References

[1] Kolodner, I.I.: Heavy rotating string - a non-linear eigenvalue problem, Comm. Pure Appl. Math., 8 (1955), 394-408.

[2] Stuart, C.A.: Spectral theory of rotating chains, to appear in Proc. Roy. Soc. Edinburgh.

[3] Rabinowitz, P.H.: Some global results for non-linear eigenvalue problems, J. Functional Anal., 8 (1971), 487-513.

[4] Courant, R. and Hilbert, D.: Methods of Mathematical Physics, I. New York: Interscience (1953).

GENERATING FUNCTIONALS AND EXTREMUM PRINCIPLES IN NONLINEAR ELASTICITY WITH APPLICATIONS TO NONLINEAR PLATE AND SHALLOW SHELL THEORY

H. Stumpf
Lehrstuhl für Mechanik II
Ruhr-Universität Bochum / BRD

1. Introduction

In finite elasticity the solution of a boundary value problem is not always unique and the associated equilibrium state of an elastic solid can be stable or unstable. A criterion for the uniqueness of a solution was derived by HILL [1]. A generalization of the dual extremum principles of linear elasticity cannot lead to extremum principles valid in the whole domain of finite deformations. In general they can only be stationary principles with extremum property in some subdomain.

In recent papers, NOBEL and SEWELL [2] and SEWELL [3] gave a unified viewpoint for the derivation of dual extremum principles in the fields of applied mathematics by introducing inner product spaces, adjoint operators and saddle functionals. For the linear elasticity the well-known dual extremum principles are considered in detail, while for the nonlinear elasticity adjoint operators and generating functionals are given for three independent continuous fields without investigating extremum principles. For the nonlinear plate theory according to von KÁRMÁN and for the nonlinear shallow shell theory dual extremum principles had been considered by the author [4 - 6].

In this paper the adjoint operators and generating functionals of [2,3] are generalized for nonlinear elastic boundary value problems by introducing complementary boundary conditions and four independent continuous fields: the unsymmetric Piola stress tensor, the displacement vector, the displacement gradient tensor and the boundary traction, associated with given surface displacements. This leads to the most general variational principle, which is equivalent to the principle of WASHIZU [7], who uses the symmetric KIRCHHOFF stress and GREEN strain tensor instead of the Piola stress and displacement gradient tensor. Special cases of the general principle are the variational statements in [3,8] and the principle of REISSNER [9].

It is shown that in nonlinear elasticity dual extremum principles can be derived from a generating functional, if this functional is saddle-shaped. The condition of saddle property leads to the uniqueness criterion of the solution and stability condition of the associated elastic state given by HILL [1]. For unique solutions the total potential energy is a convex functional generating a minimum principle, while the total complementary energy is a concave functional generating a maximum principle. Minimum and maximum values coincide and are equal to the solution value of the generating functional. With these dual extremum principles error bounds can be calculated.

For the nonlinear plate theory according to von KÁRMÁN [10] and for the nonlinear shallow shell theory according to MARGUERRE [11] the dual extremum principles are investigated in detail.

2. Variational Functionals and Extremum Principles in Nonlinear Elasticity

Consider a body B, which has in its initial virgin state the open region V, closed by an external surface S. The closure of V may be written $\overline{V} = V + S$. A fixed rectangular Cartesian coordinate system will be used with $x_j (j = 1,2,3)$ denoting the position of a particle in the initial state. The displacements will be denoted by u_j. Assume that B is deformed from the initial configuration to a deformed configuration by body forces p_j, surface tractions (dead-load type) and surface displacements. On the surface S may be given the surface tractions $P_j^{*\prime}$ and the complementary components of the surface displacements $u_j^{*\prime\prime}$ [8]. We denote by asterisk given components, by prime and double-prime components referring to prescribed components of tractions and displacements, respectively. This includes the special case, in which the surface S consists of two parts S_1 and S_2 with given surface tractions P_j^* on S_1 and given displacements u_j^* on S_2. Let the body B consist of an elastic material with the strain-energy density $U(e_{ij})$ per undeformed unit volume with e_{ij} the nonlinear GREEN strain tensor. U can also be considered as a function of the symmetric part of the deformation gradient tensor $u_{j,i}$.

The unsymmetric Piola stress tensor follows from the strain energy density $U(u_{j,i})$ by differentiation according to

$$t_{ij} = \frac{\partial U}{\partial u_{j,i}} \tag{2.1}$$

Using the Piola stress tensor the nonlinear elastic boundary value problem can be defined by the following equations:

Deformation gradient tensor	$u_{j,i} = d_{ij}$	(2.2)
Prescribed displacement boundary conditions	$u_j'' = u_j^{*\prime\prime}$	(2.3)
Equilibrium equations	$- t_{ij,i} = p_j$	(2.4)
Prescribed traction boundary conditions	$n_i t_{ij}' = P_j^{*\prime}$	(2.5)

Piola stress tensor $$0 = t_{ij} - \frac{\partial U}{\partial d_{ij}} \tag{2.6}$$

Traction boundary conditions $$0 = f''_j - n_i t''_{ij} \tag{2.7}$$

with n_i the exterior unit normal to the surface S. Indices notation with summation convention is used.

The boundary value problem (2.2) - (2.7) can be described by a matrix operator equation with two linear adjoint operators. Let E and F be two inner product spaces consisting of elements t, d, ... and u, v, f, ..., constructed as column vectors

$$t = \begin{bmatrix} t_{ij}(V) \\ t'_{ij}(S) \\ t''_{ij}(S) \end{bmatrix} \quad ; \quad u = \begin{bmatrix} u_j(V) \\ u'_j(S) \\ u''_j(S) \end{bmatrix} \tag{2.8}$$

with the inner products (,) and < , > , defined by

$$(t,d) = \int_V t_{ij} d_{ij} \, dV + \int_S (t'_{ij} d'_{ij} + t''_{ij} d''_{ij}) \, dS \tag{2.9}$$

$$\langle u,v \rangle = \int_V u_j v_j \, dV + \int_S (u'_j v'_j + u''_j v''_j) \, dS \, . \tag{2.10}$$

Two subspaces E' and F' are introduced with
E': subspace of E of continuous and single-valued tensor-functions t_{ij}, ... ;
F': subspace of F of continuous and single-valued vector-functions u_j, ...

A linear operator T : E' → F and its adjoint T* : F' → E are defined as

$$Tt = \begin{bmatrix} -t_{ij,i} \\ n_i t'_{ij} \\ 0 \end{bmatrix} \qquad T^*u = \begin{bmatrix} u_{j,i} \\ 0 \\ -n_i u''_j \end{bmatrix} \, . \tag{2.11}$$

The adjointness property can be proved by using the divergence theorem:

$$(t,T^*u) = \int_V t_{ij} u_{j,i} \, dV - \int_S n_i t''_{ij} u''_j \, dS$$

$$= - \int_V t_{ij,i} u_j \, dV + \int_S n_i t'_{ij} u'_j \, dS = \langle Tt,u \rangle \, . \tag{2.12}$$

With vectors [t,u] as elements of a Cartesian product space E x F, SEWELL [3] introduced an operator matrix with the mapping property

$$\begin{bmatrix} 0 & T^* \\ T & 0 \end{bmatrix} \; : \; E' \times F' \to E \times F \, . \tag{2.13}$$

The operator (2.13) is self-adjoint, which is shown by the adjointness property (2.12):

$$[t_1,u_1]\begin{bmatrix}0 & T^*\\ T & 0\end{bmatrix}\begin{bmatrix}t_2\\ u_2\end{bmatrix} = (t_1,T^*u_2) + < u_1,Tt_2 >$$
$$= (t_2,T^*u_1) + < u_2,Tt_1 > . \quad (2.14)$$

The self-adjoint operator (2.13), acting on the vector [t,u], can be expressed as the gradient of a bilinear functional Q

$$\begin{bmatrix}0 & T^*\\ T & 0\end{bmatrix}\begin{bmatrix}t\\ u\end{bmatrix} = \partial Q/\partial\begin{bmatrix}t\\ u\end{bmatrix} \quad (2.15)$$

with

$$Q = (t,T^*u) = < u,Tt > . \quad (2.16)$$

The gradient of a functional is defined by the GÂTEAUX differential [12].

To treat the nonlinear elastic boundary value problem, described by the equations (2.2) - (2.7), we introduce a vector [t,u,d,f] of four continuous fields : t represents the Piola stress tensor t_{ij}, u the displacement field u_j, d the displacement gradient tensor d_{ij} and f the boundary traction f_j. With an operator matrix self-adjoint in the Cartesian product space E x F x E x F, the left sides of the equations (2.2) - (2.7) can be expressed by

$$\begin{bmatrix}0 & T^* & 0 & 0\\ T & 0 & 0 & 0\\ 0 & 0 & 0 & 0\\ 0 & 0 & 0 & 0\end{bmatrix} = \partial Q/\partial\begin{bmatrix}t\\ u\\ d\\ f\end{bmatrix} . \quad (2.17)$$

We introduce a functional H [t,u,d,f] with the domain E' x F' x E' x F' by defining

$$H\,[t,u,d,f] = \int_V [t_{ij}d_{ij} - U(d_{ij}) + p_j u_j]\,dV$$
$$+ \int_S [P_j^{*\prime}\,u_j' - n_i t_{ij}''\,u_j'' + f_j''\,(u_j'' - u_j^{*\prime\prime})]\,dS . \quad (2.18)$$

It is assumed that each integrand of H and also the partial derivatives with respect to its arguments are continuous functions. The gradient of H is given by the matrices

$$\frac{\partial H}{\partial t} = \begin{bmatrix}d_{ij}\\ 0\\ -n_i u_j''\end{bmatrix};\quad \frac{\partial H}{\partial u} = \begin{bmatrix}p_j\\ P_j^{*\prime}\\ f_j'' - n_i t_{ij}''\end{bmatrix};\quad \frac{\partial H}{\partial d} = \begin{bmatrix}t_{ij} - \frac{\partial U}{\partial d_{ij}}\\ 0\\ 0\end{bmatrix};\quad \frac{\partial H}{\partial f} = \begin{bmatrix}0\\ 0\\ u_j'' - u_j^{*\prime\prime}\end{bmatrix} . \quad (2.19)$$

Using the functionals Q and H the system of governing equations (2.2) - (2.7) can be expressed by

$$\partial(Q - H)/\partial \begin{bmatrix} t \\ u \\ d \\ f \end{bmatrix} = \partial L/\partial \begin{bmatrix} t \\ u \\ d \\ f \end{bmatrix} = 0 \tag{2.20}$$

with the generating functional

$$\begin{aligned} L[t,u,d,f] &= Q - H[t,u,d,f] \\ &= \int_V [U(d_{ij}) + (u_{j,i} - d_{ij})t_{ij} - p_j u_j]\, dV \\ &\quad - \int_S [P_j^{*\prime} u_j^{\prime} + f_j^{\prime\prime}(u_j^{\prime\prime} - u_j^{*\prime\prime})]\, dS \quad . \end{aligned} \tag{2.21}$$

The domain of L is E' x F' x E' x F'.

Equation (2.20) is equivalent to the free variational principle with no subsidary conditions:

$$\delta L[t,u,d,f] = 0 \tag*{$(2.22)_1$}$$

or

$$(\delta t, \frac{\partial L}{\partial t}) + < \delta u, \frac{\partial L}{\partial u} > + (\delta d, \frac{\partial L}{\partial d}) + < \delta f, \frac{\partial L}{\partial f} > = 0 \tag*{$(2.22)_2$}$$

respectively:

$$\begin{aligned} &\int_V [(u_{j,i} - d_{ij})\delta t_{ij} - (t_{ij,i} + p_j)\delta u_j + (\frac{\partial U}{\partial d_{ij}} - t_{ij})\delta d_{ij}]\, dV \\ &+ \int_S [(n_i t_{ij}^{\prime} - P_j^{*\prime})\delta u_j^{\prime} + (n_i t_{ij}^{\prime\prime} - f_j^{\prime\prime})\delta u_j^{\prime\prime} - (u_j^{\prime\prime} - u_j^{*\prime\prime})\delta f_j^{\prime\prime}]\, dS = 0 \quad . \end{aligned} \tag{2.23}$$

The variational principle (2.22) respectively (2.23) with four independent continuous fields [t,u,d,f] is the most general variational statement in finite elasticity. A principle of this kind is considered by WASHIZU [7], who uses in contrary to (2.22) as independent fields the symmetric KIRCHHOFF stress tensor, the displacement vector, the GREEN strain tensor and the surface traction f_j.

Eliminating the surface traction f_j by introducing condition (2.7) a generating functional L[t,u,d] with the domain E' x F' x E' is obtained leading to a free variational statement with three independent fields [t,u,d] and without subsidary conditions [3,8] :

$$\delta L[t,u,d] = 0 \tag{2.24}$$

respectively

$$(\delta t, \frac{\partial L}{\partial t}) + < \delta u, \frac{\partial L}{\partial u} > + (\delta d, \frac{\partial L}{\partial d}) = 0 \tag{2.25}$$

Eliminating the surface traction f_j and the displacement gradient tensor d_{ij} by satisfying the equations (2.7) and (2.2) and introducing the complementary energy function $U_c(t_{ij})$ by the Legendre transformation

$$U(u_{j,i}) = t_{ij}u_{j,i} - U_c(t_{ij}) ,$$

a variational functional $L[t,u]$ in the domain $E' \times F'$ with two independent fields $[t,u]$ is obtained:

$$L[t,u] = \int_V [t_{ij}u_{j,i} - U_c(t_{ij}) - p_j u_j]\, dV - \int_S [P_j^{*\prime} u_j^{\prime} + n_i t_{ij}^{\prime\prime}(u_j^{\prime\prime} - u_j^{*\prime\prime})]\, dS . \qquad (2.27)$$

The variational statement

$$\delta L[t,u] = (\delta t, \frac{\partial L}{\partial t}) + < \delta u, \frac{\partial L}{\partial u} > = 0 \qquad (2.28)$$

is equivalent to the variational principle of REISSNER [9], who uses in contrary to (2.28) the symmetric KIRCHHOFF stress tensor and the displacement field as independent variables.

The considered variational statements are stationary principles without extremum property. Sharper results in the sense of calculating error bounds can be obtained by extremum principles. A necessary condition to derive dual extremum principles is a saddle property of the generating functional $L[t,u,d,f]$. The saddle property can be shown by verifying that $L[t,u,d,f]$ is jointly concave in t and f and jointly convex in u and d. Since a purely linear term may be regarded as either weakly concave or weakly convex, the saddle property of L is defined by the convexity property of the deformation energy $U(u_{j,i})$. For all large displacement problems (2.2) - (2.7), for which $L[t,u,d,f]$ is a saddle functional, the solution of (2.2) - (2.7) is unique.

Let $[t_+, u_+, d_+, f_+] - [t_-, u_-, d_-, f_-]$ are pairs of points and $\Delta L \equiv$ $\equiv L[t_+, u_+, d_+, f_+] - L[t_-, u_-, d_-, f_-]$ and $\Delta t \equiv t_+ - t_-$. Then the saddle property of L requires [2,3]:

$$\Delta L - (\Delta t, \frac{\partial L}{\partial t}\Big|_+) - < \Delta u, \frac{\partial L}{\partial u}\Big|_- > - (\Delta d, \frac{\partial L}{\partial d}\Big|_-) - < \Delta f, \frac{\partial L}{\partial f}\Big|_+ > \geq 0 \qquad (2.29)$$

for all possible distinct pairs of points . Inequality (2.29) is equivalent to the condition:

$$- (\Delta t, \Delta \frac{\partial L}{\partial t}) + < \Delta u, \Delta \frac{\partial L}{\partial u} > + (\Delta d, \Delta \frac{\partial L}{\partial d}) - < \Delta f, \Delta \frac{\partial L}{\partial f} > \geq 0 \qquad (2.30)$$

for all Δt, Δu, Δd, Δf not all zero. With (2.21) inequality (2.30) leads to the necessary condition for a saddle property of the functional L :

$$\int_V \frac{\partial^2 U}{\partial d_{ij} \partial d_{kl}} \Delta d_{ij} \Delta d_{kl} \, dV \geq 0 \, . \tag{2.31}$$

A sufficient condition for the saddle property is given by

$$\frac{\partial^2 U}{\partial d_{ij} \partial d_{kl}} \quad \text{positive definite} \tag{2.32}$$

With (2.2) and (2.3) the saddle condition (2.32) leads to a sufficient criterion for stability of an elastic solid and the criterion for a unique solution of the nonlinear elastic boundary value problem.

Analogous to the extremum principles of SEWELL [3] in linear elasticity, we formulate the following dual extremum principles of the nonlinear elasticity.

I. The solution of the nonlinear elastic boundary value problem (2.2) - (2.7) minimizes the functional

$$\begin{aligned} I &= L[t,u,d,f] - (t, \frac{\partial L}{\partial t}) - < f, \frac{\partial L}{\partial f} > \\ &= \int_V [U(d_{ij}) - p_j u_j] \, dV - \int_S p_j^{*\prime} u_j^{\prime} \, dS \end{aligned} \tag{2.33}$$

among all solutions of the subproblem, defined by (2.32) and

$$\frac{\partial L}{\partial t} = 0 = \begin{bmatrix} u_{j,i} - d_{ij} \\ 0 \\ 0 \end{bmatrix} \; ; \quad \frac{\partial L}{\partial f} = 0 = \begin{bmatrix} 0 \\ 0 \\ -(u_j^{\prime\prime} - u_j^{*\prime\prime}) \end{bmatrix} . \tag{2.34}$$

II. The solution of the nonlinear elastic boundary value problem (2.2) - (2.7) maximizes the functional

$$\begin{aligned} I_c &= L[t,u,d,f] - < u, \frac{\partial L}{\partial u} > - (d, \frac{\partial L}{\partial d}) \\ &= \int_V [U(d_{ij}) - \frac{\partial U}{\partial d_{ij}} d_{ij}] \, dV + \int_S f_j^{\prime\prime} u_j^{*\prime\prime} \, dS \end{aligned} \tag{2.35}$$

among all solutions of the subproblem, defined by (2.32) and

$$\frac{\partial L}{\partial u} = 0 = \begin{bmatrix} -(t_{ij,i} + p_j) \\ n_i t_{ij}^{\prime} - p_j^{*\prime} \\ n_i t_{ij}^{\prime\prime} - f_j^{\prime\prime} \end{bmatrix} \; ; \quad \frac{\partial L}{\partial d} = 0 = \begin{bmatrix} \frac{\partial U}{\partial d_{ij}} - t_{ij} \\ 0 \\ 0 \end{bmatrix} \tag{2.36}$$

with a unique inversion of $(2.36)_2$.

III. The minimum value of I and the maximum value of I_c for the solution of the non-linear elastic boundary value problem (2.2) - (2.7) are the same and equal to the solution value of L[t,u,d,f].

In the following sections the dual extremum principles will be considered for the nonlinear von KÁRMÁN plate theory and the nonlinear shallow shell theory according to MARGUERRE [11].

3. Extremum principles of the nonlinear von KÁRMÁN plate theory

The Langrangian description with Cartesian coordinates x_j (j = 1,2,3) will be used. The x_α (α = 1,2) plane of the coordinate system coincide with the middle plane of the undeformed plate and the x_3-axis is normal to it with the plate faces at $x_3 = \pm h/2$. Henceforth the ranges of the subscripts will be i,j = 1,2,3 and α,β = 1,2. Let u_j be the displacement vector of the middle plane of the plate. Letting $M_{\alpha\beta}$ be the Piola components of the stress moments and $N_{\alpha i}$ the Piola components of the stress resultants, we introduce the following notations:

$$N_{\alpha 3} = N_{\alpha\beta}u_{3,\beta} \qquad M_{\beta n} = n_\alpha M_{\beta\alpha} \qquad M_{ns} = \varepsilon_{3\alpha\beta}n_\alpha M_{\beta n}$$
$$N_{ni} = n_\alpha N_{\alpha i} \qquad M_{nn} = n_\alpha M_{\alpha n} \qquad Q_{n3} = M_{\alpha n,\alpha} \tag{3.1}$$

with n_α the unit normal vector to the boundary curve C of the undeformed middle plane F and $\varepsilon_{3\alpha\beta}$ the skew-symmetric permutation tensor.

Presuming the validity of KIRCHHOFF's hypothesis the strain energy density $U(u_{j,i})$, measured per undeformed middle area of the plate, is given as

$$U(u_{j,i}) = \frac{Eh}{2(1-\nu^2)}[(1-\nu)e^{(0)}_{\alpha\beta}e^{(0)}_{\alpha\beta} + \nu(e^{(0)}_{\alpha\alpha})^2]$$
$$+ \frac{Eh^3}{24(1-\nu^2)}[(1-\nu)e^{(1)}_{\alpha\beta}e^{(1)}_{\alpha\beta} + \nu(e^{(1)}_{\alpha\alpha})^2] \tag{3.2}$$

with the GREEN strain tensor

$$e^{(0)}_{\alpha\beta} = \frac{1}{2}(u_{\alpha,\beta} + u_{\beta,\alpha} + u_{3,\alpha}u_{3,\beta}) \; ; \; e^{(1)}_{\alpha\beta} = -u_{3,\alpha\beta} \; . \tag{3.3}$$

E is YOUNG's modulus and ν POISSON's ratio.

For the nonlinear von KÁRMÁN plate theory the functional (2.33) of total potential energy is defined as

$$I = \int_F [U(u_{j,i}) - pu_3]\,dF - \int_C [N^{*\prime}_{n\alpha}u'_\alpha + Q^{*\prime}_{n3}u'_3 - M^{*\prime}_{n\alpha}u'_{3,\alpha}]\,dC \tag{3.4}$$

with the lateral load $p = p(x_\alpha)$ and the energy density $U(u_{j,i})$ according to (3.2) and (3.3). To minimize the functional I, we have to introduce trial functions u_j, fulfilling the geometric boundary conditions on C:

$$u_i'' = u_i^{*''} \quad ; \quad u_{3,n}'' = u_{3,n}^{*''} \quad . \tag{3.5}$$

Criterion (2.32) is a sufficient condition for pointwise convexity of the functional (3.4). It can be shown [6] that the matrix (2.32) is positive definite only if

$$N_{11} > 0 \quad ; \quad N_{11}N_{22} - N_{12}^2 > 0 \tag{3.6}$$

holds.

Criterion (2.31) is a necessary condition for convexity of the functional (3.4). With the energy density $U(u_{j,i})$ according to (3.2) and (3.3) the convexity condition (2.31) leads to a stability limit, which is characterized by eigenvalues and eigenfunctions. This will be shown in a forthcoming paper.

Introducing the subsidary conditions (2.36) into the functional (2.35) the complementary functional I_c for the von KÁRMÁN plate theory can be given in the form [4,5]:

$$I_c = -\int_F U_c \, dF + \int_C [N_{n\alpha}'' u_\alpha^{*''} + (Q_{n3}'' + M_{ns,s}'' + N_{n3}'') u_3^{*''} - M_{nn}'' u_{3,n}^{*''}] \, dC \tag{3.7}$$

with the complementary energy density per undeformed plate area

$$U_c = \frac{1}{2Eh} [(1+\nu) N_{\alpha\beta} N_{\alpha\beta} - \nu (N_{\alpha\alpha})^2] + \frac{1}{2} N_{\alpha 3} \varphi_\alpha$$

$$+ \frac{12}{2Eh^3} [(1+\nu) M_{\alpha\beta} M_{\alpha\beta} - \nu (M_{\alpha\alpha})^2] \tag{3.8}$$

and the functions

$$\varphi_1 = \frac{N_{13}N_{22} - N_{23}N_{12}}{N_{11}N_{22} - N_{12}^2} \quad ; \quad \varphi_2 = \frac{N_{23}N_{11} - N_{13}N_{12}}{N_{11}N_{22} - N_{12}^2} \tag{3.9}$$

The subsidary conditions (2.36) lead to the necessary conditions

$$N_{\alpha\beta,\alpha} = 0 \; ; \; M_{\alpha\beta,\alpha\beta} + N_{\alpha 3,\alpha} + p = 0 \qquad \text{in } F \tag{3.10}$$

$$N_{n\alpha}' - N_{n\alpha}^{*'} = 0 \; ; \; Q_{n3}' + M_{ns,s}' + N_{n3}' - Q_{n3}^{*'} - M_{ns,s}^{*'} = 0 \; ; \; M_{nn}' - M_{nn}^{*'} = 0 \qquad \text{on } C \tag{3.11}$$

Criterion (2.31) is a necessary condition for the complementary functional I_c to be concave and to have a maximum value for the solution of the von KÁRMÁN plate theory.

4. Extremum principles of the nonlinear shallow shell theory

The dual extremum principles of section 2 will be considered for the geometric nonlinear shallow shell theory according to MARGUERRE [11], who described the form of the middle surface of the shell by a function $W(x_\alpha)$. With the usual assumptions the strain energy density $U(u_{j,i})$ is given by equation (3.2) with the Green strain tensor

$$e^{(0)}_{\alpha\beta} = \frac{1}{2}(u_{\alpha,\beta} + u_{\beta,\alpha} + u_{3,\alpha}u_{3,\beta} + u_{3,\alpha}W_{,\beta} + u_{3,\beta}W_{,\alpha}) \; ; \; e^{(1)}_{\alpha\beta} = -u_{3,\alpha\beta} \tag{4.1}$$

With the notations of the last section and with the external load $p_i (i = 1,2,3)$ the functional (2.33) is given as

$$I = \int_F [U(u_{j,i}) - p_i u_i]\, dF - \int_C [N^{*\prime}_{n\alpha} u'_\alpha + Q^{*\prime}_{n3} u'_3 - M^{*\prime}_{n\alpha} u'_{3,\alpha}]\, dC. \tag{4.2}$$

The trial functions u_i have to satisfy the geometric boundary conditions (3.5). Sufficient and necessary conditions for the convexity of (4.2) are equivalent to those for the von KÁRMÁN plate theory in the last section.

The functional (2.35) together with the subsidary conditions (2.36) lead to a complementary energy principle of the nonlinear shallow shell theory, which can be described by the complementary functional (3.7) - (3.9) [6]. The necessary conditions are the equilibrium equations

$$N_{\alpha\beta,\alpha} + p_\beta = 0 \; ; \; M_{\alpha\beta,\alpha\beta} + (N_{\alpha\beta}W_{,\beta} + N_{\alpha 3})_{,\alpha} + p_3 = 0 \qquad \text{in } F \tag{4.3}$$

and the static boundary conditions

$$\left.\begin{aligned} & N'_{n\alpha} - N^{*\prime}_{n\alpha} = 0 \; ; \; M'_{nn} - M^{*\prime}_{nn} = 0 \\ & Q'_{n3} + M'_{ns,s} + n_\alpha N'_{\alpha\beta} W_{,\beta} + N'_{n3} - Q^{*\prime}_{n3} - M^{*\prime}_{ns,s} = 0 \end{aligned}\right\} \quad \text{on } C \tag{4.4}$$

The conditions for concavity of the complementary functional I_C are the same as the conditions for convexity of the functional (4.2).

References

[1] R. HILL, On Uniqueness and Stability in the Theory of Finite Elastic Strain, J. Mech. Phys. Solids 5, 229 - 241 (1957).

[2] B. NOBEL and M. J. SEWELL, On Dual Extremum Principles in Applied Mathematics, University of Wisconsin, Math. Res. Cen. Rep. 1119 (1971) (available in J. Inst. Math. Appl. 9, 123 - 193 (1972).

[3] M. J. SEWELL, The Governing Equations and Extremum Principles of Elasticity and Plasticity Generated from a Single Functional, part I, J. Struct. Mech. 2,1, 1 - 32 (1973); part II, ibid. 2,2, 135 - 158 (1973).

[4] H. STUMPF, Die Extremalprinzipe der nichtlinearen Plattentheorie, ZAMM 55, T 110 - T 112 (1975).

[5] H. STUMPF, Dual extremum principles and error bounds in the theory of plates with large deflections, Archives of Mechanics (Archiwum Mechaniki Stosowanej) 27, 3, 485 - 496 (1975).

[6] H. STUMPF, Die dualen Variationsprinzipien mit Extremaleigenschaft in der nichtlinearen Theorie flacher Schalen, ZAMM, GAMM-Sonderheft 1975.

[7] K. WASHIZU, Variational Methods in Elasticity and Plasticity, Pergamon Press Oxford-London-New York 1968.

[8] S. NEMAT-NASSER, General Variational Principles in Nonlinear and Linear Elasticity with Applications, 214 - 261. In: Mechanics Today 1, Ed. S. Nemat-Nasser, Pergamon Press, New York 1972.

[9] E. REISSNER, On a Variational Theorem for Finite Elastic Deformations, J. Math. Phys. 32, 129 - 135 (1953).

[10] Th. v. KÁRMÁN, Festigkeitsprobleme im Maschinenbau, Enz. d. math. Wiss. 4, 348 - 352 (1910).

[11] K. MARGUERRE, Zur Theorie der gekrümmten Platte großer Formänderung, Proc. 5-th Int. Congr. Appl. Mech., 93 - 101 (1938).

[12] D. G. LUENBERGER, Optimization by Vector Space Methods, Wiley, New York, 1969.

DETERMINATION DE LA CONFIGURATION D'EQUILIBRE D'UN PLASMA

R. TEMAM

Département Mathématique - Université de Paris-Sud

91405 - Orsay, France

On étudie ici les équations qui régissent un plasma confiné dans une machine de type Tokomak. Il s'agit essentiellement des équations de la Magnétohydrodynamique (M.H.D. en abrégé), à l'équilibre, dans une géométrie de révolution.

On est ramené à un problème à frontière libre d'un type non classique. Après une formulation convenable, on démontre l'existence de solutions du problème ; ces solutions sont les valeurs critiques de certaines fonctionnelles.

1. FORMULATION DU PROBLEME.

Le Tokomak est un tore d'axe Oz ; on appelle Ω la section droite du tore, dans le demi-plan Oxz , $x>0$, et Γ frontière de Ω représente la coque. Le plasma occupe le domaine $\Omega_p \subset \Omega$, et on note Γ_p la frontière du plasma et $\Omega_v = \Omega \setminus (\Omega_p \cup \Gamma_p)$ la partie complémentaire qui est vide.

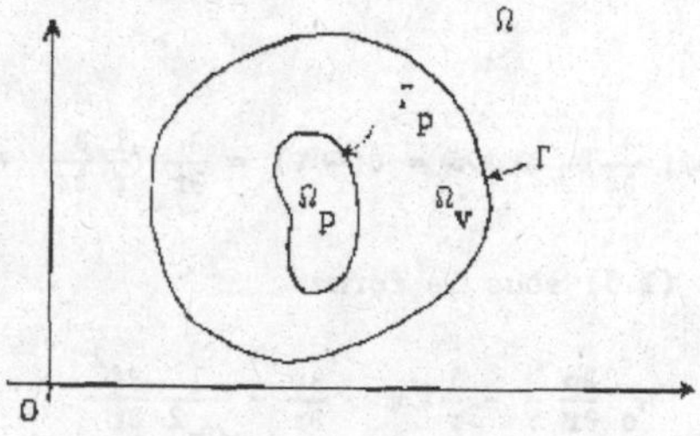

L'espace est rapporté aux coordonnées cylindriques (r,θ,z) et e_r, e_θ, e_z représente le trièdre orthonormé canonique en chaque point.

Dans le vide, on a les équations de Maxwell

(1.1) $$\text{div } B = 0 \quad \text{et} \quad \text{rot } B = 0 \quad \text{dans } \Omega_v$$

et dans le plasma, les équations de la M.H.D. se réduisent (en l'absence de mouvement macroscopique) à :

(1.2) $$\text{div } B = 0 \quad \text{et} \quad \text{rot } B = \mu_0 J \quad \text{dans } \Omega_p$$

(1.3) $$\text{grad } p = J \times B \quad \text{dans} \quad \Omega_p$$

où $B = B_r e_r + B_\theta e_\theta + B_z e_z$, $J = J_r e_r + J_\theta e_\theta + J_z e_z$, et p désignent le champ magnétique, le courant et la pression. En raison de la symétrie de révolution, ces quantités sont indépendantes de θ . En conséquence, la première équation (1.1) ou (1.2) est équivalente à

(1.4) $$\text{div } B = \frac{1}{r}\frac{\partial}{\partial r}(r\,B_r) + \frac{\partial B_z}{\partial z}$$

et on en déduit l'existence d'une fonction ψ définie localement pour l'instant et telle que

(1.5) $$B_r = \frac{1}{r}\frac{\partial\psi}{\partial z}\ , \quad B_z = -\frac{1}{r}\frac{\partial\psi}{\partial r}\ .$$

Il est commode de poser $f = f(r,z) = r\,B_\theta$. Alors, la deuxième équation (1.1) ou (1.2) devient

(1.6) $$\mathcal{L}\,\psi = 0 \quad \text{dans} \quad \Omega_v$$

(1.7) $$\mathcal{L}\,\psi\, e_\theta + \left(\frac{1}{r}\right)\nabla f \times e_\theta = \mu_o J \quad \text{dans} \quad \Omega_p$$

où

(1.8) $$\nabla = \left\{\frac{\partial}{\partial r},\ 0,\ \frac{\partial}{\partial z}\right\} \quad \text{et} \quad \mathcal{L} = \nabla\left(\frac{1}{r}\,\nabla\right) = \frac{\partial}{\partial r}\left(\frac{1}{r}\frac{\partial}{\partial r}\right) + \frac{1}{r}\frac{\partial^2}{\partial z^2}\ .$$

On écrit à présent (1.3) sous la forme

(1.9) $$\left\{\begin{aligned} \mu_o\frac{\partial p}{\partial r} &= -\frac{1}{r}\mathcal{L}\psi\ .\ \frac{\partial\psi}{\partial r} - \frac{1}{2r^2}\frac{\partial f^2}{\partial r} \\ 0 &= \frac{1}{r^2}\left(\frac{\partial\psi}{\partial z}\frac{\partial f}{\partial r} - \frac{\partial f}{\partial z}\frac{\partial\psi}{\partial r}\right) \\ \mu_o\frac{\partial p}{\partial z} &= -\frac{1}{r}\mathcal{L}\psi\ .\ \frac{\partial\psi}{\partial z} - \frac{1}{2r^2}\frac{\partial f^2}{\partial z}\ . \end{aligned}\right.$$

La seconde équation (1.9) montre que ∇f est parallèle à $\nabla\psi$ dans Ω_p , si bien que f ne dépend que de ψ

(1.10) $$f^2 = g_o(\psi)\ , \qquad \nabla f^2 = g_o'(\psi).\nabla\psi \quad \text{où} \quad g_o' = \frac{dg_o}{d\psi}\ .$$

La première et troisième équation (1.9) entraînent alors que

$$\nabla_p = -(\frac{1}{r}\mathcal{L}\psi + \frac{1}{2r^2}g_o'(\psi))\nabla\psi ,$$

De même ∇_p étant parallèle à $\nabla\psi$, p ne dépend que de ψ,

(1.11) $$p = g_1(\psi) , \nabla_p = g_1'.\nabla\psi$$

Avec (1.10) et (1.11), les équations (1.9) se réduisent à

(1.12) $$\mathcal{L}\psi = -\mu_o r\, g_1'(\psi) - \frac{1}{2r}g_o'(\psi) \text{ dans } \Omega_p .$$

En raison de (1.5), (1.7), (1.10) et (1.11) toutes les inconnues s'expriment en fonction de ψ, et ψ est solution de (1.6) et (1.12) (avec des conditions aux limites qui seront précisées). Evidemment les fonctions g_o et g_1 sont des fonctions inconnues de ψ. Elles ne peuvent être déterminées à l'aide des équations de Maxwell mais dépendent de "l'histoire" du plasma. Ce sont des équations d'état du plasma qui seront supposées données.

Conditions aux limites.

Soit ν le vecteur unitaire normal sur Γ_p ou Γ dirigé vers l'extérieur de Ω_p ou Ω et soit τ le vecteur unitaire tangent.

Les conditions aux limites et d'interface sont classiques

(1.13) $$B.\nu = 0 \text{ sur } \Gamma$$

(1.14) $$B.\nu = 0 \text{ et } B.\tau \text{ est continu sur } \Gamma_p .$$

Comme $B.\nu = -\frac{1}{r}\frac{\partial\psi}{\partial r}$ et $B.\tau = \frac{1}{r}\frac{\partial\psi}{\partial\nu}$, on a $\frac{\partial\psi}{\partial\tau} = 0$ localement sur Γ_p et Γ. Donc ψ est univoque dans tout Ω, et on peut supposer que

(1.15) $$\psi = 0 \text{ sur } \Gamma_p , \ \psi = \text{constante inconnue} = \gamma \text{ sur } \Gamma .$$

Enfin la deuxième condition (1.11) se réduit à

(1.16) $$\frac{\partial\psi}{\partial\nu} \text{ est continu sur } \Gamma_p .$$

Hypothèses complémentaires.

Des hypothèses à caractère plus physique sont habituellement imposées (cf. Mercier [4]) :

(1.17) $$p = 0 \text{ et } J_\theta = 0 \text{ sur } \Gamma_p , \ J_\theta \neq 0 \text{ dans } \Omega_p ,$$

ce qui se traduit avec (1.11) et (1.15) par

(1.18) $$g_1(0) = 0$$

et

$$\mu_o J_\theta = \mathcal{L}\psi = -\mu_o r\, g_1'(\psi) - \frac{1}{2r} g_o'(\psi) = 0 \quad \text{sur} \quad \Gamma_p$$

donne

(1.19) $$g_o'(0) = g_1'(0) = 0 .$$

Avec (1.18), $J_\theta \neq 0$ dans Ω_p se traduit par

(1.20) $$\psi \neq 0 \quad \text{dans} \quad \Omega_p .$$

Deux modèles.

Les hypothèses à faire sur g_o et g_1 nous amènent à considérer deux modèles : le modèle simplifié où g_o et g_1 sont supposées quadratiques en ψ et donc, en raison de (1.18), (1.19),

(1.21) $$f^2 = g_o(\psi) = b_o + b_2\,\psi^2 , \quad p = g_1(\psi) = a_2\psi^2 .$$

Comme f^2 et p sont $\geqslant 0$, on a bien sûr,

(1.22) $$b_o, b_2, a_2 \geqslant 0 ;$$

et les équations se réduisent à

(1.23) $$\mathcal{L}\psi = 0 \quad \text{dans} \quad \Omega_v , \quad \mathcal{L}\psi = -\lambda\, b(r)\psi \quad \text{dans} \quad \Omega_p ,$$

où $\lambda = \mu_o a_2 > 0$ et $b(r) = (2r + \frac{b_2}{\mu_o a_2 r})$.

Les équations sont linéaires, mais le problème aux limites est non linéaire (Ω_p , la forme d'équilibre du plasma est aussi inconnue).

On considèrera aussi au paragraphe 2 le modèle plus général

(1.24) $$\mathcal{L}\psi = 0 \quad \text{dans} \quad \Omega_v , \quad \mathcal{L}\psi = \frac{\partial g}{\partial \psi}(r,\psi) \quad \text{dans} \quad \Omega_p ,$$

où $g(r,\psi) = -\mu_o r\, g_1(\psi) - \frac{1}{2r} g_o(\psi)$. Les fonctions g_o, g_1 étant mal connues, il sera commode d'introduire un paramètre $\lambda > 0$ dans (1.24) et de considérer les équations

(1.24') $$\mathcal{L}\psi = \lambda \frac{\partial g}{\partial \psi}(r,\psi) \quad \text{dans} \quad \Omega_p .$$

Dans tous les cas, on suppose aussi que le courant total dans le plasma est un nombre donné

(1.25) $$I = \mu_o \int_{\Omega_p} J_\theta \, dr\, dz = \int_{\Omega_p} \mathcal{L}\psi \, dr\, dz = \int_{\Gamma_p} \frac{1}{r}\frac{\partial\psi}{\partial\nu} d\ell = \int_{\Gamma} \frac{1}{r}\frac{\partial\psi}{\partial\nu} d\ell$$

I>0 pour fixer les idées.

2. ETUDE DES PROBLEMES A FRONTIERE LIBRE.

2.1. Le Cas quadratique.

On étudie le cas où g_o et g_1 sont donnés par (1.21). On rapporte le plan R^2 aux axes Ox_1x_2 (i.e. $r = x_1$, $z = x_2$) , et donc

(2.1) $$\mathcal{L}\psi = \sum_{i=1}^{2} \frac{\partial}{\partial x_i}\left(\frac{1}{x_1}\frac{\partial\psi}{\partial x_i}\right) .$$

L'ouvert Ω de frontière Γ est situé dans une bande

(2.2) $$0 < x_* \leq x_1 \leq x_{**} < +\infty \quad , \quad \forall x = (x_1,x_2) \in \Omega ,$$

si bien que l'opérateur $\mathcal{L}$ est régulier et uniformément elliptique dans $\bar{\Omega}$.

Regroupant les équations établies au paragraphe 1, nous trouvons le problème suivant :

Soit I>0 donné ; on cherche $\lambda \in R$, $\lambda > 0$, un ouvert Ω_p , $\bar{\Omega}_p \subset \Omega$ (région occupée par le plasma), et une fonction $\psi : \Omega \longrightarrow R$, tel que

(2.3) $$\mathcal{L}\psi = -\lambda b \psi \quad \text{dans} \quad \Omega_p ,$$

(2.4) $$\mathcal{L}\psi = 0 \quad \text{dans} \quad \Omega_v = \Omega - \bar{\Omega}_p ,$$

(2.5) $$\psi = 0 \quad \text{sur} \quad \Gamma_p = \partial\Omega_p$$

(2.6) $$\frac{\partial\psi}{\partial\nu} \text{ est continue sur } \Gamma_p ,$$

(2.7) $$\psi = \text{constant} = \gamma \quad \text{sur} \quad \Gamma \quad (\gamma \text{ inconnu})$$

(2.8) $$\int_{\Gamma_p} \frac{1}{x_1}\frac{\partial\psi}{\partial\nu} d\ell = I$$

(2.9) $$\psi \text{ ne s'annule pas dans } \Omega_p .$$

Rappelons que b est donnée continue avec

(2.10) $$0 < b_o \leq b(x) \leq b_1 \quad \text{dans } \overline{\Omega} .$$

Supposant Ω_p connu, nous voyons que ψ et λ sont fonctions propre et valeur propre d'un problème de Dirichlet dans Ω_p. La condition (2.9) montre alors que λ est la première valeur propre du problème aux limites.

D'autre part $\frac{\partial\psi}{\partial\nu}$ a un signe constant sur Ω_p, et ce signe est positif en raison de (2.8). Ainsi

(2.11) $$\begin{cases} \Omega_p = \Omega_-(\psi) = \{x \in \Omega \ , \ \psi(x) < 0\} \\ \Gamma_p = \Gamma_o(\psi) = \{x \in \Omega, \ \psi(x) = 0\} \\ \Omega_v = \Omega_+(\psi) = \{x \in \Omega, \ \psi(x) > 0\} . \end{cases}$$

Soit $H^1(\Omega)$ l'espace de Sobolev d'ordre 1 rattaché à Ω, et si $\mathcal{O}$ est un ouvert $\subset \Omega$, on note

$$a_{\mathcal{O}}(\psi,\phi) = \int_{\mathcal{O}} \frac{1}{x_1} \operatorname{grad} \psi . \operatorname{grad} \phi \, dx ,$$

$$b_{\mathcal{O}}(\psi,\phi) = \int_{\mathcal{O}} b \, \psi \, \phi \, dx ,$$

et si $\Omega = \mathcal{O}$ on note $a_\Omega = a$, $b_\Omega = b$. Soit également

(2.12) $$W = \{\psi \in H^1(\Omega) \ , \ \psi = \text{constante sur } \Gamma\}$$

Une solution éventuelle de (2.3)-(2.9) est dans W et si ϕ est une autre fonction de W, alors utilisant la formule de Green on vérifie que

(2.13) $$a(\psi,\phi) - I \, \phi(\Gamma) = \lambda \, b \, (-\psi_-,\phi)$$

où $\phi(\Gamma) \in R$ désigne la valeur de ϕ sur Γ. Il est élémentaire de vérifier la réciproque: si $\psi \in W$ est assez régulière et si (2.13) a lieu $\forall \ \phi \in W$, alors ψ est solution de (2.3)-(2.9), Ω_p étant défini par (2.11) ; pour les détails, cf. [6] .

On a donc une sorte de formulation variationnelle du problème. Introduisons également les fonctionnelles k_1, k_2 de $W \longrightarrow R$ définies par

$$k_1(\psi) = \frac{1}{2} \, a(\psi,\psi) - I \, \psi(\Gamma)$$

$$k_2(\psi) = b(\psi_-,\psi_-) .$$

On vérifie que k_1 et k_2 sont bateaux différentiables sur W , de différentielles k_1' , k_2'

$$(k_1'(\psi),\phi) = a(\psi,\phi) - I\ \phi(\Gamma) \ , \quad \forall \phi \in W$$

$$(k_2'(\psi),\phi) = - b(-\psi_-,\phi) \quad , \quad \forall \phi \in W \ ,$$

si bien que l'équation (2.13) est équivalente à

$$(2.14) \qquad k_1'(\psi) = \lambda\ k_2'(\psi) \ .$$

Autrement dit : ψ <u>est point critique de</u> k_1 <u>sur les sous-ensembles</u> $k_2(\phi) =$ constante de W , <u>et</u> λ <u>est la valeur critique associée</u> (cf. [3], [5]) .

Réciproquement, toute solution de (2.14) vérifie (2.13) et, si elle est régulière, vérifie (2.3)-(2.9) : toute solution de (2.14) est solution faible de (2.3)-(2.9).

On a alors le résultat d'existence suivant :

<u>THEOREME 2.1</u>. Pour toute constante $c > 0$, la fonctionnelle k_1 est bornée inférieurement sur

$$\{\phi,\ \phi \in W,\ k_2(\phi) = \int_\Omega b(x)\ [\phi_-(x)]^2\ dx = c\} \ ,$$

et atteint son minimum en un point ψ (au moins).

La fonction ψ appartient à $W^{3,\alpha}(\Omega)$ pour tout $\alpha \geq 1$ et à $\mathcal{C}^{2+\eta}(\Omega)$ pour tout η , $0 \leq \eta < 1$.

Les ensembles Ω_p, Γ_p, Ω_v , étant définis par (2.11), ψ vérifie (2.3), (2.4), (2.5), ψ est analytique dans Ω_p et Ω_v . Au voisinage de tout point $x \in \Gamma_p$ tel que grad $\psi(x) \neq 0$, Γ_p est une courbe de classe $\mathcal{C}^{2+\eta}$ et (2.6) est vérifiée. Si Γ_p est, globalement, une courbe $\mathcal{C}^1$, alors (2.8) est aussi vérifiée.

2.2. <u>Le Cas général</u>.

Nous passons à l'étude de cas plus généraux. Nous considérons (cf. 1.24)) une fonction $g(x,\psi)$ deux fois continûment différentiable sur $\bar{\Omega} \times R$ et telle que

(2.15) $\quad \exists\ \beta > 1$, $b_1, b_2 > 0$, tels que

$$b_1(|\psi|^\beta - 1) \leq \frac{\partial g}{\partial \psi}(x,\psi) \leq b_2(|\psi|^\beta + 1) \ , \quad \forall x,\ \forall \psi$$

$$(2.16) \qquad \frac{\partial g}{\partial \psi}(x,0) = 0$$

(2.17) $$\frac{\partial g}{\partial \psi}(x,\psi) > 0 \quad \text{pour} \quad \psi < 0 \,,$$

Remplaçant (2.3) par l'équation (1.24') le problème est maintenant de trouver Ω_p avec $\overline{\Omega}_p \subset \Omega$, et $\psi : \Omega \to R$, tels que

(2.18) $$\mathcal{L}\psi = \lambda \frac{\partial g}{\partial \psi}(x,\psi) \quad \text{dans} \quad \Omega_p \,,$$

(2.19) $$\mathcal{L}\psi = 0 \quad \text{dans} \quad \Omega_v \,,$$

(2.20) $$\psi = 0 \quad \text{sur} \quad \Gamma_p = \partial\Omega_p$$

(2.21) $$\frac{\partial \psi}{\partial \nu} \quad \text{est continue sur} \quad \Gamma_p \,,$$

(2.22) $$\psi = \text{constant sur} \quad \Gamma \quad (\gamma \text{ inconnu})$$

(2.23) $$\int_{\Gamma_p} \frac{1}{x_1} \frac{\partial \psi}{\partial \nu} \, d\ell = I$$

(2.24) $$\psi \text{ ne s'annule pas dans } \Omega_p \,.$$

Le traitement est analogue au précédent. Les relations (2.11) sont encore vraies. On a encore $\psi \in W$ et si ψ vérifie (2.18)-(2.24) alors, pour tout $\phi \in W$, on a

(2.25) $$a(\psi,\phi) - I\,\phi(\Gamma) = -\lambda \int_{\Omega} \frac{\partial g}{\partial \psi}(x,-\psi_-(x))\,\phi(x)\,dx \,,$$

et réciproquement si ψ est assez régulière.

L'équation (2.25) est équivalente à

(2.26) $$k_1'(\psi) = \lambda\, k_2'(\psi) \,,$$

avec k_1 comme précédemment, et cette fois

(2.27) $$k_2(\psi) = \int_{\Omega} \left[g(x,-\psi_-(x)) - g(x,0)\right] dx \,.$$

Une forme faible de (2.18)-(2.24) est : <u>Trouver</u> $\psi \in W$ <u>qui soit point critique de</u> k_1, <u>sur les sous-ensembles</u> $k_2(\psi)$ = <u>const. de</u> W.

On a un résultat analogue au Théorème 2.1, mais dont la démonstration se heurte à une difficulté mathématique plus importante que nous signalons dans la section 2.3.

<u>THEOREME 2.2</u>. Pour toute constante $c > 0$, la fonctionnelle k_1 est bornée inférieurement sur

(2.28) $$\{\phi,\ \phi \in W\ ,\ k_2(\phi) = c\}$$

et atteint son minimum en un point ψ (au moins).

La fonction ψ est dans $W^{3,\alpha}(\Omega)$, $\forall \alpha \geqslant 1$ et dans $\mathscr{C}^{2+\eta}(\bar{\Omega})$, $\forall \eta$, $0 \leqslant \eta < 1$. Soient Ω_p, Γ_p, Ω_v , les ensembles définis par (2.11). Alors ψ vérifie (2.18), (2.19) (2.20), (2.22). (2.24) et ψ est analytique dans Ω_v . L'ensemble Γ_p est d'intérieur vide dans R^2 et, au voisinage de tout point $x \in \Gamma_p$, tel que grad $u(x) \neq 0$, Γ_p est une courbe de classe $\mathscr{C}^{2+\eta}$ et (2.21) est vérifiée. Enfin si Γ_p est une courbe $\mathscr{C}^1$ par morceaux, alors (2.23) est vérifiée.

Remarque 2.1. Il serait intéressant de connaître l'ensemble des valeurs de λ , pour lesquelles (2.26) a une solution (i.e. déterminer toutes les valeurs critiques de k_1 sur l'ensemble $\{\psi \in W,\ k_2(\psi) = c\}$.

Il serait intéressant également de préciser la régularité de Γ_p et les questions d'unicité (l'unicité est facile pour λ donné, assez petit). Les aspects numériques font l'objet d'un travail en cours cf. [2] .

Signalons aussi qu'un problème analogue au problème (2.3)-(2.9) apparaît en mécanique des fluides classiques (problème de cavitation, cf. [7]) , cf. aussi la communication de Berger et Fraenkel dans ce même volume.

2.3. Une inégalité fonctionnelle.

Dans les conditions du Théorème 2.2 il n'est pas évident que k_1 soit borné inférieurement sur les ensembles (2.28) (alors que cela est facile dans le cas du Théorème 2.1).

On démontre (cf. [6]) et utilise pour cela l'inégalité suivante :

THEOREME 3.1. Soit $\Omega \subset R^2$ un ouvert borné de classe $\mathscr{C}^2$. Pour tout $\alpha > 0$, il existe une fonction $\delta_\alpha :]0,+\infty[\longrightarrow R$, telle que

(2.29) $$|u_+|_{L^2(\Omega)} \leqslant \alpha |\text{grad } u|^2_{L^2(\Omega)} + \delta_\alpha(|u_-|_{L^2(\Omega)}) ,$$

pour tout $u \in H^1(\Omega)$ tel que $|u_-|_{L^2(\Omega)} \neq 0$.

BIBLIOGRAPHIE

[1] J. Bernstein, E.A. Friedman, N. Kruskal, R.M. Kulsrud - An energy principle for hydromagnetic stability problems. Proc. Royal Soc., A.244 (1958), p.17-40.

[2] J.P. Boujot, J. Laminie, R. Temam - A paraître.

[3] J.S. Fučik, J. Nečas, J. Souček, V. Souček - Spectral analysis of non-linear

operators.
Lecture Notes in Math., vol. 346, Springer Verlag 1973.

[4] C. Mercier - The magnetohydrodynamic approach to the problem of plasma confinment in closed magnetic configuration.
Publication Euratom - C.E.A., Luxembourg 1974.

[5] P. Rabinowitch - Cours Université de Paris VI et XI, 1973.

[6] R. Temam - A non-linear eigenvalue problem : the equilibrium shape of a confined plasma.
Arch. Rat. Mech. Anal., à paraître.

[7] Garabeddian -

ELASTIC-PLASTIC TORSION OF CYLINDRICAL PIPES*

Tsuan Wu Ting
University of Illinois at Urbana-Champaign
Urbana, Illinois, 61801, U.S.A.

1. Statement of the Problem. Let D be the cross section of a cylindrical pipe bounded externally and internally by Jordan curves C_0 and C_1, respectively. We shall assume that each C_α, $\alpha = 0,1$, has only a finite number of corners and that between any two adjacent corners, each C_α possesses continuously varying curvature which assumes only a finite number of maxima and minima there. Denote by G_α the simply connected domain enclosed by C_α and by $H^1(G_0)$ the Hilbert space of functions whose distribution derivatives are square integrable over G_0. The L^2 and the Dirichlet norms of a function u in $H^1(G_0)$ will be denoted by $\|u\|_0$ and $\|u\|_1$, respectively. We shall mainly work with the subspace $H_0^1(G_0)$ of $H^1(G_0)$, which consists of functions vanishing almost everywhere on C_0.

Let k be a given positive constant and let

$$K \equiv \{u \in H_0^1(G_0) \,\big|\, |\text{grad } u| \leq k \text{ a.e. in } G_0 \text{ and } u = \text{a constant on } G_1\}.$$

Then, K is a non-empty closed convex subset of $H_0^1(G_0)$. According to the deformation theory in plasticity, the elastic-plastic torsion problem of the cylindrical pipe with cross section D is to find the minimizer ψ in K such that

$$(1.1)\qquad J[\psi] \equiv \int_{G_0} \{|\text{grad } \psi|^2 - 4\mu\theta\psi\}dx = \inf J[u] \quad \text{over} \quad K,$$

where $dx \equiv dx_1\, dx_2$ and where μ and θ stand for the shear modulus and the angle of twist per unit length, respectively.

In passing, we note that the introduction of the stress function ψ is a consequence of St. Venant's semi-inverse method, [16], as well as the requirements for equilibrium and that it is the minimization

*Work supported in part by NSF Grant no. MPS75-07118.

principle which takes the place either of the constitutive relations in plasticity or of the Beltrami-Michell compatibility conditions in elasticity, [14, (d)]. In the meantime, the essential yield criterion and the boundary conditions have been included as side constraints in the admissibility conditions. Accordingly, all mechanical aspects of the problem have been taken into account and we would expect that the problem so formulated is a well-posed one. In fact, Poincare's inequality together with the parallelogram law ensures that every minimizing sequence of the functional J is actually a Cauchy sequence in $H_0^1(G_0)$. Moreover, the $\|\cdot\|_1$-limit ψ of a minimizing sequence is a minimizer because of the closedness of K and the bounded convergence theorem. As for the uniqueness of the minimizer, it is a direct consequence of the convexity of $J[u]$ in u. Also, such a uniqueness proof may also be interpreted as a Steiner's symmetrization, [5]. Thus, we have

Theorem 1.1. Problem (1.1) has a unique minimizer ψ which is uniformly Lipschitz continuous with the Lipschitz constant equal to k in D.

2. Imbedding of the Minimizer. Having been assured of the existence and the uniqueness of the minimizer, we proceed to derive as much information as possible from the minimizing property of ψ. However, the inequality constraint on the gradients of the admissible functions is rather restrictive on our choice of possible variations. Accordingly, we shall replace it by a majorant and a minorant function. To do this, let k_1 be the constant value of the minimizer ψ on G_1 and let

$$K_1 \equiv \{u \mid u \in K,\ u = k_1 \text{ a.e. on } G_1\}.$$

Then, K_1 is also a closed convex subset of $H_0^1(G_0)$. Of course, K_1 is non-empty, since the minimizer ψ belongs to it. Consequently, there are a unique function Φ and a unique function ϕ in K_1 such that

$$I[\Phi] \equiv \int_{G_0} \Phi(x)dx = \sup I[u] \text{ over } K_1,$$

(2.1)

$$I[\phi] \equiv \int_{G_0} \phi(x)dx = \inf I[u] \text{ over } K_1.$$

Moreover, for all functions u in K_1, $\phi \leq u \leq \Phi$ a.e. in D. In particular, $\phi \leq \psi \leq \Phi$ a.e. in D.

Now, let $K^* \equiv \{u \in H_0^1(G_0) \mid u = k_1 \text{ on } G_1, \ \phi \leq u \leq \Phi \text{ a.e. in } D\}$, and consider the problem of finding a function ψ^* in K^* such that

$$J[\psi^*] = \inf J[u] \quad \text{over } K^*. \tag{2.2}$$

Then, the same reasoning as for Theorem 1.1 ensures the truth of the following result.

Theorem 2.1. Problem (2.2) has a unique minimizer ψ^*.

Of course, the two problems in (2.1) have been so formulated that the following theorem holds.

Theorem 2.2. The minimizers ψ and ψ^* are identical in $H_0^1(G_0)$.

Although the proof of Theorem 2.2 is not a trivial one, the technique initiated by Stampacchia and successfully applied to the elastic-plastic torsion of solid bars in [2] can be directly applied to the present case.

By virtue of Theorem 2.2, we shall regard the stress function ψ as the minimizer of problem 2.2 and then derive possible informations from its variational inequalities. It should be said that problem (2.2) is not an independent one, because it depends on the existence of ψ through the constant k_1. Actually, the dependence of problem (2.2) upon the constant k_1 is to ensure that all admissible functions have the same constant values over the entire boundary, ∂D, of the multiply connected domain D as that taken by the minimizer ψ of problem (1.1). This is just what has been done for the case of solid bars, [2], except that in that case, the lower envelope is not needed because of the maximum principle.

3. The Edges of the Enveloping Surfaces. Although the upper envelope Φ and the lower envelope ϕ were introduced to replace the gradient constraint, the edges $\Gamma(\Phi)$ and $\Gamma(\phi)$ of the surfaces defined, respectively, by $\Phi(x)$ and $\phi(x)$ also furnish definite informations about the location as well as the extent of the "elastic core" in the multiply connected domain D. In fact, the set $\Lambda \equiv D \cap \Gamma(\Phi) \cap \Gamma(\phi)$ is essentially independent of the value of ψ on G_1 and hence it can be roughly located when D is given. Moreover, the elastic core is an open set in $D + \partial D$, which contains it. However, in order to derive these results, we have to know more about the set of discontinuities of $\operatorname{grad} \Phi$ and $\operatorname{grad} \phi$ in D. Fortunately, as the consequences of the restrictions on the Jordan curves C_α, $\alpha = 0,1$, we know, by the

same reasoning as for the ridge of a Jordan domain, [14, (a)], that both $\Gamma(\Phi)$ and $\Gamma(\phi)$ consist of a finite number of Jordan arcs. We say that a point γ on $\Gamma(\Phi)$ or $\Gamma(\phi)$ is a branch point if it is a common point of several Jordan arcs and that it is an end point if it is an end point of a single arc. To be precise, we state these results as

<u>Theorem</u> 3.1. Both the set of discontinuities of grad Φ and that of grad ϕ in D consist of a finite number of smooth Jordan arcs. Consequently, each of them contains, at most, a finite number of branch points and end points.

4. <u>The Existence of an Elastic Core</u>. A circular pipe may become completely plastic under a finite angle of twist per unit length. In fact, if D is a circular ring, then the set Λ is empty. On the other hand, if D possesses a non-reentrant corner on C_0, then Λ is always non-empty. Since the problem can be solved in closed form, if D is a circular ring, we shall consider those domains D for which Λ never becomes empty. By an elastic core, we mean an open neighbourhood E^* of the set Λ such that ψ is analytic and satisfies the equation, $\Delta\psi = -2\mu\theta$ in E^*. We shall see later in §6 that $|\text{grad } \psi| < k$ in E^*. Thus, the material in E^* is elastic in the sense that it obeys Hook's law and that its stress deviator stays below the yield point. Our main result is

<u>Theorem</u> 4.1. For every point x_0 on Λ, there is a positive constant ε_0 depending on x_0 such that if $D(x_0,\varepsilon)$ is a disk contained in D, centered at x_0 and with radius $\varepsilon \leq \varepsilon_0$, then the minimizer ψ of problem (2.2) is strictly greater than ϕ and strictly less than Φ in $D(x_0,\varepsilon)$. Moreover, ψ is analytic and satisfies the equation, $\Delta\psi = -2\mu\theta$, in $D(x_0,\varepsilon)$.

The proof of this theorem is to consider the function u which satisfies the equation, $\Delta u = -2\mu\theta$, in $D(x_0,\varepsilon)$ and equal to ψ on $\partial D(x_0,\varepsilon)$. Then, we show that $\phi < u < \Phi$ in $D(x_0,\varepsilon)$ if ε_0 is less than some positive number. Of course, one has to, as usual, guess this fact in advance. The crucial point for this to be the case is the following: Along any branch of the arcs in $\Lambda \cap D(x_0,\varepsilon)$, the upper and the lower enveloping surfaces defined by $\Phi(x)$ and $\phi(x)$ are, respectively, of $\wedge$-shape and v-shape. Hence, their distributional Laplacians are both Dirac measures distributed along $\Lambda \cap D(x_0,\varepsilon)$ modulo a continuous bounded function defined on $D(x_0,\varepsilon)$. Using this fact, we can transform the integral representation for u in terms of the Green function into the desired estimate, $\phi < u < \Phi$ in $D(x_0,\varepsilon)$, by choosing ε_0 small enough. Finally, we conclude from the Dirichlet principle and the uniqueness of the minimizer for the Dirichlet problem over $D(x_0,\varepsilon)$ that u and ψ are identical in $H^1(D(x_0,\varepsilon))$.

Since Theorem 4.1 holds for all points on Λ, the existence of the elastic core E^* is now assured. However, a simple way to prove that $|\text{grad}\,\psi| < k$ in E^* is to establish the smoothness of ψ in the entire domain D and then derive it as a simple consequence. This will be discussed in the next two sections.

It is interesting to note how the minimizer ψ smoothes out the edges of the enveloping surfaces. We also emphasize that in the proof we have made use of the fact that ψ is also the minimizer of problem (2.2). Although, the details of the proof are elementary but are rather involved. Since the set Λ can be roughly located once D is given, the theorem also tells us directly the position of the elastic core relative to D without knowing the precise value of k_1. Also, it is this theorem that relates the existence of the elastic core to the smoothness of ∂D.

5. <u>Continuity of the Stress</u>. For our problem, the only non-vanishing components of the stresses are the components of $\text{grad}\,\psi$. Accordingly, it suffices to establish the smoothness of ψ in the entire domain D. Since the existence of the elastic core E^* is known and since ψ is analytic in E^*, it is enough to verify the smoothness of ψ in $D\backslash E^*$. Moreover, the smoothness of a function is a local problem, we need only to show that for every point x_0 in $D\backslash E^*$, there is a constant $\varepsilon_0 > 0$, which may depend on x_0, such that if $D(x_0,\varepsilon)$ is a disk contained in $D\backslash E^*$, centered at x_0 with radius $\varepsilon \leq \varepsilon_0$, then ψ is smooth in $D(x_0,\varepsilon)$.

We carry out the above program in two steps. First, we show that ψ is smooth along the edges of the enveloping surfaces. Since ψ is smooth along $\Lambda \equiv D \cap \Gamma(\Phi) \cap \Gamma(\phi)$, it suffices to establish the smoothness along

$$\Lambda'(\Phi) \equiv (D \cap \Gamma(\Phi))\backslash\Lambda, \qquad \Lambda'(\phi) \equiv (D \cap \Gamma(\phi))\backslash\Lambda.$$

We state the results as

<u>Theorem</u> 5.1. For every point x_0 on $\Lambda'(\Phi)$, there is a constant $\varepsilon_0 > 0$, which may depend on x_0, such that if $D(x_0,\varepsilon)$ is a disk contained in D, centered at x_0 and with radius $\varepsilon \leq \varepsilon_0$, then $\psi < \Phi$ in $D(x_0,\varepsilon)$. Moreover, (i) ψ is analytic and satisfies the

equation, $\Delta\psi = 2\mu\theta$, in $D(x_0,\varepsilon)$ if $\psi(x_0) > \phi(x_0)$ and (ii) if $\psi(x_0) = \Phi(x_0)$, then $\psi(x)$ is smooth in $D(x_0,\varepsilon)$.

Completely similar statments hold for points on $\Lambda'(\phi)$. Note that if $\psi(x_0) > \phi(x_0)$, then we have a situation similar to that in Theorem 4.1 and hence the same proof applies. For the assertion (ii), we consider the variational-inequality problem for determining ψ over the disk $D(x_0,\varepsilon)$ and assume that the values of ψ are prescribed on $\partial D(x_0,\varepsilon)$. Then we may either appeal to the general regularity theorem for the solutions of variational inequalities, [1, 11, 13], or directly apply the relatively elementary arguments in [14, (c)].

Now, ψ is smooth in a complete neighbourhood N of $D \cap \Gamma(\Phi)$ and $D \cap P(\phi)$. Our next step is to establish the smoothness of ψ in $D \backslash N$. By choosing smaller N if necessary, we may assume N is a closed set. By the same reasoning as for Theorem 5.1, we have

Theorem 5.2. For every point x_0 in $D\backslash N$, there is a constant $\varepsilon_0 > 0$ such that if $D(x_0,\varepsilon)$ is a disk contained in $D\backslash N$, centered at x_0 and with radius $\varepsilon \leq \varepsilon_0$, then (i) ψ is analytic and $\Delta\psi = -2\mu\theta$ in $D(x_0,\varepsilon)$ if $\phi(x_0) < \psi(x_0) < \Phi(x_0)$ and (ii) ψ is smooth in $D(x_0,\varepsilon)$ if either $\psi(x_0) = \phi(x_0)$ or $\psi(x_0) = \Phi(x_0)$.

We have seen that the continuity of the stresses was established in three steps because of the presence of the corners on ∂D. On the other hand, such an approach provides additional informations about the elastic-plastic partition of the cross sections D, which are physically interesting. Physically, the continuity of the stresses is desirable. As was emphasized in [12], for the dynamical compatibility conditions to be satisfied across the elastic-plastic boundary, it is necessary that the stresses must be everywhere continuous across this boundary. However, to justify the fulfillment of the compatibility conditions completely, we have to show that the elastic-plastic boundary is a rectifiable curve. But this remains an open question even if D is simply connected.

Although, it has not been stated, detailed examinations of the proofs of Theorems 4.1, 5.1 and 5.2 indicate that the second derivatives of ψ are integrable over D to the power p for all $p > 1$. Hence, the first derivatives of ψ are Holder continuous in D. Thus, as far as regularity question is concerned, we have the same results as for the simply connected domains.

6. <u>Elastic-Plastic Partition of the Cross Sections.</u> Consider the sets:

$$E \equiv \{x \in D \mid \phi(x) < \psi(x) < \Phi(x)\}, \quad P \equiv \bar{D} \setminus E.$$

Then E is open and P is closed in $\bar{D} \equiv D + \partial D$. Moreover, it follows from Theorems 4.1, 5.1 and 5.2 that ψ is analytic and satisfies the equation, $\Delta\psi = -2\mu\theta$, in E. Hence, E contains the elastic core E^*. We assert that $|\text{grad } \psi| < k$ everywhere in E. In fact, the admissibility conditions together with the regularity results imply that $|\text{grad } \psi| \leq k$ everywhere in D. In particular, $|\text{grad } \psi| \leq k$ on the boundary of E. On the other hand, direct computation shows that $\Delta(|\text{grad } \psi|^2) > 0$ in E. It follows from these facts and the strong maximum principle for the solutions of elliptic differential inequalities that $|\text{grad } \psi|$ is strictly less than k in E. This shows that the material in E obeys Hook's law and that the modulus of the stress deviator in E stays below the yield point. Thus, E and hence E^* are elastic zones of the cross section D in the usual sense. It may be noted that for solid bars, the elastic zone E is a simply connected open set containing the ridge of the cross section, [14, (c)]. However, this is not so, if D is multiply connected. In fact, the set $\Lambda \equiv D \cap \Gamma(\Phi) \cap \Gamma(\phi)$ is, in general, disconnected.

Consider now the set P. For every point x in P, $\psi(x)$ is either equal to $\phi(x)$ or equal to $\Phi(x)$. Hence, if the interior of P is non-empty, then $|\text{grad } \psi| = k$ everywhere there. Thus, P is the plastic zone of D in the sense that the modulus of the stress deviator equals to k everywhere in its interior. As we shall see, there exist a warping function and a Lagrange's multiplier such that the stress-strain relations in P are given by the deformation theory in plasticity.

The most interesting and important property of P is that its intersection with every inward normal to ∂D is a single segment which may be degenerated into a single point on ∂D. This segment always has one end point on ∂D. To see this, we first note that the inward normals to ∂D, that meet along $D \cap \Gamma(\Phi)$, completely covers D without overlap. Similar statements also holds for the inwards to ∂D, that meet along $D \cap \Gamma(\phi)$. Consider the variations of ψ along the inward normals to ∂D, if the intersection property of P should not hold, then the mean-value theorem in differential calculus would lead to a contradiction to the fact that $|\text{grad } \psi| \leq k$ everywhere in D. Thus the plastic zone P always adheres to the boundary of the cross section. This implies that as the angle of twist per unit length increases from zero, yielding always begins on the lateral surfaces of the pipe.

7. The Elastic-plastic Boundary. By elastic-plastic boundary we mean the set of points $\Sigma \equiv \partial P \cap D = \partial E \cap D$. We proceed to show how Σ can be decomposed into disjoint union of a finite number of Jordan arcs. Since the minimizer ψ is equal ϕ or Φ on ∂P, Σ is the disjoint union of Σ^+ and Σ^-, where $\psi = \Phi$ on Σ^+ and $\psi = \phi$ on Σ^-. Since the inward normals to ∂D, that meet along $D \cap \Gamma(\Phi)$ completely covers D without overlap and since $D \cap \Gamma(\Phi)$ possesses, at most, a finite number of branch points and end points, $D \setminus \Gamma(\Phi)$ consist of a finite number of components, say $D_1, \ldots, D_n$. Then $\Sigma^+ \cap D_j$, $j = 1,2, \ldots, n$, are disjoint sets, because $\Sigma^+ \cap (D \cap \Gamma(\Phi)$ is empty. Now, restrict our attention to, say, $\Sigma^+ \cap D_1$. For every point $x \equiv (x_1,x_2)$ on $\Sigma^+ \cap D_1$, there is, at least, one point s on ∂D such that

$$R(s) \equiv \operatorname{dist}(x,s) = \operatorname{dist}(x,\partial D).$$

Let $x_1 = f(s)$, $x_2 = g(s)$ be the equations defining ∂D, where s also stands for the arc length of ∂D. The intersection property of the plastic zone P leads to the following parametric presentation for $\Sigma^+ \cap D_1$:

$$x_1 = f(s) + R(s)n_1(s), \qquad x_2(s) = g(s) + R(s)n_1(s),$$

where n_1, n_2 are the components of the unit inward normal to ∂D at s. Since the functions f, g, n_1 and n_2 are all continuous, the functions x_1 and x_2 will be continuous if $R(s)$ is continuous. Actually, $R(s)$ is a continuous function. Moreover, $\Sigma^+ \cap D_1$ consists of a finite number of components. The proof of the last statements is similar to those in [14, (b)].

Of course, similar conclusions hold for each $\Sigma^+ \cap D_j$, $j = 2, \ldots, n$. Also, Σ^- has similar decomposition. Thus, we have

Theorem 7.1. The elastic-plastic boundary consists of a finite number of Jordan arcs each of which is a continuous image of a portion of ∂D.

8. <u>The Existence of the Warping Function</u>. Since the torsion problem has been formulated in terms of the stress function, i.e., since only the dual problem has been considered, [3], we are obliged to consider the primary problem for determining the corresponding displacement field so as to check whether any fracture occurs during the process of loading. It is a characteristic feature of the deformation theory that the problem of determining the displacements corresponding to a given stress field with the modulus of the stress deviator equal to the yield constant is to find a Lagrange's multiplier so that the compatibility conditions will be satisfied. Although, this problem remains open for the general case, [3], it has been treated for elastic-plastic torsion of solid bars, [1]. A crucial fact for the existence of the Lagrange multiplier for the solid bars is the intersection property of the plastic zone P with the inward normals to the boundary of the simply connected domain. Since this intersection property is still preserved even if D is multiply connected, all the results in [1] can be applied to the present case. Also, the elegant geometrical construction of the warping function given in [12] is applicable too.

REFERENCES

[1] Brezis, H., Problemes unilateraux, J. Math. Pures and Appl., 51 (1972), pp. 1-68; Multiplicateur de Lagrange en torsion "elasto-plastique", Arch. Rat. Mech. Anal. 49(1972), pp. 32-40.

[2] Brezis, H. and M. Sibony, Equivalence de deux inequalities variationelles et applications, Arch. Rat. Mech. Anal. 41(1971), pp. 254-265.

[3] Duvaut, G. and J. L. Lions, Les inequations en mecanique et en physique, Dunod, Paris, 1972, Chapter 5.

[4] Fichera, G., Existence theorems in elasticity, Handbuch der Physik, via/2, pp. 347-389, Springer-Verlag, New York, 1972.

[5] Garabedian, P. R., Proof of uniqueness by symmetrization, Studies in Math. Anal. and Selected Topics, Standford Univ., Calif. 1962.

[6] Gerhardt, G., Regularity of the solutions of non-linear variational inequalities with a gradient bound as constraint (preprint).

[7] Germain, P., Remarks on the theory of partial differential equations of mixed type and applications to the study of transonic flow, Comm. Pure and Appl. Math. 7(1954), pp. 117-144.

[8] Lanchon, H., Torsion elasto-plastique d'un cylindrique de section simplement on multiplement connexe, J. Mech., 13(1974), pp. 267-320.

[9] Lanchon, H. and R. Glowinski, Torsion elasto-plastique d'un barre cylindrique de section multiconnexe, J. Mech., 12(1973), pp. 151-171.

[10] Lanchon, H. and G. Duvant, Sur la solution du probleme de torsion elasto-plastique dune barre cylindrique de section quelconque, C. R. Acad. Sc. t. 264, serie A, 1967, pp. 520-523.

[11] Lions, J. L., Quelques methods de resolution des problemes aux limites non lineaires, Dunod, Paris, 1969.

[12] Prager, W. and P. G. Hodge Jr., Theory of perfectly plastic solids, John Wiley and Sons, New York, 1951, Chapter 4.

[13] Stampacchia, G., Regularity of solutions of some variational inequalities, Non-linear Functional Analysis (Proc. Symp. Pure Math. Vol. 18), Part I, A.M.S., Providence, RI, pp. 271-281.

[14] Ting, T. W., (a) The ridge of a Jordan domain and completely plastic torsion, J. Math. Mech. 15(1966), pp. 15-48. (b) Elastic-plastic torsion of convex cylindrical bars, J. Math. Mech. 19(1969), pp. 531-551. (c) Elastic-plastic torsion of simply connected cylindrical bars. (d) St. Venant's compactibility conditions

and basic problems in elasticity (to appear in Rocky Mountain J. Math.).

[15] Truesdell, C. and W. Noll, Handbuch der Physik, Vol. III/3, Springer-Verlag, New York, 1965.

[16] St. Venant, B. De, Memoire sur la torsion des prisme, Mem. div. Sov. Acad. Sci. 14(1856), pp. 233-560.